ELECTRONIC PROPERTIES OF NOVEL MATERIALS— PROGRESS IN MOLECULAR NANOSTRUCTURES

ELECTRONIC PROPERTIES OF NOVEL MATERIALS— PROGRESS IN MOLECULAR NANOSTRUCTURES

XII International Winterschool

Kirchberg, Tyrol, Austria   March 1998

EDITORS
Hans Kuzmany
Universität Wien, Austria

Jörg Fink
*Institut für Festkörper- und Werkstoff-Forschung
Dresden, Germany*

Michael Mehring
Universität Stuttgart, Germany

Siegmar Roth
*Max-Planck-Institut für Festkörperforschung
Stuttgart, Germany*

AIP CONFERENCE
PROCEEDINGS 442

American Institute of Physics   **Woodbury, New York**

Editors:

Hans Kuzmany
Universität Wien, Institut für Materialphysik
Strudlhofgasse 4
A-1090 Wien AUSTRIA

Email: kuzman@pap.univie.ac.at

Jörg Fink
Institut für Festkörper- und Werkstoff-Forschung
Postfach 270016
D-01171 Dresden GERMANY

Email: jfink@ifw-dresden.de

Michael Mehring
Universität Stuttgart
2. Physikalisches Institut
Pfaffenwaldring 57
D-70569 Stuttgart GERMANY

Email: m.mehring@physik.uni-stuttgart.de

Siegmar Roth
Max-Planck-Institut für Festkörperforschung
Heisenbergstr. 1
D-70569 Stuttgart GERMANY

Email: roth@klizix.mpi-stuttgart.mpg.de

Authorization to photocopy items for internal or personal use, beyond the free copying permitted under the 1978 U.S. Copyright Law (see statement below), is granted by the American Institute of Physics for users registered with the Copyright Clearance Center (CCC) Transactional Reporting Service, provided that the base fee of $15.00 per copy is paid directly to CCC, 222 Rosewood Drive, Danvers, MA 01923. For those organizations that have been granted a photocopy license by CCC, a separate system of payment has been arranged. The fee code for users of the Transactional Reporting Service is: 1-56396-808-8/ 98 /$15.00.

© 1998 American Institute of Physics

Individual readers of this volume and nonprofit libraries, acting for them, are permitted to make fair use of the material in it, such as copying an article for use in teaching or research. Permission is granted to quote from this volume in scientific work with the customary acknowledgment of the source. To reprint a figure, table, or other excerpt requires the consent of one of the original authors and notification to AIP. Republication or systematic or multiple reproduction of any material in this volume is permitted only under license from AIP. Address inquiries to Office of Rights and Permissions, 500 Sunnyside Boulevard, Woodbury, NY 11797-2999; phone: 516-576-2268; fax: 516-576-2499; e-mail: rights@aip.org.

L.C. Catalog Card No. 98-86576
ISBN 1-56396-808-8
ISSN 0094-243X
DOE CONF- 980379

Printed in the United States of America

Contents

Preface xiii
Table of Previous Kirchberg Winterschools xiv
Sponsors xv

NANOTUBES: PREPARATIONS

Production of Carbon Single Wall Nanotubes Versus Experimental Parameters 3
    C. Journet, V. Micholet, P. Bernier, W. K. Maser, A. Loiseau,
    M. Lamy de la Chapelle, S. Lefrant, R. S. Lee, and J. E. Fischer

Laser-Assisted Production of Multi-Walled Carbon Nanotubes from Acetylene 12
    H. Kanzow, A. Schmalz, and A. Ding

Effect of Hydrogenation on Catalytically Produced Carbon Nanotubes 16
    P. Piedigrosso, J.-F. Colomer, A. Fonseca, and J. B. Nagy

The Role of Catalyst Support in Carbon Nanotube Synthesis 20
    A. Siska, K. Hernadi, I. Kiricsi, I. Rojik, and J. B. Nagy

Pyrolysis of C_{60}-Thin Films Yields Ni-Filled Sharp Nanotubes 25
    N. Grobert, M. Terrones, A. J. Osborne, H. Terrones, W. K. Hsu,
    S. Trasobares, Y. Q. Zhu, J. P. Hare, H. W. Kroto, and D. R. M. Walton

Nanotechnology of Nanotubes and Nanowires: From Aligned Carbon Nanotubes to Silicon Oxide Nanowires 29
    N. Grobert, J. P. Hare, W. K. Hsu, H. W. Kroto, A. J. Pidduck, C. L. Reeves,
    H. Terrones, M. Terrones, S. Trasobares, C. Vizard, D. J. Wallis,
    D. R. M. Walton, P. J. Wright, and Y. Q. Zhu

Bulk Properties of Crystalline Single Wall Carbon Nanotubes: Purification, Pressure Effects, and Transport 34
    J. E. Fischer, R. S. Lee, H. J. Kim, A. G. Rinzler, R. E. Smalley,
    S. L. Yaguzhinski, A. D. Bozhko, D. E. Sklovsky, and V. A. Nalimova

Chromatographic Purification and Size Separation of Carbon Nanotubes 39
    G. S. Duesberg, J. Muster, V. Krstic, M. Burghard, and S. Roth

Assembling Techniques for Micellar Dispersed Carbon Single-Walled Nanotubes 44
    M. Burghard, J. Muster, G. Duesberg, G. Philipp, K. Krstic, and S. Roth

Intercalation Reactions in Catalytic Multiwall Carbon Nanotubes 51
    K. Metenier, L. Duclaux, H. Gaucher, J.-P. Salvetat, P. Lauginie,
    S. Bonnamy, and L. Beguin

NANOTUBES: EXPERIMENTS

Electrical Resistivity of Single-Wall Carbon Nanotubes Obtained by the Arc-Discharge Technique 57
    E. Jouguelet, C. Mathis, P. Petit, C. Journet, and P. Bernier

Electrical Transport in Carbon Nanotubes 61
    K. Liu, S. Roth, G. S. Duesberg, G.-T. Kim, and M. Schmid
Electrical Properties of Single Carbon Nanotubes 65
    A. Bachtold, C. Strunk, C. Schönenberger, J.-P. Salvetat, and L. Forró
Transport and Localization in Single-Walled Carbon Nanotubes. 69
    M. S. Fuhrer, U. Varadarajan, W. Holmes, P. L. Richards, P. Delaney,
    S. G. Louie, and A. Zettl
**Transmission Electron Microscopy and Electrical Transport Investigations
Performed on the Same Single-Walled Carbon Nanotubes** 74
    G. Philipp, M. Burghard, G. S. Duesberg, S. Roth, and K. V. Klizting
**Thermoelectric Power and Thermal Conductivity of Single-Walled
Carbon Nanotubes** 79
    J. Hone, I. Ellwood, M. Whitney, M. Muno, C. Piskoti, and A. Zettl
**Electronic Structure Studies of Single-Wall Carbon Nanotubes Using
Electron Energy-loss Spectroscopy in Transmission** 83
    M. S. Golden, T. Pichler, M. Knupfer, J. Fink, A. G. Rinzler, and R. E. Smalley
^{13}C-NMR in Carbon Nanotubes 87
    Y. Maniwa, M. Hayashi, Y. Kumazawa, H. Tou, H. Kataura, H. Ago, Y. Ono,
    T. Yamabe, and K. Tanaka
**Observation of Elastic Deformations in Single-Walled Carbon Nanotubes
by Scanning Tunneling Microscopy** 92
    W. Clauss, D. J. Bergeron, and A. T. Johnson
**Spring-Like Behaviour of Carbon Nanotubes Observed by Transmission
Electron Microscopy** 97
    W. H. Knechtel, G. S. Duesberg, and W. J. Blau
**Resonance Raman Scattering of the Radial Breathing Mode in
Single Wall Carbon Nanotubes** 101
    J. Kürti, H. Kuzmany, B. Burger, M. Hulman, A. G. Rinzler, and R. E. Smalley
**Resonant Raman Scattering and the Zone-Folding Effects in Single-Wall
Nanotubes** 107
    A. Kasuya, M. Sugano, Y. Saito, Y. Tani, K. Tohji, H. Takahashi,
    and Y. Nishina
**Resonant Raman Scattering from Single-Wall Nanotubes of Diameters
Between 1.1 nm and 1.4 nm** 112
    M. Sugano, A. Kasuya, K. Tohji, Y. Saito, and Y. Nishina
Molecular Dynamics of Single Wall Nanotubes 116
    E. Anglaret, J. L. Sauvajol, S. Rols, C. Journet, T. Guillard, L. Alvarez,
    E. Muñoz, A. M. Benito, W. K. Maser, M. T. Martinez, G. F. de la Fuente,
    D. Laplaze, and P. Bernier
**Mid-Infrared Investigations and Spatially Resolved Raman Spectra
of Singlewalled Carbon Nanotubes** 123
    C. Thomsen, H. Jantoljak, U. Kuhlmann, C. Journet, and P. Bernier
Raman Studies of Singlewalled Nanotubes 128
    M. Lamy de la Chapelle, S. Lefrant, C. Journet, P. Bernier, E. Muñoz,
    A. M. Benito, W. K. Maser, M. T. Martinez, G. F. de la Fuente, and D. Laplaze
Resonance Effects in Raman Spectra of Carbon Nanotubes 132
    E. D. Obraztsova, J.-M. Bonard, and V. L. Kuznetsov

**Micro-Raman Investigation of the Low-Energy Mode of Multiwalled
Carbon Nanotubes** 136
    H. Jantoljak, J.-P. Salvetat, L. Forró, and C. Thomsen

NANOTUBES: THEORY

Electronic Structure and Transport in Nanotube Ropes 143
    C. L. Kane and E. J. Mele
Correlation Effects in Single-Wall Nanotubes 147
    R. Egger and A. O. Gogolin
**Non Conventional Screening of the Coulomb Interaction in C_{60}
and in Carbon Nanotubes** 152
    J. van den Brink and G. A. Sawatzky
**Tight-Binding Calculation of the Elastic Properties of Single-Wall
Nanotubes** 156
    E. Hernández, C. Goze, P. Bernier, and A. Rubio
**Self-Assembly and Electronic Structure of Bundled Single-
and Multi-Wall Nanotubes** 159
    D. Tománek
**Simulation of STM Images of 3D Objects and Comparison with
Experimental Data: Carbon Nanotubes** 164
    G. I. Márk, L. P. Biró, and J. Gyulai
**Electronic Processes in Scanning Tunneling Microscopy of
Carbon Nanotubes** 168
    V. Meunier and Ph. Lambin
C_{60}-Based Molecular and Electronic Nanostructures 172
    E. V. Buzaneva, L. A. Bulavin, V. E. Pogorelov, V. N. Yashchuk,
    Y. I. Prilutski, S. S. Durov, A. V. Nazarenko, Y. A. Astashkin,
    T. Yu. Ogul'chansky, G. V. Andrievsky, and P. Scharff

FULLERENES

Computing Most Stable Fullerenes 179
    Z. Slanina, X. Zhao, and E. Ōsawa
The First Stable Lower Fullerene: C_{36} 183
    C. Piskoti and A. Zettl
Simulated Behavior of Fullerenes at High Temperatures 186
    N. Kurita and E. Ōsawa
Contact Dependence of the Conductance Through C_{60} 190
    M. Paulsson and S. Stafström
**Intermolecular Bond Stability of C_{60} Dimers and 2D Pressure-
Polymerized C_{60}** 194
    P. Nagel, V. Pasler, S. Lebedkin, C. Meingast, B. Sundqvist, T. Tanaka,
    and K. Komatsu
Energetics of Two-Dimensionally Polymerized C_{60} Materials 198
    S. Saito and S. Okada

Structural Properties of a C_{120} Crystal..203
    Ch. Laforge, P. Senet, and Ph. Lambin

Electrosynthesis and Characterization of Dimers of Mono Alkyl Adducts of C_{60}..207
    K. Noworyta, P. Kuran, L. Dunsch, and W. Kutner

Catalytic C_{120}: Vibrational Spectra and Stability..211
    B. Burger, H. Kuzmany, and K. Komatsu

C_{120} and $C_{120}O$: Vibrational Spectroscopy and PM3 Calculations..215
    H.-J. Eisler, F. H. Hennrich, and M. M. Kappes

High Pressure Synthesis and Thermal Properties of C_{60} Dimers..219
    Y. Iwasa, K. Tanoue, T. Mitani, A. Izuoka, T. Sugawara, and T. Yagi

Film Growth of $C_{59}N$ on Layered Materials..223
    B. Pietzak, C. Sommerhalter, A. Weidinger, B. Nuber, U. Reuther, and A. Hirsch

Investigations of Fullerenes and their Derivatives Synthesized in the Plasma Chemical Reactor..227
    S. G. Ovchinnikov, G. N. Churilov, L. A. Solovyov, E. A. Petrakovskaya, Ya. N. Churilova, O. V. Chupina, Ya. Koretz, N. V. Bulina, Ya. I. Pouhova, A. A. Savchenko, and V. V. Fefelova

Synthesis and Properties of Novel Fullerene Derivatives..232
    T. Da Ros, M. Prato, D. Guldi, E. Alessio, L. Valli, M. Carano, F. Paolucci, P. Ceroni, and S. Roffia

Phase Diagram of the C_{60}/C_{70} System..237
    D. Havlik, M. Steinmetz, P. Huber, W. Schranz, M. Enderle, and K. Knorr

Catalytic Reduction of 1,2-Dihaloethanes by Electrochemically Generated C_{60}^n (n = 2 and 3)..241
    F. D'Souza, J-P. Choi, Y. Y. Hsieh, K. Shriner, and W. Kutner

**Fullerene Radicals, Electrochemistry, and Electron Spin Resonance:
Part A: Anomalous Rotational Dependence of the ESR Signals of Single Crystals of $[C_{70}][I][(C_6H_5)_4P]_2$
Part B: Electrochemical and ESR Characterization of Mono-Anionic Radicals of Four Minor Isomers of C_{84}: $[84]C_1$, $[84]C_s(V)$, $[84]D_{2d}(I)$, and $[84]D_2(III)$**..245
    J.-A. Azamar, R. Clérac, C. Coulon, J. Dennis, H. Shinohara, and A. Pénicaud

In Situ ATR-FTIR Spectroscopic Investigations during Electrochemical Reduction of Fullerene Thin Films..249
    H. Neugebauer, C. Kvarnström, and N. S. Sariciftci

High-Resolution Vibronic Spectroscopy of Fullerenes in Shpolskii Systems........253
    A. N. Starukhin, B. S. Razbirin, A. V. Chugreev, Yu. S. Grushko, V. N. Zgonnik, E. Yu. Melenevskaya, M. Happ, and F. Henneberger

Studies of Photoinduced Charge Transfer in Conjugated Polymer-Fullerene Composites by Light-Induced ESR..257
    V. Dyakonov, G. Zoriniants, M. Scharber, C. J. Brabec, R. A. J. Janssen, J. C. Hummelen, and N. S. Sariciftci

Optical Properties and Heat of Solution of Fullerenes Taking Account the Aggregation of Fullerenes In Solutions..261
    V. N. Bezmelnitsyn, A. V. Eletskii, and M. V. Okun

A Spectroscopic Study of the Fullerene/Metal Interface in
C_{60}-Al Multilayers 265
    T. Pichler, T. Böske, M. S. Golden, M. Knupfer, M. Sing, J. Fink, Ch. Jung,
    C. Hellwig, and W. Frentrup

FULLERIDES

The Electronic Structures of Doped Fullerenes Studied Using High
Energy Spectroscopy 271
    T. Pichler, M. S. Golden, M. Knupfer, and J. Fink

Rb_1C_{60} as a 3D Electronic System 277
    M. Fally and H. Kuzmany

On a Description of the Normal-State Resistivity in Alkali-Metal Doped C_{60} .. 283
    Yu. Huang, A. Ito, and K. Tanaka

Mott Transition and Superconductivity in Alkali-Doped Fullerides 287
    O. Gunnarsson and E. Koch

Dielectric Screening in Doped Fullerides 292
    E. Koch, O. Gunnarsson, and R. M. Martin

Frequency Dependent ESR Study of the Magnetic Phase Transition
in $NH_3K_3C_{60}$ 296
    F. Simon, A. Jánossy, Y. Iwasa, H. Shimoda, G. Baumgartner, and L. Forró

Antiferromagnetic Resonance in Rb_1C_{60} 300
    M. Bennati, R. G. Griffin, S. Knorr, A. Grupp, and M. Mehring

Electronic Correlations and Magnetic Ordering in CsC_{60} 305
    K.-F. Thier, M. Mehring, and F. Rachdi

Local Symmetry Distortion Evidenced by ^{133}Cs NMR in the Rapidly
Quenched Cubic CsC_{60} 310
    V. Brouet, H. Alloul, F. Quéré, and L. Forró

$^{133}CsNMR$ Study Under Pressure in Linear Polymers $(CsC_{60})_n$ 314
    B. Simovic, P. Auban-Senzier, D. Jerome, G. Baumgartner, and L. Forró

Low-Frequency Modes of CsC_{60} Phases 318
    J. L. Sauvajol and E. Anglaret

Fingerprints of Solid-State Chemical Reactions in the Dynamics
of Fullerenes 322
    B. Renker, H. Schober, R. Heid, and B. Sundqvist

Fullerides Polymerisation at Ambient and Elevated Pressure 327
    S. Margadonna, C. M. Brown, A. Lappas, K. Kordatos, K. Tanigaki,
    and K. Prassides

First-Principles Study of Polymerized Alkali-Fullerene Compounds 331
    T. Ogitsu, K. Prassides, K. Tanigaki, K. Kusakabe, and S. Tsuneyuki

Two Dimensional Fulleride Polymers 335
    M. Christensen and S. Stafström

Ferromagnetic Resonance and High Field ESR in a TDAE-C_{60}
Single Crystal 340
    D. Arčon, P. Cevc, A. Omerzu, R. Blinc, M. Mehring, S. Knorr, A. Grupp,
    A.-L. Barra, and G. Chouteau

Perdeuteration of TDAE in [TDAE]C$_{60}$: A Study by ESR and NMR 344
    A. Schilder, W. Bietsch, J. Gmeiner, and M. Schwoerer
Searching for the Reactions of Fullerenes with Fe Compounds 348
    P. Byszewski, E. Kowalska, J. Radomska, Z. Kucharski, R. Diduszko,
    A. Huczko, H. Lange, R. Kochkanjan, A. Zaritowskij, A. Bondarenko,
    and V. Chabanenko
On the Structure of Iron Fullerene Complexes 353
    E. Kowalska, Z. Kucharski, and P. Byszewski
Spectroscopy of C$_{60}$ and C$_{70}$ Complexes 357
    D. V. Konarev, N. V. Drichko, V. N. Semkin, Y. M. Shul'ga, A. Graja,
    and R. N. Lyubovskaya

ENDOHEDRALS

Study of N@C$_{60}$ and P@C$_{60}$ 363
    A. Weidinger, B. Pietzak, M. Waiblinger, K. Lips, B. Nuber, and A. Hirsch
Production, HPLC Separation, and UV-vis Spectroscopy of Li@C$_{70}$ 368
    N. Krawez, A. Gromov, R. Tellgmann, and E. E. B. Campbell
Heterogeneous Electron Transfer at Endohedral Fullerenes 373
    L. Dunsch, P. Kuran, and M. Krause
**Far- and Mid-Infrared Transmission for Two Isomers of the Endohedral
Metallofullerene Sc$_2$@C$_{84}$** 379
    M. Hulman, M. Inakuma, H. Shinohara, and H. Kuzmany
**Electron Paramagnetic Resonance Investigation of Phosphorus and Nitrogen
in [60] Fullerene** 383
    C. Knapp, A. Adla, N. Weiden, H. Käß, K.-P. Dinse, B. Pietzak, M. Waiblinger,
    and A. Weidinger
Thermal Stability of N@C$_{60}$ 388
    M. Waiblinger, B. Pietzak, K. Lips, and A. Weidinger
Stabilisation of Atomic Elements Inside C$_{60}$ 392
    H. Mauser, T. Clark, and A. Hirsch
EPR Studies of N@C$_{60}$ and its Adducts 396
    A. Gruß, C. Knapp, N. Weiden, K.-P. Dinse, E. Dietel, A. Hirsch, B. Pietzak,
    M. Waiblinger, and A. Weidinger

CLUSTERS

**Giant Gold-Cluster Compounds—Gaps in Optical and Charging Spectra,
and an Electronic Origin of Abundance Anomalies** 403
    R. L. Whetten, M. M. Alvarez, T. Bigioni, J. T. Khoury, B. E. Salisbury,
    T. G. Schaaff, M. N. Shafigullin, and I. Vezmar
Proton NMR of a Manganese-Ion Cluster Nanomagnet 407
    F. Milia, R. Blinc, R. M. Ashey, and N. S. Dalal
**Deuteron Magnetic Resonance and Relaxation in a Manganese-Ion Cluster
Nanomagnet** 412
    R. Blinc, J. Dolinšek, T. Apih, D. Arčon, R. M. Achey, and N. S. Dalal

Synthesis and Characterization of Divalent Metal Containing Mesoporous
Silicas by an Ionic Templating Route 416
    D. Petridis and M. A. Karakassides
Carbon Clusters—Size Effects of Properties 420
    G. Seifert, K. Vietze, and P. W. Fowler
Laser Induced Emission Spectroscopy of Carbon Clusters in Solid Argon 425
    I. Čermák, M. Förderer, S. Kalhofer, I. Čermáková, and W. Krätschmer
Carbon Onions Produced by Ion-Implantation 430
    T. Cabioc'h, M. Jaouen, M. F. Denanot, J. P. Rivière, J. Delafond,
    and J. C. Girard
Molecular Dynamics Study of Carbon Structures 435
    I. László
Plasmon Excitations in Carbon Onions: Model vs. Measurements 439
    Th. Stöckli, Z. L. Wang, J.-M. Bonard, P. A. Stadelmann, and A. Châtelain
Highly Symmetric Borane Clusters as Fullerene Analogs 443
    Z. Szekeres and P. R. Surján
Pulse ESR of Triplet States of Large Molecular π Systems 447
    S. Knorr, A. Grupp, M. Mehring, M. Wehmeier, P. Herwig, V. S. Iyer,
    and K. Müllen
Metallic, Insulating, and Superconducting States in κ-ET$_2$X Systems,
Where ET is the BEDT-TTF (bis(ethylenedithio)tetrathiafulvalene)
Molecule 451
    V. A. Ivanov, E. A. Ugolkova, and M. Ye. Zhuravlev
STM Studies of Synthetic Peptide Monolayers 456
    D. J. Bergeron, W. Clauss, D. L. Pilloud, P. L. Dutton, and A. T. Johnson
Self-Assembly of Ropes of Cyanine Dye Molecules 460
    S. Blumentritt, M. Burghard, and S. Roth

APPLICATIONS

Physical Properties of Carbon Nanotubes 467
    J.-P. Salvetat, J.-M. Bonard, R. Basca, Th. Stöckli, and L. Forró
Hydrogen Storage in Carbon Materials—Preliminary Results 481
    L. Jörissen, H. Klos, P. Lamp, G. Reichenauer, and V. Trapp
Pressure Dependent ^{1}H-NMR-Measurements of Activated Carbon
and Carbon Nanofibers 485
    W. Schütz and H. Klos
Encapsulation of Ferromagnetic Metals into Carbon Nanoclusters 489
    S. Seraphin, J. Jiao, C. Beeli, P. A. Stadelmann, J.-M. Bonard, and A. Châtelain
Fullerene Incorporated Nanocomposite Resist Systems for Practical
Nano-Fabrication 494
    T. Ishii, T. Shibata, H. Nozawa, and T. Tamamura
Cluster Structure and Elastic Properties of Superhard and Ultrahard
Fullerites 499
    V. D. Blank, S. G. Buga, N. R. Serebryanaya, G. A. Dubitsky, V. M. Prokhorov,
    M. Yu. Popov, N. A. Lvova, V. M. Levin, and S. N. Sulyanov

Electronic Properties of Nanotube Junctions 504
    Ph. Lambin and V. Meunier

Phthalocyanine-C_{60} Composites as Improved Photoreceptor Materials? 509
    B. Kessler, C. Schlebusch, J. Morenzin, and W. Eberhardt

New Metallilc Alloys Incorporating Fullerenes and Carbon Nanotubes 515
    R. J. Doome, A. Fonseca, and J. B. Nagy

Realization of Large Area Flexible Fullerene—Conjugated Polymer Photocells: A Route to Plastic Solar Cells 519
    C. J. Brabec, F. Padinger, V. Dyakonov, J. C. Hummelen, R. A. J. Janssen, and N. S. Sariciftci

Fullerenes and Nanostructured Plastic Solar Cells 523
    J. Knol and J. C. Hummelen

Semiconductor Device Structure Based on Fullerene: Ag/C_{60} Thin Film Schottky Barrier 527
    E. A. Katz, D. Faiman, S. Shtutina, B. Mishori, and Y. Shapira

APPENDIX

Scientometrics on Fullerenes and Nanotubes 533
    W. Marx, M. Wanitschek, and H. Schier

Author Index 545

Preface

The present book contains the proceedings of the 12th International Winterschool on Electronic Properties of Novel Materials in Kirchberg, Tyrol, Austria. The series of these schools started in 1985. Originally the school was held every second year and was devoted to conducting polymers. After the discovery of high temperature superconductors the periodicity changed to an annual format and the topic alternated between conjugated polymers and superconductors. Since fullerenes are both conjugated compounds and (in some cases) superconductors, it was tempting to choose fullerenes as topic for the Kirchberg schools. The evident extension of this topic is carbon nanotubes and so the title changed from Fullerenes via Fullerene Derivatives and Fullerene Nanostructures to Molecular Nanostructures. This gradual change enables us to keep a fairly large interdisciplinary scientific community together and to stimulate numerous international cooperations. A compilation of the previous Kirchberg Winterschools will be presented in the table at the end of this preface.

The term Molecular Nanostructures implies the "bottom-up" (synthetic) approach, as opposed to the "top-down" (lithography and etching) techniques in semiconductor technology. As for the physics, we are in a field where solid state physics and molecular physics overlap. This is nicely seen on the example of carbon nanotubes: Their diameter is in the order of a few nanometers and thus perpendicular to their axis nanotubes are molecular (different diameters lead to different electronic structures), along their axis they are extended solids.

Most oral contributions to the winterschool were on carbon nanotubes, and there are also three chapters of the proceedings and a large part of the chapter on applications devoted to this topic. The majority of the posters was on fullerenes and fullerides. A special chapter deals with endohedrally "doped" fullerenes, and in the chapter on "clusters" an attempt is made to point to common aspects in carbonaceous and in inorganic nanometer-sized structures.

The meeting could not have taken place without financial support from the Bundesministerium für Wissenschaft und Forschung in Austria, the US Army Research Development and Standardization Group in London, UK, and the Verein zur Förderung der Internationalen Winterschulen in Kirchberg, Austria, as well as from numerous industrial sponsors. Without their contribution, all the enthusiasm and dedication could be wasted and so we express our gratitude to the sponsors and supporters.

Finally, we are indebted, as ever, to the managers of Hotel Sonnalp, Herr Gradnitzer, and Frau Jurgeit, and to their staff for their continuous support and for their patience with the many special arrangements required during the meeting.

H. Kuzmany
J. Fink
M. Mehring
S. Roth

Table of Previous Kirchberg Winterschools

Year	Title	Published by
1985	Electronic Properties of Polymers and Related Compounds	Springer Series in Solid-State Sciences 63
1987	Electronic Properties of Conjugated Polymers	Springer Series in Solid-State Sciences 76
1989	Electronic Properties of conjugated Polymers III - Basic Models and Applications	Springer Series in Solid-State Sciences 91
1990	Electronic Properties of High-T_c Superconductors and Related Compounds	Springer Series in Solid-State Sciences 99
1991	Electronic Properties of Polymers - Orientation and Dimensionality of Conjugated Systems	Springer Series in Solid-State Sciences 107
1992	Electronic Properties of High-T_c Superconductors	Springer Series in Solid-State Sciences 113
1993	Electronic Properties of Fullerenes	Springer Series in Solid-State Sciences 117
1994	Progress in Fullerene Research	World Scientific Publ., 1994
1995	Physics and Chemistry of Fullerenes and Derivatives	World Scientific Publ. 1995
1996	Fullerenes and Fullerene Nanostructures	World Scientific Publ., 1996
1997	Molecular Nanostructures	World Scientific Publ., 1998

Sponsors

AVL LIST GmbH, Kleiststraße 48, A-8020 Graz, Austria

BRUKER Analytische Meßtechnik GmbH, Wikingerstraße 13, D-7500 Karlsruhe 21, Germany

CREDITANSTALT BANKVEREIN, Nußdorferstraße 2, A-1090 Wien, Austria

DIGITAL INSTRUMENTS, Janderstraße 9, D-68199 Mannheim, Germany

HOECHST AG, Brüningstraße 50, D-6230 Frankfurt am Main 80, Germany

Fa. E. MILLER, Mariahilferstraße 93, A-1060 Wien, Austria

PIRELLI, Viale Sarca 222, 20126 Milano, Italy

SGL CARBON, W. v. Siemens Straße, 86404 Meitingen, Germany

VAKUUM- und SYSTEMTECHNIK GmbH (ALCATEL), Hohenauergasse 10, A-1190 Wien, Austria

The financial assistance from the sponsors and from the supporters is greatly acknowledged.

NANOTUBES: PREPARATIONS

Production of Carbon Single Wall Nanotubes Versus Experimental Parameters

C. Journet[1], V. Micholet[1], P. Bernier[1], W.K. Maser[1*], A. Loiseau[2], M. Lamy de la Chapelle[3], S. Lefrant[3], R. Lee[4], J.E. Fischer[4]

1 Groupe de Dynamique des Phases Condensées, Université Montpellier, 34095 Montpellier, France
* Present adress: Instituto de Carboquimica, CSIC, 50015 Zaragoza, Spain
2 Laboratoire de Physique de Solide, ONERA, 92322, Châtillon cedex, France
3 Laboratoire de Physique Cristalline, IMN, Université de Nantes, 44072 Nantes cedex 3, France
4 Laboratory for Research on the Structure of Matter, University of Pennsylvania, Philadelphia, PA 19104, USA

Abstract. Bundles of carbon single wall nanotubes (SWNTs) are produced by sublimating selected metal mixtures and carbon in an inert atmosphere during an electric arc [1]. Various experimental parameters such as the nature and relative proportions of metallic catalysts [1] or the kind and pressure of gas can influence the quantity and geometry of bundles produced by the arc process. In this paper, we particularly focus on the role of the nature and pressure of gas used. Systematic studies have been made and we present the results obtained by Scanning Electron Microscopy (SEM), High Resolution Transmission Electron Microscopy (HRTEM), X-Ray Diffraction (XRD) and High Resolution Raman Spectroscopy (HRRS).

INTRODUCTION

Since their observation for the first time [2] as a by-product of fullerene materials formed by the arc technique, carbon nanotubes have attracted a great interest from a fundamental point of view and as well for future applications. Even if some production techniques have been improved [1, 3], the main point remains obtaining large quantities of pure nanotubes. Only a systematic study of the production parameters and their influence on the growth of carbon nanotubes can enhance the quantity and quality of nanotubes produced. In the following sections, we will study the effect of the nature and pressure of gas used during the production of carbon SWNTs by the arc technique.

Experimental

The synthesis are carried out in a water-cooled experimental chamber [4] first evacuated and then filled with an inert gas. Two graphite rods (also cooled with water) are used as electrodes. One is pure graphite while the other one is drilled and filled with a mixture of Ni: Y: C. This composite rod, acting as the anode, is moved towards the fixed cathode until the distance between them is so small that a current (100 A) passes through them, producing an electric arc discharge. The temperature of the plasma created in the interelectrode region is extremely high (on the order of 3000 K) and carbon sublimates. By controlling the voltage (around 35-40 V), the distance between rods can be controlled and kept constant, in order to reduce the fluctuations of the plasma.

Various pressures of both helium and argon ranging from 100 to 1500 mbar have been tested. The macroscopic products obtained are very different depending on the gas and pressure. At low pressures (from 100 to 300 mbar) of helium, a small (Å1g, 1.5 cm long) hard grey deposit grows at the surface of the cathode. Around this deposit is formed a small (Å200 mg), black and crumbly collaret. Crumbly soot is also collected from the reactor walls. Increasing the helium pressure between 400 and 800 mbar, we observe a very long (Å2.5 g, 5.5 cm long) deposit at the surface of the cathode and a very large (Å600 mg) and spongious black collaret similar to a soft belt around it. A lot of web-like structures can be found growing from the cathode to the reactor walls where rubbery soot is now collected in large pieces. At higher pressures (900-1500 mbar), a medium (Å2g, 3.5 cm long) deposit is grown on the cathode. When formed (under 1300 mbar), the collaret is very small (Å20 mg). A few webs are found and the soot on the reactor walls is again crumbly. When using argon instead of helium, we find, at low pressures (100-300 mbar), a medium (Å2g, 3 cm long) deposit, a very small collaret (Å50 mg), no webs and a very few of crumbly soot. At intermediate pressures (400-800 mbar), the deposit is bigger (Å3g, 6 cm long) but the collaret is always small (Å20 mg) and no webs are formed. A very small quantity of crumbly soot is collected from the reactor walls. Between 900 and 1500 mbar, the deposit is quite long (Å3g, 5 cm long), the collaret is becoming smaller and smaller (Å10 mg), the soot is crumbly and no webs are observed.

Known to contain the highest density of SWNTs among the products formed by the arc technique [1], the collaret was studied in details using SEM, HRTEM, XRD and HRRS for each of the experimental conditions described above.

Results

SEM pictures obtained at low (100-300 mbar) (fig.1) and high (900-1200 mbar) (fig.3) pressures of helium show a low density of fibrilar structures demonstrated to be bundles of SWNTs [1]. On the contrary, at medium (400-800 mbar) pressures (fig.2), a large amount of such fibrilar structures curved and crossing each other is observed. The diameter of these fibriles is between 10 and 20 nm. Their length is always more than 1mm.

These results are well correlated with HRTEM studies. At low pressures, we observe a low density of short and small bundles of carbon SWNTs covered with amorphous carbon. Increasing the pressure, we increase the density, the length and the size of bundles of carbon SWNTs. In a bundle, a few tens individual SWNTs are found to be very well organized in a two-dimensional triangular lattice with a parameter of about 17 Å. These bundles are free of amorphous carbon at medium pressures while they are covered with at higher pressures. In this last case, also isolated tubes are observed.

XRD studies (fig.7) on the samples obtained at intermediate helium pressures show an intense peak at $Q=0.43$ Å^{-1} and smaller ones in the low Q region (from 0 to 2.5 Å^{-1}) while no peaks are observed at low and high pressures. These peaks have already been explained as related to the organization of SWNTs in a two-dimensional triangular lattice having a parameter of 17 Å [1, 3] which is consistent with the HRTEM observations. At the high Q values and for each pressure, we observe two peaks (3.1 and 3.6 Å^{-1}) which correspond to pure (111) and (200) FCC nickel respectively.

HRRS experiments (fig.8) show no effect of the pressure on the diameter of the SWNTs. Indeed whatever the pressure is, the obtained spectra are exactly the same, presenting a narrow distribution of tube diameters between 10 and 20 Å.

Concerning the results obtained by varying the argon pressure, preliminary SEM pictures show that the highest density of fibriles structures is observed at low pressures (fig.4,5,6). Compared to helium, the optimal pressure for the formation of carbon SWNTs with argon seems to be lower. With the arc technique, results obtained with argon are quite different from those obtained with helium while they are exactly the same using helium or argon with the laser ablation method.

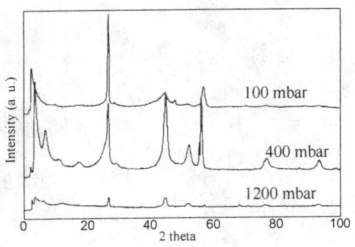 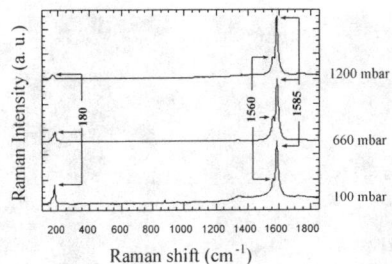

Fig.7: XRD spectra at various pressures   **Fig.8:** Raman spectra at various pressures

Discussion

From the observations obtained by varying the helium pressure, it seems that pressure forces the individual SWNTs to arrange themselves in bundles. From 400 mbar of helium, the SWNTs are very well organized in a two-dimensional triangular lattice while this is not observed at lower presures. There seems to be an optimal pressure around 400-600 mbar where the bundles are very long, very large and not covered with amorphous carbon.

From the observations obtained by varying the argon pressure, we can say that the optimal pressure for the formation of carbon SWNTs is shifted to lower pressures. It seems that helium is more favorable to the production of carbon SWNTs than argon as the density of fibrilar structures is higher in the case of helium than in the case of argon but we need more characterization experiments.

Summary

Only systematic studies of the influence of the experimental parameters on the production of carbon SWNTs can lead to optimal conditions and can give a better understanding of their growth mechanism.

ACKNOWLEDGEMENTS

The Montpellier and Nantes groups thank the European Economic Community for financial support through its Training and Mobility of Researchers Programme under contract n° ERBFMRX-CT96-0067.

REFERENCES

1. Journet C. et al., *Nature* **388**, 756-758 (1997)
2. Iijima S., *Nature* **354**, 56-58 (1991)
3. Thess A. et al., *Science* **273**, 483-487 (1996)
4. Zahab A., *PhD thesis*, Université Montpellier II (1992)

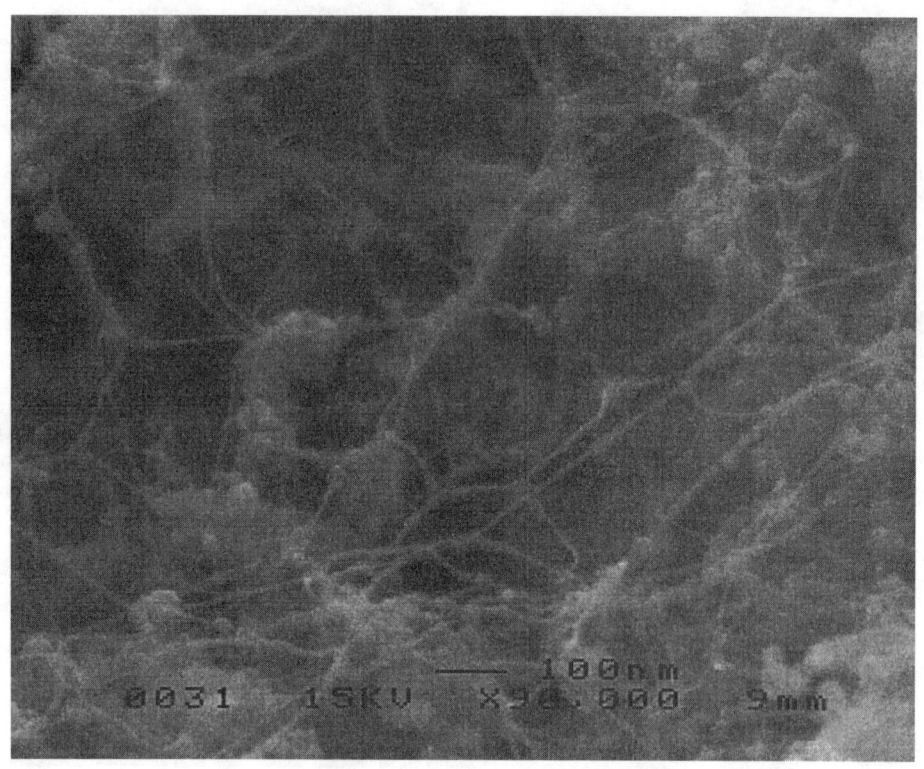

Fig.1: 100 mbar of helium

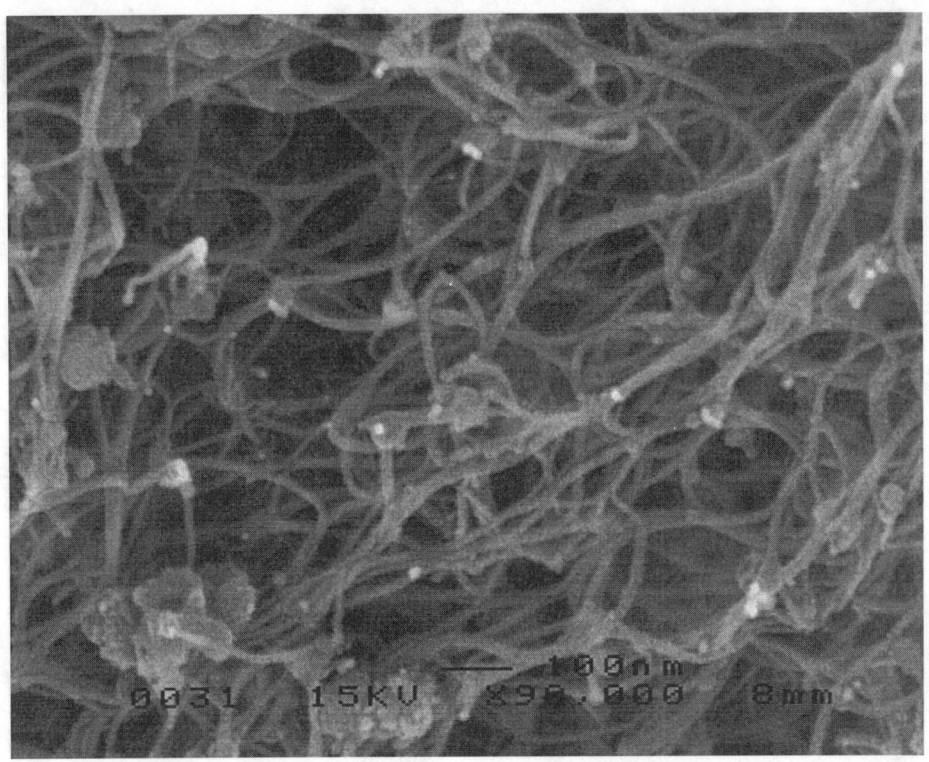

Fig.2: 400 mbar of helium

Fig.3: 1000 mbar of helium

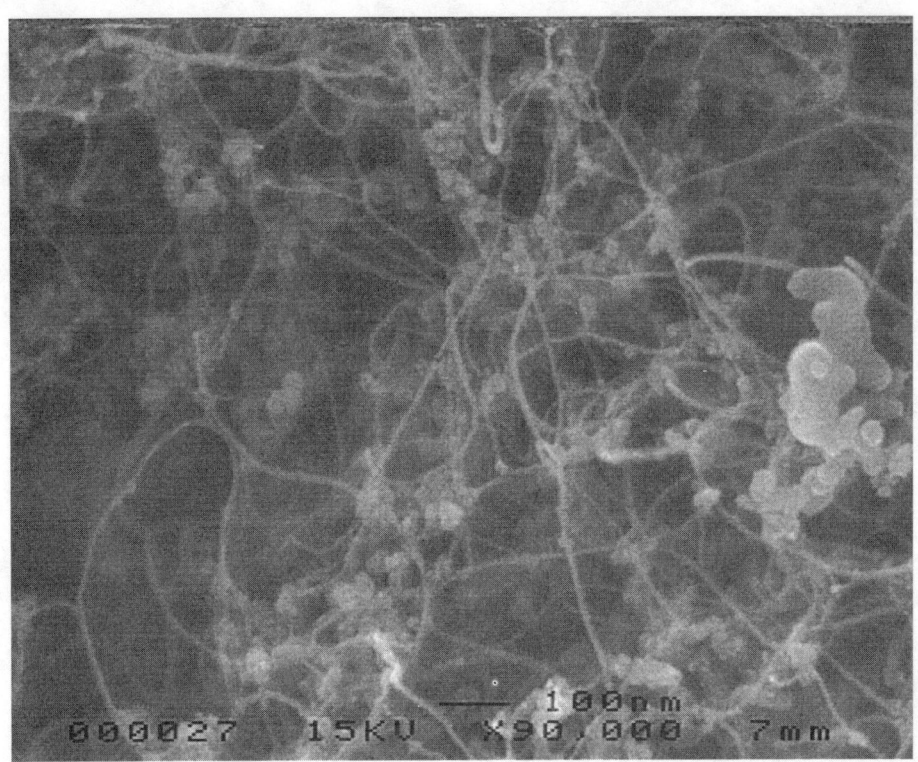

Fig.4: 100 mbar of argon

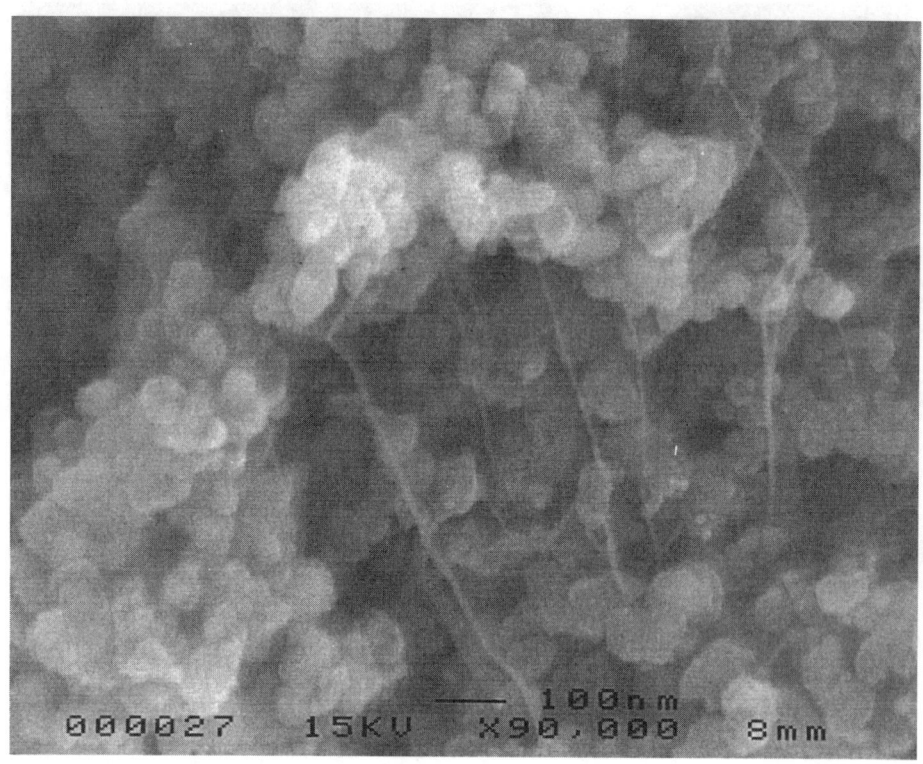

Fig.5: 500 mbar of argon

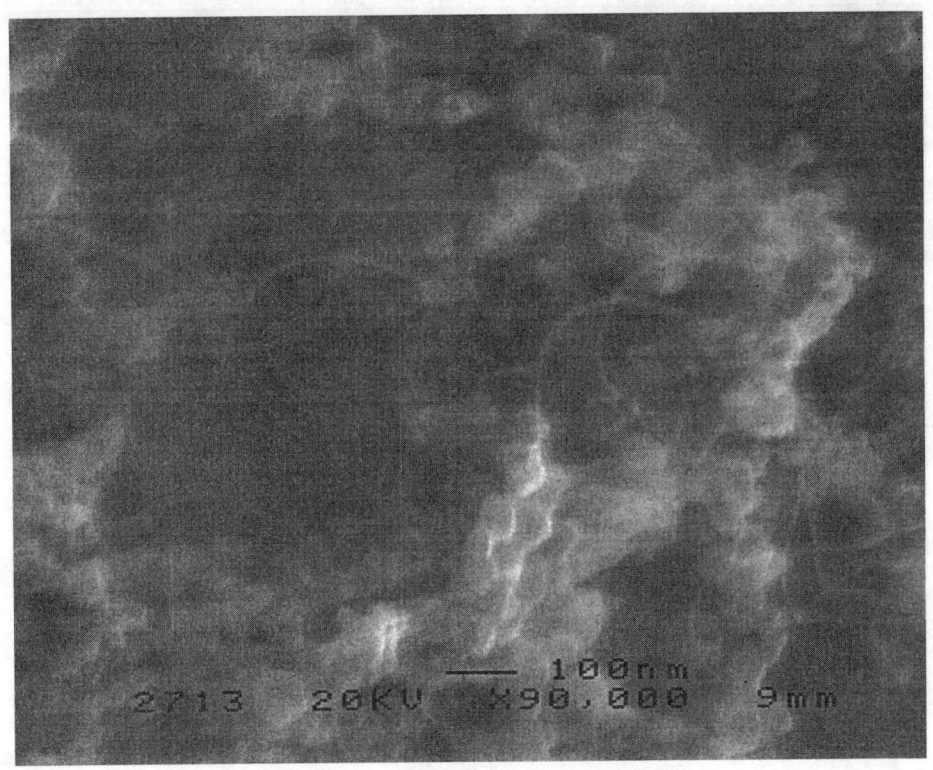

Fig.6: 1200 mbar of argon

Laser-assisted Production of Multi-walled Carbon Nanotubes from Acetylene

Henning Kanzow, Andrea Schmalz and Adalbert Ding

Technische Universität Berlin, Optisches Institut, Sekr. P 1-1, Str. d. 17. Juni 135, D-10623 Berlin

Abstract. A batch method for the gas phase production of carbon nanotubes by the catalytic decomposition of acetylene has been developed. Metal clusters are formed by laser vaporization of solid nickel in situ in the hot reaction tube containing a mixture of acetylene and argon. The laser generated clusters act as catalysts for the growth of nanotubes. With this method multi-walled tubes were obtained with inner diameters of 3 - 10 nm and outer diameters of 10 - 100 nm, respectively. TEM analysis shows that the tubes are well graphitized.

INTRODUCTION

For at least four decades the chemical vapour deposition (CVD) has been known as a technique suitable for the production of filamentous carbon by the catalytic decomposition of carbon monoxide or hydrocarbons on metal catalysts (1-7).
During the past years new methods for the production of carbon fibrils have been developed which are based on the co-vaporization of graphite and a metal catalyst either in an electric arc discharge (8-10) or by high power laser irradiation (11).
We have combined the laser vaporization technique with the CVD method. In our set-up the laser is used to vaporize the solid catalyst and generate catalytic metal aggregates in situ in the hot reaction tube while the carbon is being introduced as a gaseous compound. In this way it should be possible to achieve higher absolute yields, since the supply of carbon is not limited by the vaporization of solid graphite. The growth of tubes take place in the gas phase as well as at the walls of the reaction tube. Most of the process parameters can be controlled independently.

EXPERIMENTAL

The experiments have been performed in a reaction chamber originally designed for the laser vaporization of graphite and modified according to Figure 1. A nickel target attached to a water cooled rod was placed in the center of the tube. The quartz reaction chamber was filled with a mixture of argon and acetylene at a pressure of 300 mbar (P(Ar) = 260 mbar, P(C2H2) = 40 mbar). The focussed laser beam of the high power laser was scanned over the target in a circle using a rotating mirror. Experimental results were obtained with two pulsed Nd:Yag-lasers with λ=1064 nm, ν=20 Hz, pulse length 5 ns or 30 ns and pulse energy 700 mJ or 1000 mJ respectively. The laser was operated for 2 to 4 minutes generating highly excited nickel aggregates.
After cooling down the reaction chamber the reaction products were removed from the target holder and from the walls of the chamber. The samples were examined using scanning electron microscopy (SEM), trans-mission electron microscopy (TEM) and energy-dispersive X-Ray spectroscopy (EDX).

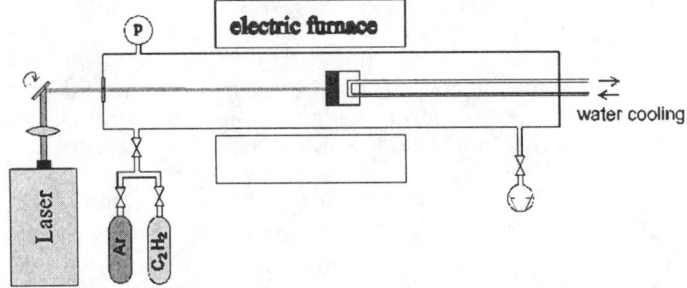

Figure 1. Experimental set-up

RESULTS

The laser parameters were not optimized with respect to the particle size. From TEM pictures of our samples the size of the nickel aggregates can be estimated to be between less than 1 nm and 100 nm.

The reaction product removed from the chamber contains parts that consists almost entirely of multi-walled nanotubes as shown in Figure 2. These tubes have outer diameters of 10 to 40 nm, in some cases 50 to 100 nm. Inner diameters are between 3 and 10 nm. High resolution TEM clearly demonstrate an almost perfect crystalline graphitic structure of the walls. Some of the tubes have a nickel aggregate at its end. Others have open ends, possibly by a loss of the catalyst particle. Most of the metal aggregates found at the end of the tubes are rather elongated than spherical. The size of these particles is 15 to 25 nm in the smaller dimension. Clusters incorporated in or between tubes vary between 5 and 10 nm in diameter. Other tube containing parts of the samples are covered with a significant amount of amorphous carbon. Very small nickel clusters are often embedded in this material between the tubes.

Figure 2. TEM / Multi-walled Nanotubes

GROWTH MODEL

The growth model for CVD was first proposed from Baker et al (4) who found a remarkable correlation of the directly measured activation energies for the growth of the fibrils with those for the diffusion of carbon through the corresponding metals. The model describes the formation of carbon nanotubes in four stages:

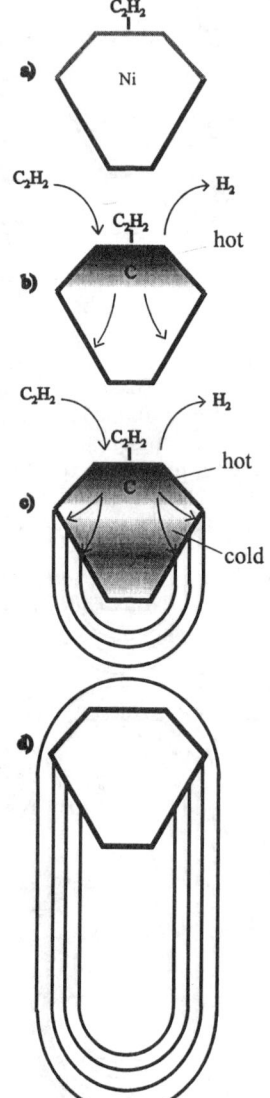

1. Hydrocarbon molecules are adsorbed at the surfaces of small metal aggregates. Its surfaces have different catalytic activity. (Figure 3a)

2. The active metal surfaces crack the carbon – hydrogen bonds. The carbon diffuses into the bulk material. For unsaturated hydrocarbons this process is highly exothermic, leading to a heating of the active sides. (Figure 3b)

3. When the saturation limit for carbon at the cooler surfaces of the particle is reached, the carbon is precipitated from the metal surface which is an endothermic process. The resulting temperature gradient from the active surfaces to the other sides within the particle is the driving force for the carbon diffusion through the particle. The rather slow diffusion determines the overall reaction rate. To avoid energetically unfavorable dangling bonds a carbon tube with a closed cap is formed. (Figure 3c)

4. In the hot environment the acetylene also reacts with itself to form larger molecules, which condense at the surfaces of the tubes and particles. If this excess carbon does not diffuse fast enough into the metal particles, the metal particles become completely encapsulated and thus stopping further growth. (Figure 3d)

Figure 3. Mechanism of Carbon Nanotube Formation

DISCUSSION

In our experiment the laser generated nickel aggregates react with the acetylene either directly in the gas phase or after deposition on the quartz walls of the reaction chamber. The smaller the particles are the faster should be the growth rate. For big particles the reaction should be too slow due to the long diffusion lengths and the short reaction times. This explains why there is almost no contribution of aggregates larger than 25 nm under our reaction conditions.

We believe that very small particles are also not useful for the growth of tubes. To avoid dangling bonds during the growth process the whole tube end should be in contact with the metal surface. But tubes of very small diameters are not favorable because of its inner strain. Therefore the particle's diameter should be at least as large as the tube's ones.

The tube growth in the gas phase according to the model should only take place, if there are surfaces of different reactivity. Theoretical studies (12) suggest that icosahedral cluster structures are favored for small clusters with less than a few thousand atoms, while single crystals are estimated to occur at 17000 atoms corresponding to a minimum particle size of about 7 nm.

ACKNOWLEDGEMENT

This work has been funded by the German Ministry of Science and Technology (BMBF) under contract no. 13N6705.

The SEM pictures were taken by J. Nissen, Central Institute for Electron Microscopy (ZELMI), Technical University Berlin.

TEM was performed by T. Link, Technical University Berlin, and N. Pfänder, Fritz-Haber-Institut Berlin.

The lasers were provided by the Laser and Medicine Technology Berlin (LMTB) gGmbH.

REFERENCES

1. L.V.Radushkevich, V.M.Luk´yanovich, Zhur.Fiz.Khim., 26, 88 (1952), C.A., 47, 6210 (1953)
2. L.J.E.Hofer et al., J.Phys.Chem. 59 (1955) 1153-55
3. P.A.Tesner et al., Carbon 8 (1970) 435-42
4. R.T.K.Baker et al., J.Catal. 26 (1972) 51-62
5. J.S.Speck et al., J. Cryst. Growth 94 (1989) 834
6. V.Ivanov et al., Chem.Phys.Lett. 223 (1994) 329-335
7. M.Terrones et al., Nature, 388 (1997) 52-55
8. S.Iijima, Nature, 354 (1991) 56-58
9. S.Iijima,T.Ichihashi, Nature, 363 (1993) 603-5
10. D.S.Bethume et al.,Nature, 363 (1993) 605-7
11. T.Guo et al., Chem.Phys.Lett. 243 (1995) 49-54
12. C.L.Cleveland, U.Landman, J.Chem.Phys. 94,11 (1991) 7377-96

Effect of Hydrogenation on Catalytically Produced Carbon Nanotubes

P. Piedigrosso, J.-F. Colomer, A. Fonseca and J. B.Nagy

Laboratoire de Résonance Magnétique Nucléaire,
Facultés Universitaires Notre-Dame de la Paix,
61, Rue de Bruxelles, B-5000, Namur, Belgium

Abstract. Carbon nanotubes can now be produced in large scale by catalytic decomposition of acetylene in the presence of various supported metal catalysts. However, purification of the tubes obtained by this process is very difficult because it requires the separation both from the catalyst (support and metal particles) and the amorphous carbon which is a product of the thermal decomposition of hydrocarbons. In this work, separation of nanotubes from the catalyst has been carried out using an acidic treatment. On the other hand, the elimination of amorphous carbon by hydrogenation has also been investigated. The quality of the nanotubes was characterized by means of transmission electron microscopy and the yield of pure nanotubes was also determined.

INTRODUCTION

Since 1993, the catalytic decomposition of acetylene in the presence of various supported transition metal catalysts has been widely investigated in our laboratoty (1-3). In these studies, it was shown that the catalytic decomposition of acetylene is an effective method to synthesize carbon nanotubes of different forms and having lengths up to 60 μm. However, the tubes produced by this method are always associated with a significant amount of amorphous carbon which comes from the thermal decomposition of the hydrocarbon. In this work, elimination of amorphous carbon by a hydrogenation treatment was investigated.

EXPERIMENTAL

Carbon nanotubes were produced by catalytic decomposition of acetylene over supported catalyst Co / zeolite NaY containing about 2.5% weight of metal. The catalyst was prepared by the impregnation method described earlier (4). The synthesis of nanotubes was carried out in a fixed-bed flow reactor (quartz tube of 50 mm diameter, 80 mm in length in a Carbolite horizontal reactor) at 600°C with a time of contacting of acetylene (reaction time) of 60 min. The acetylene and nitrogen flow were 20 ml/min and 110 ml/min, respectively. For the characterization of the samples obtained by this catalytic process, transmission electron microscopy (Philips CM 20) was used.

From figure 1a, it can be seen that catalyst particles are covered by a large amount of carbon nanotubes. Separation of the nanotubes from the catalyst (support and metal particles) was performed using fluorhydric acid (4-5) with stirring during 24 hours. According to low resolution TEM observations, amorphous carbon (dark spots in Figure 1b) is released from the inner pores of the zeolite support during the acidic treatment.

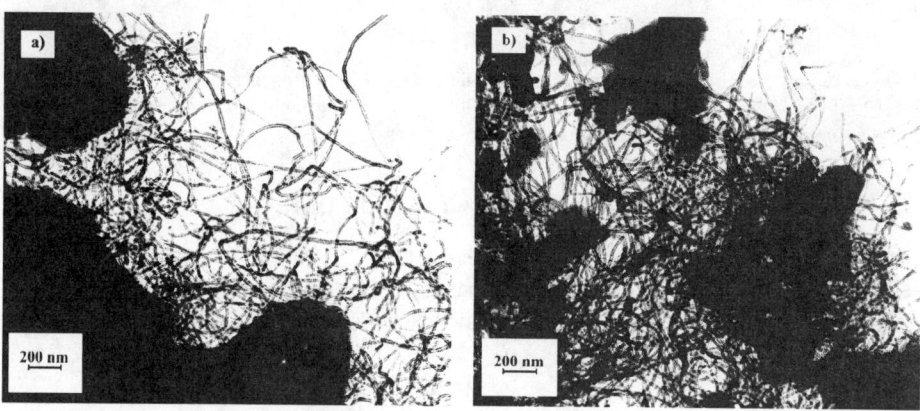

FIGURE 1. Low magnification TEM images of carbon nanotubes produced in the decomposition of acetylene at 600°C over Co / NaY catalyst. a) As made nanotubes oover the support b) Nanotubes recovered by filtration after dissolution of the catalyst in HF.

Elimination of amorphous carbon by hydrogenation was then investigated under different conditions of temperature, reaction time and gas flows. For this purification process, each time 50 mg of carbon deposit obtained after the HF treatment (Figure 1b) hydrogenated in a horizontal reactor.

RESULTS AND DISCUSSION

Influence of hydrogenation temperature

The reaction temperature was varied between 750 and 1000°C under different conditions of hydrogenation. It was observed that by varying the temperature, the carbon yield obtained after the hydrogenation treatment is unchanged (See graph 1). Concerning the quality and purity of hydrogenated nanotubes, low magnification TEM images showed that amorphous carbon could be partially removed from the initial sample during hydrogenation (Figure 2a) but nanotubes were also consumed (Figure 2b).

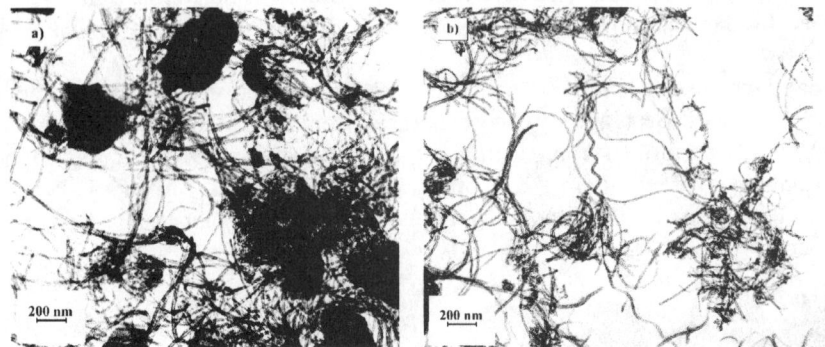

FIGURE 2. Low magnification images of carbon nanotubes after hydrogenation in the conditions: H_2 flow 10 ml/min, N_2 flow 10 ml/min, temperature 900 °C. a) Amorphous carbon is partially removed from the sample. b) Long and mainly short carbon nanotubes.

Influence of the reaction time

The reaction time was investigated between 5 min to 8 hours. As a result, the carbon yield decreased when the reaction time increased (Graph 2), but no selectivity was observed in the hydrogenation process: amorphous carbon but also nanotubes were partially eliminated from the initial sample.

Influence of the hydrogen and nitrogen flows

The hydrogen and nitrogen flows were both varied between 10 ml/min to 50 ml/min. These series of experiments showed no variation of the carbon yield and hand no selectivity between nanotubes and amorphous carbon in the hydrogenation process: they were consumed in the same time.

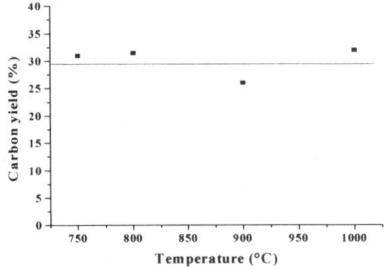
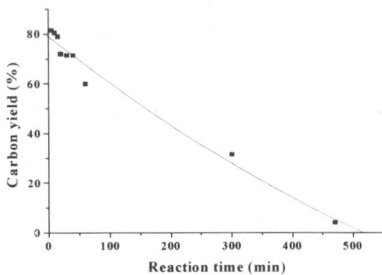

GRAPH 1. Carbon yield variation versus temperature in the following conditions: H_2 flow 50 ml/min, N_2 flow 50 ml/min, reaction time 270 min.

GRAPH 2. Carbon yield variation versus reaction time in the following conditions: H_2 flow 10 ml/min, N_2 flow 10 ml/min, 900 °C.

CONCLUSION

Although hydrogenation can eliminate part of the amorphous carbon contained in the samples produced by catalytic decomposition of acetylene, this treatment is not selective and destroys also the nanotubes. The lack of selectivity is probably due to the high temperature required for the hydrogenation process (> 700°C).

ACKNOWLEDGMENTS

This work was fully supported by the European Commission, TMR Program, " Program Research Network " (Contract n° ERBFMRX 96-0067). This text presents research results of the Belgian Programme on Inter University Poles of Attraction initiated by the Belgian State, Prime Minister's Office of Science Policy Programming (PAI/IUAP n°4).

REFERENCES

1. V. Ivanov, A. Fonseca, J. B.Nagy, A. A. Lucas, D. Bernaerts and X. B. Zhang, Carbon 33, 1727-1738 (1995).
2. K. Hernadi, A. Fonseca, J. B.Nagy, D. Bernaerts, J. Riga and A. A. Lucas, Synth. Met. 77, 31-34 (1996).
3. V. Ivanov, J. B.Nagy, Ph. Lambin, A. A. Lucas, X. B. Zhang, X. F. Zhang, D. Bernaerts, G. Van Tendeloo, S. Amelincks and J. Van Landuyt, Chem. Phys. Lett. 223, 329 (1994).
4. K. Hernadi, A. Fonseca, J. B.Nagy, D. Bernaerts, A. Fudala and A. A. Lucas, Zeolites 17, 416-423 (1996).
5. A. Fonseca, K. Hernadi, J. B.Nagy, D. Bernaerts and A. A. Lucas, J. Molec. Catal. A.: Chemical 107, 159-168 (1996).

The Role of Catalyst Support in Carbon Nanotube Synthesis

Andrea Siska[1], Klara Hernadi[1], Imre Kiricsi[1], Imre Rojik[2] and Janos B.Nagy[3]

[1] *Applied Chemistry Department, JATE, H-6720 Szeged, Rerrich B. tér 1., Hungary*
[2] *Department of Comparative Physiology, JATE, H-6726 Szeged, Középfasor 52., Hungary*
[3] *Laboratoire de R.M.N., FUNDP, 61 rue de Bruxelles, B-5000 Namur, Belgium*

Abstract. Acetylene decomposition over supported cobalt (or iron) catalysts proved to be an effective method for the preparation of well-graphitized carbon nanotubes. Compared to other techniques, catalytic synthesis is operated under relatively mild reaction conditions (700°C, atmospheric pressure) and experimental apparatus is very simple.
In order to improve catalyst performance, we try to understand the reaction mechanism. Catalysts were prepared by the impregnation method using different materials as catalyst support. Physico-chemical characterization of the samples were carried out by XRD, IR, etc. Surface acidity was measured by pyridine adsorption technique. Catalyst samples were tested in the decomposition of acetylene in a fixed bed flow reactor at 722°C. The quantity of carbon deposit was weighted (catalyst activity). The quality of carbon nanotubes produced was characterized by means of transmission electron microscopy.

INTRODUCTION

Acetylene decomposition over supported cobalt (or iron) catalysts proved to be an effective method for the preparation of well-graphitized carbon nanotubes. Catalytic synthesis is operated under relatively mild reaction conditions (~700°C, atmospheric pressure) and experimental apparatus is very simple. Some of our previous investigations suggested that beside catalyst particles, catalyst support also has a significant role in activity and carbon nanotube selectivity.

Unsupported catalyst samples showed neither activity nor selectivity in carbon nanotube synthesis [1]. Previous results were obtained mainly using silica gel or zeolite as catalyst support [1-4].

In order to improve catalyst performance, we try to understand the reaction mechanism. Various materials with different crystallinity, structure, surface area, acidity have been applied as catalyst support, and tested in carbon nanotube synthesis.

EXPERIMENTAL

Catalysts were prepared by the impregnation method using different oxide materials (different silica and alumina materials, zeolite-type materials, TiO_2, SnO_2, and MgO) as catalyst support. Cobalt was deposited on the surface from a slightly basic solution, and the final catalysts contain 1 w% of cobalt, respectively. Before reaction, the samples were calcined in air at 450°C for 4.5 hours. Physico-chemical characterization of the samples were carried out by XRD and density measurements. Surface acidity was measured by pyridine adsorption technique.

Catalyst samples (approx. 60 mg in each reaction) were tested in the decomposition of acetylene (20 ml/min in 80 ml/min nitrogen flow) in a fixed bed flow reactor at 722°C (Thermolyne F21130-26 oven). The quantity of carbon deposit was weighted. Catalyst activity is given as a ratio of carbon deposit and initial catalyst (g/g). The nature of carbon nanotubes produced was characterized by means of transmission electron microscopy (Tesla BS-500). For EM sample preparation copper grid and the glue technique was applied.

RESULTS

Table 1 summarizes results of XRD and density measurements and carbon yields in catalytic test. Crystallinity and surface acidity of support have no significant effect

sample	cryst.	carbon yield/(g/g)	$\varrho/ (g/cm^3)$*	$\varrho/(g/cm^3)$**
Co/MgO	+	1.10	0.2126	1.4279
Co/TiO$_2$ (rutile)	+	0.29	0.4856	3.9844
Co/TiO$_2$ (anatase)	+	0.35	0.5295	3.6166
Co/SnO$_2$	+	-	1.2929	4.5148
Co/Bentolit-H	+	0.19	0.5009	2.0448
Co/alumina	+	0.20	0.8385	3.8840
Co/alumina (Katalco)	-	1.23	0.4118	1.5879
Co/alumina, activated basic	-	0.36	0.9403	3.5202
Co/fumed alumina	-	0.95	0.0281	3.5943
Co/kieselgel	-	0.82	0.5043	2.020
Co/silicagel 15-40µ	-	0.99	0.4063	1.5760
Co/silicagel 5-25µ	-	1.07	0.2683	1.9880
Co/Cab-O-Sil	-	1.43	0.0506	1.6278

* apparent density
** true density

Table 1 Cristallinity, carbon yield and density of catalyst samples

on carbon nanotube synthesis. A considerable difference was found in apparent density values. Relevant surface area could not be measured for materials having pore diameter larger than a few nanometers. Many catalysts produced carbon yield of 1.0 g/g or higher value, but the quality of carbon deposit was very altering, as it is illustrated on electron microscopy images.

MgO-supported Co catalyst has high activity but absolutely no tubular carbon was found on the surface. Both forms of TiO_2, namely rutile and anatase were used for catalyst synthesis. Neither activity nor selectivity was remarkable. Co/SnO_2 sample did not give special results. After having good experiences with zeolites [3], carbon fibers formed over montmorillonite-supported sample were a kind of regress.

Carbon yield of alumina-supported catalyst was quite low, and the quality of the tubes was satisfactory. Using activated basic alumina improved both catalyst activity and selectivity. Prominent results were obtained with fumed alumina support. Carbon nanotubes were produced in a good quality with a yield of about 1.0 g/g. While some regions are similar to other samples, carbon nanotubes can also be found separately from the catalyst particles as it can be seen on Figure 1.

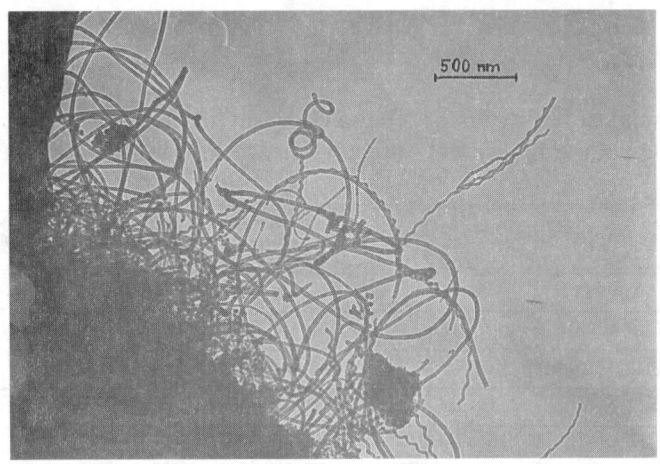

Figure 1   Carbon nanotubes on the surface of Co/fumed alumina

SiO_2 materials with different morphology were also used for catalyst synthesis. Pure quartz sand is not suitable for support. Kiesel gel has an average activity and produced tubes with varying quality. Certain silica gel proved to be a good support with high carbon yield and well-graphitized nanotubes. Catalysts on Cab-O-Sil (fumed silica) support gave the best results. Carbon nanotubes with regular spirals can be seen on the TEM image of Figure 2.

Figure 2   Carbon nanotubes on the surface of Co/Cab-O-Sil

DISCUSSION

From our investigations it can be concluded that both activity and selectivity is very sensitive to the nature of catalyst support.

SiO_2 and Al_2O_3 provides a great variety of possible structures. Microamorphous silica materials are composed of ultimate particles or structural units <1 µm. So-called pyrogenic silicas are formed at high temperature by condensation of SiO_2 from the vapor phase. They have high surface area and very low apparent density. These features provide very high outer surface for cobalt dispersion.

Catalysts using support with low apparent density showed not only high activity but produced well-graphitized carbon nanotubes. With a view to high yield and good separability, fumed materials (mainly silica and alumina) seem to be the most advantageous catalyst supports.

ACKNOWLEDGEMENTS

This work was financed by the Ministry of Education (FKFP 0972), and the National Science Foundation of Hungary (OTKA T025246). This text presents research results of the Belgian Programme on Inter University Poles of Attraction initiated by the Belgian State, Prime Minister's Office of Science Policy Programming (PAI/IUAP n° 4).

REFERENCES

1. K. Hernadi, A Fonseca, J. B.Nagy, D. Bernaerts, A. Lucas: Carbon, **34**, 1249 (1996)
2. A. Fonseca, K. Hernadi, J. B.Nagy, D. Bernaerts, A. Lucas: J. Mol. Catal. A: Chemical Vol. **107**, 159 (1996)
3. K. Hernadi, A Fonseca, J. B.Nagy, A. Fudala, D. Bernaerts, A. Lucas: Zeolites, **17**, 416 (1996)
4. K. Hernadi, A Fonseca, P. Piedigrosso, J. B.Nagy, D. Bernaerts, J. Riga, A. Lucas: Catal. Lett., **48**, 229 (1997)

Pyrolysis of C_{60}-thin films yields Ni-filled sharp nanotubes

N. Grobert [a], M. Terrones [a,b], A. J. Osborne [a], H. Terrones [c],
W. K. Hsu [a], S. Trasobares [a], Y. Q. Zhu [a], J. P Hare [a],
H. W. Kroto [a] & D. R. M. Walton [a]

[a] *School of Chemistry, Physics and Environmental Science, University of Sussex, Brighton, BN1 9QJ, U.K.*
[b] *Materials Research Laboratory, University of California, Santa Barbara, CA 93106, USA*
[c] *Instituto de Física, UNAM, Apartado Postal 20-364, 01000 México, D. F.*

Abstract. Highly graphitised needle-like elongated carbon nanostructures containing encapsulated Ni, are produced by heating alternating thin films of C_{60} and Ni on a silica plate at 950 °C. High Resolution Transmission Electron Microscopy (HRTEM) studies reveal that these tapering structures are fully filled with Ni and are closed at both ends. The diameters of the needles (*ca.* 2-5 μm in length) range between 10-20 nm at one end and 30-200 nm at the other. A surprisingly high degree of graphitisation was observed for the nanostructures. They may prove to be useful as Scanning Tunnelling Microscope (STM) tips due to their shape.

INTRODUCTION

Chemical/wetting and capillarity methods are used to fill the inner cores of carbon nanotubes with pure metals or metal oxides (1-3). However, these processes only result in partial incorporation of the metal/oxide. The modified Krätschmer-Huffman apparatus, in conjunction with graphite/metal mixtures as anodes also generates nanotubes partly filled with metals (*e.g.* Se, Sb, Ge, Mn) (4,5) and metal carbides (*e.g.* ZrC, TaC, HfC, *etc.*) (6) in addition to encapsulated carbon particles. Condensed-phase methods are now available which permit efficient introduction of low melting point metals (*e.g.* Bi, Sn, Pb) into carbon nanotubes, which exhibit moderately disordered graphite domains (7). In this context, we describe the generation of carbon nanotubes, fully filled with Ni using C_{60} as a pure carbon source. The graphitisation of the tubes is remarkably high and, due to their ferromagnetic properties, they may prove to be useful as magnetic data storage devices and novel STM probes.

EXPERIMENTAL

Alternating thin (*ca.* 20-30 nm) films of C_{60} and Ni (5 in total; see Fig. 1a) were successively deposited on a silica substrate (5 mm wide; 25 mm long; 1 mm thick). In the following way. First, C_{60} powder was sublimed from a tungsten boat on to the silica plate. Then an electron beam (15 A, 4-5 kV) was used to sublime Ni onto the C_{60}-film. The procedure was repeated, and a final C_{60} layer was deposited on top of the second Ni-film (Fig.1a). At this stage the upper C60 film was a shiny dark. The coated substrate was inserted into a silica tube (600 mm long; 5 mm OD), which was placed inside a furnace fitted with a temperature controller. An Ar flow (40 ml/min) was passed through the tube in the direction of the furnace in order to remove oxygen from the system. The furnace was then heated to 950 °C, and was then moved towards the cold (room temperature) substrate. After 2-5 min, the furnace was removed and the substrate, upon cooling, taken out of the silica tube. (see Fig.1b) The upper C_{60} surface of the substrate was now a matt black. The silica plate was sonicated in acetone (2 cm^3) for 5 min and a few drops of the resulting suspension was transferred onto a holey carbon grid (400 mesh) for TEM and HRTEM observations (JEM400 at 400 keV, Hitachi 7100 at 125 keV).

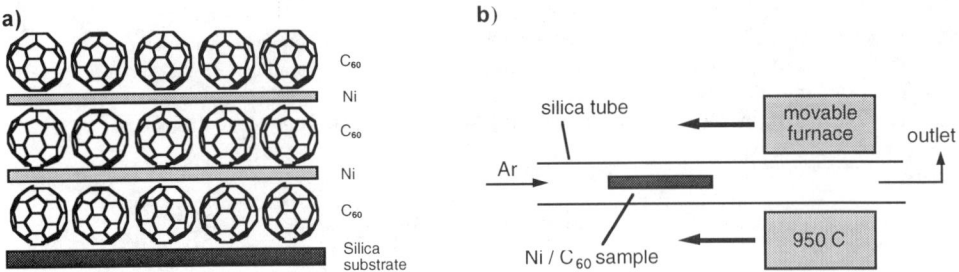

FIGURE 1. (a) Alternating thin (ca. 20-30 nm) films (5 in total) of C_{60} and Ni deposited on a silica plate; (b) thermolysis apparatus.

RESULTS

A TEM inspection of the material revealed the presence of needle-like structures (fully Ni encapsulated nanotubes) (Fig.2) and multi-walled carbon nanotubes. Both types of structure were highly graphitic. The quality of graphitisation is unusual for pyrolytic material produced at 950 °C (see Fig.3a) (the temperature associated with arc discharge methods is much higher (2000-4000°C)). The Ni-filled needles (2-5µm long taper from and 200-30nm to 20-10 nm OD Fig.3b,c).

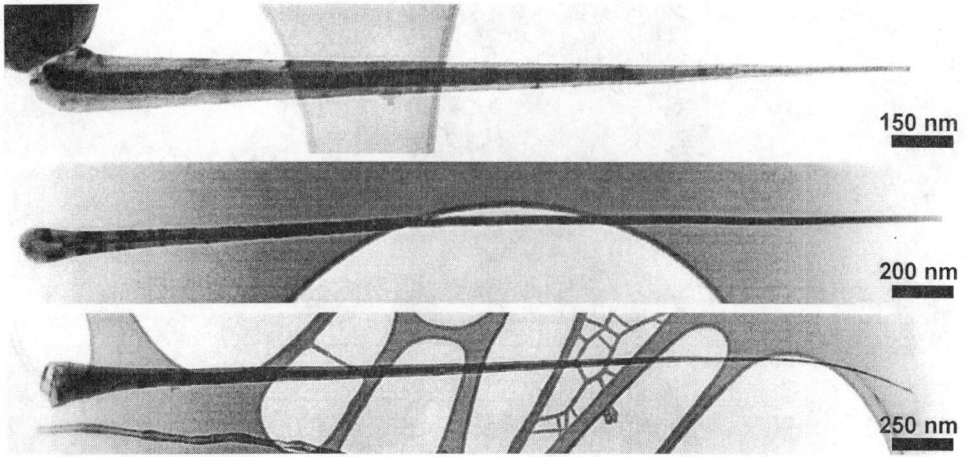

FIGURE 2. TEM images of typical Ni-filled needles.

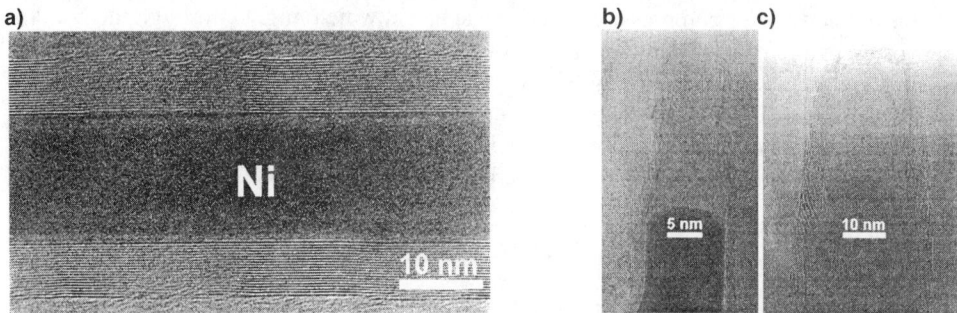

FIGURE 3. (a) HRTEM image of a needle section showing the Ni-filled inner core and highly graphitised tube walls; (b) needle tip showing part filling with Ni. The number of graphitic layers decreases from eight to three when approaching the tip; (c) needle tip void of Ni. The number of layers in this tip decreases gradually from twelve to nine.

In some cases, the encapsulated Ni creates steps along the needle axis (Fig.4). We believe that gaseous carbon (arising from the C_{60} fragmentation at high temperature) diffuses through the metallic Ni and subsequently is extruded as graphite sheets, thus creating needles and/or tubes. It is noteworthy that in all the observed cases, Ni single crystals are contained into the inner core of the nanostructures.

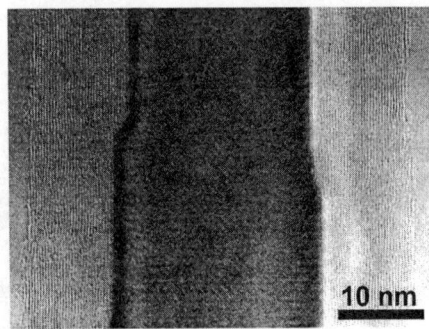

FIGURE 4. HRTEM image showing very clearly the stepped tube axis.

CONCLUSIONS

Pyrolysis of a layered Ni/C_{60} mixture yields novel needle like carbon structures, resembling drawn out capillaries, which are mostly Ni filled. Such needles are more ordered than hollow carbon nanotubes formed during the same process. The latter may be due to extrusion and/or compression of graphene layers which migrate through the Ni. It is noteworthy that pyrolysis of hydrocarbons or other organic precursors over catalysts (*e.g.* Ni, Fe, Co) never lead to such highly graphitic material, possibly due to the presence of hetero-atoms (*e.g.* H, N, etc.) within the carbon network. Therefore, the use of C_{60} as a pure carbon source during thermolysis experiments proves to be advantageous and to offer a new route to ferromagnetic metal-filled carbon nanotubes.

Acknowledgements

We are grateful to J. Thorpe, D. P. Randall, L. Rendon and M. I. Heggie for useful discussions and for providing Electron Microscope facilities. We thank the Royal Society (London), BOC gases, the DERA (NG), CONACYT-Mexico (HT), DGAPA-UNAM IN 107-296 (HT), CAI-CONSI+D (ST) and the EPSRC for financial support.

References

1. Ajayan, P. M., Iijima, S., *Nature* **361**, 333 (1993)
2. Tsang, S, C., Chen, Y. K. , Harris, P. J. F., Green, M. L. H. *Nature* **372**, 159 (1994)
3. Loisseau, A.., Pascard, H.., *Chem. Phys. Lett.* **256**, 246 (1996)
4. Dujardin, E., Ebbesen, T. W., Hiura, H., Tanigaki, K., *Science* **265**, 1850 (1994)
5. Terrones, M., Hsu, W. K., Schilder, A., Terrones, H., Grobert, N., Hare, J. P., Zhu, Y. Q., Schwoerer, M., Prassides, K., Kroto, H. W., Walton, D. R. M., *Appl. Phys. A*. **66**, 307 (1998)
6. Yosida, Y., *Appl Phys Lett* **62**, 3447 (1993)
7. Hsu, W. K., Terrones, M., Terrones, H., Grobert, N., Kirkland, A. I., Hare, J. P., Prassides, K., Townsend, P. D., Kroto, H. W., Walton, D. R. M. *Chem. Phys. Lett.* **284**, 177 (1998)

Nanotechnology of Nanotubes and Nanowires: From Aligned Carbon Nanotubes to Silicon Oxide Nanowires

N. Grobert [a], J. P. Hare [a], W. K. Hsu [a], H. W. Kroto [a], A. J. Pidduck [b], C. L. Reeves [b], H. Terrones [c], M. Terrones [a], S. Trasobares [a], C. Vizard [b], D. J. Wallis [b], D. R. M Walton [a], P. J. Wright [b], Y. Q. Zhu [a]

[a] *School Of Chemistry, Physics and Environmental Science, University of Sussex Brighton BN1 9QJ, England, U.K.*

[b] *Defense Evaluation & Research Agency (DERA), St. Andrews Road, Malvern Worcestershire, WR14 3PS, UK.*

[c] *Instituto de Fisica UNAM, Apartado Postal 20-364, Mexico 01000, Mexico*

Abstract. Laser etching of Co, Ni and Fe films, in conjunction with the pyrolysis of solid organic precursors (*e.g.* aminodichlorotriazine, melamine, *etc.*) generates aligned carbon nanotube bundles and films of uniform length (< 200 μm) and diameter (30 Å - 80 Å). However, nanotube alignment strongly depends upon laser etching conditions (*e.g.* laser power, pulse duration and focus distance). Additionally, condensed-phase techniques, using mixtures of molten LiCl and soft metals (*e.g.* Bi, Pb, *etc.*) as electrolytes, generate high yields of metallic nanowires (< 45 % overall material and < 2 μm in length, < 100 nm OD). Finally, it is shown that novel 3-D flower-like silica nanostructures are produced by a simple and surprising solid-phase approach. It is observed that single catalytic nanoparticles act as nucleation sites, leading to unusual morphologies of silicon oxide nanofibres (20-120 nm OD and < 200 μm). The latter structures may be useful in the context of catalysis, 3-D composite materials, and optoelectronic devices, thus breaking new ground in nanowire and nanofibre technology.

INTRODUCTION

Nanometer scale materials research continues to accelerate and, as a result, novel structures with remarkable mechanical and transport properties have been developed. Some of these nanostructures may prove to be very useful as microscope probes (1), field emission sources (2), gas storage materials (3), and as super-strong carbon-fibre-reinforced materials. In this context, we describe new results obtained using pyrolytic, electrolytic and gas-solid methods, which lead to aligned carbon nanotubes, metal nanowires and silicon oxide nanofibres respectively.

ALIGNED NANOTUBES

The generation of aligned nanotube arrays, was first achieved by cutting thin slices (50-200 nm thick) of a nanotube-composite polymer (4). The tubes appeared to be well separated as a result of mechanical deformation suffered by their being embedded in the polymer. However, the alignment was impractical for larger areas (4). In this context, two recent reports describe the large scale synthesis of aligned carbon nanotubes, of uniform length and diameter, by pyrolysis of organic precursors over templated/catalyst supports (5,6). These methods are by far superior as compared to plasma arcs, since other graphitic structures such as polyhedral particles, encapsulated particles and amorphous carbon are notably absent.

Metal thin films (ca. 10-100 nm) were deposited on silica plates using sublimation techniques. The plates were then exposed to air and etched with *single* laser pulses (Nd:YAG 266nm and 355nm; 5-20 mJ per pulse) using cylindrical lenses (65 mm focal length), thus creating linear tracks (width 1-20 mm; length ≤ 5mm; Fig. 1) where the metal had been ablated.

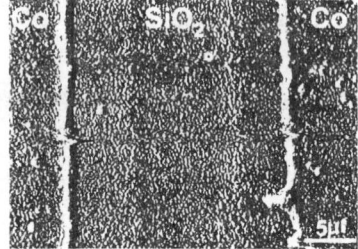

FIGURE 1. AFM image of an etched nanotrack (*laser in focus on the film*) showing how the uniform channel is made. It is believed that along this channel/track uniform metal and/or metal oxide nanoparticles (≤ 50nm OD) are generated evenly by the laser striking the thin film.

In particular, the pyrolysis of organic precursors containing nitrogen (*e.g.* 2-amino-4,6-dichloro-s-triazine, melamine, *etc.*) over laser-etched thin metal films (*e.g.* Ni, Co, Fe) yields aligned nanotubes that grow perpendicularly to the catalytic substrate 'only' in the etched regions (Fig. 2).

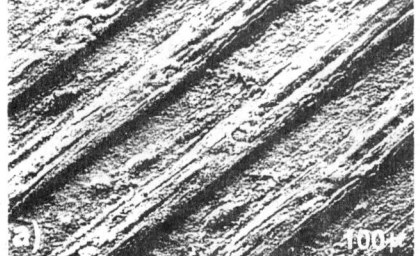 

FIGURE 2. SEM images of aligned nanotube films generated by pyrolysis of melamine over laser etched Fe thin films (*laser out of focus, thus creating wider tracks*): (a) low magnification, in which films of nanotubes grow along the laser etched tracks; (b) higher magnification of a single track showing aligned nanotubes of uniform length (20 μm) and diameter (30-80 nm).

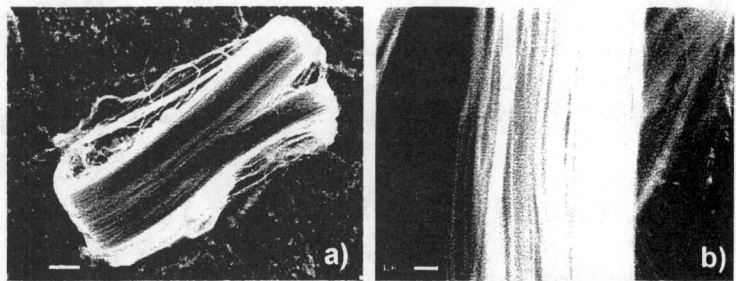

FIGURE 3. SEM images of aligned nanotubes produced by pyrolysing aminodichlorotriazine over laser-etched Fe thin films (*laser in focus*): (a) low magnification of an area showing well-aligned nanotubes; (b) higher magnification of (a) exhibiting aligned nanotubes with uniform dimensions. HRTEM images of this sample showed extremely good graphitisation within the nanotube walls.

Pyrolysis experiments were conducted with aminodichlorotriazine and Fe laser-etched thin films, generated long aligned nanotube bundles (< 200 μm, 20-40 nm OD; Fig. 3). It is important to note that laser etching energies > 10 mJ per pulse or double laser shots on the thin films, frustrate (almost totally) the nanotube growth. Therefore, careful control of the laser energy should be excercised when generating the catalytic substrate.

NANOWIRES

Condensed phase processes offer alternative routes to nanotubes and encapsulated compounds (7,8). Electrolysis experiments, performed using pure LiCl as electrolyte in conjunction with graphite electrodes, produced mainly nanotubes (30-35 nm OD.; 5-7 nm id.) and graphite-like nanostructures (7). Interestingly, with 0.5-1% $SnCl_2$ in LiCl, fully-filled carbon nanotubes (termed nanowires) were generated (30-50 % of extracted material) (8). Electron and X-ray powder diffraction studies indicated that the encapsulated material was β-Sn. Furthermore, the presence of small quantities of Pb or Bi (< 1% by weight) in LiCl also yielded nanowires of the respective metals (Fig. 4).

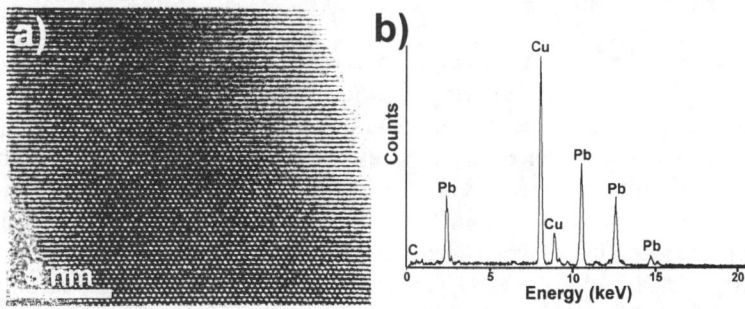

FIGURE 4. (a) HRTEM image of a Pb nanowire generated by electrolysis of graphite in molten Pb+LiCl mixtures (1:100 ratio) at 600 °C. The nanowire walls appear to be essentially graphitic, but fairly disordered; (b) Energy Dispersive X-ray (EDX) spectrum of (a), exhibiting Pb and C signals (Cu siganls arise from the TEM grid).

It is important to note that the epitaxial relationship between the metal crystals and carbon (graphitic) shells, which is usually observed in arc discharge material (produced in the presence of high melting point metals), was absent. In general, the carbon nanowire walls appear to be essentially graphitic, but fairly disordered (see Fig. 4).

SILICON OXIDE NANOSTRUCTURES

Novel flower-like nanostructures consisting of silicon oxide nanofibers, radially attached to a single catalytic particle, are generated in bulk quantities by a simple yet effective solid-gas-solid method. In this process, a mixture of SiC and Co powders, deposited on silica substrates and heated under a Ar atmosphere at *ca.* 1500°C, produced material with unusual 3-dimensional (3D) networks of nanofibres of uniform diameter (*ca.* 20 - 120 nm) and length (*ca.* 10 - 250 μm).

SEM, HRTEM and X-ray powder diffraction analyses reveal that the generated nanofibres are amorphous and consist only of silicon oxide. The formation of the nanostructures is catalysed by large Co particles, which act as nucleation centres and templates for the 3D growth (see Fig. 5)

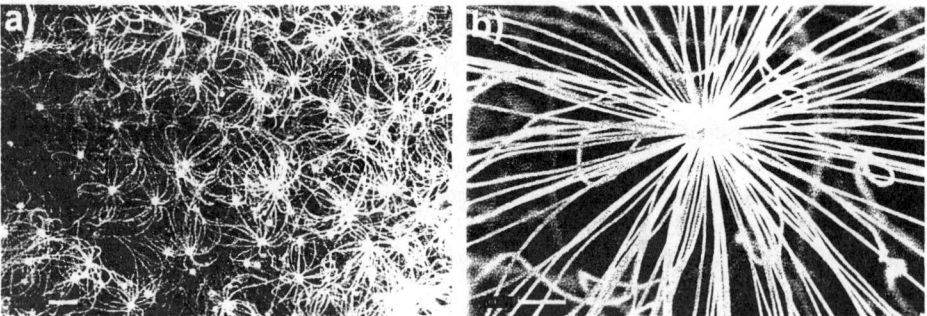

FIGURE 5. SEM images: (a) surface of the bulk material covered by silicon oxide flower-like nanostructures; (b) higher magnification of a nanoflower exhibiting pronounced 3-dimensional features radiating from a spherical core.

It is possible that smaller Co particles act as aggregation sites for fiber growth, and may prevail at the growing ends during nanoflower formation. It is noteworthy that the number of arms and the size of the central particles within the nanoflowers are intimately related. For instance, as the size of the central core decreases, the more radial grown nanofibres appear. HRTEM and EDX analyses confirmed that the single spherical particle, located in the radial centers of the nanoflowers and surrounded by silicon oxide, is metallic Co. Experiments using Si_3N_4 and Si in conjunction with other catalysts (*e.g.* Fe, Ni and CoO) yield similar results and confirm that the resulting SiO_x fibres display virtually unique and remarkable radial growth starting from single metal particles.

CONCLUSIONS

Matrices consisting of aligned nanotube bundles may be useful as novel mechanically strong composite materials and as ultra-fine field emission sources. However, further research is needed in this direction. The novel electrolytic technique for the production of fully-filled metallic nanowires (*e.g.* Pb, Bi, Sn) has been discussed. Note that in the arc discharge method, the encapsulated material inside carbon nanotubes is usually a metal carbide and is not fully located within the nanotube. Finally, the elegant silicon oxide (flower-like) nanostructures may prove to be useful in nanoscale devices, optoelectronic sensors, 3-dimensional composite materials, novel catalytic supports and biological microfilters and, in particular, in situations where network growth is important.

ACKNOWLEDGMENTS

We are grateful to J. Thorpe, D. P. Randall, S. Tehuacanero, R. Hernández, R. Guardián and L. Rendón for providing Electron Microscope facilities. We thank the Royal Society (London), BOC gases, the DERA (NG), CONACYT-México (HT), DGAPA-UNAM IN 107-296 (HT), CAI-CONS+D (ST) and the EPSRC for financial support.

REFERENCES

1. Dai, H. J., Hafner, J. H., Rinzler, A. G., Colbert, D. T., Smalley, R. E. *Nature* **384**, 147 (1996).
2. de Heer, W. A., Chatelain, A., Ugarte, D. *Science* **270**, 1179 (1995); Rinzler, A. G., Hafner, J. H., Nikolaev, P., Lou, L., Kim, S. G., Tomànek, D., Nordlander, P., Colbert, D. T., Smalley, R. E. *Science* **269**, 1550 (1995); Collins, P. G., Zettl, A. *Appl. Phys. Lett.* **69**, 1969 (1996); Saito, Y., Hamaguchi, K., Hata, K., Uchida, K., Tasaka, Y., Ikazaki, F., Yumura, M., Kasuya, A., Nishima, Y. *Nature* **389**, 555 (1997).
3. Gadd, G. E., Blackford, M., Moricca, S., Webb, N., Evans, P. J., Smith, A. M., Jacobsen, G., Leung, S., Day, A., Hua, Q. *Science* **277**, 933 (1997); Dillon, A. C., Jones, K. M., Bekkedahl, T. A., Kiang, C. H., Bethune, D. S., Heben, M. J., *Nature* **386**, 377 (1997).
4. Ajayan, P. M., Stephan, O., Colliex, C., Trauth, D. *Science* **265** 1212 (1994).
5. Terrones, M., Grobert, N., Olivares, J., Zhang, J. P., Terrones, H., Kordatos, K., Hsu, W. K., Hare, J. P., Townsend, P. D., Prassides, K., Cheetham, A. K., Kroto, H. W., Walton, D. R. M. *Nature* **388** 52 (1997).
6. Li, W. Z., Xie, S. S., Qian, L. X., Chang, B. H., Zou, B. S., Zhou, W. Y., Zhao, R. A., Wang, G. *Science* **274**, 1701-1703 (1996); Terrones, M., Grobert, N., Zhang, J. P., Terrones, H., Olivares, J., Kordatos, K., Hsu, W. K.,. Hare, J. P., Prassides, K., Cheetham, A. K., Kroto, H. W., Walton, D. R. M. *Chem. Phys. Lett.* **285**, 299 (1998).
7. Hsu, W.K., Hare, J. P., Terrones, M., Harris, P. J. F., Kroto, H. W. Walton, D. R. M. *Nature* **377**, 687. (1995); Hsu, W. K., Terrones, M., Hare, J. P., Terrones, H., Kroto, H. W., Walton, D. R. M. *Chem. Phys. Lett.* **261**, 161 (1996).
8. Hsu, W. K., Terrones, M., Terrones, H., Grobert, N., Kirkland, A. I., Hare, J. P., Prassides, K., Townsend, P. D., Kroto, H. W., Walton, D. R. M. *Chem. Phys. Lett.* **284**, 177 (1998).
9. Zhu, Y.Q., Hsu, W. K., Terrones, M., Grobert, N., Terrones, H., Hare, J. P., Kroto, H. W., Walton, D. R. M to be published (1998).

Bulk Properties of Crystalline Single Wall Carbon Nanotubes: Purification, Pressure Effects and Transport

J. E. Fischer[1], R. S. Lee[1], H. J. Kim[2],
A. G. Rinzler[3], R. E. Smalley[3], S. L. Yaguzhinski[4],
A. D. Bozhko[4], D. E. Sklovsky[4] and V. A. Nalimova[4]

[1] *University of Pennsylvania, Philadelphia, PA 19104-6272 USA*
[2] *Hallym University, South Korea*
[3] *Rice University, Houston TX 77251 USA*
[4] *Moscow State University, Moscow 119899 Russia*

Abstract. Pulsed laser ablation (PLA) has been scaled up to yield several grams/day of single-walled nanotubes. Annealed, purified material is highly crystalline, essentially free of amorphous carbon, fullerenes and catalyst residues, and about 3 times denser than the highly porous, as-grown product. In principle the interactions between tubes in a rope, and/or between rope crystallites, may be "tuned" by 3 different approaches - chemical doping, hydrostatic pressure, or purification/annealing, all of which have a dramatic effect on the temperature dependence of resistivity. In particular, we suggest that the crossover from positive to negative dR/dT at low temperature is a 3D effect and not an intrinsic property of isolated neutral SWNT.

Large quantities of single-walled carbon nanotubes are produced in good yield by laser ablation (1) or carbon arc (2). A purification scheme involving acid reflux, cross-flow filtration and vacuum annealing has been developed (3). Combined with an enlarged graphite target and 10 cm. diameter quartz flow tube, this scheme is capable of producing grams/day of high purity randomly-oriented SWNT material. A primary goal of this work is to characterize this material.

Purification begins with a 45 hr. reflux in 2-3 M nitric acid, followed by centrifugation. The sediment still contains substantial trapped acid which is removed by repeated washing in de-ionized water. Electron micrscope observation of the sediment showed that the SWNT were coated with acid decomposition products, which could be removed by vacuum filter washing with pH 11 NaOH solution and hollow fiber cross-flow filtration (CFF). The use of a surfactant (Triton X-100, non-ionic, Aldrich, Milwaukee, WI) helps to maintain the SWNT in suspension during this procedure.

High resolution TEM images at this stage show the material to still contain a significant quantity of impurities. In order to remove these, our approach has been to use successively more oxidizing acid treatments. These are sufficiently reactive to attack the SWNT from the sides so the reaction times are kept much shorter. The first of these is a (3:1) mixture of sulfuric (98%) and nitric (70%) acids (typically 500 ml) stirred and maintained at 70 C in an oil bath for 20-30 min. This is followed by another CFF cycle as described above. The final polish is done with a 4:1 mixture of sulfuric acid (98%) and hydrogen peroxide (30%) following the same procedure as with the sulfuric/nitric mixture. TEM imaging at this stage reveals the SWNT to be largely free of impurities.

X-ray diffraction was performed in order to determine the effects of chemical treatment and annealing on the nanotubes and their crystalline organization, to follow the evolution of impurity phases throughout the process, and to obtain another estimate of the diameter distribution. Figure 1 shows the evolution of diffraction profiles from the as-grown material to the final product. Crystallinity is largely destroyed by the acid treatment and is restored and enhanced by vacuum annealing. Surprisingly the intermediate product also exhibits peaks attributable to crystalline C_{60} which must have been present as isolated molecules or small aggregates in the initial material. Peaks associated with small crystallites of Co/Ni catalyst are essentially gone after purification, although the material still responds (weakly) to a permanent magnet.

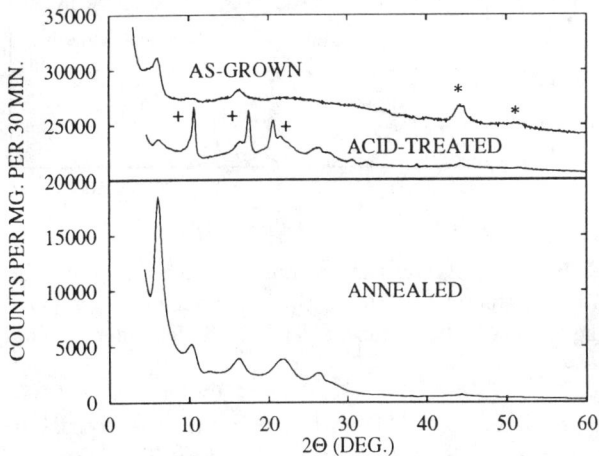

Figure 1. Raw x-ray data for 2 inch SWNT material. TOP (zero shifted to 20000 for clarity): as-grown and acid-treated material. (*) indicate peaks from Ni/Co catalyst and (+) are indexable as fcc C_{60}. BOTTOM: after final vacuum anneal.

The directionally averaged resistivity $<\rho>$ was measured on rectangular pieces 2-4 mm wide and 8 mm long. In Figure 2, the temperature dependence of $<\rho>$ from 90 K to 550 K shows some unexpected evolution with purification. The combination of chemical processes reduces the absolute value of $<\rho>$ by a factor of 25, which we attribute to the removal of non-conducting impurities and some degree of compaction which improves interparticle contacts. Subsequent vacuum annealing *increases* $<\rho>$ by about 4 times. The major difference between these two states of the material is the more highly developed inter-tube ordering in the latter (cf. Figure 1). More dramatically, the temperature dependence changes sign after annealing. $<\rho>$(T) is strongly metallic for the as-grown material, with positive $d\rho/dT$ persisting down to at least 90 K; previous experiments (4,5) on non-purified, vacuum-annealed samples showed a crossover to negative $d\rho/dT$ at 200 K. $<\rho>$(T) remains metallic after acid treatment albeit with a smaller slope, but the slope changes sign after vacuum-annealing the purified material. The x-ray data suggests that the latter should come closest to the intrinsic behavior of well-ordered, randomly oriented, rope crystallites while the as-grown and acid-treated samples no doubt contain large fractions of isolated tubes or very small crystals.

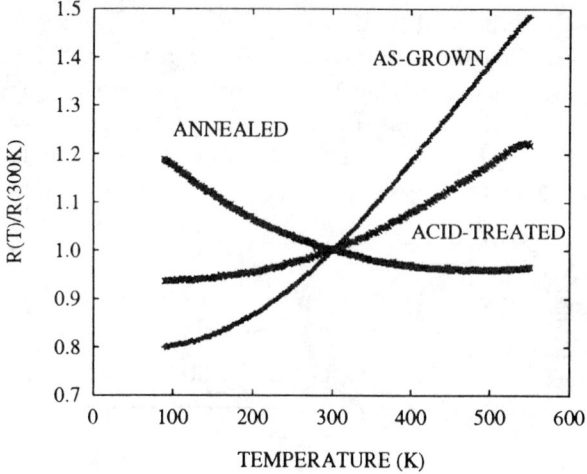

Figure 2. Four-probe resistivity vs. temperature for 2 inch material. The acid purification results in a 25-fold reduction in rho with little change in temperature dependence. Vacuum annealing increases rho somewhat and the T dependence changes from metallic to non-metallic (see text).

The sign change of $d<\rho>/dT$ with crystallinity is consistent with an effect we observed upon potassium doping of highly crystalline but unpurified material. Before the development of the acid/filtration process, the usual procedure was to vacuum-anneal the raw product to remove fullerenes and some of the metal impurities; the resulting material was used for our initial x-ray studies of the triangular rope lattice (1). After vapor-transport doping with potassium in sealed glass tubes at 400°C, the major effect on diffraction was an overall loss of Bragg intensity suggesting a reduction in inter-tube correlations (6). The corresponding effect on $<\rho>$ is again dramatic - a 40-fold overall reduction, probably associated with charge transfer (7,8) and again a sign change of $d<\rho>/dT$. Figure 3 shows the normalized temperature dependence before and after doping. The pristine sample shows the typical shallow minimum near 200K while the doped sample shows no sign of the low-T divergence. The latter curve suggests classical Matthiesen's rule behavior: $\rho = \rho_{defects} + AT^n$ with $1 < n < 2$. The undoped sample is more crystalline than the doped one, so this trend in $<\rho>(T)$ correlates with inter-tube interactions in the same manner as described above. Similar results were obtained after doping with bromine, an electron acceptor.

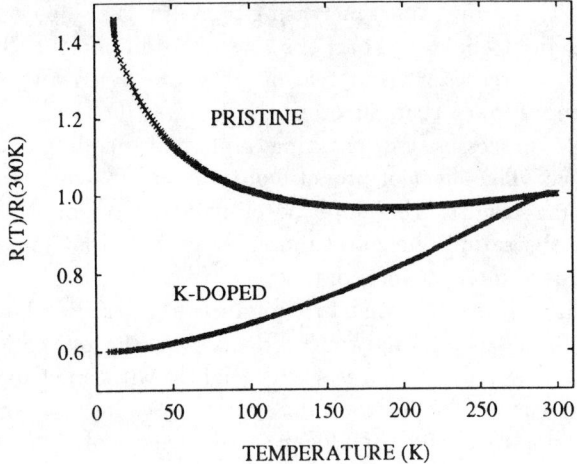

Figure 3. Four-probe resistivity vs. temperature for unoriented unpurified material, before and after doping. Approximate composition of the doped material corresponds to KC_8.

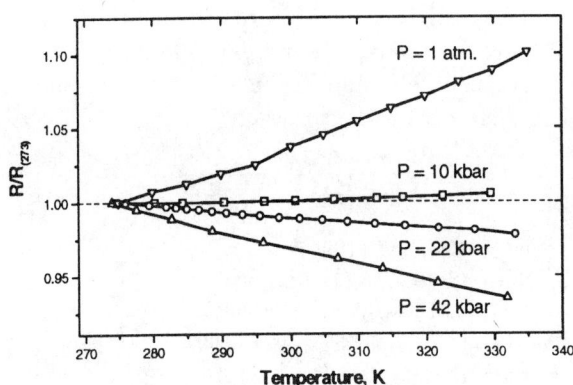

Figure 4. Temperature dependence of resistance (2-probe) measured at 4 fixed pressures. Data at one atm. from Ref. 5; the rest obtained from the lenticular cell. Note especially the sign change in the derivative between 10 and 22 kbar.

Hydrostatic pressure provides a direct probe of inter-tube interactions. Experiments were performed in an anvil apparatus up to 90 kbar with no pressure-transmitting fluid. Measurements of V(P) showed that 2-3 cycles up to 27 kbar yielded a stable compact pellet for which subsequent P cycles yielded reproducible and reversible data (9). Figure 4 shows the evolution of $<\rho>(T)$ with increasing pressure, measured over a limited temperature range. At low pressure the slope remains positive despite a very large volume reduction - the density increases from a few mg/cm^3 to 1-2 g/cm^3. With increasing pressure, the slope decreases continuously, changing sign at about 12 kbar; the crossover temperature T*, from positive to negative slope, increases with increasing pressure. Even after compaction the modulus is 2-3 times lower than theoretical estimates for the inter-tube contribution, so some other processes (e.g. crushing, kinking or bending of tubes) dominate the volume change. The effect of pressure on $<\rho>(T)$ is consistent with the annealing and doping results: the slope becomes negative (or T* shifts to higher temperature) as the sample becomes more 3D, either from increased inter-tube coupling or improved interparticle contacts.

Several ideas have been proposed to explain the differences between transport in isolated SWNT and crystalline ropes. Backscattering by torsional shape fluctuations is believed to reduce the resistivity slightly with no change in slope (10), contrary to observations. The loss of some symmetry elements upon crystallization was shown to open a 0.1-0.2 eV pseudogap in the density of states (11); this would appear to be far too large to explain the crossover phenomenon which can occur as low as 50 K in an oriented single-rope measurement (5). The dependence of the low-T resistivity on T and electric field was shown in pristine material to be consistent with localization and 3D variable range hopping (12). Here the ~ 1 μm length scale is attributed to defects within a tube or rope rather than interparticle contacts. If this model is correct, we need to explain the extreme reduction in length scale upon doping (or conversely, its extreme increase with annealing after acid treatment). Detailed field- and T-dependent measurements on doped and annealed samples may help clarify the situation.

REFERENCES

1. A. Thess *et al.*, Science **273**, 483 (1996).
2. C. Journet *et al.*, Nature **388**, 756 (1997).
3. A. G. Rinzler *et al.*, Appl. Phys. A (June 1998).
4. J. E. Fischer *et al.*, Phys. Rev. B **55**, R4921 (1997).
5. C. L. Kane *et al.*, Europhys. Letters (in press).
6. Milder doping conditions reveal a coherent expansion of the lattice: T. Pichler *et al.*, this volume.
7. R. S. Lee *et al.*, Nature **388**, 255 (1997).
8. A. Rao *et al.*, Nature **388**, 258 (1997).
9. S. L. Yaguzhinski *et al.*, (submitted).
10. C. L. Kane and E. J. Mele, Phys. Rev. Lett. **78**, 1932 (1997).
11. P. Delaney *et al.*, Nature **391**, 466 (1998).
12. A. Zettl *et al.*, this volume.

Chromatographic purification and size separation of carbon nanotubes

G. S. Duesberg*, J. Muster, V. Krstic, M. Burghard, S. Roth

Trinity College, Dublin, Ireland
Max-Planck-Institut für Festkörperforschung, Heisenbergstr. 1, D-70569 Stuttgart, Germany

Abstract. The efficient purification of single-wall and multi-wall carbon nanotubes (NTs) by columnar size exclusion chromatography (SEC) is reported. In this process, carbon nanospheres (polyhedra), amorphous carbon and metal particles are removed from aqueous surfactant-stabilised dispersions of NT raw material. TEM and AFM investigations revealed that more than 40-50% of the purified material consists of individual tubes. In addition, length separation of the tubes is achieved.

1 INTRODUCTION

Several large scale synthesis routes for carbon nanotubes have been reported (1, 2) which, however, lead to by-products of other carbon species. In addition to NTs the soot of a conventional arc discharge experiment contains fullerenes, carbon polyhedra, and amorphous carbon, which are interconnected to a dense network. In the growth of single-wall nanotubes (SWNTs) by arc discharge (3) or laser ablation (4) a catalyst is essential (Fe, Ni, Co) which leads to an additional contamination by metal particles. For a number of proposed applications of NTs including hydrogen storage (5), field emission (6), or the fabrication of electronic devices (7, 8, 9), efficient purification and size separation of NTs is very important.

In destructive purification methods like oxidation, the smaller particles are decomposed while some tubes remain. However, the caps of the tubes are opened and chemical functionalities are introduced on the tube surface (10, 11). Non-destructive methods like filtering or flocculation have been reported but their efficiency is limited (12, 13, 14, 15), and especially in filtration techniques blocking of pores is a severe problem. Another disadvantage is that a number of successive filtration steps are required to achieve satisfactory purity and that size selection of the tubes is not easily attained.

Here we report the purification and size separation of multi-wall as well as single-wall NTs through size exclusion chromatography (SEC). The success of this procedure strongly relies upon an appropriate stationary phase like controlled porous glass (CPG) which is available with large, defined pore sizes and is characterised by a high chemical stability. SEC is known as a powerful tool for the separation of large molecules, e.g., biological macromolecules or virus particles (16).

2 EXPERIMENTAL

Dispersions of NTs in water were prepared with the aid of sodium dodecylsulfate (SDS). For that purpose, 10 mg of MWNT raw material (prepared via a conventional arc discharge experiment) or 1 mg of SWNT raw material (produced by arc discharge or laser ablation with Ni/Y catalyst) were added to 2 ml of a 1wt% aqueous SDS solution and sonicated for 5 min with an ultrasonic tip. After a settling time of 15 min a black supernatant and a sediment of some undispersed aggregates were obtained. The supernatant, which is stable for days, was directly subjected to chromatography.

Two successive columns were used for the fractionation of the MWNTs. The purpose of the first column was to remove the majority of small particles and fullerenes. The packing (7 cm x 2 cm^2) of this column consisted of controlled pore glass (CPG) with an average pore size of 140 nm (CPG 1400 Å, Fluka). The column was loaded with 1.5 ml of the supernatant and eluted with a 0.25 wt% aqueous SDS solution buffered at pH 7. The flow rate was adjusted to 9 ml/h. After a void volume of 16 ml, two fractions of 6 ml were collected. The first fraction was concentrated to 1.5 ml by addition of 100 mg of polyacrylamide adsorbent gel (Fluka) followed by continuous shaking for 30 min. The concentrated first fraction was loaded onto the second column (33 cm x 1 cm^2) filled with CPG with an average pore size of 300 nm (CPG 3000 Å, Fluka). Elution was performed similar as described above with a flow rate of 5 ml/h. After a void volume of 16 ml, eight fractions of 1.5 ml were collected.

For the SWNT purification, the dispersions were directly loaded onto the second type of column. Subsequent to fractionation, the SDS concentration was raised to 1wt%, and the fractions were centrifuged at 8000 rpm for 30 min (Eppendorf Centrifuge 5417 C). Treatment of the Si wafers and the NT adsorption were performed as described recently (17).

3 RESULTS AND DISCUSSION

Generally, the dispersed tube material moved completely through the column, except for a number of larger particles (roughly 10wt%) which precipitated on top of the column. These particles consist of network fragments that could not be disintegrated during the ultrasonic treatment.

The particles in the different fractions were characterised by transmission and scanning electron microscopy (TEM/SEM). Fig. 1 shows a representative TEM micrograph of fraction 3 of purified MWNTs. Individual MWNTs are clearly recognised, demonstrating that the network of the soot was successfully disintegrated by the surfactant. In fraction 1 from the second column, aggregates of NTs and other carbon species were found while fractions 7 and 8 contained mainly spherical particles and a few short tubes, both smaller than 0.1 µm. Fractions 2 - 6 consisted of individual nanotubes and, for the later fractions, a low content of spherical particles. A similar composition of the fractions was found for the SWNTs. In this case, however, centrifugation was additionally performed subsequent to the chromatographic separation. This step removes the majority of the tubes with attached catalyst particles, but also lowers the overall yield of SWNTs.

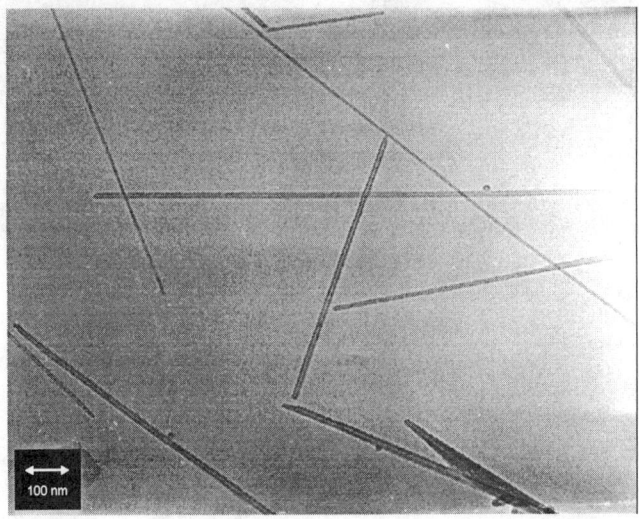

FIGURE 1. Transmission electron micrograph of multi-wall nanotubes purified by size exclusion chromatography (fraction 3).

In order to evaluate the NT sizes in the different fractions, NTs adsorbed onto amino-functionalised Si-wafers were investigated by SEM and atomic force microscopy (AFM). As a representative example, Fig. 2 displays AFM images of fractions 3 and 5 obtained from a chromatographic SWNT separation. Only a small number of catalyst particles, which are attached to the tubes, can be detected.

Detailed studies have shown that the purified material is composed to about 50% of both, ropes of SWNTs and individual SWNTs (as deduced from a height of between 1 and 2 nm). It is further apparent from the AFM images that the average tube lengths are different in the two fractions. Such length separation was observed for both MWNTs and SWNTs. A statistical evaluation was performed with NT lengths determined from a number of SEM and AFM micrographs. The histograms of the SWNT length distribution in fraction 3 and 5 are shown in Fig. 3a and b, respectively.

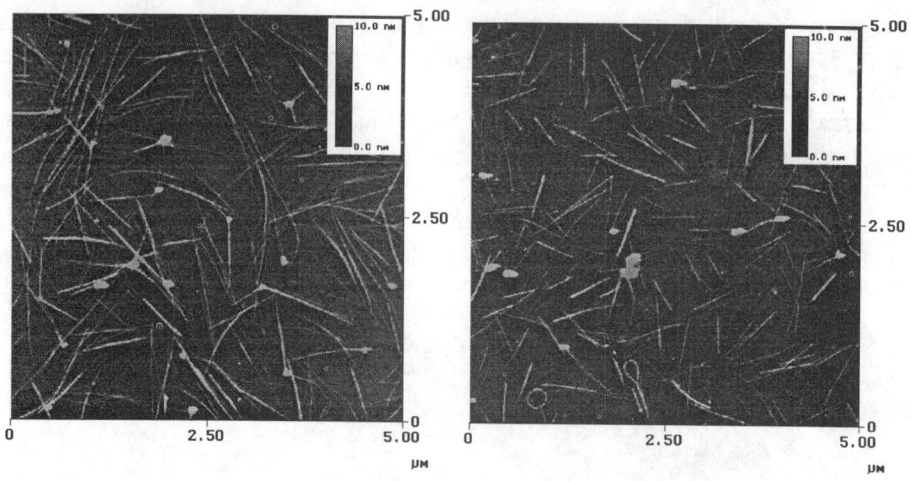

FIGURE 2. AFM images of single-wall carbon nanotubes adsorbed on a chemically modified Si wafers after chromatographic separation. The average length of the tubes is larger in fraction 3 (left) than in fraction 5 (right).

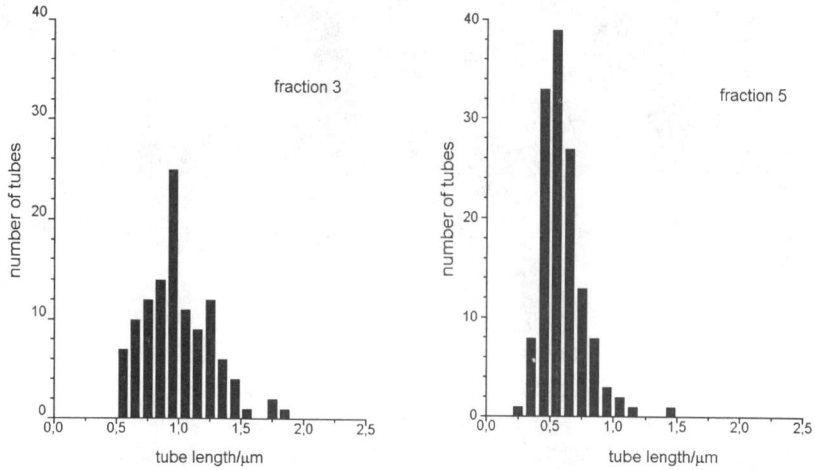

FIGURE 3. Histograms of single-wall nanotube lengths in fraction 3 (left) and fraction 5 (right).

There is a significant difference in the length distribution of the SWNTs: the average length in fraction 3 was calculated to be 1 μm and in fraction 5 to be 0.6 μm. Similar results were obtained for the MWNTs (17).

CONCLUSION

Size exclusion chromatography is an effective, non-destructive method for purification and size separation of carbon NTs. Moreover, the method can easily be scaled up, and by the use of more sophisticated chromatographic systems like high performance liquid chromatography (HPLC), the size separation might be improved.

ACKNOWLEDGMENTS

This work was supported by the European Community TMR project NAMITECH (ERBFMRX-CT96-0067 (DG12-MIHT)). The authors are grateful to C. Journet, P. Bernier (University of Montpellier), and W. Maser (University of Zaragoza) for supplying SWNT samples as well as to S. Morlang and P. Redlich for helpful discussions.

REFERENCES

1. Ebbesen,T.W. and Ajayan, P.M., *Nature* **358**, 220 (1992).
2. Hernadi, K., Fonseca, A., B.Nagy, J., Bernaerts, D., Riga, J., and Lucas, A., *Synth. Met.* **77**, 31 (1996).
3. Journet, C., Maser, W. K., Bernier, P., Loiseau, A., de la Chapelle, M. L., Lefrant, S., Deniard, P., Lee, R., and Fischer, J. E., *Nature* **388**, 756 (1997).
4. Thess, A., Dai, H., and Smalley, R., *Science* **273**, 483 (1996).
5. Dillon, A.. C., Jones, K. M., Bekkedahl, M. A., Kiang, C. H., Bethune, D. S., and Heben, M. J., *Nature* **386**, 377 (1997).
6. de Heer, W. A., Bonard, J.-M., Fauth, K., Chatelain, A., Forro, L., and Ugarte, D., *Adv. Mater.* **9**, 87 (1997).
7. Chico, L., Crespi, V. H., Benedict, L. X., Louie, S. G., and Cohen, M. L., *Phys. Rev. Lett.* **76**, 971 (1996).
8. Saito, S., *Science* **278**, 77 (1997).
9. Tans, S. J., Verschueren, A. R. M., and Dekker, C., *Nature* **393**, 49 (1998).
10. Hiuara, H., Ebbesen, T. W., and Tanigaki, K., *Adv. Mater.* **7**, 275 (1995).
11. Ebbesen, T. W., Ajayan, P. M., Hiura, H., and Tanigaki, T, *Nature* **367**, 519 (1994).
12. de Heer, W. A., Bacsa, W. S., Chatelain, A., Gerfin, T., Humphrey-Baker, R., Forro, L., and Ugarte, D., *Science* **268**, 845 (1995).
13. Bonard, J.-M., Stora, T., Salvetat, J.-P., Maier, F., Stöckli, T., Duschl, C., Forro, L., de Heer, W.A., and Chatelain, A., *Adv. Mater.* **9**, 827 (1997)
14. Shelimov, K. B., Rinat, R. O., Rinzler, A. G., Huffmann, C. B., and Smalley, R. E., *Chem. Phys. Lett.* **282**, 429 (1998).
15. Bandow, S., Rao, A. M., Williams, K. A., Thess, A., Smalley, R. E., and Eklund, P. C., *J. Phys. Chem. B* **101**, 8839 (1997).
16. Wu, C.-S., *Handbook of Size Exclusion Chromatography*, New York: Marcel Dekker, 1995.
17. Duesberg, G. S., Burghard, M., Muster, J., Philipp, G., and Roth, S., *Chem. Commun.*, 435 (1998).

Assembling techniques for micellar dispersed carbon single-walled nanotubes

M. Burghard, J. Muster, G. Duesberg*, G. Philipp, V. Krstic, S. Roth

Max-Planck-Institut für Festkörperforschung, Heisenbergstr. 1, D-70569 Stuttgart, Germany
**Trinity College, Dublin 2, Ireland*

Abstract. Surfactant-stabilised aqueous dispersions of carbon single-walled nanotubes (SWNTs) provide attractive possibilities for different types of assembling processes. The adsorption behaviour of chromatographically purified, micellar suspended SWNTs on silica substrates and metal electrodes is presented. Chemical modifications of the substrate surface allow to control the adsorption kinetics and the fraction between adsorbed individual SWNTs and bundles of SWNTs. Tube alignment occurs presumably due to flow effects upon removal of the surfactant. As a second assembling technique, we describe the preparation of Langmuir-Blodgett films consisting of SWNTs embedded in a surfactant matrix.

1 INTRODUCTION

Disintegration of single-walled carbon nanotube (SWNT) raw material in an aqueous surfactant solution provides the basis for purification methods like filtration (1, 2) or size exclusion chromatography (3). Especially with the latter technique, length separation of the tubes can be achieved simultaneously to the removal of amorphous carbon, carbon nanospheres and catalytic particles. It is further expected that no or only very little defects are introduced to the tubes by above techniques as compared to purification by oxidation (4) or by filtration during excessive sonication (5). Therefore, chromatographically purified SWNT dispersions represent attractive starting materials for the investigation of electric or magnetic properties of SWNTs.

If an ionic surfactant is used, the micelles incorporating the tubes expose charged groups which in principle allow the tubes to be guided to oppositely charged surface regions. Such properties of surfactant-coated SWNTs open the possibility of controlling tube assembly to a degree which is not accessible through simple drying of tube suspensions or spin-coating. One important goal is to gain control over the density of adsorbed tubes by appropriate adsorption times, the concentration of SWNTs, and surface chemistry. High densities of SWNTs would be desired for, e.g., surface enhanced Raman spectroscopy, whereas only a few tubes per μm^2 are needed for electrical measurements on individual tubes. In the present contribution, we describe studies of tube adsorption from purified, micellar suspended SWNTs onto (chemically modified) silica surfaces and electrode arrays. In addition, the organisation of SWNTs at the gas/liquid interface and the transfer of the mixed SWNT/surfactant monolayer onto solid substrates are presented.

2 SWNT ADSORPTION ON (PATTERNED) SUBSTRATES

Assembly of the tubes was performed with SWNTs dispersions purified by size exclusion chromatography similar to the procedure described previously for multi-walled nanotubes (3), except for the use of the lithium dodecyl sulfate (LDS) as surfactant instead of the sodium salt (SDS). Compared to the latter, LDS is somewhat more water soluble, which is favourable for the removal of surfactant subsequent to tube adsorption. After chromatography, the LDS concentration was raised to 1wt%. After a few days, the dispersions usually contain some precipitated aggregates which can easily be re-dispersed through short ultrasonic agitation.

Adsorption kinetics and composition of the deposited material were observed to sensitively depend on the chemical nature of the surface. Closely arranged SWNTs were obtained in case of adsorption on silica with positive surface charges. For that purpose, ammonium groups were attached to Si/SiO_2 wafers via silanization (6). Fig. 1 displays a representative AFM image of particles adsorbed onto such type of surface.

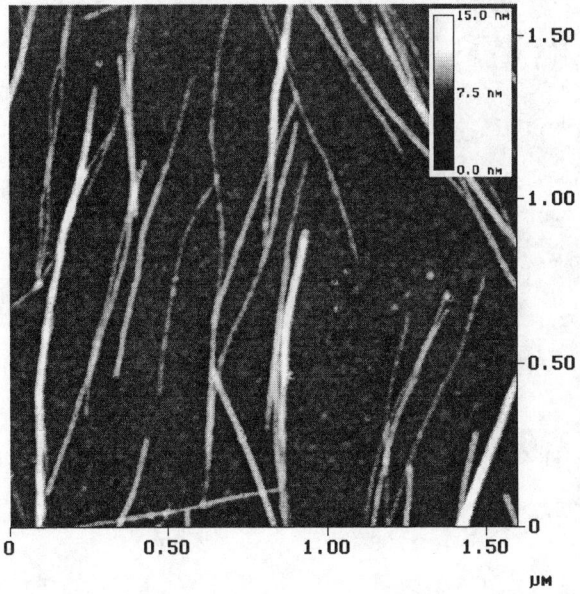

FIGURE 1. AFM image of SWNTs adsorbed from a surfactant-stabilised, chromatographically purified dispersion on an amino-silanized silica surface. The substrate was rinsed with water to remove the surfactant.

Due to the electrostatic attraction between the negatively charged LDS molecules around the tubes and the amino-functionalised SiO_2 surface, high coverages are achieved by a relatively short adsorption time of 20 min. Subsequent to the adsorption step, excess surfactant can be removed by rinsing with pure water. It is, however, not possible to deduce from the AFM results whether some traces of surfactant remain between tube and surface.

The particles observed in Fig. 1 consist of individual SWNTs (with a height of between 1 nm and 2 nm) as well as ropes of SWNTs. Inspection of a larger number of samples revealed that about half of the particles in the purified material are individual tubes. The situation changes if the SWNT adsorption is performed on a pure SiO_2 surface. Chemically unmodified silica exhibits negative surface charges at not too low pH values. Thus the repulsive interactions between the negatively charged LDS surfactant shell and the SiO_2 surface slow down the adsorption process. As a consequence, much lower tube densities than in Fig. 1 are obtained even after several hours of adsorption. Remarkably, the fraction of adsorbed individual tubes is found to increase from about ½ to more than ¾. This result can be explained by a lower kinetic stability of the surfactant/individual tube complex as compared to a surfactant-stabilised thicker rope. On the contrary, this difference does not seem to play a role in the time scale of the much faster adsorption onto the positively charged SiO_2 surfaces.

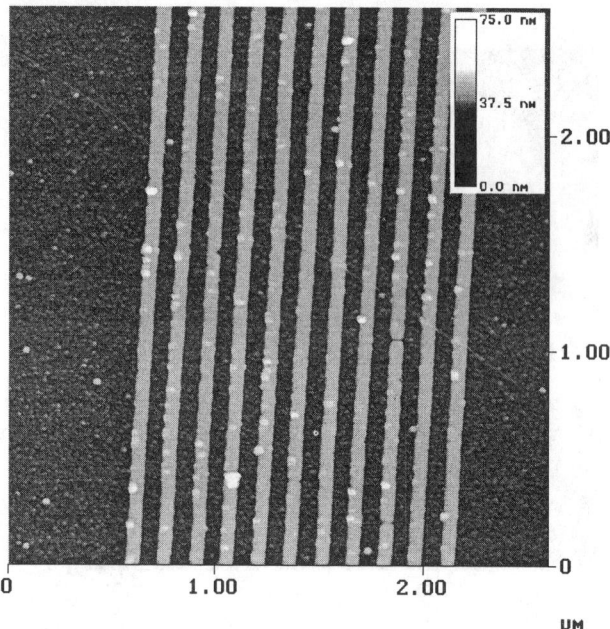

FIGURE 2. AFM image of an individual SWNT deposited on an array of eleven electrode lines (electrode distance ≈ 75 nm). The measured height of the tube is about 1.6 nm.

It turned out that for SWNT assembly on noble metal electrodes, no specific chemical surface modification is needed to attain good adhesion of the tubes. This observation is in accordance with the known affinity of noble metal surfaces to π-conjugated molecular systems. Adsorption of SWNTs from surfactant-stabilised dispersions allows an easy and reproducible access to tubes which are deposited over extended electrode arrays. As an example, Fig. 2 shows an individual SWNT crossing eleven electrode lines. The Au/Pd lines are 75 nm in width, 14 nm high, and are separated by about 75 nm. By using a fixed adsorption time of 30 min, a similar number of adsorbed tubes per area as in Fig. 2 is obtained with different substrates. Again the question arises whether a surfactant monolayer separates the tube from the electrode surface. Room temperature electrical measurements on the deposited tubes revealed a resistance of maximally a few 10 MΩ. Because surfactant molecules with a length of between 1.5 nm and 2 nm (as for LDS) should lead to highly insulating behaviour (7), the presence of an intact surfactant monolayer is unlikely.

In the adsorption experiments, tube orientation was observed if after the appropriate adsorption time the dispersion was removed by exposing the surface to a fast stream of water. An example is given in Fig. 3 which shows SWNTs adsorbed close to an electrode array with a different geometry from that shown in Fig. 2. As a striking feature in Fig. 3, the tubes are aligned within two regions denoted as 1 and 2, respectively.

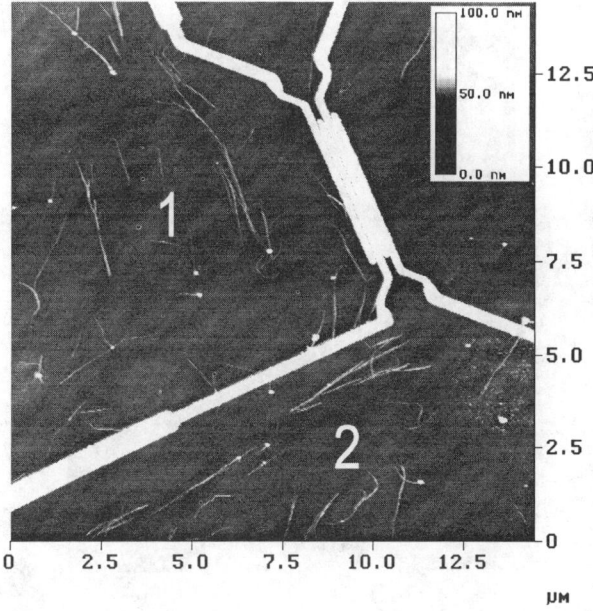

FIGURE 3. AFM image of SWNTs adsorbed in vicinity of an electrode array. The tubes are aligned in two different, almost perpendicular directions within the regions 1 and 2, respectively.

Within region 1 the tubes are oriented parallel to the array consisting of four electrode lines, whereas the tubes in region 2 follow the same direction as the lower left lead. These observations can be explained by assuming that the electrodes with a height of 25 nm act as mechanical barriers for the tubes which exert flow forces during removal of their surfactant shell. Tube orientation effects of similar extent were found for electrode arrays like the one depicted in Fig. 2.

3 SWNTs IN LANGMUIR-BLODGETT FILMS

Mixed SWNT/LDS surfactant monolayers can be prepared at the gas/liquid interface by directly spreading the aqueous SWNT dispersions onto a subphase in a Langmuir trough. For that purpose, the subphase must consist of an aqueous solution of a positively charged polyelectrolyte (e.g., poly-allylamine, PAA). This leads to the formation of an insoluble complex between the positively charged polymer and the oppositely charged LDS molecules and prevents the SWNTs from being transferred into the subphase. After compression of the monolayer to the Langmuir film, horizontal dipping allows the transfer of the latter onto various types of substrate. In Fig. 4, an AFM image of a SWNT/LDS monolayer deposited on a hydrophobic silicon substrate is displayed.

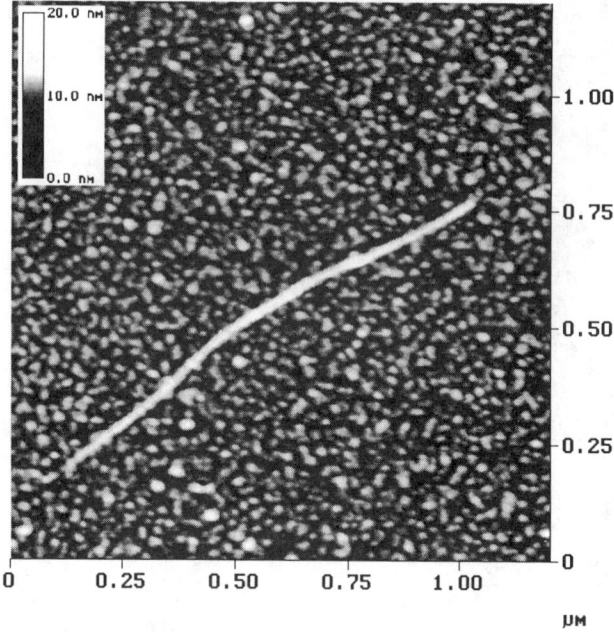

FIGURE 4. AFM image of a mixed SWNT/surfactant Langmuir-Blodgett monolayer transferred onto a hydrophobic silicon wafer. Shown is a rope of SWNTs with a measured height of ≈ 5 nm.

The image shows a SWNT rope surrounded by a matrix composed of an ionic LDS/PAA complex. The density of tube particles (individual SWNTs or ropes of SWNTs), which are homogeneously dispersed in the Langmuir-Blodgett films without agglomeration, typically ranges between 1 and 5 per μm^2. The film matrix is not continuous but consists of dot-like islands with a height of about 3 nm. These islands are surrounded by regions exposing the bare substrate surface. It is assumed that upon transfer to the hydrophobic surface, the LDS/PAA system undergoes a structural change to almost spherical micelle-like aggregates.

4 CONCLUSIONS

Surfactant-stabilised SWNT dispersions are very useful in the assembly of carbon nanotubes on different substrates. The density and composition of deposited particles can be controlled via chemical surface modifications. In addition, the surfactant can also play the role of a film matrix in Langmuir-Blodgett films incorporating SWNTs. Such films are promising for the realisation of ultrathin composite layers containing SWNTs which are oriented by flow effects originating from the dipping of the substrate.

ACKNOWLEDGMENTS

The authors are grateful to C. Journet and P. Bernier (University of Montpellier) for supplying SWNT raw material. This work was financially supported by the network NAMITECH of the European community.

REFERENCES

1. Bandeau, S., Rao, A.M., Williams, K.A., Thess, A., Smalley, R.E., and Eklund, P.C., *J. Phys. Chem. B* **101**, 8839 (1997).
2. Bonard, J.-M., Stora, T., Salvetat, J.-P., Maier, F., Stöckli, T., Duschl, C., Forro, L., de Heer, W.A., and Chatelain, A., *Adv. Mater.* **9**, 827 (1997).
3. Duesberg, G.S., Burghard, M., Muster, J., Philipp, G., and Roth, S., *Chem. Commun.* 435 (1998).
4. Hiura, H., Ebbesen, T.W., and Tanigaki, K., *Adv. Mater.* **7**, 275 (1995).
5. Shelimov, K.B., Rinat, R.O., Rinzler, A.G., Huffmann, C.B., and Smalley, R.E., *Chem. Phys. Lett.* **282**, 429 (1998).
6. Ulman, A., *Introduction to Ultrathin Organic Films*, San Diego: Academic Press, 1991.
7. Boulas, C., Davidovits, J.V., Rondelez, F., and Vuillaume, D., *Phys. Rev. Lett.* **76**, 4797 (1996).

INTERCALATION REACTIONS IN CATALYTIC MULTIWALL CARBON NANOTUBES

K. METENIER, L. DUCLAUX, H. GAUCHER, J.P. SALVETAT*, P. LAUGINIE,
S. BONNAMY and F. BEGUIN
CRMD, CNRS - Université, 1b rue de la Férollerie, 45071 Orléans Cedex 2, France
*Ecole Polytechnique Fédérale de Lausanne, IGA/DP 1015 Lausanne, Suisse

ABSTRACT

Heat-treated catalytic multiwall carbon nanotubes (MWNTs) were intercalated by K and $FeCl_3$ in vapor phase, using the two-bulb technique. A first stage KC_9 intercalation compound was formed with potassium. After elimination of potassium, the tubular morphology is still preserved showing that intercalation is a reversible phenomenon. In the case of $FeCl_3$, the saturated compound is less rich than with graphite. However, well defined in plane hk bands prove the intercalation. Due to the position of the 002 line at 0.345 nm, it is likely that intercalation is incomplete and that the material is a mixture of intercalated and non intercalated zones. A model of catalytic nanotubes is presented which accounts for the reversibility of the intercalation reactions.

1. Introduction

Various kinds of carbon nanostructures such as filaments, nanofibers...[1] and MWNTs [2] have been already obtained by the catalytic decomposition of hydrocarbons. Catalytic MWNTs were synthezized in a large scale through the catalytic decomposition of acetylene [3].

Due to their tubular morphology, nanotubes are a very attractive 1D-type host lattice for insertion reactions [4,5]. Catalytic MWNTs have been already applied for lithium insertion [6,7].

In the present work, they have been used in reactions with potassium and iron (III) chloride, in order to modify their electronic properties. On the other hand, intercalation-deintercalation is a way to better elucidate the dilemma of Russian doll and scroll structures.

2. Experimental

Cobalt nitrate supported on silica (1 g) was placed in a graphite crucible at the centre of a vertical flow furnace, and reduced at 500°C by hydrogen diluted in nitrogen (1:9 volume ratio). Acetylene diluted in nitrogen (1:9 volume ratio) was decomposed at 900°C on the catalyst during two hours, after what the system was cooled down to room temperature under nitrogen. Subsequent chemical treatments allowed to dissolve the silica substrate in 73% hydrofluoric acid and the main part of cobalt in 3N nitric acid. In a typical experiment, 600 mg of purified nanotubes were obtained starting from 1 gram of catalyst.

The catalytic nanotubes were heat-treated between 2500 and 2800°C in a graphitization furnace under argon flow during 15 minutes; the selected temperature was reached at 20°C/min heating rate. Heat-treatments enable to improve the structural and microtextural characteristics of the nanotubes and to eliminate the encapsulated metallic cobalt.

Intercalation reactions on the heat-treated MWNTs were performed in the vapor phase by using the two-bulb technique [8]. The reagent (K or $FeCl_3$) and the nanotubes were respectively placed at the opposite ends of a pyrex glass tube. After being carefully evacuated under vacuum (10^{-5} mbar), the tube was sealed and placed four days in twin furnaces at two temperatures, respectively $T_{nanotubes}=350°C$ and $T_K=300°C$ for the reaction with K and $T_{nanotubes}=300°C$ and $T_{FeCl_3}=270°C$ in the case of the reaction with $FeCl_3$.

For their characterizations by X-ray diffraction (INEL CPS 120) and Electron Spin Resonance (BRUKER ESP 300E spectrometer), the samples were transferred into a quartz capillary (diameter = 1 mm) in a glove box under high purity argon atmosphere.

3. Results and discussion
3.1 Electron Spin Resonance (ESR) on heat-treated catalytic MWNTs

The ESR line recorded at 300 K is symmetrical (figure 1); the g factor and the linewidth ΔH_{pp} are temperature-dependent with a maximum at about 100 K (figure 2). The susceptibility given by the intensity of the signal shows a Curie-Pauli behaviour with temperature (figure 3), suggesting the existence of free carriers and of a small amount of localized paramagnetic centers. Its quasi-constancy over the range 50-300 K is in favor of a true metallic character.

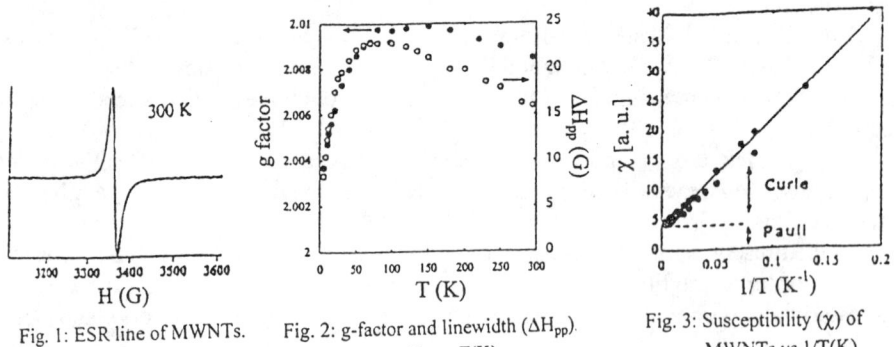

Fig. 1: ESR line of MWNTs. Fig. 2: g-factor and linewidth (ΔH_{pp}) of MWNTs vs T(K). Fig. 3: Susceptibility (χ) of MWNTs vs 1/T(K).

3.2 Intercalation-deintercalation of K in MWNTs

The brown colour of the sample after reaction is a proof of a charge transfer and consequently of modified electronic properties for the nanotubes. The weight uptake (35%) gives the composition $KC_{9\pm1}$, which is close to the usual MC_8 stoichiometry of the 1st stage GICs with heavy alkali metals.

The X-ray diffraction data (figure 4b) are those of a 1st stage compound. Indeed, after reaction, the 00l reflections confirm an identity period of 0.54 nm, whereas the hk0 lines fit well with the 2*2 superstructure. The asymmetry of the hk bands show that the arrangement of the carbon layers is still turbostratic.

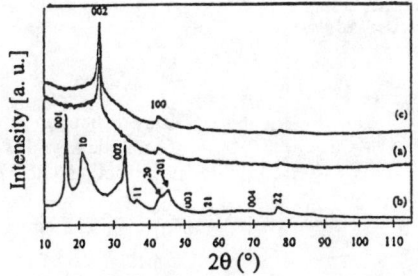

Fig. 4: X-ray diffraction patterns of catalytic MWNTs heat-treated at 2800°C (a), intercalated with K(b), after deintercalation (c).

$\lambda_{CuK\alpha} = 0.15405$ nm.

Intercalation is well supported by the ESR measurements (figure 5) showing a broad and asymmetrical (asymmetry ratio≈2.3) line very comparable to that of the KC_8 GIC. The g factor and the linewidth (figure 6) are nearly temperature-independent. As compared to the pristine nanotubes and similarly to the parent graphite-potassium intercalation compound, the susceptibility is increased by a factor 40 and is almost independent of temperature (figure 7). This Pauli-like dependence is the finger print of a a metallic behaviour. The effective number of spins is in agreement with a 1st stage intercalation compound.

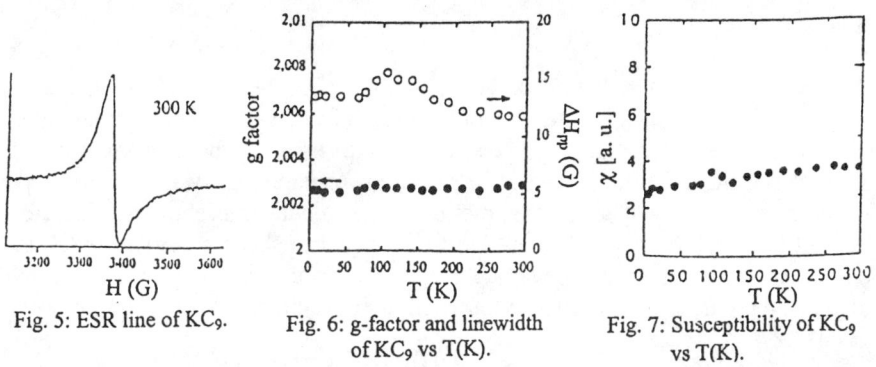

Fig. 5: ESR line of KC_9.

Fig. 6: g-factor and linewidth of KC_9 vs T(K).

Fig. 7: Susceptibility of KC_9 vs T(K).

In order to check the reversibility of the reaction, the compound was introduced into a two-bulb tube under high vacuum and heat-treated at 450°C during two days with one end of the tube at room temperature. After deintercalation, the diffractogram (figure 4c) is similar to that of the starting nanotubes (figure 4a). Moreover, Transmission Electron Microscopy observations have shown that the tubular morphology is still existing, however with some additionnal defects.

3.3 Intercalation of $FeCl_3$ in MWNTs

After the reaction with $FeCl_3$, the sample remains black. The weight uptake is only 25% while it is usually 55% for the 1st stage saturated $FeCl_3$-GIC. The diffractogram (figure 8) exhibits well defined in plane *hk* bands coming from the bidimensionnal superstructure of the intercalate. The *00l* lines are interpreted by a mixture of 3rd and 4th stages derivatives with an identity period of about 18 Å. The still important reflection of the pristine nanotubes present at $d_{002} = 3.44$ Å suggests that the intercalation is incomplete and leads to a mixture of intercalated and non-intercalated zones.

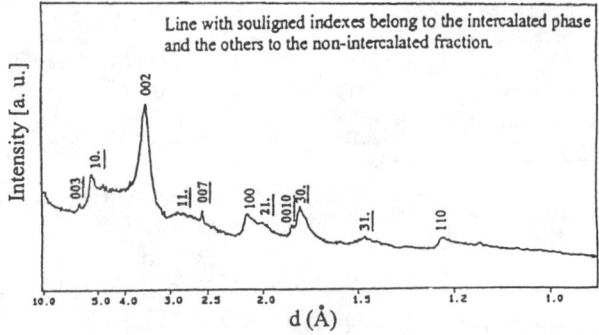

Fig. 8: X-ray diffraction pattern of catalytic MWNTs after reaction with $FeCl_3$.

$\lambda_{MoK\alpha}=0.7093$ nm.

We could not detect any signal by ESR on this derivative. This is not surprising, since $FeCl_3$-GICs present an extremely broad (325 G) asymmetrical line. We could suspect that nanotubes intercalated with $FeCl_3$ could give an even broader signal which would be undetectable.

4. Conclusion

The fact that the catalytic carbon nanotubes can be reversibly intercalated by potassium has to be related with the numerous defects still existing even after high temperature treatment. Continuous layers, as one can imagine them in the Russian doll arrangement, would not allow an easy intercalation reaction. Anyway, if such a process would occur, it would lead to the destruction of the tubular morphology. Therefore, we suggest that the catalytic nanotubes are built of concentric shells separated by boundaries through which the reagent can penetrate (figure 9). In the case of $FeCl_3$, it cannot be excluded that there is, at least partly, an insertion in the microtexture or in the canal.

Fig. 9: "Shell" model of catalytic MWNTs.

This work has been supported by E. C. TMR program : NAMITECH, ERBFMRX-CT96-0067 (DG 12-MIHT).

References

1. R.T.K. Baker and P.S. Harris, in *Chemistry and Physics of Carbon*, vol. 14, ed. P.L. Walker and P.A. Thrower (Marcel Dekker, New York, 1978) pp. 83-165.
2. M. José-Yacaman, M. Miki-Yoshida, L. Rendon and J.G. Santiesteban, *Appl. Phys. Lett.*, 62 (1993) 202.
3. A. Hamwi, H. Alvergnat, S. Bonnamy and F. Béguin, *Carbon*, 1997, 35, 723.
4. S. Suzuki and M. Tomita, *J. Appl. Phys.*, 1996, 79(7), 3739.
5. V.Z. Mordkovich, M. Baxendale, S. Yoshimura and R.P.H. Chang, *Synth. Metals*, 1996, 34, 1301.
6. V.A. Nalimova, D.E. Slovsky, G.N. Bondarenko, H. Alvergnat, S. Bonnamy and F. Béguin, *Synth. Metals*, 1997, 88, 89.
7. E. Frackowiak, S. Gautier, H. Gaucher, S. Bonnamy and F. Béguin, *Carbon*, 1998, in press.
8. A. Hérold, *Bull. Soc. Chim. Fr.*, 1955, 999.

NANOTUBES: EXPERIMENTS

Electrical Resistivity of Single-wall Carbon Nanotubes Obtained by the Arc-Discharge Technique

E. Jouguelet*, C. Mathis*, P. Petit*, C. Journet[†] and P. Bernier[†]

*Institut Charles Sadron, 6, rue Boussingault, 67000 Strasbourg, France
[†]GDPC, Université Montpellier II, 34095 Montpellier cedex 5, France

Abstract. Electrical resistivity studies performed on single-wall carbon nanotubes obtained by the arc discharge method are reported in the present paper. The experiments have been performed on both doped and undoped non-purified bulk samples and the results are compared to those reported for single-wall carbon nanotubes obtained by the laser vaporisation technique. Except the fact that we never observed a "normal" metallic behaviour, both kind of materials exhibit the same electrical properties.

INTRODUCTION

Nonchiral single-wall carbon nanotubes (SWNT) are extensively studied since they are produced in high quantity. The first method leading to a high yield of production is the laser vaporisation technique[1]. The bulk resistivity of the "as grown" material ranges in the milli-Ohm.cm scale and its thermal behaviour is characteristic of a metal ($d\rho/dT>0$) above a sample dependent temperature T* of about 100-250 K[2,3]. Below T*, $d\rho/dT<0$ while the ESR signal characteristic of a highly conducting system remains[4]. Moreover, the resistivity is found to be non-linear below T* and quantum effects, like weak localisation with a large coherence length, are invoked to interpret the low-T behaviour of the resistivity[5,6]. SWNT have also been shown to be easily intercalated either by Potassium or Bromide leading to a drop of the room temperature resistivity by a factor 30 and 15 respectively[7]. The second synthesis method is the arc-discharge technique leading to a high density of SWNT in a "collaret" around the cathod[8]. In this paper, we report electrical resistivity measurements on the bulk "as grown" material obtained by this technique. Both pristine and doped samples are studied and the results are compared to those performed on SWNT obtained by the laser vaporisation method. We also report the possibility of intercalating SWNT using a chemical route.

EXPERIMENTAL

SWNT are produced using the arc discharge method described in reference 8. From published SEM images[1,8], the main *apparent* differences between samples obtained by this technique as compared to those produced by laser vaporisation are a weaker mean diameter of the ropes and a larger density of ropes.

Br_2 or I_2 intercalated SWNT were obtained by exposure to halogen vapour, following the procedure described in ref. 7.

For the chemical doping, the procedure is as follows: a THF solution of naphthalene radical anion with a potassium contre-ion (N^-K^+) is prepared under high vacuum in a sealed glass ampoule equipped with a break-seal. The SWNT sample, mounted on a glass support with two pressed electrical contacts, is iserted in a separate glass apparatus fitted with airtight outlets allowing resistivity measurements. The glass ampoule containing N^-K^+ is then sealed onto this apparatus, and is later air evacuated on a vacuum line and sealed.

Both kind of doping were performed on non-pressed samples taken from the soot.

RESULTS AND DISCUSSION

The electrical resistivity of the as grown material in the temperature range 4-300 K is shown on figure 1. On the whole temperature range, we observe a negative slope of $\rho(T)$. This constitutes the main difference between this material and the one obtained by laser vaporisation which shows a metallic behaviour above a sample dependent temperature of about 150 to 250 K. However, we measured the microwave resisitivity on ten different samples of different shapes and found that the room temperature resistivity of each sample is characteristic of a metallic system[9,4] and equal to the dc resistivity (50 to 400 mΩcm, depending on the sample). As the upturn in the thermal behaviour of the resistivity in the samples obtained by laser vaporisation is shifted to very high temperature when the sample is exposed to hydrostatic pressure[10,11], we argue that the difference between both kind of materials is related to a difference in the ropes density[4].

By exposure to I_2 and Br_2 vapours, the room temperature resistivity drops by a factor 12, close to what is reported for SWNT obtained by laser vaporization[7]. However, we found that opening the apparatus to lab. air leads to an increase of the resistivity which stabilises after a few hours to a value of 1/4 that of the undoped material. The temperature dependence of the resistivity of the doped samples still exhibits a negative slope, but the increase of ρ with decreasing temperature is very much weaker than for the undoped material, using the same applied current (1mA), as can be seen on figure 2.

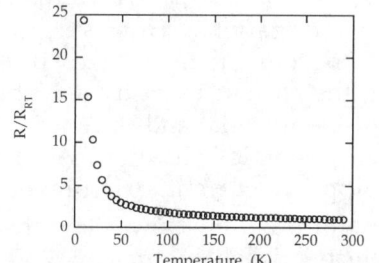 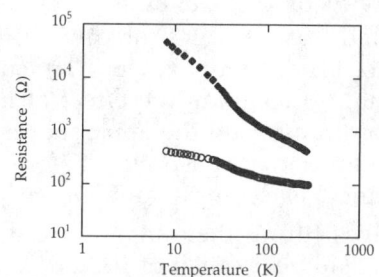

FIGURE 1. Temperature dependence of the normalised resistance of SWNT (I=1mA)

FIGURE 2. Temperature dependence of the resistance of undoped (♦) and doped SWNT. The experiment has been performed on the same sample (I=1mA).

For the chemical doping, the breakseal separating both part of the glass apparatus is broken and the N⁻K⁺ solution is poured over the sample and the evolution of the resistance recorded. A fast decrease of the resistance is observed indicating that electron transfer occurs as soon as the sample is immersed in the N⁻K⁺ solution (figure 3). The discontinuity around 1000 sec in the R(t) curve results probably from changes in the soot structure, linked to the sample texture. However, it can be seen that the reaction goes on up to, at least, 30000 s.

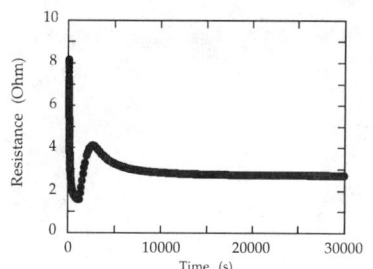 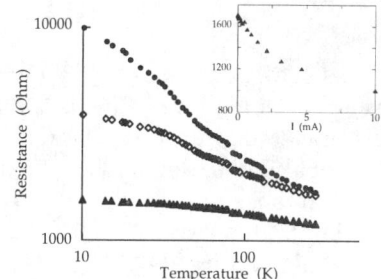

FIGURE 3. Time dependence of the resistance upon chemical doping.

FIGURE 4. Thermal variation of undoped SWNT for different field intensities. (▲ 4.6mA, ◊ 0.46mA, • 0.046mA). Inset: resistance vs. I at room temperature.

Taking into account the accident in this preliminary experiment, a total drop of the resistivity of the same amplitude as observed for gas vapour doping can be expected. The interest in this route of doping is that the amount of transferred electrons can be directly derived from the decrease of the concentration of N-K+. However, due to the small amount of sample used in the present study (less than 1 milligram), the total consumption of N⁻K+ was too small to be estimated.

We also studied the effect of the electrical field intensity on the resistivity of this material and found the same non-linearity as already reported for samples obtained by laser vaporization[5]. However, we observed non-linearity in I(V) curves up to room temperature. It is tempting to compare the effect of increasing the current intensity and the electron transfer on the apparent resistivity of the material and to conclude that charge injection occurs before reaching the ohmic behaviour. As the resistance is strongly dependent on the applied current, we prevent ourselves fitting the temperature dependence of ρ by any power law that could lead to a misinterpretation of the electrical behaviour of SWNT. According to us, magnetoresistivity measurements are better suited to check weak localisation effects.

CONCLUSION

From the above results, we conclude that SWNT synthesised by both techniques have the same electrical properties. This is supported by the disappearance of the upturn in ρ(T) for purified samples obtained by laser vaporisation for which the density is three times that of the pristine material.

REFERENCES

1. A. Thess et al, Science **273**, 483, (1996).
2. J. E. Fischer et al., Phys. Rev. B **55**, R4921, (1997).
3. C. L. Kane et al., Europhys. Lett., in press, (1998).
4. P. Petit et al, Phys. Rev. B **56**, 9275, (1997).
5. A. Zettl, this issue.
6. K. Liu et al., this issue.
7. R. Lee et al., Nature **388**, 255, (1997).
8. C. Journet et al., Nature 388, 756, (1997) and C. Journet et al., this issue.
9. L. I. Buranov and I. F. Shchegolev, Prib. Tek. Eksp. **2**, 171, (1971) [Instrum. Exp. Tech. **14**, 258, (1971).]
10. J. E. Fischer et al., this issue.
11. A. G. Rinzler et al., Applied Phys. A, in press.

Electrical transport in carbon nanotubes

K. Liu, S. Roth, G. S. Duesberg*, G. -T. Kim, M. Schmid

Max-Planck-Institut für Festköperforschung, Heisenbergstr. 1, D-70569, Germany
*University of Dublin, Trinity College, Physics Department, Dublin 2, Ireland

Abstract. Network samples of different purified multi-walled carbon nanotubes (MWCNTs) and single-walled carbon nanotubes (SWCNTs) are prepared for electrical property studies. Classic electrical properties of tubes are responsible for electron transport in the high temperature region, while quantum behaviors control electron transport in the low temperature region. In particular, metallic behavior is observed in the high temperature region (T > 104 K). Based on experimental studies and theoretical predictions, the metallic behavior is attributed to scattering of electrons by twistons in entangled tubes.

1 INTRODUCTION

Since the discovery of carbon nanotubes (CNT) by S. Iijima (1), electrical properties of this new material have drawn more and more attention (2) due to their implications of fundamental physics and potential applications in electronic devices. So far, most experimental studies have been performed on either bulk samples or single tube/bundle samples. Bulk samples are easy to prepare, but impurity components such as amorphous carbon, fullerenes, metal catalysts, and carbon polyhedra particles still exist in the sample, which makes experimental analysis more difficult. On the other hand, experiments on single tubes or bundles require complicated e-beam techniques, which sometimes modifies the electrical properties of samples (3). In this paper, network sample of purified carbon nanotubes are used for electrical studies. Experiments on a series of network samples of purified multi-walled carbon nanotubes (MWCNTs) and single-walled carbon nanotubes (SWCNTs) have been performed, classic and quantum transport behaviors are observed.

2 SAMPLE PREPARATION AND EXPERIMENT

The MWCNTs samples were synthesized by cracking of acetylene with the help of Co catalysts (4) and purified by the liquid oxidation method. The SWCNTs are produced by the *d.c.* arc-discharge method and purified by the size exclusion chromatography method (5) and centrifugation method (6). After purification, pure

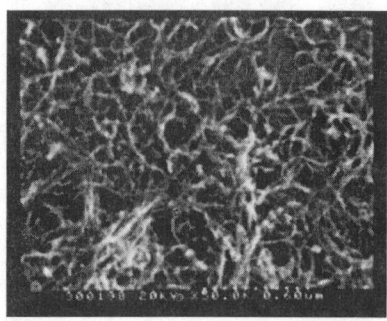 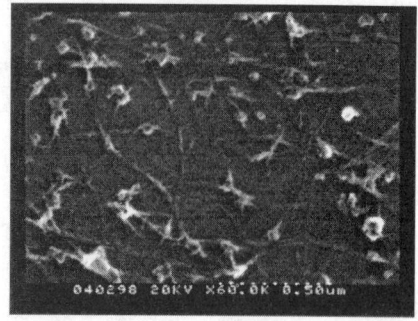

FIGURE 1a. SEM image of MWCNT network on electrode array.

FIGURE 1b. SEM image of MWCNT network on electrode array.

MWCNTs and SWCNTs are obtained with very few catalyst impurities. With the help of sodium dodecyl sulfate surfactant (SDS), tubes are uniformly dispersed in suspensions that are stable for a long time (at least one week). Then, suspensions are dropped over an electrode array on a SiO_2(1μm)/Si chip, and washed with pure water after drying in order to remove extra surfactant. In this way, a series of MWCNT and SWCNT network samples with good contacts to electrodes is obtained. The electrode array has 6 parallel electrodes separated by a 3 μm gap, the electrode width is 2 μm. The insulation resistance between neighboring electrodes is larger than 10^{13} Ω before covering with the tube network, and about 10~300 Ω at room temperature after covering. Figs. 1a and 1b are typical SEM images of the MWCNT and SWCNT tube networks, respectively. The networks are composed of bundles of pure tubes. In this paper, six network samples are studied, three are MWCNTs and three SWCNTs. A home built four-probe measuring system is used to investigate electrical transport properties.

3 EXPERIMENTAL RESULTS AND DISCUSSION

Fig. 2a and Fig. 2b present the temperature dependent resistance R(T) of the six network samples from 4.2 K to 300 K. The resistance of each sample has been normalized to its room temperature resistance R(T = 300 K). For each sample, high and low currents are applied for the R(T) measurements. In both MWCNTs and SWCNTs, it is found that Ohm's law (linear current/voltage characteristics) holds till a very low temperature T^O ($T^O \approx 50$ K), T^O is called „Ohmic temperature". In the Ohmic conduction region (T > 50 K), the temperature dependence of the resistivity is different in different MWCNT samples, but almost the same in different SWCNT samples. Since the electron thermal energy k_BT is very high (> 4 meV) in the Ohmic region, electrical transport in tubes is „classic", i.e. determined by the energy band structures of the tubes and influenced by sample morphologies (disorder, strain, coupling among

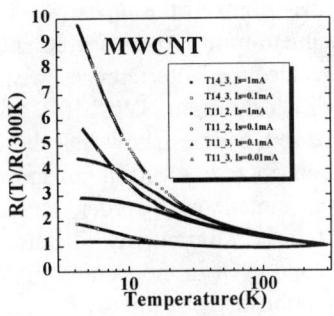
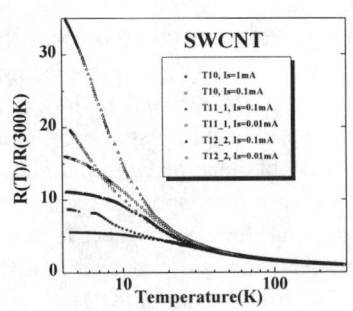

FIGURE 2a. Temperature dependent resistance of different MWCNT networks

FIGURE 2b. Temperature dependent resistance of different SWCNT networks

tubes). Considering that the energy band structures of tubes are determined by tube configurations (diameter, length, helicity, tube layers), the experimental observations indicate that, (I) the deviation of the tube configurations is large in the MWCNT samples and small in the SWCNT samples, (II) the morphology influence is strong in the MWCNT samples and weak in the SWCNT samples, which is in agreement with the SEM investigations (see Fig. 1a and 1b). Among the six network samples, five show semiconducting behavior (i.e. the resistivity increases on cooling) and one MWCNT sample (T14_3) metallic behavior in the high temperature region (i.e. the resistivity decreases on cooling).

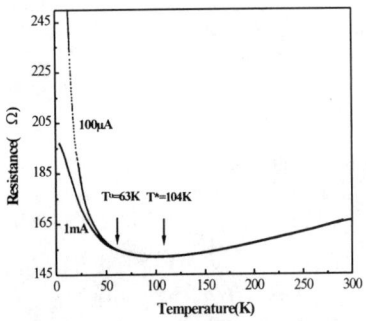

FIGURE 3. Temperature dependent resistance in one of the MWCNT network samples.

The metallic behavior of the sample T14_3 is plotted in Fig. 3. The resistance reaches a minimum at T* of about 100 K on cooling. In another serious of MWCNT samples, T* is found to vary between 18 K (4) to 250 K depending on sample preparation process. The temperature dependence observed here is similar to that observed previously on single-walled nanotubes. The resistance decrease on cooling is consistent with the freeze-out of phonons (like in ordinary metals) or with the freeze-out of twistons, as proposed by Kane et al. (7). The upturn of the resistance below T* shows that the metallic behavior in the sample differs from that in normal metals. Twistons are long wavelength, low energy torsional modes, which strongly interacts with conduction electrons. The twiston spectrum depends strongly on defects, strains and network entanglement and therefore the strong dependence of T* on sample preparation process is taken as an evidence of the importance of twistons in the sample.

A remarkable feature, as seen in Fig. 2 and 3, is that the resistance in the non-Ohmic regime depends on the electric current through the sample. Some authors propose that in this regime charge- or spin- density waves form in the tubes (8). To investigate the transport properties in this regime magneto-transport measurements have been carried out. Electron-electron interactions, electron interference, weak localization (two-dimensional in MWCNTs and one-dimensional in SWCNTs) and antilocalization induced by spin-orbit coupling are observed. Studies show that these quantum behaviors take place in individual tubes, and contribute to electron transport. Since the quantum transport are very sensitive to local situations (disorder, strain, coupling among tubes) in tube layers as well as tube configurations, different conduction behaviors are expected in different samples and even in the same sample with applying different currents. Detailed studies will be published elsewhere (9).

4 CONCLUSION

A series of network samples of purified MWCNTs and SWCNTs has been prepared for electron transport studies in tubes. It is found that in the high temperature region classic behavior dominates the electron transport in the tubes, while in the low temperature region quantum behavior controls the electron transport. Metallic behavior is observed in one of the MWCNT samples, which is attributed to electron scattering by twistons in the tubes.

REFERENCES

1. Iijima, S., *Nature* **354**, 56 (1991).
2. Song, S. N., Wang, X. K., and Chang, R. P. H., *Phys. Rev. Lett.* **72**, 697 (1994);
   Langer, L., Bayot, V., Grivei, E., and Issi, J.-P., *Phys. Rev. Lett.* **76**, 479 (1996);
   Fischer, J. E., Dai, H., Thess, A., Lee, R., Hanjani, N. M., Dehaas, D. L., and Smalley, R. E., *Phys. Rev. B* **55**, R4921 (1997).
3. Burghard, M., et al. (private communication).
4. Mukhopadhyay, K., Colomer, J.-F., Konya, Y., Piedigrosso, P., and Popa, D. (to be published).
5. Duesberg, G. S., Burghard, M., Muster, J., Philipp, G., and Roth, S., *Chem. Commun.*, 435 (1998).
6. Bandow, S., Rao, and A. M., Williams, K. A., *Chem. B* **101**, 8839 (1997).
7. Kane, C. L., Mele, E. J., and Lee, R. S., *Europhys. Lett.* **41**, 683 (1998).
8. Balen, L. and Fischer, M. P. A., *Phys. Rev B* **55**, R11973 (1997).
9. Liu, K., Roth, S., Duesberg, G. S., Popa, D., Mukhopadhyay, K., Doome, R., and B.Nagy, J., (submitted to *Phys. Rev. B*).

Electrical properties of single carbon nanotubes

A. Bachtold[1], C. Strunk[1], C. Schönenberger[1], J.-P. Salvetat[2], L. Forró[2]

1 Institut für Physik, Universität Basel, Klingelbergstr. 82, CH-4056 Basel, Switzerland
2 Institut de Génie Atomique, Ecole Polytechnique Fédérale de Lausanne, CH-1015 Lausanne, Switzerland

Abstract. Magnetoresistances of individual multiwall nanotubes have been measured at low temperatures in a four-probe configuration with low-ohmic contacts. A negative magnetoresistance is found which is in agreement with weak-localisation theory in one dimension. A large coherence length $L_\phi > 200$nm is deduced at T=4.2K. Measurements on different nanotubes show only minor variations in the value of the resistance and in its temperature dependence.

INTRODUCTION

Carbon nanotubes are attracting increasing interests as a new class of nanoscale material. One distinguishes between multi-wall nanotubes (MWNTs), consisting of a series of coaxial graphitic cylinders, and single-wall nanotubes (SWNTs). The latter have been predicted the remarkable property to be metallic or semiconducting depending on how the graphite layer is wrapped into a cylinder.[1] Wildöer et al.[2] have been able to verify this prediction on single SWNTs with scanning tunnelling spectroscopy. For MWNTs, there is at present no consensus whether such drastically different behaviours are observable. Using four-probe measurements, Ebbesen et al.[3] reported metallic as well as semiconducting temperature dependences. Resistivies, measured at T=300K, varied by six orders of magnitudes. In contrast, experiments by Dai et al.[4] showed a much smaller variation in electrical resistivities at 300K. Here, a brief account of our (four-probe) electrical resistance measurements on single MWNTs with low-ohmic contacts is presented.

CONTACTING ONE SINGLE MWNT

MWNTs are prepared by arc-discharge evaporation and purified by centrifugation and sedimentation.[5] Transmission electron microscopy reveals an outer diameter of 9±5nm and an inner diameter of ≈2nm for these nanotubes. The deduced

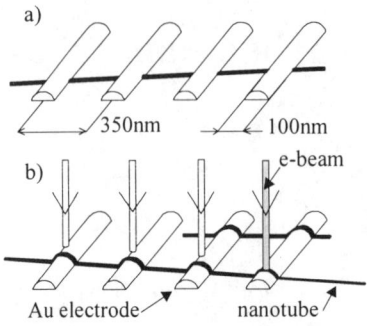
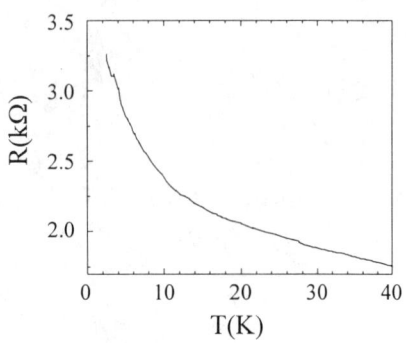

FIG. 1. Two techniques to attach low-ohmic contacts to MWNTs: in (a) Au electrodes are fabricated after the deposition of nanotubes, while in (b) the sequence is reversed. In (b) the e-beam is used to "spot-weld" the nanotube to the electrode.

FIG. 2. Temperature dependence of the resistance for a single MWNT.

number of shells is 10±7. Four metallic Au electrodes are attached to single MWNTs by two methods (figure 1): in method (a), the nanotubes are deposited onto an oxidised Si wafer first. Then, an arbitrarily positioned array of 200 similar electrode-structures, each consisting of four Au fingers, are fabricated. Inspection with the scanning-electron microscope (SEM) allows to identify those structures that have only one single nanotube contacted. Contact resistances as low as 300Ω are found. In the second method (b), the Au electrodes are fabricated before the deposition of the nanotubes. In contrast to method (a), measured contact resistances are initially very large (1MΩ...1GΩ), but can substantially be reduced to values of 4kΩ...30kΩ by locally irradiating the nanotube-Au contacts with a large electron dose in the SEM (the nanotubes are "spot-welded" to the contacts).[6] Method (b) has the advantage that even in situations for which more than one nanotube happen to bridge the four electrodes, a single nanotube can be selected by local electron-beam exposure. We mention, that imaging in the SEM and the local electron-beam exposure does not seem to cause any damage to the nanotubes (at least for doses used in our work), since the four-probe resistances of nanotubes are not observed to change.

ELECTRICAL PROPERTIES OF INDIVIDUAL MWNTS

Using both contact methods, the four-probe resistances of single MWNTs at T=300K lie in a narrow range between 0,35kΩ and 2,6kΩ (more than 30 samples). At T=4K, the resistances vary between 1,0kΩ and 5,2kΩ (11 samples). A representative resistance versus temperature dependence R(T) is shown in figure 2. All measured R(T) curves have in common a negative slope ($\partial R/\partial T<0$) describing a rather moderate increase of R of at most a factor of 3.5 on lowering the temperature from 300K to 4K.

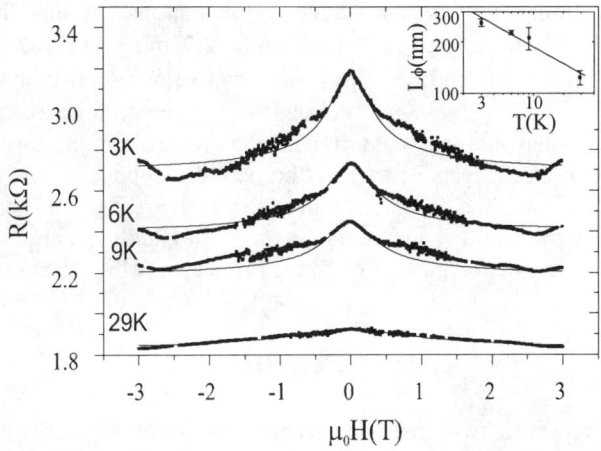

FIG. 3. Magnetoresistance for a MWNT perpendicular to the field. The full lines are obtained from 1D WL theory with the condition to fit the curvature at H=0. The inset shows the temperature dependence of the coherence length deduced from the 1D WL fits.

Since variable-range hopping or semiconducting (activated) behaviour cannot account for this increase, the MWNTs are believed to be metallic, although $\partial R/\partial T<0$. Figure 3 shows the magneto-resistance (MR) of one MWNT aligned perpendicular to the magnetic field H. It is found to be negative. In addition, there is a field range (0.7T...1.7T) in which large temporal fluctuations are observed, giving rise to a strong scatter in the data. Furthermore, at higher fields (>2T) oscillations appear which are possibly caused by universal conductance fluctuations. In this brief paper, only the negative MR will be discussed further. The negative MR is interpreted in the framework of weak-localisation[7] in one dimension (1D) from which a consistent picture emerges. A theoretical 1D-WL curve is fitted to the curvature of the measured R(H) data around zero field. These curves (thin solid in figure 3) are seen to represent the overall magnitude of the MR in the full magnetic-field range of 0...3T in a satisfactory way. Furthermore, the deduced phase-coherence lengths L_ϕ of order 200nm are larger than the circumference of the nanotube in agreement with the requirement for one-dimensional behaviour. Applying a similar procedure, but using 2D-WL[8] theory, either the magnitude of the MR or the curvature around zero field cannot accurately be fitted. Even worse, however, is the deduced coherence length, which is of order 500nm, inconsistent with the assumed dimensionality. 1D-WL together with a large coherence length for our MWNTs is strikingly different to previous measurements on MWNTs by Langer et al,[9] who found 2D-WL behaviour and a short coherence length of only 10nm. This difference presumably arises from the fact that our nanotubes are less disordered.

The temperature dependence of the phase-coherence length $L_\phi(T)$ can be described by $L_\phi(T)=A\cdot T^{-1/3}$, where A is a constant (see inset of figure 3). This power law is in agreement with the model of Altshuler,[10] in which L_ϕ is destroyed by the interaction of electrons with electromagnetic fluctuations. In this theory, A only depends on the diffusion constant D. We obtain D≈0.03m^{2}s^{-1}. From D=$v_f^2\tau/2$, where τ is the elastic-scattering time and v_f≈10^{6}m/s the Fermi velocity for turbostratic graphite, an elastic mean-free path of l≈60nm is derived. This estimation gives a first indication that MWNTs are quasi-ballistic one-dimensional conductors at low temperatures. Moreover, quantum-size effects caused by the periodic boundary condition along the tube circumference should be observable. In figure 3, the resistance is seen to increase on lowering the temperature even in large magnetic fields. Therefore, there is an additional field-independent contribution to the resistance, which is not yet understood.

CONCLUSION

We have reported two different techniques to attach low-ohmic electrical contacts to single multi-wall nanotubes. Only minor differences are found in the electrical resistance and its temperature dependence when comparing several measured nanotubes. The magnetoresistance at low temperature can consistently be interpreted by weak-localisation in one dimension. Both the coherence length and the elastic-mean free path are larger than the tube diameter, placing multiwall nanotubes into the exciting category of one-dimensional quasi-ballistic quantum-wires.

---

[1] J. W. Mintmire et al., Phys. Rev. Lett. **68**, 631 (1992); N. Hamada et al., Phys. Rev. Lett. **68**, 1579 (1992); R. Saito et al., Appl. Phys. Lett. **60**, 2204 (1992).
[2] J. Wildöer et al., Nature **391**, 59 (1998).
[3] T. W. Ebbesen et al., Nature **382**, 54 (1996).
[4] H. Dai et al., Science **272**, 523 (1996).
[5] J. M. Bonard et al, Advanced Materials **9**, 827 (1997).
[6] A. Bachtold, J. P. Salvetat, J. M. Bonard, M. Henny, C. Terrier, C. Strunk, L. Forro and C. Schönenberger (submitted).
[7] B. L. Altshuler and A. G. Aronov, Pis'ma Zh. Eksp. Theor. Fiz. [JETP Lett. **33**,499 (1981)].
[8] G. Bergmann, Phys. Rep. **107**, 1 (1984).
[9] L. Langer et al., Phys. Rev. Lett. **76**, 479 (1996).
[10] B. L. Altshuler et al., J. Phys. C. **15**, 7367 (1982).

Transport and Localization in Single-Walled Carbon Nanotubes

M. S. Fuhrer, U. Varadarajan, W. Holmes, P. L. Richards,
P. Delaney, S. G. Louie, A. Zettl

Department of Physics, University of California, Berkeley, and Materials Sciences Division, Lawrence Berkeley National Laboratory, Berkeley, CA 94720 USA

We have measured the electrical transport properties of mats of single-walled carbon nanotubes (SWNT) as a function of applied electric and magnetic fields. We find that at low temperatures the resistance as a function of temperature R(T) follows the Mott variable range hopping (VRH) formula for hopping in three dimensions. Measurement of the electric field dependence of the resistance R(E) allows for the determination of the Bohr radius of a localized state a ≈ 650nm. The magnetoresistance (MR) of SWNT mat samples is large and negative at all temperatures and fields studied, and can be qualitatively described by theories of MR for VRH systems. The Hall coefficient R_H is positive and nearly temperature-independent. The sign of R_H agrees with the sign of the thermopower. The small magnitude of R_H suggests a large carrier density, but may be the result of cancellation of electron and hole terms.

Introduction

The electrical transport properties of single-walled carbon nanotubes are currently the subject of much debate. Individual achiral (armchair) SWNT are predicted to be metallic[1-3]. Resistance measurements on bulk samples (mats) of such tubes indeed show metallic behavior (positive dR/dT) at high temperature, but negative dR/dT at low temperature[4]. Transport measurements[5,6] on individual tubes or bundles of tubes show some samples to be conducting to low temperatures, but the quantized electron energy levels which lead to such spectacular phenomena as Coulomb blockade cloud interpretation of the temperature dependence of the resistivity. It is then important to determine whether the properties of the mat samples are intrinsic to the nanotubes or are a product of impurities or poor inter-tube connections.

The samples used in this study were produced by catalyst-assisted arc-vaporization[7] or laser-vaporization[8] of a graphite source, yielding similar results. The resulting material was purified in HNO₃ to remove catalyst impurities and amorphous carbon, and then washed and filtered. The samples are observed by transmission electron microscopy to consist of bundles of tens or hundreds of tubes arranged in a triangular lattice. The mats consist of a random network of these bundles, connected on a length scale of approximately 100nm. Electrical contact to the bundles was made using silver paint in a standard four probe or Hall bar configuration. Sample resistances at room temperature ranged from 100 to 3000 Ohms.

Nonlinear Resistance

The R(E) measurements used a standard pulsed current technique: 10μsec current pulses were applied to the sample. The current and voltage were monitored simultaneously as a function of time using an oscilloscope, allowing the intrinsic resistance to be separated from (time-dependent) self-heating effects.

Figure 1 shows the R(T) behavior of a SWNT mat at various applied electric fields. The resistance of SWNT mats shows a striking dependence on E; a modest electric field can completely suppress the resistance upturn and recover metallic behavior from 2.2K to 300K. The inset of figure 1 shows the metallic high temperature resistance.

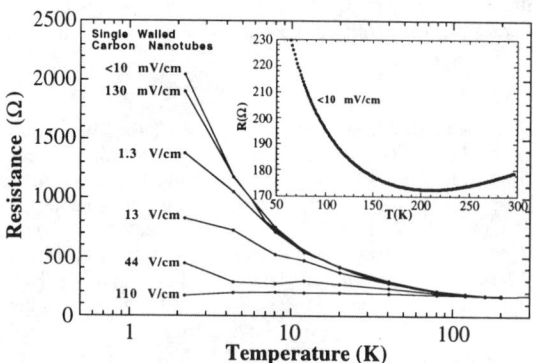

Figure 1. Resistance versus Temperature of SWNT mat at several applied electric fields. Inset shows high temperature R(T) at low electric field.

In order to clarify the nature of the non-linear resistance, measurements of R(T) and R(E) were repeated on the cold stage of an adiabatic demagnetization refrigerator, to temperatures as low as 227mK. Figure 2 shows the low-temperature R(T) data plotted semi-logarithmically as a function of $T^{-1/4}$; the data indicate $R \propto \exp(-T^{1/4})$, the Mott variable range hopping formula for three dimensions.

Temperature and electric field both have a similar delocalizing effect on the carriers. The two energy scales for temperature and electric field are kT and eEa, respectively, where k is Boltzmann's constant, e the electronic charge, and a the Bohr radius of an electronic state. It is found that[9]:

$$R(kT,0) = \frac{R}{2}(0, \frac{3}{8}eEa),$$

valid for finite values of R. Our R(T) data are measured in the Ohmic regime, and thus give R(kT,0). We measured R(0,E) at 227mK. The data are plotted

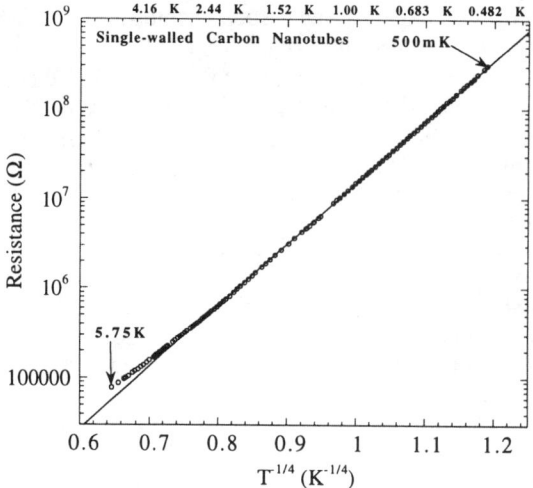

Figure 2. Resistance versus $T^{-1/4}$ of SWNT mat showing three-dimensional variable range hopping.

together in Figure 3. The temperature in Kelvin is scaled by the factor 3.5 in order to match the electric field in V/cm. The scaling of the two curves indicates a ≈ 650nm.

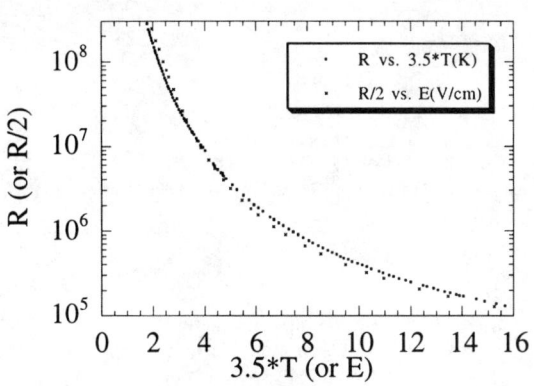

Figure 3. R(E) at 227mK and R(T) in the Ohmic regime for a SWNT mat. The scaling determines the radius of the localized states, 650nm.

SWNT mats are localized electronic systems where the electronic states are three-dimensional, with a radius of approximately 650nm, much greater than the average distance between bundle contacts observed by TEM. This result is inconsistent with any model of localization in which the electrons are confined to an individual tube or bundle. The dimensionality in Mott's VRH formula indicates that the number of available states for hopping is proportional to the distance r from the localized state raised to the d power. Hopping confined to a single tube or to nearest-neighbor bundles is one-dimensional, since the number of available states scales linearly with r along the tube. Only if the localized state extended across many tube crossings would the number of available states be proportional to r^3. We conclude that the localization is not due to poor junctions between tubes or bundles, but rather due to some intrinsic disorder, either in the individual tubes or bundles, or in the morphology of the network, which is inherently random.

Magnetoresistance

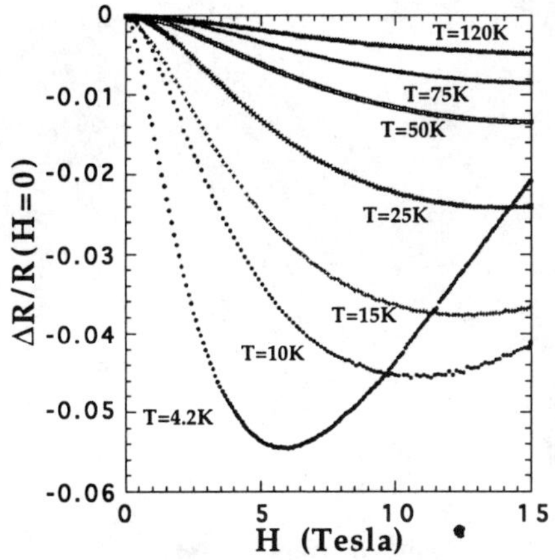

Figure 4 shows the magnetoresistance (MR), defined as the change in resistance divided by the resistance at zero field, of a SWNT mat sample at various temperatures. The MR is negative at all fields and temperatures measured, but shows a minimum, increasing at higher fields.

Figure 5 shows the MR data replotted on a double logarithmic scale. The lines on the plot indicate H^2 dependence of the MR, in agreement with the variable range hopping model. A tendency to saturation is also observed at high field, but the positive high field slope is unexplained.

In hopping systems, the effect of magnetic field can be modeled by considering the interference between all different hopping paths between two states[10]. Magnetic field induces a phase difference

Figure 4. Magnetoresistance of a SWNT mat as a function of magnetic field at various temperatures.

in the different paths, which has a net delocalizing effect. In this model MR in variable range hopping systems is predicted to be negative, with a quadratic dependence at low magnetic field, saturating to a constant value at high field.

Hall Effect

SWNT mat samples were prepared with a standard Hall bar geometry in order to measure the Hall effect. In order to circumvent the uncertainty in thickness of the SWNT films caused by their unusual morphology, the mass and area of each SWNT mat were measured. The experimentally accessible quantity is $\Delta R_{xy}/\Delta H$, the change in Hall resistance (R_{xy}) with magnetic field. In a single band model,

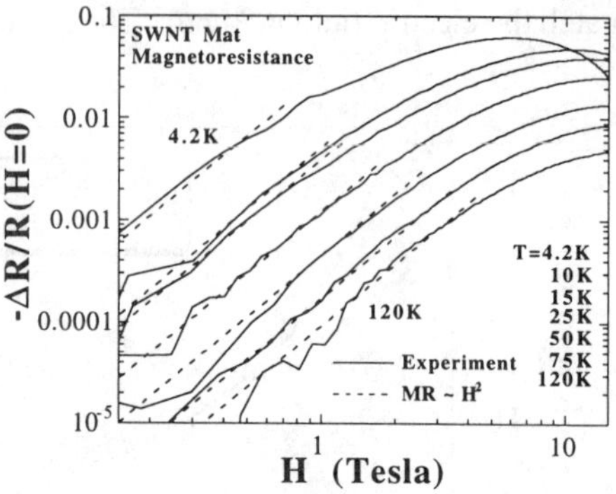

Figure 5. Negative magnetoresistance of a SWNT mat is plotted on a double-logarithmic scale. Solid lines represent the experimental data; straight dotted lines indicate H^2 behavior.

$$\frac{\Delta R_{xy}}{\Delta H} = \frac{R_H}{t} = \frac{1}{nec}$$

where R_H is the Hall coefficient, t is the thickness of the sample, and n is the density of carriers per *area*. Knowing the mass per area of the film allows for a calculation of the number of carriers per carbon atom. Figure 6 shows R_H/t and n for a SWNT mat sample. The Hall coefficient is positive and nearly temperature independent. The positive sign of R_H/t agrees with the sign of the thermopower[11]. The magnitude of R_H/t indicates a carrier density of approximately 0.1 holes/atom in a single band picture. For a metallic (10,10) tube this corresponds to about 4 holes per 40 atom unit cell, assuming all tubes in the sample are metallic, and contributing to the Hall effect. The presence of some percentage of semiconducting tubes would increase this number.

Band structure calculations for the armchair nanotubes show 2 states at the Fermi level, with an electron-hole symmetry. The electron-

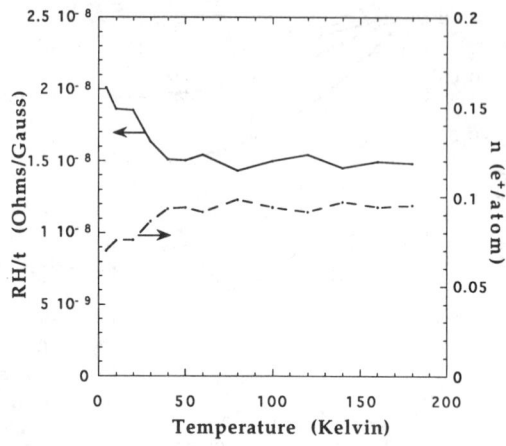

Figure 6. Hall coefficient divided by sample thickness and carrier density of SWNT mat plotted as a function of temperature. The solid line (Hall coefficient) corresponds to the left scale, the dotted line (carrier density) corresponds to the right.

hole symmetry should lead to zero Hall effect and thermopower, but significant positive Hall effect and thermopower are observed. Furthermore, the apparent n from the single band picture is much too large. It has been predicted[12] that interactions between tubes in a bundle will significantly perturb the states near the Fermi level leading to the formation of electron-like and hole-like pockets of the Fermi surface, much as are found in graphite. We suggest that these electron and hole terms incompletely cancel, leading to a small R_H/t, and apparent large n.

We note that although a small downturn of the carrier density is observed at low temperatures, this cannot account for the large upturn in resistance. This is further evidence that the resistance upturn is a reduction of the mobility of the carriers, as would be expected in the case of localization, not in their number, as would be the case if a gap in the Fermi surface were causing the resistance upturn.

Conclusions

We have measured the electrical transport properties of SWNT mats in electric and magnetic fields. Although composed of one-dimensional wires, mats of single-walled carbon nanotubes exhibit three-dimensional metallic behavior, as evidenced by their metallic resistivity and the Hall effect. At low temperatures, the carriers are localized, leading to a 3D variable range hopping form of the resistance. Comparison of the temperature and electric field dependencies of the resistance allows for the extraction of the Bohr radius of localized states a $\approx$ 650nm, consistent with three-dimensional states. The magnetoresistance of SWNT mats is consistent with theories of MR for variable range hopping, although the high-field positive component of the MR remains unexplained. The Hall effect is positive in agreement with TEP results and small in magnitude, which would indicate an unusually large hole concentration in a single band model. The small Hall coefficient likely results from the cancellation of electron and hole terms, supporting the idea that tube-tube interactions in a bundle perturb the Fermi surface significantly.

We thank Prof. R. E. Smalley for providing some of the samples used in this study. This work was supported in part by the Director, Office of Energy Research, Office of Basic Energy Sciences, Materials Sciences Division of the U. S. Department of Energy under Contract No. DE-AC03-76SF00098, and by the National Science Foundation, Grant No. DMR95-20554.

References

1. J. W. Mintmire, B. I. Dunlap and C. T. White, Phys. Rev. Lett. **68**, 631 (1992).
2. N. Hamada, S. Sawada and A. Oshiyama, Phys. Rev. Lett. **68**, 1579 (1992).
3. R. Saito, M. Fujita, G. Dresselhaus and M. S. Dresselhaus, Appl. Phys. Lett. **60**, 2204 (1992).
4. J. E. Fischer, et al., Phys. Rev. B **55**, R4921 (1997).
5. S. J. Tans, et al., Nature **386**, 474 (1997).
6. M. Bockrath, et al., Science **275**, 1922 (1997).
7. C. Journet, et al., Nature **388**, 756 (1997).
8. A. Thess, et al., Science **273**, 483 (1996).
9. N. Apsley and H. P. Hughes, Philosophical Magazine **31**, 1327 (1975).
10. U. Sivan, O. Entin-Wohlman and Y. Imry, Phys. Rev. Lett. **60**, 1566 (1988).
11. J. Hone, et al., Physical Review Letters **80**, 1042 (1997).
12. P. Delaney, H. J. Choi, J. Ihm, S. G. Louie and M. L. Cohen, Nature **391**, 466 (1998).

Transmission Electron Microscopy and Electrical Transport Investigations performed on the same Single-Walled Carbon Nanotube

G. Philipp, M. Burghard, and S. Roth

Max-Planck-Institut für Festkörperforschung, Heisenbergstr. 1, D-70569 Stuttgart, Germany

Abstract. Electrical transport measurements and high resolution transmission electron microscopy performed on the same (rope of) single-walled carbon nanotube(s) (SWCNTs) allow to establish links between structural and electronic properties of the tubes. The tubes are deposited on electron transparent ultrathin Si_3N_4-membranes bearing Cr/AuPd-electrodes defined by electron beam lithography. TEM-micrographs of the setup reveal mostly ropes consisting of 2-3 tubes which also appear on a scanning force microscope image of the same area. A current-voltage trace of the ropes at 4.2 K is also presented.

I INTRODUCTION

Transmission electron microscopy (TEM) is a powerful tool for investigating structural properties of single-walled carbon nanotubes like packing or helicity [1-3]. The obtainable resolution is better than that of the scanning force microscope (SFM), where true resolution (but not the lattice resolution) is limited by the tip radius. Scanning tunneling microscopy (STM), in contrast, allows to resolve the atomic structure of the tubes and to study electronic properties normal to the tube axis [4,5]. So far electronic properties of SWCNTs along the tube axis were studied with the structural information obtained from SFM imaging [6].

II SWCNTS ON MEMBRANES

The approach presented here allows to study electron transport along the tube axis and to perform high resolution transmission electron microscopy on the same tube. To achieve this it is necessary to have metal electrodes for electrically contacting the tubes defined on a substrate that is extremely flat and also transparent for electrons. Suitable substrates for this purpose are Si_3N_4 membranes that can be prepared using conventional silicon processing techniques [7].

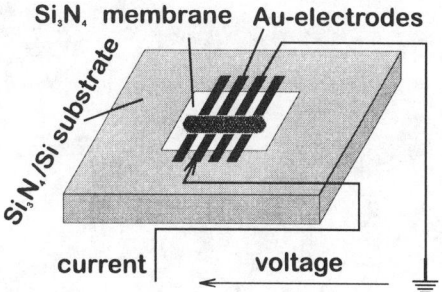

FIGURE 1. Schematic of the device. The tubes are deposited on a electron transparent Si_3N_4 membrane with interdigitated Cr/AuPd electrodes. The corresponding TEM micrograph is shown in FIG. 2

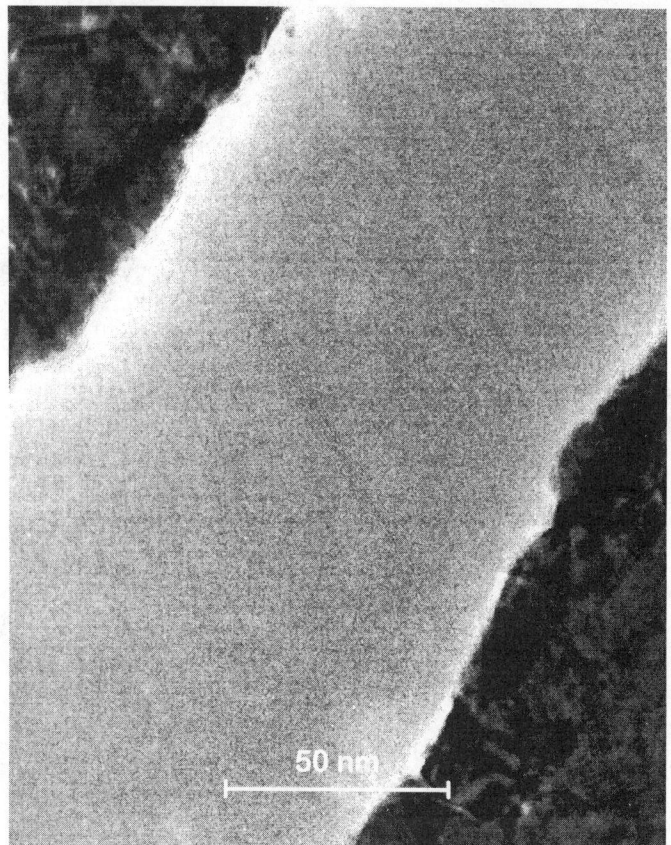

FIGURE 2. TEM micrograph of a carbon nanotube rope bridging two electrode fingers of the device shown in FIG. 1. The three tubes making up the rope can be recognized.

FIGURE 3. TEM micrograph of an area of the membrane where nanotubes are adsorbed.

The obtained Si_3N_4 membranes have a uniform thickness of 20 nm over an area of 200 μm x 200 μm with a roughness of less than 0.5 nm and can be easily handled using normal tweezers. To define electrodes (5nm Cr, 15 nm AuPd) on such membranes a standard bilayer PMMA resist electron beam lithography process was used.

Fig. 1 shows schematically an electrode array with deposited SWCNTs. Before deposition the single-walled carbon nanotube raw material was disintegrated with the aid of a surfactant and the resulting dispersions purified by size exclusion chromatography [8]. Prior to tube adsorption the substrate was exposed to UV ($\lambda = 254$ nm) radiation in order to improve the wetting of the surface by aqueous media. Subsequently, the purified tubes were deposited onto the otherwise untreated substrate as described in [9].

The TEM micrograph in Fig. 2 shows a rope consisting of three carbon nanotubes bridging two electrode fingers of the device shown in Fig. 1.

III   COMPARISON SFM - TEM

In addition to TEM, scanning force microscopy (SFM) can be applied to the electron transparent silicon nitride membranes. This allows to investigate the same area on the membrane by both methods and thus obtain a better assessment of the features in the SFM image. In Fig. 3 and Fig. 4 an TEM micrograph and a SFM image of the same area are shown. The structures seen in the SFM image can be clearly identified in the transmission image Fig. 3. Zooming in easily allows to establish the number of tubes in a rope and also their structural arrangement.

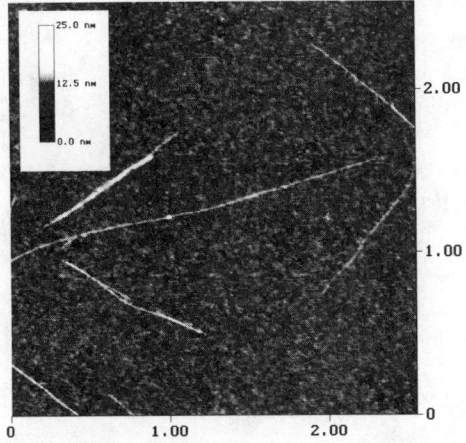

FIGURE 4. SFM image of the same membrane area as in Fig. 3 with the same nanotubes

IV ELECTRICAL TRANSPORT

A current-voltage characteristic at T=4.2 K of the device sketched in Fig. 1 is shown in Fig. 5. The nonlinear behaviour is indicative of Coulomb blockades, which will be investigated in more detail when experiments are continued to lower temperatures.

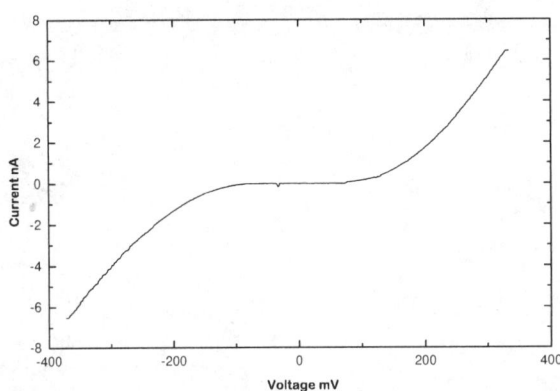

FIGURE 5. Current-voltage trace at T = 4.2 K of a rope similar to that shown in Fig. 2

ACKNOWLEDGMENTS

The presented work was supported by the Bundesministerium für Bildung und Forschung under grant 03N1012E8. Fruitful discussions with G. Duesberg and J. Muster, the expert help of M. Kelsch with TEM specimen preparation and the supply of SWNT raw material by C. Journet and P. Bernier (Université de Montpellier) are gratefully acknowledged.

REFERENCES

1. Bernaerts, D., Amelinckx, S., van Tendeloo, G. and van Landuyt, J.: *J. Phys. Chem. Solid.* **58**, 1807, (1997)
2. Bernaerts, D., Zettl, A., Chopra, N. G., Thess, A. and Smalley, R. E.: *Solid State Commun.*, **105**, 145, (1998)
3. Cowley, J. M., Nikolaev, P., Thess, A. and Smalley, R. E.: *Chem. Phys. Lett.*, **265**, 379, (1997)
4. Wildoer, J. W. G., Venema, L. C., Rinzler, A. G., Smalley, R. E.and Dekker, C.: *Nature*, **391**, 59, (1998)
5. Odom, T. W., Huang, J.-L., Kim, P. and Lieber, C. M.: *Nature*, **391**, 62, (1998)
6. Tans, S. J., Devoret, M. H., Dai, H., Thess, A., Smalley, R. E., Geerligs, L. J. and Dekker, C.: *Nature*, **386**, 474, (1997)
7. Jacobs, J. W. M. and Verhoeven, J. F. C. M.:*J. Microscopy*, **143**, 103, (1986)
8. Duesberg, G. S., Burghard, M., Muster, J., Philipp, G. and Roth, S.: *Chem. Commun.*, 435, (1998)
9. Burghard, M., Duesberg, G. S., Philipp, G., Krstic, V. and Roth, S.: *present proceedings*

Thermoelectric Power and Thermal Conductivity of Single-Walled Carbon Nanotubes

J. Hone, I. Ellwood, M. Whitney, M. Muno, C. Piskoti, and A. Zettl

Department of Physics, University of California, Berkeley, and Materials Science Division, Lawrence Berkeley National Laboratory, Berkeley CA 94720 USA

We have measured the temperature-dependent thermopower (TEP) and thermal conductivity (κ) of bulk samples of single-walled nanotube (SWNT) bundles. The TEP of SWNT's approaches zero as T→0, indicating a metallic density of states at the Fermi level in spite of their non-metallic resistivity behavior. At moderate temperatures, the TEP is large and positive, while a single metallic tube should have electron-hole symmetry and thus a zero thermopower. The measured data can be fit using a model comprising hole-like metallic tubes and electron-like semiconducting tubes in parallel. The thermal conductivity of SWNT's is found to be large, and dominated by phonons at all temperatures. At low temperature, $\kappa(T)$ is linear. The data can be fit by a single-scattering-time model; the model confirms that the low-temperature linear behavior is due to the thermal conductivity of a single one-dimensional phonon subband, and that the phonon mean free path is of order 100 nm.

Introduction

Most investigations into the properties of carbon nanotubes have focused on either their electrical properties or their mechanical properties. However, their thermoelectric and thermal properties are equally interesting. Thermopower is a sensitive probe of the bandstructure of materials, and in particular can discriminate between different possible mechanisms for the low-temperature resistivity upturn seen in bulk single-walled nanotube (SWNT) samples. The thermal concuctivity of nanotubes has been predicted to be quite large--similar to that of diamond or graphite; such a large thermal conductivity could represent a significant advantage for nanotubes in electronic device applications. In addition, nanotubes are an ideal system in which to study low-dimensional thermal conduction effects.

The samples used in this study are 'mats' of SWNT bundles. The bundles contain tens to hundreds of nanotubes in a triangular lattice, and can be microns in length. They were synthesized by an arc-vaporization method[1]; samples synthesized by the laser-vaporization method[2] were used to check the results. A portion of the tubes were 'sintered' by resistively heating them under pressure in an attempt to improve the contacts between tubes. It was found that the electrical conductivity of these sintered tubes is in fact lower than that of the as-grown samples, and that the conductivity shows nonmetallic temperature dependence at all temperatures.

Thermoelectric Power

Figure 1 shows the measured thermoelectric power (represented as the solid squares) of a representative SWNT mat, taken at a number of fixed temperature points from 300 K to 4 K. The TEP of other samples, including the sintered sample, shows similar behavior. Thus we believe that the measured TEP represents the intrinsic properties of the SWNT bundles composing the mat. The measured thermopower is large and positive, indicating hole-like conduction, and approaches zero as T approaches zero. Because a system with a gap at the Fermi level should display a (1/T)-dependent thermopower, we conclude that the opening of a gap is not the cause of the nonmetallic resistivity observed in SWNT mats at low temperature.

We first examine the predicted one-dimensional bandstructure of a metallic armchair nanotube to examine whether it is consistent with the observed TEP. The Mott formula for the thermopower of a metal reduces in one dimension to

$$S = -\frac{\pi^2 k_B^2 T}{3e}\left(\frac{v'}{v} + \frac{\tau'}{\tau}\right), \qquad (1)$$

where v is the band velocity, τ is the electron relaxation time, the derivatives are with respect to energy, and the expression is evaluated at the Fermi points. An armchair nanotube is expected to have two one-dimensional bands which cross at the Fermi points[3]. These bands are highly linear, and thus should exhibit electron-hole symmetry and a thermopower close to zero; we calculate a maximum contribution of 0.5 μV/K for the room temperature contribution from Eq. (1). This is two orders of magnitude smaller than the observed TEP. Likewise, the electron-hole symmetry of this tube should produce a small phonon drag contribution to the TEP. Therefore we conclude that the expected contribution from an armchair tube is insufficient to account for the measured thermopower.

Since nanotube bundles likely contain tubes with a range of chiralities, we consider the possibility that semiconducting tubes, in parallel with metallic tubes, could be producing the measured TEP. We do this by employing a two-band model. By assuming simple functional forms for the thermopower and conductivity of the metallic and semiconducting contributions, we obtain

$$S = AT + (B\lambda + CT)\exp\left(\frac{-\lambda}{T}\right). \qquad (2)$$

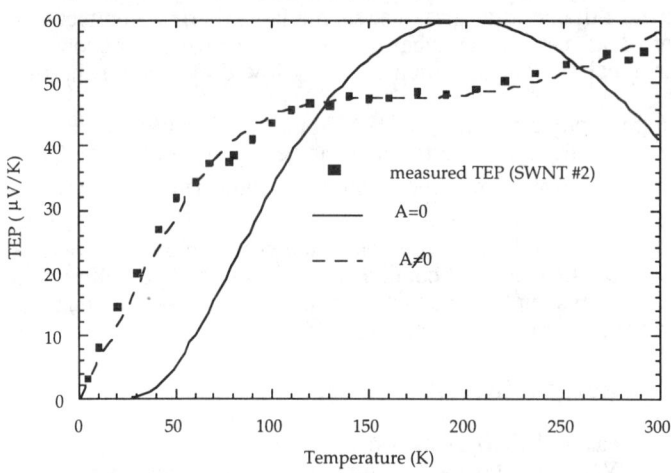

Figure 1. Thermopower of SWNT's. The solid line is a fit to the measured TEP data to Eq. (2) with A=0, representing metallic tubes with electron-hole symmetry; the dashed line represents a fit with an arbitrary contribution from the metallic tubes.

In this formula, the first (linear) term represents the contribution of the metallic channel, while the second represents the contribution of the semiconducting channel. We first assume that the metallic tube retains its electron-hole symmetry, and thus A=0. The solid line in Fig. (1) represents an attempt to fit the measured data to Eq. 2 using A=0. This clearly does not produce a good fit--

the observed low-temperature behavior cannot come from semiconducting tubes, because their contribution freezes out at low temperature. Therefore we conclude that there must be at least some metallic tubes with broken electron-hole symmetry. We model this by allowing the A parameter to vary arbitrarily. The dashed line in Fig. (1) represents this fit to the measured data. The high quality of this fit leads us to the conclusion that a model comprising hole-like metallic tubes and electron-like semiconducting tubes is a viable explanation for the magnitude and temperature dependence of the measured thermopower.

Thermal Conductivity

Figure (2) represents the thermal conductivity $\kappa(T)$ of a representative SWNT sample as a function of temperature from 350 K to 8 K. The same temperature dependence was observed in as-grown samples and the sintered sample; and therefore we conclude that it reflects the intrinsic $\kappa(T)$ of SWNT's. The thermal conductivity is near-linear at all temperatures, in contrast to that of graphite and carbon fibers, which display a T^2-like $\kappa(T)$. The inset to figure (2) represents the low-temperature behavior of κ. It is strictly linear in T, and extrapolates to zero as T→0. This linear behavior is usually associated with electron thermal conductivity; however, we will demonstrate below that it is in fact the signature of one-dimensional phonon thermal conductivity.

We first attempt to deduce the magnitude of the thermal conductivity of SWNT's from the measured data. Using just the dimensions and measured thermal conductance of the sample, we obtain a room-temperature thermal conductivity of 0.7 W/m-K. Because the filling fraction of the tubes in a mat is only about 2%, we correct for this factor to obtain a value of 36 W/m-K for the thermal conductivity of a dense-packed SWNT mat. However, even a dense-packed mat will have a disordered structure: the longitudinal thermal conductiviy of a single tube or rope will likely be significantly larger. To deduce this value, we note that the longitudinal electrical conductivity of a single rope has been measured to be 50-150 times that of the bulk mat[4]. If the same holds true for the thermal conductivity, then the longitudinal thermal conductivity of a single rope is 1800-6000 W/m-K (comparable to or greater than that of graphite or diamond) at room temperature.

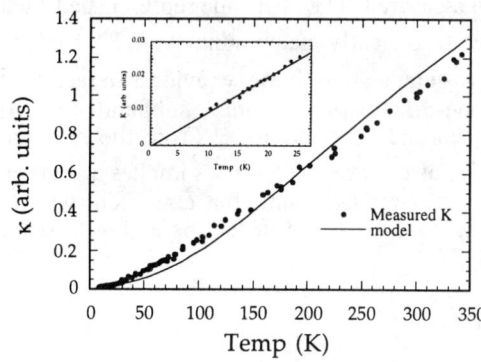

Figure 2. Thermal conductivity of SWNT's. The inset highlights the low-temperature behavior.

We now address the observed temperature dependence of the thermal conductivity. We first seek to determine whether κ is due to electrons or phonons. To do this, we measure the thermal and electrical conductivity of each sample using identical contact geometries to determine the Lorentz ratio $\kappa/\sigma T$. All samples displayed a Lorentz ratio which ranged from $2\text{-}6\times 10^{-6}$ W-Ω/K^2 over the measured temperature range; this value is two orders of magnitude larger than what would be expected for electrons. The thermal conductivity of SWNT's is dominated by phonons at all temperatures.

We next attempt to model $\kappa(T)$ by calculating the phonon thermal conductivity of a single tube. The diagonal term of the thermal conductivity tensor is given by

$$\kappa_{zz} = \sum C v_z^2 \tau, \qquad (3)$$

where C is the heat capacity and v is the sound velocity of a given phonon state; the sum is over all phonon states. We consider only the four acoustic modes of a tube (one longitudinal, one twist, and two transverse) as contributing significantly κ, and use estimates of 2, 1, and 0.8×10^6 cm/s for their sound velocities. In addition, we carefully consider the circumferential quantization of the phonon wavevector: this has the effect of splitting up the phonon modes into multiple 'subbands,' with $\Delta k_x = 1/R_{tube}$[5]. Figure 3 shows the results of Eq. (3), evaluated separately for all of the subbands of the 'twist' mode, for a 1.4 nm diameter tube. The zero-order subband passes through the center of the Brillouin zone; as a true one-dimensional acoustic mode, it provides a linear $\kappa(T)$ at low temperature. Higher subbands do not contribute to κ at low temperature; the first subband begins to contribute significantly near 30 K, just the point where the measured $\kappa(T)$ diverges from linearity.

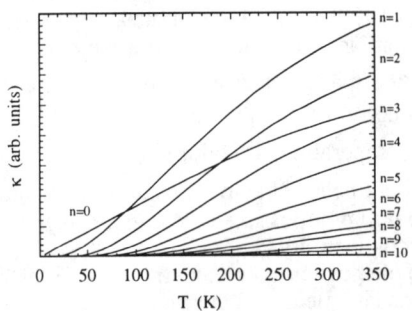

Figure 3. Calculated contribution (Eq. (3)) to the thermal conductivity from each subband of the 'twist' acoustic mode

The solid line in Fig. (2) shows the total expected contribution to $\kappa(T)$ from all four phonon branches, scaled to best fit the measured data. This line represents a fit with only one free parameter, the scattering time τ. It clearly matches the overall behavior of the measured $\kappa(T)$, and confirms that the low-temperature linear behavior is in fact due to the thermal conductivity of a single one-dimensional phonon subband--the first observation of one-dimensional phonon thermal conductivity. From the measured magnitude of $\kappa(T)$, we can extract an estimate of τ; its value of $\approx 10^{-11}$ s implies a scattering length of ≈ 100 nm, smaller than the length of a single tube, but consistent with the distance between inter-tube contacts; these contacts could in fact be acting to scatter phonons.

---

[1] C. Journet, et al., Nature **388**, 756 (1997).

[2] A. Thess, et al., Science **273**, 483 (1996).

[3] N. Hamada, S.-I. Sawada and A. Oshiyama, Phys. Rev. Lett. **68**, 1579 (1992).

[4] J. E. Fischer, et al., Phys. Rev. B **55**, R4921 (1997).

[5] L.X. Benedict, S.G. Louie, and M.L. Cohen, Solid State Comm. **100**, 177 (1996)

Electronic structure studies of single-wall carbon nanotubes using electron energy-loss spectroscopy in transmission

M. S. Golden[1], T. Pichler[1], M. Knupfer[1], J. Fink[1], A. Rinzler[2] and R.E. Smalley[2]

[1] *Institut für Festkörper- und Werkstofforschung Dresden, Postfach 270016, D-01171 Dresden.*
[2] *CNST, Rice Quantum Institute, Rice University, Houston, Tx 77251, USA.*

Abstract. We present momentum-dependent electron energy-loss measurements of bulk samples of purified single-wall carbon nanotubes. The π and the $\pi + \sigma$ plasmons have dispersion relations similar to those of the graphite plane, and thus signal the graphitic nature of the electronic system along the tube axis. In contrast, the interband plasmons at low energy, which are related to transitions between the characteristic singularities in the nanotube density of states, have a vanishingly small dispersion in momentum space and can be used to show that these samples contain a significant fraction of semiconducting nanotubes.

INTRODUCTION

Carbon nanotubes are a promising new member of the growing family of novel fullerene-based materials, and represent model building blocks for nanoengineering as a result of their special electronic [1] and mechanical [2] properties. Nanotubes can be envisaged as rolled-up graphene sheets which are capped with fullerene-like structures. Their electronic properties are predicted to vary depending upon the wrapping angle and diameter of the graphene sheet, thus giving either metallic or semiconducting behavior [3]. Single wall nanotubes (SWNTs) are the best system in which to investigate the intrinsic properties of this new material class. However, macroscopic nanotube samples generally contain a distribution of tubes with different diameters and chirality and thus present the experimentalist with an averaged picture of their properties. One approach which has been successfully used to overcome this problem is to study individual SWNTs. Consequently transport measurements [4] and scanning tunneling spectroscopic (STS) and topographic (STM) studies of *single* nanotubes [5] have done much in the recent months to advance our knowledge regarding the properties of SWNTs, for example by experimentally verifying the remarkable relationship between nanotube geometry and their electronic properties. In addition, spatially-resolved electron energy-loss spectroscopy (EELS) has been performed on individual multi-wall nanotubes [6] or on a single bundle of SWNTs [7]. In this contribution, we report recent EELS in transmission results from bulk samples of purified SWNTs, recorded with both a high energy and momentum resolution. We show that SWNTs support two general types of electronic excitations and that in particular the low energy excitations can be used to distinquish between semiconducting and metallic SWNTs in bulk samples.

SWNTs were produced by the laser vaporization of graphite [8]. The material consists of up to 60 % SWNTs with approximately 1.4 nm mean diameter and was purified as described in Ref. [9]. Free-standing films for EELS with an effective thickness about 1000 Å were prepared on standard copper microscopy grids via vacuum filtration of a nanotube suspension in a 0.5 % surfactant (Triton X100) solution in de-ionised water, with a SWNT concentration of $\sim$ 0.01 mg/ml. The surfactant was then rinsed off and the grid was transferred into the spectrometer. EELS was carried out in a purpose-built high-resolution spectrometer [10] which combines both good energy *and* momentum resolution. For the data shown here, an energy and momentum resolution of 115 meV and 0.05 Å^{-1} were chosen. Unlike many electron spectroscopies, EELS in transmission is volume sensitive, which in the context of the inhomogenous nature of bulk samples of SWNTs is a crucial point. All spectra were recorded at room temperature.

RESULTS AND DISCUSSION

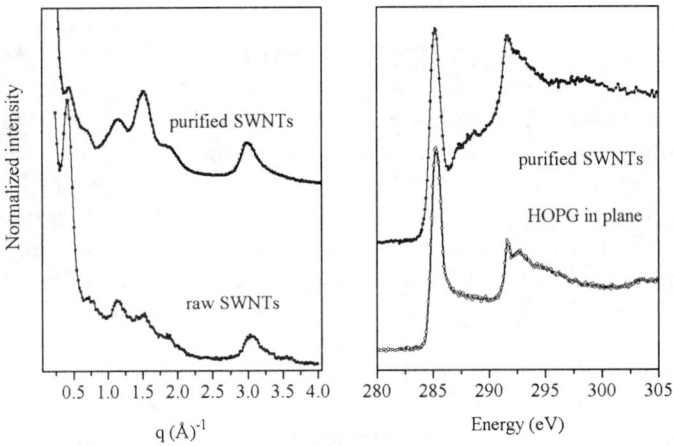

FIGURE 1. Left panel: electron diffraction profiles raw and purified SWNT material. Right panel: C1s excitation spectra of purified SWNTs and HOPG (in plane).

By setting the energy-loss to zero, we can carry out electron diffraction experiments in-situ in the EELS spectrometer. The left-hand panel of Fig. 1 shows electron diffraction data for the raw and purified SWNT material. The data are consistent with the published x-ray diffraction results which were interpreted in terms of a triangular lattice formed by the individual SWNTs in the ropes [8]. It

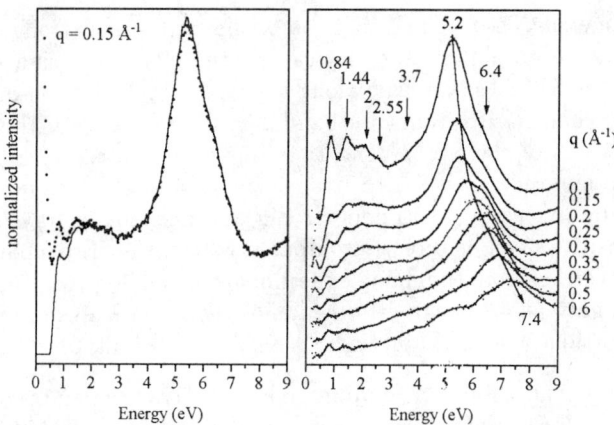

FIGURE 2. Left panel: the loss function of purified SWNTs for q=0.15 Å^{-1}, before (dots) and after (solid line) subtraction of the contributions from the quasi-elastic line. Right panel: the loss function of purified SWNT as a function of momentum transfer, **q**.

can be seen that the essential features of the electron diffraction profile of the raw SWNTs occur at the same **q**-values in the purified material. The rest of the data presented in this paper are from purified SWNT, but the untreated material gave qualitatively similar data in all cases.

The right-hand panel of Fig. 1. shows the EELS C1s excitation spectrum of purified SWNT compared to that of HOPG (measured with the momentum transfer in the graphite plane). The spectrum of the SWNTs strongly resembles that of graphite. By comparison with directional-dependent C1s excitation measurements [11,12], we see that the SWNT spectrum most closely resembles an average of the in-plane and out-of-plane graphite spectra. In graphite the maxima at the π^* [13] and the σ^* [12] onsets are both excitonic in nature, and it is reasonable to assume that the same holds for the SWNTs.

The left-hand panel of Fig. 2 shows the loss function (Im(-1/$\epsilon(\mathbf{q}, E)$)) of purified SWNTs for q=0.15 Å^{-1}, plotted both before and after subtraction of the quasi-elastic line. In contrast to the EELS data of nanotubes published up to now [6,7], we can measure reliably down to energies as low as 0.5 eV, which is a direct result of the superior momentum resolution of the dedicated, non-spatially resolving spectrometer used here. In the right-hand panel of Fig. 2 we show the **q**-dependence of the loss function, after the subtraction of the quasi-elastic line and the effects of multiple scattering from the raw data. At low momentum transfer, features in the loss function are visible at about 0.84, 1.44, 2.0, 2.55, 3.7, 5.2 and 6.4 eV. These peaks in the loss function are, by definition, plasmons, and originate from transitions between the occupied and unoccupied π electronic levels of the SWNTs. Two distinct **q**-dependences are observed: the π plasmon disperses strongly from 5.2 eV at q=0.1 Å^{-1} to 7.4 eV at q=0.6 Å^{-1}, whereas all the other peaks have a vanishingly small dispersion in **q**. The dispersion of the π plasmon strongly resembles that observed in the graphite plane and can be therefore be assigned to the response function of the SWNTs polarized along tube axis [14]. The non-dispersive features on the other hand arise from interband transitions between the characteristic DOS singularities of the SWNT and thus represent the response function perpendicular to the nanotube axis. After transformation of the loss function to give the optical conductivity via a Kramers-Kronig analysis, the energy positions of the interband transitions can be directly compared with the energy separation of the DOS singularities either from tunneling experiments on single tubes [5], or from theory [3]. Such an analysis has been performed and clearly indicates the presence of a significant proportion of semiconducting SWNT in these bulk samples [14].

T.P. thanks the European Union for funding under the 'Training and Mobility of Researchers' program. The work at Rice was supported by the NSF (DMR9522251), the Advanced Technology Program of Texas (003604-047) and the Welch Foundation (C-0689).

REFERENCES

1. S. Saito, Science **278**, 77 (1997); L. Chico et al., Phys. Rev. Lett. **76**, 971 (1996).
2. E.W. Wong, P.E. Sheehan and C.M. Lieber, Science **277**, 1971 (1997).
3. J. W. Mintmire, B. I. Dunlop and C. T. White, Phys. Rev. Lett. **68**, 631 (1992); N. Hamada, S. I. Sawada and A. Oshiyama, Phys. Rev. Lett. **68**, 1579 (1992); M. S. Dresselhaus, G. Dresselhaus and P. C. Eklund, *Science of Fullerenes and Carbon Nanotubes*, (Academic Press Inc., San Diego, 1996).
4. S. J. Tans et al., Nature **386**, 474 (1997); C. Dekker et al. in this volume.
5. J. W. G. Wildör et al., Nature **391**, 59 (1998); C. Dekker et al. in this volume.
6. R. Kuzuo, M. Terauchi and M. Tanaka, Jpn. J. Appl. Phys. **31**; L1484 (1992); P. M. Ajayan, S. Ijima and T. Ichihashi, Phys. Rev. **B 49**, 2882 (1994); A. Bursill et al., Phys. Rev. **B 49**, 2882 (1994).
7. R. Kuzuo et al., Jpn. J. Appl. Phys. **33**, L1316 (1994).
8. A. Thess et al., Science **273**, 483 (1996).
9. A. G. Rinzler et al., Appl. Phys **A** in press, J. Liu et al., unpublished
10. J. Fink, Adv. Electron. Electron Phys. **75**, 121 (1989) and references therein.
11. P. E. Batson, Phys. Rev. **B 48**, 2608 (1993).
12. Y. Ma et al., Phys. Rev. Lett. **71**, 3725 (1993).
13. P. Brühwiler et al., Phys. Rev. Lett. **74**, 614 (1995).
14. T. Pichler et al., Phys. Rev. Lett. in press.

^{13}C-NMR in Carbon Nanotubes

Y. Maniwa, M. Hayashi, Y. Kumazawa, H. Tou and H. Kataura,
*Department of Physics, Tokyo Metropolitan University, Minami-osawa, Hachi-oji,
Tokyo 192-0397, JAPAN*

H. Ago, Y. Ono, T. Yamabe and K. Tanaka
*Department of Molecular Engineering, Graduate School of Engineering, Kyoto University, Sakyo-ku,
Kyoto 606-01, JAPAN.*

Abstract. ^{13}C-NMR studies of multi-wall carbon nanotube (MWNT), which was purified by oxidation technique, are reported. Temperature dependence of ^{13}C-spin lattice relaxation time, T_1, clearly shows a metallic behavior, T_1T=const law, at 9.4T between 4.2K and 77K. This suggests that there are metallic tubes in the sample at 9.4T. The field dependence of T_1T is also discussed.

INTRODUCTION

Physical properties of carbon nanotube (CN) [1] strongly depend on structural parameters, like diameter and chirality [2]. For example, metallic and semiconducting tubes are distinguished by chairal vector. About one third of the chairal vectors lead to metallic tubes, and the others are semiconducting tubes. Another striking feature of CN is behavior of the electronic states in a magnetic field. According to Ajiki & Ando [3], the energy band gap oscillates as a function of magnetic flux passing through the cross section of tube with a period of flux quantum, ϕ_0.

The aim of the present study is to clarify the electronic states of CN, especially as a function of magnetic field, by ^{13}C-NMR technique. We expect that this technique can be one of powerful tools to study local structures and electronic properties of CN, as in other carbon-based materials. Here, we present preliminary results on multi-wall carbon nanotube (MWNT) and the related materials.

SAMPLE PREPARATIONS

MWNT was prepared as cathode deposit in arc discharge of carbon rod at inert gas atmosphere [4]. In order to purify the crude materials that include a lot of impurity carbon forms such as nanoparticles and amorphous carbon, we employed oxidation technique in air [4]. Samples used are summarized below.
  Sample 1 : crude MWNT (cathode deposit)
  Sample 2 : heated in air at 940C for 25min.
      Abundance of MWNT is ~80%.

Sample 3 : heated in air at 900C for 20min.
An average outer diameter is 32nm.
Sample 4 : heated in air at 900C for 27min.
An average outer diameter is 25nm.

Sample 4 is sample 3 after additional heating for 7min at 900C. The samples after heat-treated still included significant amount of nanoparticles as shown in TEM picture in Fig. 1. The crude materials for samples 2 ~ 4 were purchased from Vacuum Metallurgical Co., LTD.

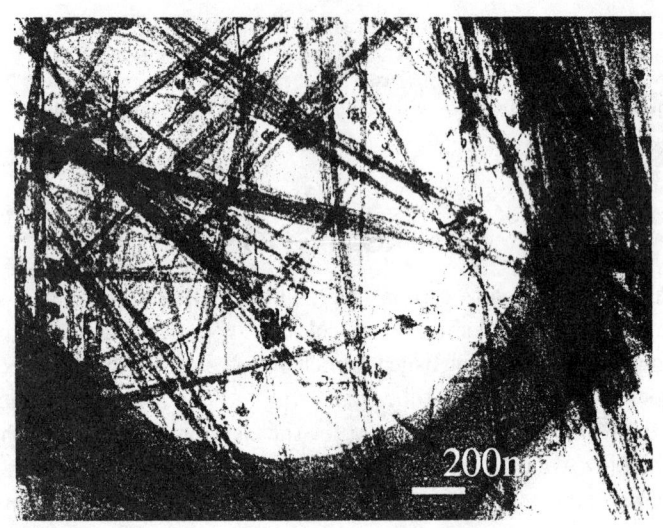

FIGURE 1. TEM image of MWNT (sample 4).

^{13}C-NMR SPECTRA

We show ^{13}C-NMR *powder spectra* of MWNT and the related materials (graphite powder, diamond powder, solid C_{60}, and pressure induced C_{60} polymers) in Fig. 2. Diamond powder exhibits typical powder spectrum of sp^3 carbon, which is sharp (isotropic) and located around 35ppm. On the other hand, solid C_{60} gives typical spectrum of sp^2 carbon that is nearly axial symmetric with a width about 170ppm. Graphite powder also shows axial symmetric powder spectrum, but the anisotropy is much larger than those of typical sp^2 carbons [5, 6]. This corresponds to a large diamagnetic susceptibility for the external magnetic field perpendicular to the graphite plane. The powder spectra of C_{60} polymers are interpreted as superposition of those of sp² and sp³ carbon atoms [7]. Therefore, we can obtain, from the ¹³C-NMR spectra, information on diamagnetic susceptibility and local structures of carbon atom such as the coordinate number.

From the ¹³C-NMR powder spectra in Fig. 2, we know that the sp² carbon is major constituent of MWNT. The anisotropic line width for MWNT is comparable to that of graphite and much larger than in C_{60} solid, suggesting that the diamagnetic susceptibility is very large when the magnetic field is applied perpendicular to the tube axis. This is consistent with a theoretical prediction [3].

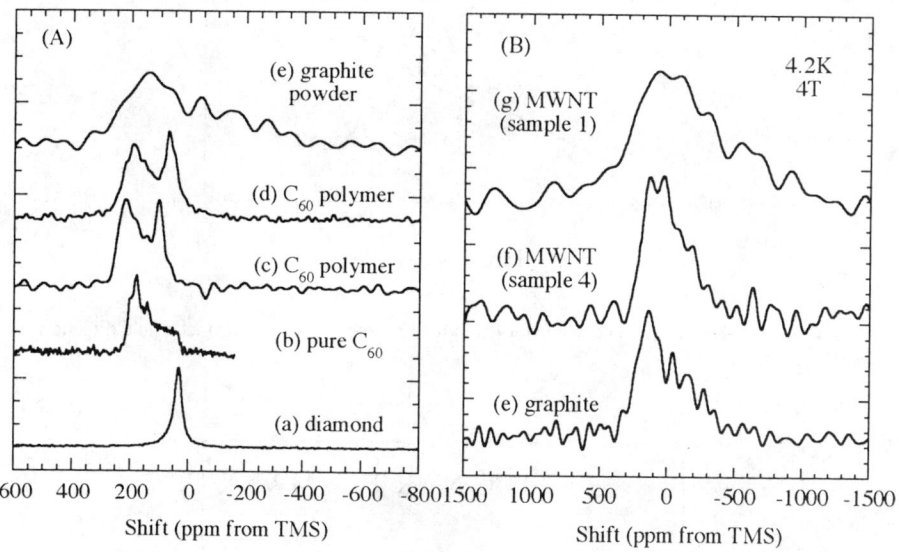

FIGURE 2. ¹³C-NMR spectra of MWNT and related materials (diamond powder (a), solid C_{60}(b), and pressure induced C_{60} polymers (c, d), and graphite powder(e)).

¹³C-NMR SPIN-LATTICE RELAXATION TIME T_1

In order to obtain more detailed information on the electronic states, we performed ¹³C-T_1 measurement. The results measured at 9.4T for sample 2 are shown in Fig. 3, where the temperature dependence of T_1 in graphite (HOPG) [8, 9] and K_3C_{60} [10] is also indicated. It is found that MWNT, as well as graphite and K_3C_{60}, follows a metallic Korringa like relation, T_1T=const, suggesting that there are metallic tubes at least 9.4T. If we assume that the observed $(1/T_1T)^{0.5}$ is simply proportional to the electronic density of states at the Fermi level, $N(E_F)$, we can estimate $N(E_F)$ of metallic MWNT, which is very close to $N(E_F)$ of HOPG.

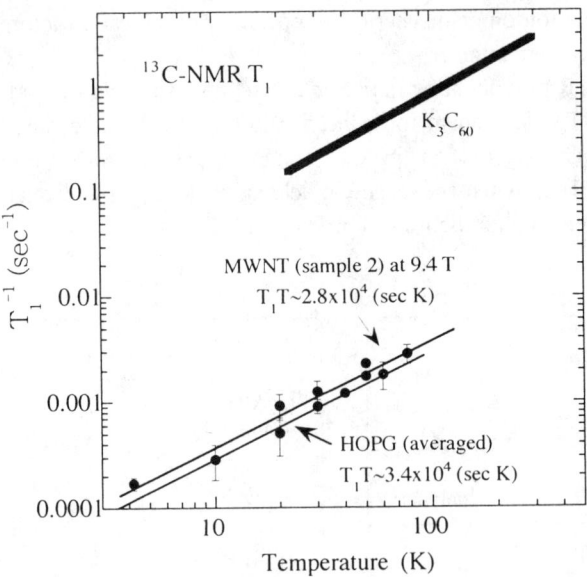

FIGURE 3. ^{13}C-NMR T_1. The value for graphite is averaged over the magnetic field direction.

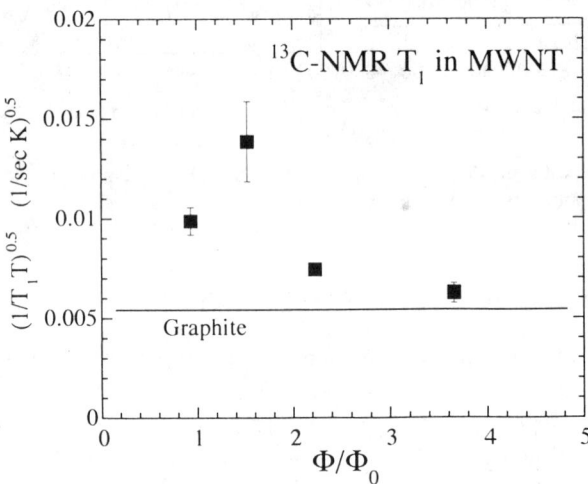

FIGURE 4. Field dependence of ^{13}C-T_1 in randomly oriented MWNT. ϕ *is* the magnetic field times the average outer cross section of the tube.

However, the electronic state of carbon nanotube is theoretically expected to depend on the external magnetic field. Metallic (insulating) tubes can be converted to insulating (metallic) tubes by changing the magnetic flux passing through the cross section of tube. Therefore, we tried to measure T_1 at different fields (4T and 9T) and for the samples with different diameters (sample 3 and sample 4). The magnetic field strength and the sample diameter can change the flux passing the tube cross section. These measurements suggested that the value for $(1/T_1T)^{0.5}$ increases with a decrease of the magnetic flux (Fig. 4). This is quite different from the cases of HOPG, GIC and K_3C_{60} where the T_1 is field-independent. Although the observed field dependence may be ascribed to the Aharonov-Bohm (AB) effect in carbon nanotube predicted by Ajiki & Ando [3], more detailed experiments on better quality samples should be required to identify the origin.

REFERENCES

1. S. Iijima, Nature **354**, 56 (1991).
2. N. Hamada et al., Phys. Rev. Lett. **68**, 1679(1992).
3. Ajiki and Ando, J. Phys. Soc. Jpn. **62**, 2470 (1993).
4. T.W. Ebbsen, Physics Today p26, June1996.
5. Y. Hiroyama and K. Kume, Solid State Commun. **65**, 617 (1988).
6. P. Lauginie et al., Synthetic Metals **55-57**, 3002 (1993)
7. Y. Maniwa et al., Carbon **34**, 1287 (1996).
8. K. Kume et al., Synthetic Metals **12**, 307 (1985).
9. Y. Maniwa et al., J. Phys. Soc. Jpn. **54**, 666 (1985).
10. Y. Maniwa et al., in *Electrochemical Society Proceedings*, **Vol. 97-14**, 1191 (1997), edited by K.M. Kadish and R.S. Ruoff.

Observation of Elastic Deformations in Single-walled Carbon Nanotubes by Scanning Tunneling Microscopy

Wilfried Clauss*, David J. Bergeron, and Alan T. Johnson

*Department of Physics and Astronomy, University of Pennsylvania,
Philadelphia, Pennsylvania 19104*
**On leave from: Institute of Applied Physics, University of Tuebingen,
72074 Tuebingen, Germany*

Abstract: Scanning Tunneling Microscopy is used to obtain atomically resolved images of single-walled carbon nanotubes, in ropes of several tens to hundreds of tubes. The images confirm that in this environment strong elastic deformations of the tube lattice occur frequently. In particular, bent and twisted tubes have been identified. The observed distortions could play an important role in explaining the electronic transport properties of nanotubes.

INTRODUCTION

Since it has become possible to synthesize single-walled carbon nanotubes (SWNT) in macroscopic quantities, the interest in the structural, mechanical and electrical properties of this new carbon material has grown considerably. Due to its very nature as an inherently perfect, one-dimensional quantum system, many interesting features have been found theoretically and experimentally[1,2,3]. In previous reports[4,5] the atomic structure of individual tubes has been related to their electronic properties and experimental evidence for the relation between the semiconducting/metallic behavior and the chirality[1,2] has been established. On the other hand, transport measurements at various temperatures on bulk SWNT samples[6,7,8] show a complicated behavior of the transport characteristics varying greatly depending on the method of sample preparation. As a reason for these large variations the influence of doping, tube-tube interactions and the mixture of tubes with different chiralities has to be taken into account. Moreover, theoretical works[9] have shown that elastic deformations of tubes, in particular a twist along the tube axis, can strongly change the electronic density of states (DOS), causing for example a metallic armchair tube to become semiconducting. In this situation, it is important to get a more detailed picture of the local structure of real SWNTs which can only be obtained by real-space imaging techniques with spatial resolution below 0.1 nm. Scanning tunneling microscopy is one of the outstanding techniques for this goal with the additional capability to get local electronic information by tunneling spectroscopy.

EXPERIMENTAL SETUP

In order to get well-defined, atomically flat substrates for high-resolution STM imaging, we evaporated gold onto mica substrates at 300° C. After annealing in a gas flame, atomically flat Au(111) terraces several hundred nanometers in size were obtained. As-grown nanotube material which was produced by laser ablation[10] was sonicated for 30 minutes in dichloroethane, then a small drop of the solution was put on the gold surface and blown dry in a nitrogen stream. Alternatively, small pieces of solid nanotube material were pressed onto the substrate. Using appropriate forces, a small amount of the material remains on the gold if the piece is removed. Measurements were taken in air with commercial Pt-Ir tips and an STM of the Besocke design (the Beetle STM from Omicron, Inc.). Although the instrument has a rather small scan size (about 2µm), it allows a reliable coarse lateral movement with steps in the range of 50 to 200 nm, making it possible to find useful sample regions in a few minutes, even with low SWNT density. Lateral dimensions were calibrated to an estimated accuracy better than 15 % using atomically-resolved images of Au(111), highly-oriented pyrolytic graphite (HOPG), and decanethiol monolayers on gold. The vertical scale was calibrated at monoatomic steps between terraces of Au(111). Images were obtained with tunneling voltages in the range of 700-900 mV and setpoint currents between 350 and 900 pA. No significant influence of the voltage polarity was noticed.

RESULTS

Figure 1 shows two examples of tube arrangements as they are found in typical STM images[11]. In Fig.1(a), a bundle of ropes is visible, each consisting of several tens of individual tubes. The ropes exhibit bends with typical radii below 100 nm. Part (b) shows a single, large rope with a very pronounced, layered structure. This image illustrates the character of ropes as single- or polycrystalline arrangements of tubes. As can be seen from the inspection of the edge, the order is not perfect, and individual tubes can be elastically deformed on a length scale of only a few nanometers, especially if they change their position between different rope layers. Figure 2 shows the atomic structure of a tube which is embedded in a rope. In order to accentuate the atomic corrugation, the surface is displayed flattened. The direction of the tube axis was derived from larger scale images. The honeycomb lattice was reconstructed by connecting the centers of hexagons along all three possible directions as shown in the

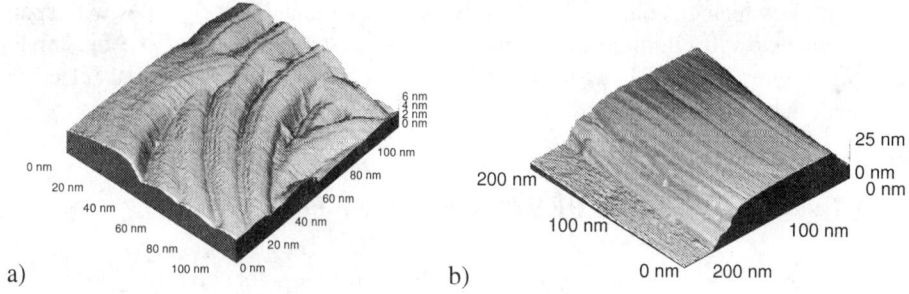

FIGURE 1: STM images of ropes of single-walled nanotubes. The rope morphology can vary considerably; part (a) shows several ropes of similar diameter which are crossing each other, while in part (b) one large rope with a distinct, but not perfect layer structure can be seen.

sketches drawn for the right and the left part of the image. This image clearly shows that the STM can resolve not only the average lattice structure but also a deviation which shows up as a bend of the tube with an angle of about 8° and a radius of only few nanometers. Moreover, while the angle between the lattice base vectors is nearly correct for the left part, it differs by about 20° from the undistorted direction for the right part, indicating a strong elastic shear deformation along the tube axis.

The second example of a strongly distorted tube is shown in Fig. 3(a), which is taken on top of the rope of Fig.1(a). Figure 3(b) shows a line scan perpendicular to the tube axis as indicated by the white line in part (a). From this cross section we derive a tube diameter of about 1.5 nm. Based on this diameter and the angle of 4° (0.07 rad) between tube axis and the direction of the most pronounced rows of hexagons, a wrapping vector of (12,9) could be assigned to this tube. But a more detailed analysis[12]

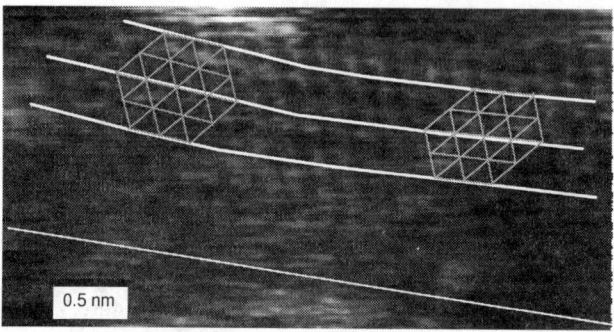

FIGURE 2. High resolution image of the surface of a single tube. The three thick lines are drawn by connecting the appropriate centers of hexagon cells. The lower line visualizes the direction of the tube axis. From the reconstructions of the lattice in the left and the right part the strong variation of the elastic distortion of the tube can be seen.

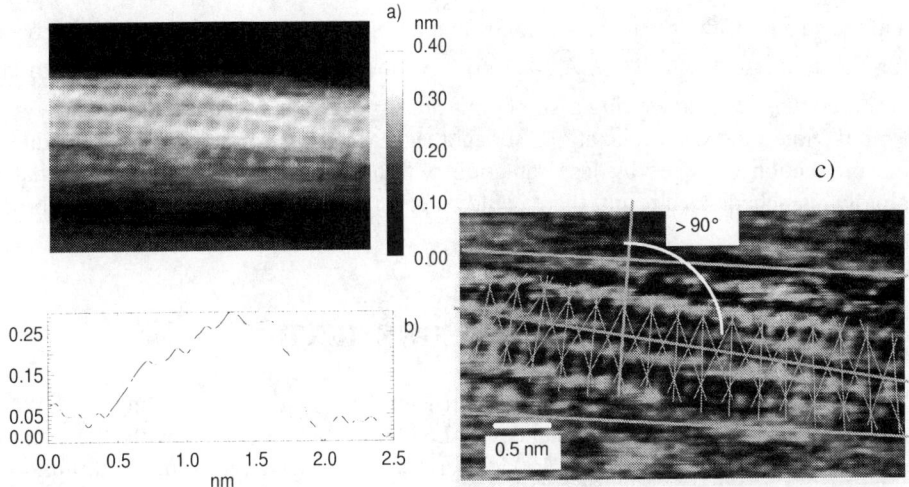

FIGURE 3. Constant current image of an individual tube embedded into amorphous background. The upper left part shows the height shape with a line scan shown below. The right part shows the same area flattened in order to accentuate the atomic structure. The reconstruction of the lattice is described in the text.

shows that if the angles between different armchair and zig-zag directions are taken into account, it is more consistent with the data to interpret this image as an armchair tube which is twisted. This twist results in an angle between zig-zag direction and tube axis of 4°. Although this value is large, it is consistent with what is known about the mechanical properties of SWNTs. They can be modeled rather well as if the tube were a continuous cylinder with a wall thickness of 0.34 nm, the interplane spacing in graphite[13]. A tube in a rope will try as much as possible to align its hexagons with those of neighboring tubes as in the AB stacking of graphite. In a rope containing tubes of differing helicities, frustration will induce tube twists. Countering this tendency is the energy cost associated with an elastic twist distortion. Little is known about the details of this energy balance and possible energy barriers that define local energy minima.

We can estimate the magnitude of twist that could be induced by interactions between tubes of differing helicities. Mentally unrolling the SWNT reveals that a twist distortion of a tube is equivalent to a shear[12] of the underlying single plane whose modulus we take from bulk graphite as $M_b = 0.45$ TPa. Similarly, a rotation of the tube away from the equilibrium alignment with its neighbors in the rope likely has a modulus similar to that of a shear distortion along the graphite c-axis, $M_c = 0.004$ TPa. For a (10,10) tube, which is invariant under a rotation of $\pi/5$, the maximum angular deviation from ideal alignment that can occur is $\pi/10$. Since the tube radius is 0.75 nm, the equivalent shear for two graphite planes separated by 0.34 nm is about $e_{max} = 0.67$. The energy stored per unit volume under a shear is the product of the relevant modulus and the square of the shear. This implies that an upper limit on the twist that can occur for the tube is $e_{max}/\sqrt{M_c/M_b} = 0.067$. Although this agrees very well with the observed twist, the underlying assumptions are strongly overestimating the energy gain from the intertube shear relaxation. But another reason for strong twists could be due to the production of ropes by laser ablation within a high temperature plasma that is rapidly quenched. As a result, there could be substantial population of non-equilibrium geometries.

ACKNOWLEDGEMENTS

We thank R. Smalley and A. Rinzler from Rice University for providing the SWNT material used in these experiments, J. Lefebvre for assistance with the sample preparation. C. L. Kane and E. J. Mele are acknowledged for valuable discussions about the electronic and mechanical properties of tubes. This work was supported by Penn's Laboratory for Research on the Structure of Matter. W. C. thanks the Deutsche Forschungsgemeinschaft for support. A. T. J. is supported by a David and Lucile Packard Foundation Fellowship.

REFERENCES:

[1] N. Hamada, S. Sawada and A. Oshiyama, Phys. Rev. Lett. **68**, 1579 (1992).
[2] R. Saito, et al., Appl. Phys. Lett. **60**, 2204 (1992).
[3] Science of Fullerenes and Carbon Nanotubes, M. S. Dresselhaus, G. Dresselhaus, P. C. Eklund (Eds.), San Diego 1996.
[4] J. W. G. Wildöer, et al., Nature **391**, 59 (1998).
[5] Teri Wang Odom, et al., Nature **391**, 62 (1998).
[6] J. E. Fischer, H. Dai, A. Thess, R. Lee, N. M. Hanjani, D. L. Dehaas, R. E. Smalley, Phys. Rev. B **55**, R4921 (1997).
[7] C. L. Kane, E. J. Mele, R. S. Lee, J. E. Fischer, P. Petit, H. Dai A. Thess, R. E. Smalley, A. R. M. Verschueren, S.J. Tans, and C. Dekker, to appear in Europhys. Lett.
[8] J. Hone, I. Ellwood, M. Muno, A. Mizel, M. L. Cohen, A. Zettl, A. G. Rinzler, R. E. Smalley, Phys. Rev. Lett. **80**, 1042 (1998).
[9] C. L. Kane, E. J. Mele, Phys. Rev. Lett. **78**, 1932 (1997).
[10] T. Guo et.al., Chem. Phys. Lett. **243**, 49 (1995), A. Thess et.al. Science **273**, 483 (1996)..
[11] J. Lefebvre, R. Antonov, A. T. Johnson., to appear in Appl. Phys. A.
[12] W. Clauss, D. J. Bergeron, A.T. Johnson, submitted to Phys. Rev. Lett.
[13] Jian Ping Lu, Phys. Rev. Lett. **79**, 1297 (1997).
  B. I. Yakobson, C. J. Brabec, and J. Bernholc, Phys. Rev. Lett. **76**, 2511 (1996).

Spring-like Behaviour of Carbon Nanotubes Observed by Transmission Electron Microscopy

Walter H. Knechtel, Georg S. Düsberg and Werner J. Blau

Trinity College Dublin, Dublin 2, Ireland

Abstract. Multi-wall carbon nanotubes can be bent by changing the current density of the electron beam in a Transmission Electron Microscope (TEM). The effect could be observed in a small fraction of nanotubes in all of the investigated samples. The bending can be varied continuously, is reversible and highly reproducible. On removing the force which make them bend, they relax to their originally straight shape without any damage, thus exhibiting spring-like behaviour. We estimate the force which is applied to such a *bending tube*.

1 INTRODUCTION

Many theoretical reports predict that carbon nanotubes (1) have an extraordinary strength-to-weight ratio and particularly, high stiffness and axial strength (2,3,4). Experiments confirmed these predictions and have also shown that carbon nanotubes are at the same time very flexible. Investigations of thermal vibrations (5) of nanotubes have shown that nanotubes have higher Young's moduli than steel or even diamond. Atomic Force Microscopy observations using carbon nanotube probes (6) showed that nanotubes bend whenever the tip is "crashed" into a hard surface, and then snap back to its original straight position when the tip is withdrawn. A recently published paper shows that multi-wall carbon nanotubes can be bent repeatedly through large angles using the tip of an atomic force microscope (7). A range of responses to this high-strain deformation was observed, which together suggests as well that nanotubes are remarkably flexible and resilient. Wong *et al.* (8) used atomic force microscopy to determine the mechanical properties of individual multi-wall nanotubes that were pinned at one end.
Bends in nanotubes have been studied by TEM (9,10), but the deformations in the investigated nanotubes were either defects already present or permanently introduced during specimen handling. We report the manipulation of multi-wall carbon nanotubes with the electron beam of a TEM.

2 EXPERIMENTAL

TEM investigations were performed using a JEOL JEM-2000FX microscope with an acceleration voltage up to 200 kV. TEM samples were prepared by sonicating a tiny amount of the raw soot produced by the conventional arc discharge method (11) in about 1ml ethanol. A single drop of this suspension was put on top of a polyvinyl formate (formvar) coated copper TEM grid. Dry prepared samples were also used: A tiny amount of the raw soot was placed on a velin tissue, a second tissue was placed on top of it and the sample was ground between them. The TEM grid was then carefully pressed onto the sample. A previously unreported effect was observed for both methods of sample preparation.

3 OBSERVATION

A small number of nanotubes can be found, which bend by increasing the current density (brightness) of the electron beam and/or by moving the centre of the electron beam relative to the sample. *Bending tubes* are usually attached to carbon clusters, thus probably not directly attached to the formvar film on the TEM grid. The bending of a tube can always be varied continuously between its straight position and its maximum bending position. The bending is always reversible and highly reproducible. The *bending effect* does not show any hysteresis, i.e. for a certain current density and for a certain position of the electron beam, a *bending tube* always shows instantaneously a certain curvature. By reducing the current density below a threshold value, where no bending is occurring, the tube reverts back to its original straight shape, thus exhibiting *spring-like* behaviour. Thus the geometry of a tube does not change and no permanent defects are induced. However if the beam intensity is increased to values much higher than necessary to observe the effect, some of the bending tubes jump away and cannot be found anymore.

Figures 1 and 2 show two different tubes which show the described bending behaviour. It was found that the maximum bending position and the threshold value of the current density varies from tube to tube. For some tubes bending could even be achieved in different directions. The threshold value for the current density varied for different tubes from 80 to about 220 pA/μm^2. Bending could be observed for different electron energies between 100 and 200 keV.

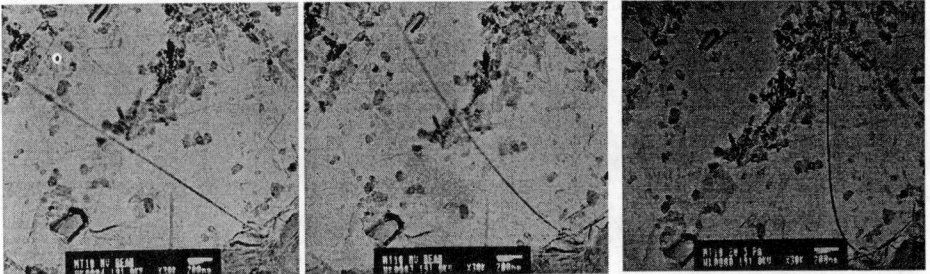

Figure 1. Left: A *bending tube* in its original (straight) position. It is clamped at one end in a big carbon cluster (right lower corner). **Middle, Right**: Above a threshold value of current density the tube can be bent continuously by moving the centre of the electron beam relative to the sample. The strongest bending takes place in a small part of the tube close to the fixed end. The radius of curvature at this position is about 500 nm.

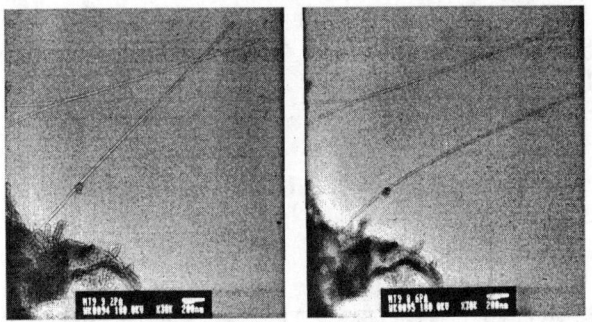

Figure 2. Another bending tube in its original position (left) and its maximum bending position (right).

4 ESTIMATION OF THE FORCE

Even though we do not fully understand their origin, we estimate the strength of the applied force using calculations from solid state mechanics. The force which acts on a bending rod can be calculated by the change in geometry of the rod if the modulus of Elasticity (Young's modulus) is known. Consider for example an applied force perpendicular to the original direction (*y*-direction) of the tube (see Figure 3). Then, after Landau and Lifshitz (12) :

$$f = \frac{2EI}{x_0^2} \cos\theta_0$$

where f is the force, E is the Young's modulus, I is the moment of inertia of the tube, x_0 is the x-coordinate of the free end and θ_0 is the angle between the tangent of the rod at the free end and the y-axis.

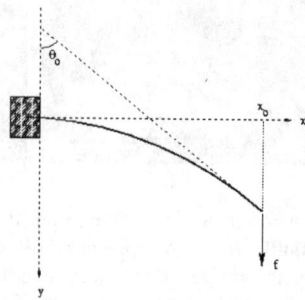

Figure 3: A rod clamped at one end and a force applied on the other end

For a rod with a circular cross-section the hollow core of the inspected nanotube is negligible), $I = \frac{1}{4}\pi R^4$, where R is the radius of the tube. From Fig. 2 the following quantities are obtained: $x_0 \approx 2.3$ μm, $\theta_0 \approx 70°$, $R \approx 12$ nm. Considering a Young's modulus of 1.3 TPa (4,8) yields a force of $f = 2.7 \cdot 10^{-9}$ N.

ACKNOWLEDGEMENTS

This work was carried out in University College Dublin (UCD) with the generous assistance of Dr. David Cottell. It was supported by the European Community through TMR contract NAMITECH, ERBFMRX-CT96-0067 (DG12-MIHT).

REFERENCES

1. Iijima, S., *Nature* **354**, 56 (1991).
2. Lu, J.P., *Phys. Rev. Lett.* **79**, 1297 (1997).
3. Ruoff, R.S., and Lorents, D.C., *Carbon* **33**, 925 (1995).
4. Hernández, E., Goze, C., Bernier, P., and Rubio, A., *Phys. Rev. Lett.* **80**, 4502 (1998).
5. Treacy, M.M.J., Ebbesen, T.W., and Gibson J.M., *Nature* **381**, 678 (1996).
6. Dai, H., Hafner J.H., Rinzler A.G., Colbert D.T., and Smalley R.E., *Nature*, **384**, 147 (1996).
7. Falvo, M.R., Clary G.J., Taylor R.M., Chi, V., Brooks F.P., Washburn S., and Superfine R., *Nature* **389**, 582 (1997).
8. Wong, E.W., Sheehan P.E., and Lieber, Ch.M., *Science* **277**, 1971 (1997).
9. Weldon, D.N., Blau, W.J., and Zandbergen, H.W., *Chem. Phys. Lett.* **241**, 365 (1995).
10. Iijima., S., Barabec, Ch., Maiti A., and Bernholc J., *J. Chem. Phys.* **104**, 2089 (1996).
11. Ebbesen, T.W., and Ajayan, P.M., *Nature* **358**, 220 (1992).
12. Landau L.D., and Lifshitz, E.M., *Theory of Elasticity*, Pergamon Press, 3rd Edition, Oxford, 1986.

RESONANCE RAMAN SCATTERING of the RADIAL BREATHING MODE in SINGLE WALL CARBON NANOTUBES

Jenő Kürti[1], Hans Kuzmany[2], Bernhard Burger[2], Martin Hulman[2],
A.G. Rinzler[3] and Richard.E. Smalley[3]

[1] *Department of Biological Physics, Eötvös University,
Puskin u. 5-7, H-1088 Budapest, Hungary,*
[2] *Institut für Materialphysik, Universität Wien,
Strudlhofgasse 4, A-1090 Wien, Austria,*
[3] *Department of Chemistry, Rice University, Houston TX 77005 USA*

Abstract. The radial breathing mode of single wall carbon nanotubes was used for the analysis of conformational and electronic properties of the tubes. We report resonance excitation of this mode with 12 different laser lines in the visible and near IR. Very sharp resonances were observed and compared to calculated cross sections. The position of the modes were evaluated from a high level first principles calculation which allows for the first time to identify the tube geometries from observed frequencies.

INTRODUCTION

Since the discovery of a technique to prepare single wall carbon nanotubes (SWCNT) [1] the interest in such materials increased dramatically. However, the interpretation of many experiments such as, for example, spectroscopic measurements, turned out to be difficult since samples were shown to consist of a distribution of tubes with various diameters and chiralities.

One of the most powerful analytical methods is Raman spectroscopy. Besides testing the normal mode frequencies it is also sensitive to the electronic structure of the tubes. The characteristic excitation energies of the SWCNTs extend from the near IR to the UV spectral range which results in a resonance enhancement for the modes with strong coupling to the electronic system. There are 15 or 16 Raman active vibrational modes in the SWCNTs depending on the type of the tubes but independent from their diameter [2]. Only four of these modes are strongly resonance enhanced from the experiments [3, 4]. Three of them with symmetry A, E_1 and E_2 are located around 1600 cm^{-1}. They are derived from from the C-C stretch mode of the graphene sheet. The fourth strong band around 200 cm^{-1} is the radial breathing mode (RBM), where all atoms are moving in phase perpendicular to the tube axis. We focus on the RBM because it is unique for the tubes, without any counterpart in graphene. Increasing the tube diameter, the frequency of the RBM approaches zero as this mode becomes a pure translation for zero curvature.

The various nanotubes are characterized by their Hamada vector (n,m) which describes how the graphene sheet is rolled up to establish the tube. n and m are integer numbers. Tubes with $n = m$ are called armchair, tubes with either $n = 0$ or $m = 0$ are called zigzag and all other tubes are called chiral [2].

From x-ray measurements it was claimed that the majority of the tubes had a diameter of 1.38 nm, corresponding to (10,10) tubes. Recent calculations [5] also pointed out that (10,10) is the energetically most stable form, although there are several other tubes with similar diameter which have almost the same stability. Tubes with significantly smaller diameter are less favourable due to the larger folding energy, whereas tubes with significantly bigger diameter are less favourable because of the larger edge energy. On the other hand several different armchair, zigzag and chiral tubes were found in the same type of material as investigated here by means of topographic and spectroscopic scanning probe microscopy [6].

Another evidence for a size distribution of the tubes comes from preliminary Raman analyses, which showed that both bands, the one from the graphitic mode and the one from the breathing mode, depend dramatically on the exciting wavelength [3, 4, 7, 8]. This dispersion is a consequence of the photoselective resonance Raman effect combined with the distribution of various kinds of tubes in the samples, similarly to the dispersion effect observed in conjugated polymers [9]. The dispersion effect was recently used to analyze different types of SWCNT material [10].

EXPERIMENT

SWCNTs were prepared by the laser evaporation process using two collinear laser pulses [1]. Raman spectra in the range of the RBM were excited in backscattering with 12 different laser lines extending from 1.064 μm to 0.458 μm. Relative scattering intensities were derived from a calibration of the instrument with the resonance of the F_{2g} mode in Si. The response for the various lasers was dramatically different with respect to line shape and intensity. An example for three selected spectra is shown in Fig. 1. A careful analysis of all 12 spectra revealed 14 components with Voigtian line-shapes in a frequency range between 140–230 cm^{-1}. As described in more detail in [11, 12] the positions of these components do

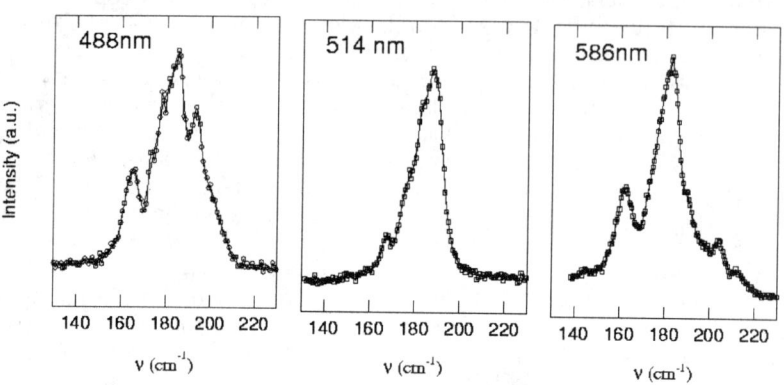

Figure 1: High resolution Raman spectra of single wall nanotubes as excited with various lasers.

Table 1: Vibrational frequencies used to fit the experimentally observed Raman lines of the radial breathing mode of single wall carbon nanotubes in comparison to calculated results.

observed lines	147	156	162	167	172	177	182	185	190	195	199	206	211	230
calculated for armchair tubes	146 (12)	159 (11)				175 (10)				195 (9)			219 (8)	
calculated for zigzag tubes		158 (19)		167 (18)		178 (17)		187 (16)			200 (15)	214 (14)		231 (13)
chiral tubes	.	.	.	.	172	.	182	.	190	.	.	206	.	.

not change by varying the wavelength of the exciting laser and can be assigned to the contributions of the various tubes to the overall response. Table 1 lists the frequencies together with some results from a first principle calculation [13]. Strongest response was recorded for the lines at 177, 185, and 195 cm^{-1}, respectively.

EVALUATION of MODE FREQUENCIES and INTENSITIES

The value of the RBM frequency as a function of diameter and chirality is crucial for the evaluation of the experiments. Frequencies calculated from semiempirical methods have been reported several times [2, 3]. Considering a force constant model for the graphene sheet with interactions up to fourth nearest neighbors the RBM frequency was found inversely proportional to the radius of the tube. From this the frequency for the (10,10) tube was 156 cm^{-1}[2] or 165 cm^{-1}[3], respectively. In contrast the maximum intensity in the Raman spectra was recorded for 177, 185, 195 cm^{-1} as discussed above.

The evaluation of the mode frequencies was performed with a well adapted first principles LDA calculation. Achiral tubes with diameters between 10 and 16 Å were considered which means armchair tubes from (8,8) to (12,12) and zigzag tubes from (13,0) to (20,0). The number of atoms was less then 100 even for the largest tubes investigated. For the evaluation the Vienna Ab initio Simulation Package (VASP) was used which already proved to be very accurate for determining the phonon dispersion relations of diamond and graphite [14]. The calculations were carried out at the Alpha Cluster of the Vienna University. The starting geometry of a tube was obtained by simply wrapping up a graphene sheet – according to (n,m) – with a carbon-carbon distance $a_{C-C} = 1.4$ Å. The number of carbon atoms in the unit cell is 4n for both (n,n) and (n,0) tubes. The calculations were performed for a tetragonal lattice but are related to single tubes since the lattice constant was chosen 20 Å.

The evaluation of the system was performed in two steps. First the geometry was optimized in two consecutive procedures. In the first procedure the carbon atoms were allowed to relax both perpendicular and parallel to the tube axis. This was a procedure with slow convergence because the unit cell – and so the basis set – was changing, at least in the tube axis direction. The final optimization was done with fixed unit cell parameters, which means the carbon atoms were only relaxed perpendicular to the tube axis. The Cartesian components of the forces on the individual atoms in the relaxed geometry were less than 0.03 eV/Å. The carbon-carbon bond lengths turned out to be nonuniform. In armchair tubes they were 1.401 Å and 1.417 Å with the shorter bond perpendicular to the tube axis. In zigzag tubes they were 1.408 Å and 1.413 Å with the shorter bond parallel to the tube axis. Details of the evaluation procedure are described in [13].

In the second step the frequency of the radial breathing mode was calculated by the frozen phonon method. The tube was blown up and pressed together perpendicular to the long axis that is the tube radius was increased and decreased by 1-2% around the relaxed value. The total energy was calculated at each configuration. From a quadratic fit the harmonic force constant was obtained, which, taking into account the mass of the unit cell resulted in the RBM frequency. A small shift of less than 0.1% in the tube radius corresponding to the minimum of the quadratic fit is a measure either for not perfect relaxation or for possible anharmonicity.

As a test for the accuracy of the calculation we evaluated the radial breathing mode of the C_{60} molecule with the same procedure. The calculated frequency was 494 cm^{-1} which is in excellent agreement with the experimental value of 493 cm^{-1} [15].

The calculated frequencies are listed in Tab. 1 for several armchair and zigzag tubes. The agreement with the experiment is surprisingly good and strongly suggest to assign the mode observed at 177 cm^{-1} to the RBM of the (10,10) tube.

The other calculated frequencies lye indeed very well on a $1/R$ curve as it was predicted from the semiempirical calculation [2]. The fit yields $\nu[\text{cm}^{-1}] = A/R[\text{Å}]$ with $A = 1181$ for armchair and $A = 1152$ for zigzag tubes, respectively. With this the assignment for all other observed modes is straight forward and as shown in the table. Even the frequencies for chiral tubes may be evaluated using the $1/R$ law and an average value for A.

There are several modes which could not be assigned to either armchair or zigzag tubes. It is very probable that they correspond to chiral species. Chiral species may as well overlap some of the armchair or zigzag tubes.

The other important quantities for the interpretation of the resonance Raman spectra are the one electron energy levels. The band structure near the Fermi level was evaluated as well by LDA-VASP for several tubes and found in very good agreement – at least within a few eV interval around the Fermi level – with tight binding calculations using a tight binding parameter $\gamma = 2.5$ eV.

From the band structure the density of states $g(\epsilon)$ can be evaluated and, in turn, the Raman cross section [16]. A comparison of such calculations with the experiments for the various modes revealed reasonable good agreement with the experiment, at least for the strongest observed lines, for a fitting parameter of $\gamma = 2.6$ eV. An example is given in Fig. 2a for the RBM mode of the (16,0) tube with a RBM at 185 cm^{-1}. The full drawn line is as calculated for constant transition matrix elements and a damping factor (Γ) for the excited state of 0.1 eV. Even though most of the experimental results fit well to the calculations there are some points in the center of the spectrum which deviate considerably. It turns out there are a few chiral modes with RBM frequencies close to 185 cm^{-1}. However, only one of them exhibits a resonance transition at the energy which matches the experiment. This is the tube (13,5) with a chiral angle of 15.6°. A selected number of branches in the dispersion relation for this tube are depicted in Fig. 2b. The

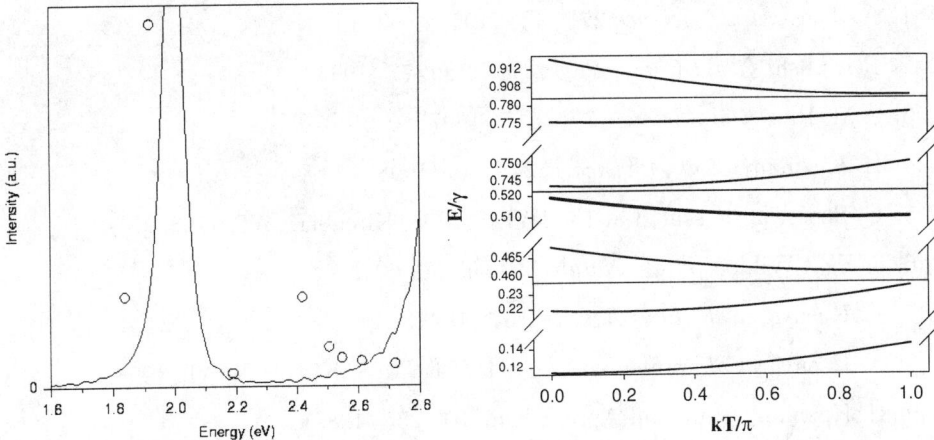

Figure 2: Experimentally observed Raman intensities for the mode at 185 cm^{-1} (o) and calculated cross section for the same tube (–) (a) and selected dispersion relations for a chiral tube with almost the same frequency for the breathing mode (b).

selection was performed to retain those branches which exhibit a divergence in the density of states. There are only 19 such branches. The seven with the lowest energy are shown (without the symmetrical counterparts below the Fermi energy). The branch, which is responsible for the deviation of the experimental results in Fig. 2a from the smooth curve is plotted as bolt in Fig. 2b. From the low scattering intensity for this tube species its concentration can be estimated to be rather low. This is remarkable since its diameter is close to the diameter of species which appear with high frequency.

SUMMARY

We have demonstrated, that the Raman response of SWCNTs grown from laser deposition exhibit a strong dispersion effect. First principle calculations of the frequency for the RBM allowed to assign the observed modes to tubes with well defined helicity. The RBM for the (10,10) tube was assigned 177 cm^{-1}. From a comparison of the observed Raman intensities with calculations the tight binding parameter γ was determined to be 2.6 eV and evidence for chiral tubes was found.

ACKNOWLEDGEMENTS

This work was supported by grants OTKA T022980 and AKP 96/2-462 2,2 in Hungary and by the FFWF, project P-11943, in Austria. One of the authors (JK) gratefully acknowledges financial support from the Soros Foundation. Valuable discussions with G. Kresse, E. Mele, C. Cane and J.C. Charlier are gratefully acknowledged.

References

[1] A. Thess et al., *Science* **273**, 483 (1996).

[2] R.A. Jishi et al., *Chem. Phys. Lett.* **209**, 77 (1993).

[3] A.M. Rao et al., *Science* **275**, 187 (1997).

[4] H. Kuzmany et al., *Physica B* **244**,186 (1998).

[5] G. Scuseria, presented at the IWEPNM98, Kirchberg/AT, 1998.

[6] J.W.G. Wildöer et al., *Nature* **391**, 59 (1998).

[7] A. Kasuya et al., *Phys Rev. B* 1 **57** (1998).

[8] A. Kasuya et al., presented at the IWEPNM98, Kirchberg/AT, 1998.

[9] H. Kuzmany, Pure and Appl. Chem. **57**, 235 (1985).

[10] M. Lamy del la Chapelle et al., presented at the IWEPNM98, Kirchberg/AT, 1998.

[11] H. Kuzmany et al., *Carbon*, in press.

[12] H. Kuzmany et al. *Europhys. Lett.*, submitted.

[13] J. Kürti et al., *Phys. Rev. B*, submitted.

[14] G. Kresse et al., *Europhys. Lett.* **32**, 729 (1995).

[15] H. Kuzmany et al., *Advanced Materials* **6** 731 (1995).

[16] R.M. Martin and L.M. Falicov, *Topics in Applied Physics* **8** 79 (1983).

Resonant Raman Scattering and the Zone-Folding Effects in Single-Wall Nanotubes

A. Kasuya[a], M. Sugano[a], Y. Saito[b], Y. Tani[b], K. Tohji[c], H. Takahashi[c] and Y. Nishina[a]

[a]*Institute for Materials Research, Tohoku University, Sendai, 980-77 Japan*
[b]*Department of Electrical and Electronic Engineering, Mie University, Tsu, 514 Japan*
[c]*Department of Geoscience and Technology, Tohoku University, Sendai, 980-77 Japan*

Raman scatterings from optical phonons in single-wall nanotubes of mean radius between 0.55 nm and 1.0 have been measured in the wavelength of incident laser beam between 450 nm and 800 nm. The observed multiple splittings of optical phonon peaks and their resonant enhancement at the wavelength in the vicinity of 690 nm show the definite presence of zone-folding effect in both phonon and electronic systems. The results are well accounted for by the two-dimensional graphite dispersion relations. These measurements, therefore, provide direct experimental evidences on the diameter-dependent properties induced by the cylindrical symmetry of nanotubes.

INTRODUCTION

Carbon nanotube has a unique feature that its basic physical properties may be analyzed in terms of the zone-folding effect based on the two-dimensional graphite (1-8). Both electronic and vibrational states can be deduced from the graphite dispersion relationships which have been studied in details both theoretically and experimentally. Because of the cylindrical symmetry of nanotubes, the translational symmetry present in the single layer of graphite remains the same along the tube axis but no longer exists around the circumference which turns to a cyclic one. As a result, the wavevectors of both electrons and phonons remain to take continuous values along the direction corresponding to the tube axis in the Brillouin zone, but can take only sets of discrete values $q_n = n/r$, (r: tube radius, n: 0,1,2,...) given by the cyclic period around the circumference. Hence, the Brillouin zone of naotubes consists of a set of equally spaced lines separated by $1/r$ along the circumference direction in the hexagonal Brillouin zone of a graphite layer.

The zone-folding effect in the electronic structure induces one-dimensional $E^{-1/2}$ singularities in the density of states at energies with their wavevectors equal to q_n in the band structure of two-dimensional graphite(6). Many of these critical points lie in the energy range of 1 eV and may play decisive roles in the electronic as well as optical properties of nanotubes derived from graphite. It is, therefore, essential to identify and characterize the low-lying electronic states in understanding the basic size-dependent properties of nanotubes.

In the phonon system, energies in the graphite dispersion relation at wavevectors

q_n correspond to vibrational energies of normal modes in nanotubes. There are 15 Raman active modes in general chiral tubes of a given diameter(8) in contrast to only one in the two-dimensional graphite.

The zone-folding method is applicable so long as the finite curvature effect of the graphite sheet is negligible. This paper presents our recent investigation on resonant Raman scattering to show that the zone-folding method is a good approximation for nanotubes of radii larger than 0.5 nm.

EXPERIMENTAL RESULTS AND DISCUSSION

Single-wall nearly mono-size nanotubes were synthesized by an arc discharge method with the positive graphite electrode mixed with different metals such as Ni&Fe, Co and La. Our electron microscopic observation shows that the mean radii of these samples prepared with Ni&Fe, Co and La are 0.55 nm, 0.65 nm and 1.0 nm, respectively. The radius distribution is measured under electron microscope and is rather sharp and its half-width is less than 0.1 nm. for each sample. Details of sample preparation process are given in our previous publications(10, 11).

The inset in Fig. 1 shows our Raman scattering spectrum measured in a back scattering geometry with 514.5 nm line of Ar-ion laser on nanotubes prepared with the positive graphite electrode filled with Ni and Fe (mean tube radius 0.55 nm). It shows three well resolved peaks at 1590.9 cm^{-1}, 1567.5 cm^{-1} and 1549.2 cm^{-1} indicated

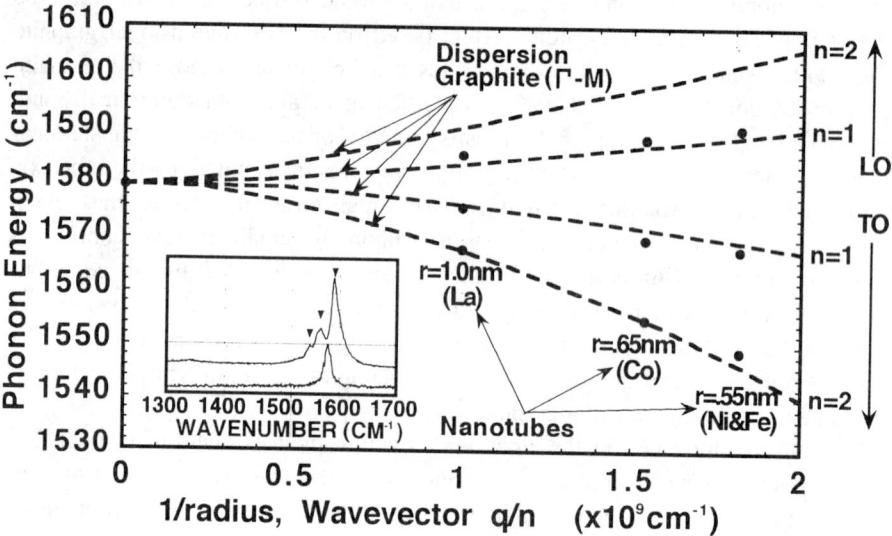

FIGURE 1. Plot of observed energy positions of multiply split peaks denoted by triangles in the spectrum in the inset vs 1/r for nanotubes prepared with Fe&Ni (r=.55nm)), Co (r=.65nm)) and La (r=1.0nm). Dashed lines are theoretical LO and TO modes in graphite along the Γ-M. The ordinate, 1/r, corresponds to the phonon wavevector q for the curves with n=1 (fundamental vibration), and wavevector q/2 with n=2 (first harmonic vibration).

by triangles. Our measurements show that the spectral splittings become narrower in going from the samples prepared with Ni&Fe to Co to La, and merge to the peak position at 1580 cm^{-1} of E_{2g} (stretching) mode in graphite shown as the lower spectrum in the inset.

Figure 1 shows a plot of observed energy positions of multiply split peaks(2) of samples of radius 0.55 nm (Fe&Ni), 0.65 nm (Co), and 1.0 nm (La). Dashed lines in Fig. 1 are optical phonon dispersions of longitudinal (LO) and transverse (TO) modes in graphite calculated along the Γ-M. (12) The abscissa, 1/r, corresponds to the phonon wavevector q for the curves with n=1, and q/2 for the curves with n=2.

Observed peak positions coincide well with n=1 or n=2 dispersion curve, showing the evidence of size-dependent discrete dispersions. The peaks (indicated by triangles in the inset), therefore, may be assigned, in order from the highest energy, to LO phonons with n=1 (denoted by L1), TO phonons with n=1 (T1) and TO phonons with n=2 (T2), respectively. Those peaks fitted on the n=1 curves correspond to optical phonons of the fundamental vibrations with their wavelength equal to the circumference of the tube, and n=2 the first harmonic vibrations with their wavelength equal to 1/2 of the circumference. These peaks with n=1 and n=2 are assigned to the Raman active E_1 and E_2 modes, respectively, for general chiral tubes (2) represented by the symmetry group $C_{N/\Omega}$. Other E_n modes with n>2 are Raman inactive (9) and may contribute to the tails of the multiply-split peak. The good agreement of experimental results with graphite dispersion shows that nanotubes with radii from 0.5 nm to 1.0 nm exhibit diameter-dependent discrete dispersion which is the unique feature of nanotubes not present in any other condensed matter. The zone-folding effect is present clearly in the phonon system in nanotubes and well accounted for by the dispersion relation in graphite. I.

To investigate the zone-folding effect in the electronic states, the resonant Raman scattering is measured by varying the wavelength of incident photons in the visible range(4). Figure 2 shows our Raman scattering spectra in the wavenumber region from 1400 cm^{-1} to 1700 cm^{-1} measured on the sample of radius 0.55 nm (Fe&Ni) at the wavelength from 488 nm to 783 nm. As the wavelength increases from 488nm to 690nm, the intensity of L1 peak decreases, whereas those of T1 and T2 increase. They become comparable to each other at around 690 nm. From 690 nm toward 800 nm, they return to their values observed at 488 nm. This behavior shows that the incident photons of wavelength 690 nm (1.8 eV) is resonant with the electronic transition between zone-folded electronic structures shown in the inset in Fig. 2. Our optical transmission measurement also shows a dip around the same wavelength(3). The recent measurement on scanning tunneling spectroscopy finds similar values for the zone-folded electronic state(13).

According to the calculation(2), the transition energy, G1 between the pair of singularities nearest from the Fermi energy (=0) is 0.76 eV, and G2, the next nearest is 1.55 eV for r= 0.55 of zigzag tubes (14,0) with γ_0=3.14, the nearest neighbor overlapping integral. For the armchair tube with radius 0.54 nm of type (8,8), G1 =1.91 eV for γ_0=2.5 estimated by a local density functional calculation(2). These values of G1 and G2 lie in the energy range of resonance in our measurement.

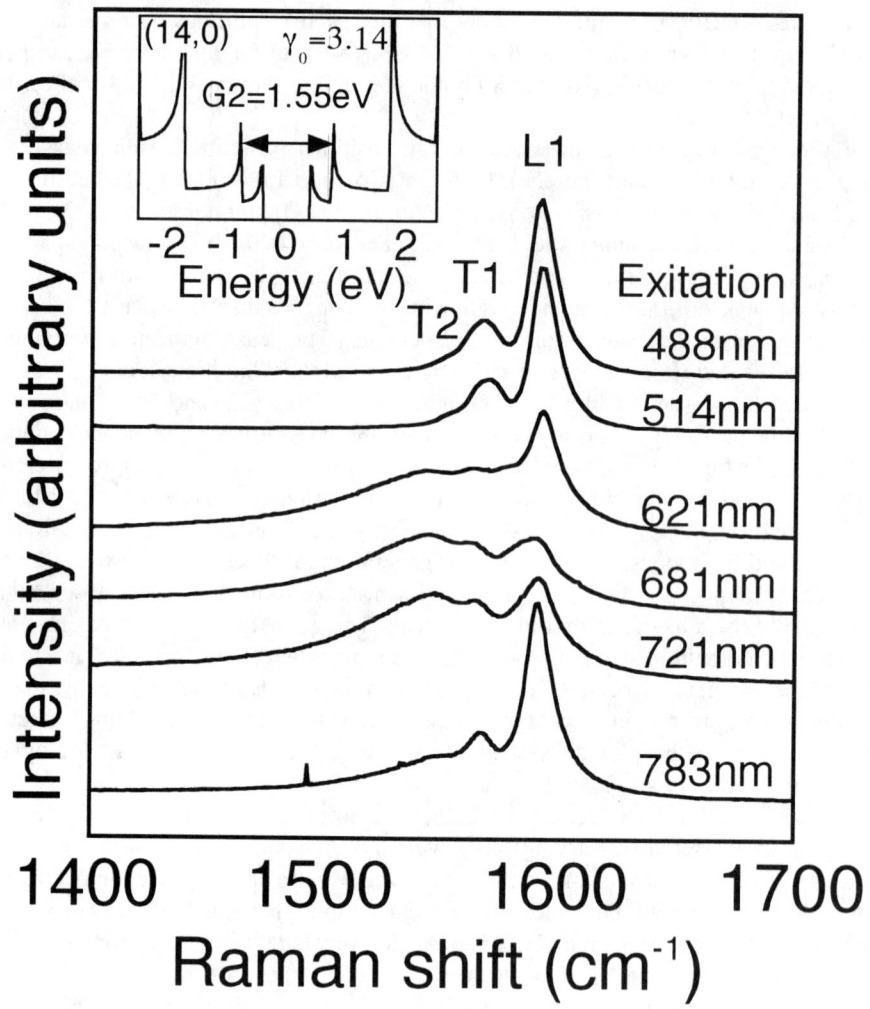

FIGURE 2. Raman scattering spectra between 1400cm^{-1} and 1700 cm^{-1} from single-wall nanotubes of radius 0.55 nm excited by the incident laser of wavelengths from 488 nm to 783 nm. The inset shows schematics of zone-folded electronic structures in nanotubes.

Our experimental results, therefore, show that the zone-folding effect is present in both vibrational and electronic states in nanotubes and the diameter-dependent properties are explained well in terms of the two-dimensional graphite.

ACKNOWLEDGEMENTS

This work was supported by CREST of Japan Science and Technology Institute.

REFERENCES

1. S. Iijima and T. Ichihashi, *Nature* **363**, 603 (1993).
2. A. Kasuya, Y. Sasaki, Y. Saito, K. Tohji and Y. Nishina, *Phys. Rev. Lett.* **78**, 4434 (1997).
3. A.M Rao, E. Richter, S. Bandow, B. Chase, P. C. Ecklund, K.A. Williams, S.Fang, K. R. Subbaswamy, M. Menon, A.Thess, R. E. Smalley, G. Dresselhaus and M.S. Dresselhaus, *Science* **275**, 187 (1997).
4. A. Kasuya, M. Sugano, T. Maeda, Y. Saito, K. Thoji, H. Takahashi, Y. Sasaki, M. Fukishima, C. Horie and Y. Nishina, *Phys Rev.* B **57**, 4999 (1998).
5. A. Kasuya, Y. Saito, Y. Sasaki, M. Fukushima, T. Maeda, C. Horie and Y.Nishina, *Science and Technology of Atomically Engineered Materials*, ed. by P. Jena, S.N.Khanna and B.K. Rao, Singapore: World Scientific, 1996, p. 431.
6. N. Hamada, S. Sawada and A. Oshiyama, *Phys. Rev. Lett.* **3**, 863 (1992).
7. R. Saito, T. Takeya, T. Kimura, G. Dresselhaus and M.S. Dresselhaus, *Phys. Rev.* B **57**, 4145 (1997).
8. E. Richter and K.R. Subbaswamy, *Phys. Rev. Lett.*, **79**, 2738 (1997).
9. R.A. Jishi, D. Inomata, K. Nakao, M.S. Dresselhaus and G. Dresselhaus, *J. Phys. Soc. Jpn.* **63**, 2252 (1994).
10. Y. Saito, M. Okuda and T. Koyama, *Surface Review and Lett.* **3**, 863 (1966).
11. K. Tohji, T. Goto, H. Takahashi, Y. Shinoda, N. Shimizu, B. Jayadevan, I. Matsuoka, Y. Saito, A. Kasuya, T. Ohsuna, K. Hiraga and Y. Nishina, *Nature* **383**, 679 (1996).
12. P. Lespade, R. A. Jishi, and M.S. Dresselhaus, *Carbon* **20**, 427 (1982).
13. J. W. G. Wildoer, L. C. Venema, A. G. Rizzler, R. E. Smalley and C. Dekker, *Nature* **391**, 59 (1998).

Resonant Raman Scattering from Single-wall Nanotubes of Diameters between 1.1 nm and 1.4 nm

M. Sugano[a], A. Kasuya[a], K. Tohji[b] Y. Saito[c] and Y. Nishina[a]

[a]*Institute for Materials Research, Tohoku University, Sendai*
[b]*Department of Geoscience and Technology,
Tohoku University, Sendai*
[c]*Department of Electrical and Electronic Engineering, Mie University, Tsu*

Abstract. Resonant Raman scattering has been measured on single-wall nanotubes of diameters from 1.1 nm to 1.4 nm. The spectra show groups of peaks near 180 cm^{-1}, 360 cm^{-1}, 440 cm^{-1}, 600 cm^{-1}, 720 cm^{-1}, 880 cm^{-1}, 1070 cm^{-1}, 1320 cm^{-1} and 1580 cm^{-1}. Observed resonant intensity enhancements show that peaks in each group belong to the same vibrational mode of different diameters, and hence different dependencies on the incident photon energy. These measurements, therefore, determine the diameter dependent vibrational modes and zone-folded electronic structures, yielding mode assignments that groups of peaks near 600 cm^{-1}, 859 cm^{-1}, 1070 cm^{-1} and 1580 cm^{-1} belong to optical modes and the rest to acoustic ones in graphite in the zone-folding scheme. Peaks near 360 cm^{-1}, 720 cm^{-1} and 880 cm^{-1} are harmonics of 180 cm^{-1} or 440 cm^{-1}.

INTRODUCTION

Raman scattering measurements have provided considerable information on the basic nanometer-scale properties of carbon nanotubes, not only in the phonon system but also in electronic structures by tuning the photon energy of incident laser beam to particular energy levels (1-5). Raman spectrum, for example, reveals diameter-dependent discrete dispersion of optical phonons and singularities in the low lying electronic states predicted in the zone-folding scheme based on graphite. This paper presents our investigation on the diameter-dependent vibrational modes and electronic structures by resonant Raman scattering measurements on single-wall nanotubes of diameter from 1.1 to 1.4 nm.

EXPERIMENTAL RESULTS AND DISCUSSION

Samples were prepared by arc discharging of graphite electrodes (6) in He gas of pressure 600 Torr with the positive electrode was filled in the center with Fe and Ni powders of equal atomic weight. The samples were purified by the hydrothermal treatment (7). Our electron microscopic observation indicates that the diameter of our samples ranges from 1.1 nm to 1.4 nm. The Raman scattering was measured in a back scattering geometry at room temperature with a single monochromator connected to a liquid nitrogen cooled CCD camera. Ar ion laser together with a dye laser and a Ti-Sapphire laser are used as excitation sources.

Figure 1-(a), 1-(b), and 1-(c) display our Raman spectrum excited by laser beams of wavelengths at 488 nm, 726 nm and 780 nm, respectively. Figure 1(a) shows dominant peaks centered at 1580 cm^{-1}. These peaks are assigned to the vibrational modes originating from the optical phonons in graphite split into longitudinal (LO) components to higher energies and transverse (TO) ones to lower energies (1,3). Figure 1-(a) also shows peaks near 180 cm^{-1}. This group of peaks have been assigned to the breathing mode (radial acoustic vibration) of tubes of different diameters (2). As the wavelength of laser beam increases to 800 nm, peaks of both 1580 cm^{-1} and 180 cm^{-1} change their intensities significantly, and many other groups of peaks appear and disappear between them. Dominant ones are those in the vicinities of 360 cm^{-1}, 440 cm^{-1}, 600 cm^{-1}, 720 cm^{-1}, 880 cm^{-1}, 1070 cm^{-1} and 1320 cm^{-1}.

Figure 1 reveals that the group of peaks near 180 cm^{-1} consists at least of four peaks at 162 cm^{-1}, 173 cm^{-1}, 185 cm^{-1} and 198 cm^{-1}. With a single laser wavelength, not all of

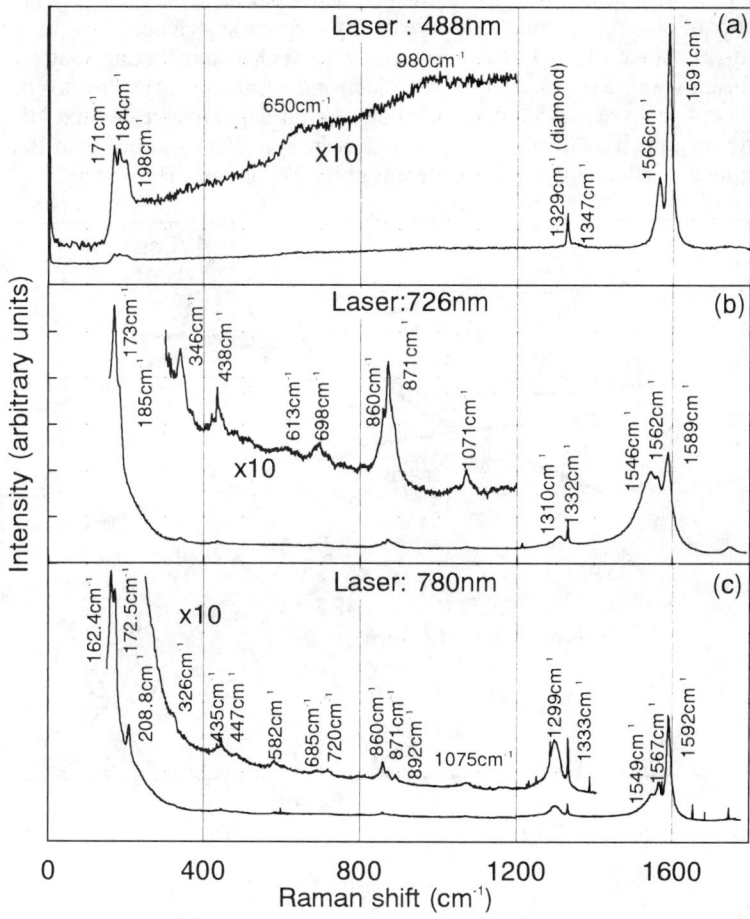

FIGURE 1. Raman scattering spectra of single-wall nanotubes measured with the laser beam of wavelengths at 488 nm (a), 726 nm (b) and 780 nm (c).

them can be observed in a spectrum. The peaks near 440 cm^{-1} consist of three major peaks at 424 cm^{-1}, 438 cm^{-1} and 446 cm^{-1}, and those near 880 cm^{-1} consist of 849 cm^{-1}, 859 cm^{-1}, 871 cm^{-1} and 891 cm^{-1}. Based on the spectral positions and intensities, groups of peaks near 360 cm^{-1} and 720 cm^{-1} are identified as the second and fourth harmonics of the peaks near 180 cm^{-1}. Similarly, peaks near 880 cm^{-1} (849 cm^{-1}, 871 cm^{-1} and 891 cm^{-1}) are the second harmonics of group 440 cm^{-1} (424 cm^{-1}, 438 cm^{-1} and 446 cm^{-1}). The peaks near 1340 cm^{-1} are rather sharp but resemble a broad peak around 1350 cm^{-1} commonly observed in various graphitic materials with structural imperfections(8).

Figure 2 shows intensity profiles of each peak in the wavelength range from 450 nm to 800 nm. The profile of LO components denoted as LO has a broad minimum and those of TO denoted as TO1 and TO2 have broad maxima around 700 nm (3). Our optical absorption measurement (3) also shows a broad peak in the same region indicating the presence of electronic structures not existing in graphite. Hence, they are interpreted as an anti-resonance for LO and a resonance for TO in the Raman scattering of the incident beam via one of the zone-folded electronic structures (4). Profiles of 859 cm^{-1} and 1070 cm^{-1} peaks are also broad and show similar dependencies to that of TO.

The intensity profiles of other Raman peaks in Fig. 2 are sharper than those of TO and show their resonant maxima at different photon energies. These maxima, however,

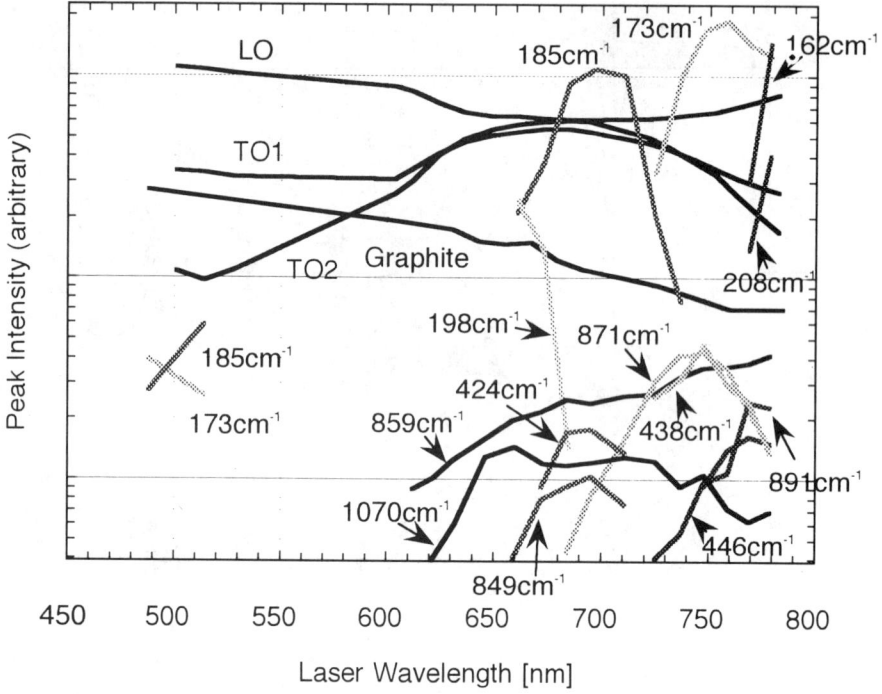

FIGURE 2. Intensity variations of Raman scattering peaks vs wavelength of incident laser beam. Optical phonon peaks for longitudinal is denoted as LO, and transverse as TO1 and TO2. Other peaks are denoted by their energies. Optical mode for graphite is also shown.

appear roughly at either one of three different wavelengths. The first is around 700 nm for a set of peaks at 185 cm^{-1}, 424 cm^{-1} (849 cm^{-1}, the second harmonics). The second is around 750 nm for peaks at 173 cm^{-1}, 438 cm^{-1} (871 cm^{-1}) and the third is around 780 nm for peaks at 162 cm^{-1}, 446 cm^{-1} (891 cm^{-1}). Since each set of peaks exhibit their maxima at a common photon energy, they form a Raman spectrum of nanotubes of a common type (diameter and chirality) having the same zone-folded electronic structure for the resonance. Hence each group belongs to one of the vibrational modes in nanotubes and the lowest energy group near 180 cm^{-1} is assigned to the breathing mode.

Figure 2 shows that the wavelength of resonant maxima is shorter for peaks with higher vibrational energies in the group of peaks near 180 cm^{-1}. This experimental fact is consistent with our assignment that this group belongs to the breathing mode. The energy of breathing mode is nearly proportional to the inverse of the diameter (5), as the zone-folded electronic structure (9). Hence the shorter the wavelength of resonant maximum, the higher the breathing mode energy and the smaller the tube diameter.

In the group near 440 cm^{-1}, the wavelength of resonant maximum is longer for a higher peak energy as oppose to the group near 180 cm^{-1}, breathing mode. This result can be explained if this group belongs to an acoustic mode near the M-point in the graphite Brillouin zone. The energy of the lowest acoustic mode in two-dimensional graphite is zero at the Γ-point and increases to 447 cm^{-1} at the M-point (10). In nanotubes, the vibrational modes become Raman active only for those having their wavevectors in the circumference direction equal to the inverse (and its integer multiples) of the tube radius. The allowed wave vector measured from the M-point towards Γ-point is larger for smaller diameter tubes, Hence the energy of vibration becomes smaller for smaller diameter tubes.

ACKNOWLEDGEMENT

Authors would like to thank Prof. C. Horie, Prof. T. Maeda and Prof. M. Fukushima for valuable discussion. This work of supported by CREST of JST.

REFERENCES

1 A. Kasuya, Y. Sasaki, Y. Saito, K. Tohji and Y. Nishina, *Phys. Rev. Lett.* **78**, 4434 (1997).
2 A.M Rao, E. Richter, S. Bandow, B. Chase, P. C. Ecklund, K.A. Williams, S. Fang, K. R. Subbaswamy, M. Menon, A.Thess, R. E. Smalley, G. Dresselhaus and M. S. Dresselhaus, *Science* **275**, 187 (1997).
3 A. Kasuya, M. Sugano, T. Maeda, Y. Saito, K. Thoji, H. Takahashi, Y. Sasaki, M. Fukishima, C. Horie and Y. Nishina, *Phys Rev.* B **57**, 4999 (1998).
4 E. Richter and K.R. Subbaswamy, *Phys. Rev. Lett.*, **79**, 2738 (1997).
5 R. Saito, T. Takeya, T. Kimura, G. Dresselhaus and M.S. Dresselhaus, *Phys. Rev.* B **57**, 4145(1997).
6 Y. Saito, M. Okuda and T. Koyama, *Surface Review and Lett.* **3**, 863 (1966).
7 K. Tohji, T. Goto, H. Takahashi, Y. Shinoda, N. Shimizu, B. Jayadevan, I. Matsuoka, Y. Saito, A. Kasuya, T. Ohsuna, K. Hiraga and Y. Nishina, *Nature* **383**, 679 (1996).
8 F. Tuinstra and J. L. Koenig, *J. Chem. Phis.* 53, 1126 (1970).
9 N. Hamada, S. Sawada and A. Oshiyama, *Phys. Rev. Lett.* **3**, 863 (1992).
10 P. Lespade, R. A. Jishi, and M.S. Dresselhaus, *Carbon* **20**, 427 (1982).

Molecular dynamics of single wall nanotubes

E. Anglaret[*], J.L. Sauvajol[*], S. Rols[*,#], C. Journet[*], T. Guillard[£]
L. Alvarez[*], E. Muñoz[+], A.M. Benito[+], W.K. Maser[+],
M.T. Martinez[+], G.F. de la Fuente[$], D. Laplaze[*] and P. Bernier[*]

[*] *Groupe de Dynamique des Phases Condensées, Université Montpellier II, Montpellier, France*
[#] *Institut Laue-Langevin, Grenoble, France*
[£] *Institut de Sciences et de génie des Matériaux, CNRS Odeillo, Font-Romeu, France*
[+] *Instituto de Carboquimica, CSIC, Zaragoza, Spain*
[$] *ICMA, CSIC, Zaragoza, Spain*

Abstract. We report on a Raman scattering study of single wall carbon nanotubes (SWNT) prepared by three different routes : laser ablation (LA), electric arc discharge (EA) and solar energy (SE). We emphasize some differences in the Raman spectra that we discuss in terms of distribution of tubes diameters and helicoidal pitches. Nanotubes prepared via EA are found to be quite monodisperse in opposite to samples prepared via LA or SE. The signature of unusually large tubes is observed for LA samples and confirmed by neutron diffraction measurements. Some infrared data are also presented.

INTRODUCTION

Many physical (mechanical, electronic) and chemical (gas adsorption) properties of single wall carbon nanotubes (SWCN) are expected to depend both on their molecular structure and intermolecular interactions (1). Mass production techniques are nowadays available (2,3), and this is an actual challenge to probe the structure of SWCN both at the molecular scale (diameter and polydispersity, helicoidal pitch) and at the bundle scale (cell parameter, bundle size) in order to relate the structural informations to the intrinsic properties. This can be achieved in the direct space using microscopic techniques but only local informations can be derived. In the reciprocal space, diffraction techniques (X-ray, neutrons, electrons) are useful tools to probe the packed crystalline structure of the SWCN bundles as well as the tubes symmetries. In the frequency space, vibrational spectroscopic techniques are well designed to characterize the molecular structure since the vibrational density of states depends both on the tubes diameter and helicoidal pitch (4,5). Raman, and especially micro-Raman, scattering has been first widely used (6-8) to characterize SWCN because of the small quantites required and the intense resonant signals recorded (9). Unfortunately, no quantitative analysis of the tubes size distribution can be easily derived in Raman precisely because of the resonant character of the signal. Infrared and inelastic neutron scattering (INS) studies do not suffer this inconvenience but only a few FTIR studies (10) and no INS have been reported up to now because of the small signals recorded (and the large mass required for INS).

In this report, we present a vibrational spectroscopy study of SWCN prepared with three different techniques : laser ablation (LA) (2,11), electric arc discharge (EA) (3) and solar energy (SE) (12) and various catalysts mixtures. Some differences are emphasized in the micro-Raman spectra and discussed in terms of polydispersity of diameters and helicoidal pitches. Differences at the microscopic scale are correlated to "bulk" differences observed in neutron diffraction (ND) and mid infrared absorption (MIR). Preliminary INS data have been presented at the winterschool and will be analyzed in details elsewhere (13).

SAMPLES DESCRIPTION AND EXPERIMENT

The LA samples were prepared using a cw CO_2 laser operating at 10.6μm in a 530 mbar Ar atmospher using the following catalysts mixtures : *Ni-Y (2/0.5% at.)* and *Ni-Co (0.6/0.6% at.)*. Details can be found elsewhere (11). The EA sample was prepared under a 600 mbar He atmospher using a catalysts mixture *Ni-Y (4.2/1% at.)* (3). SE samples were synthesized under *a 0.25 m^3/h Ar* flux in a 400 mbar atmospher using Ni-Y *(4.2/1% at.)*, Ni-Co *(2/2% at.)* and Ni-La *(2/2% at.)* as catalysts (8,12). At the time of this study, only limited quantities of SE samples were available, this is why they have been studied only in Raman.

The micro-Raman spectra have been recorded on a Jobin-Yvon T64000. The ND spectra were measured on the G6-1 spectrometer at Laboratoire Léon Brillouin, Saclay, France using incident neutrons with wavelength λ=4.743 Å. The infrared absorption spectra were recorded in the MIR on a Bruker IFS 113V FTIR spectrometer (SWCN were mixed with *KBr* to prepare pellets).

RESULTS AND DISCUSSION

Raman results

As stated above, a micro-Raman spectrum only gives a local picture of the material and a detailed study requires to present different spectra recorded on various spots of the samples. Here, we will only present some typical spectra, *i.e.* frequently observed for a same sample, and emphasize typical differences between the samples.

Figure 1 is a zoom in the low frequency range that was operated in a triple grating configuration to achieve an otpimal resolution and with two incident wavelenghts to illustrate the resonance effect. The number of individual peaks in the low frequency bunch appears to be very sample dependent. It was shown that each peak corresponds to a different tube diameter (no significant helicoidal pitch dependence is expected) (4) and thus one can identify three kinds of samples :

i) the EA and Ni-Co, SE samples appear rather monodiperse,

ii) at least five peaks can be resolved in the LA samples and Ni-La sample spectra indicating a significant polydispersity, including some larger tubes (peaks at low frequencies (3)),

iii) the low frequency bunch of sample Ni-Y, SE covers an unusually broad frequency range, indicating a huge polydispersity, including unusually small SWCN.

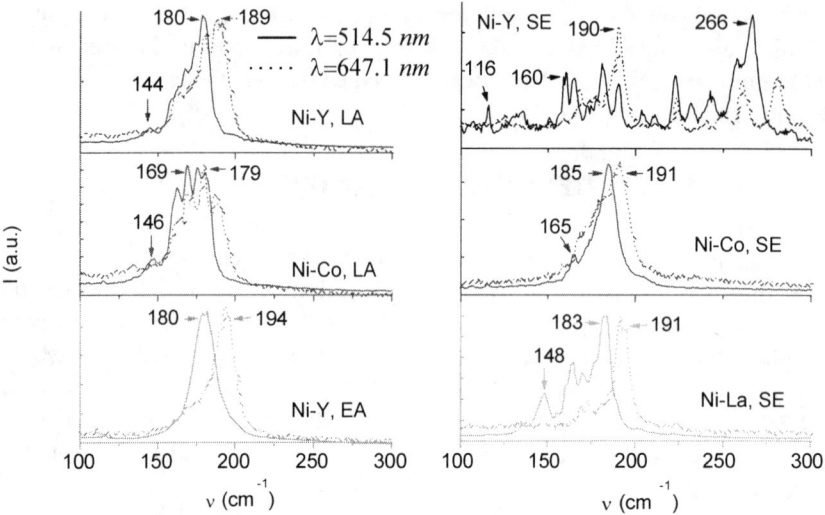

Figure 1. Low-frequency Raman spectra for various SWNT.

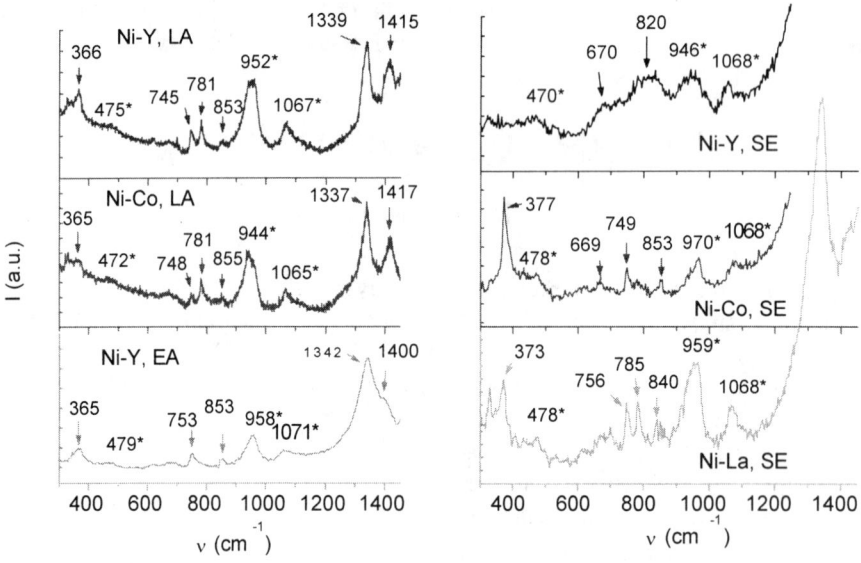

Figure 2. Mid-frequency Raman spectra for various SWNT. $\lambda=514.5nm$. Peaks assigned to harmonics or combinations are labelled with a star.

Only peaks of rather small intensities are recorded in the intermediate intramolecular frequency range (figure 2). Only a weak diameter-dependence is expected in this range while significant shifts are expected for tubes of different helicoidal pitches (4,8). This is interesting to find out that some differences in the spectra exist in the range 750-780 cm^{-1} between the same three groups of samples than above : a single peak is observed for EA and Ni-Co, SE, two peaks for LA and Ni-La, SE and a very broad ill-defined peak for Ni-Y, SE. This is another signature of polydispersity, possibly in terms of helicoidal pitch (4,8). One also notes the various shapes and intensities of the bunch at 1350 cm^{-1}, assigned to the D-line of graphite, that indicate different degrees of samples purity.

Neutron diffraction results

Figure 3 (left) present some ND spectra for LA and EA samples. The main feature is the difference in the position of the main peak, that shifts from 0.43Å^{-1} for EA to 0.34 or 0.32 Å^{-1} for LA, and slightly broadens. This peak corresponds to the (1 0) Bragg reflexion on the hexagonal bidimensional network formed by the tubes (2). This indicates that the mean cell parameter of the bundles is significantly larger for LA samples. The mean diameter of the tubes in LA is thus certainly larger and the broader linewidth indicates a larger size distribution. This is a "bulk" confirmation of the microscopic features observed in Raman.

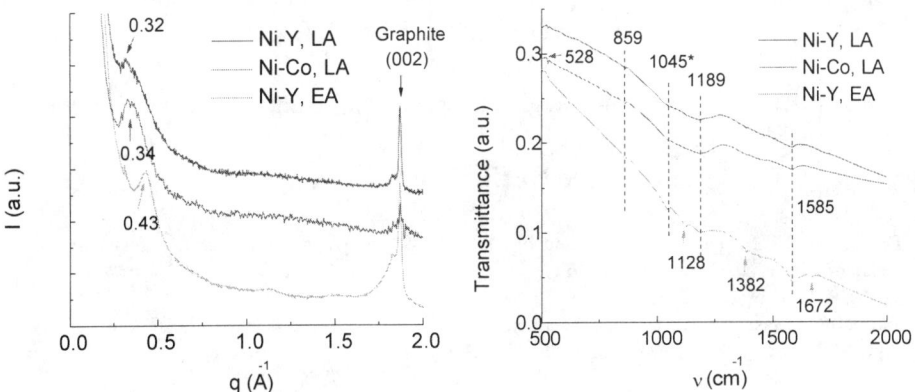

Figure 3. Left : raw neutron diffraction spectra for various SWCN.
Right : transmission FTIR spectra for various SWCN (in KBr pellets). Lines tentatively assigned to harmonics or combinations are labelled with a star.

FTIR results

The FTIR transmission spectra of LA and EA samples are presented in figure 3 (right). Some peaks (859, 1045, 1189 and 1585 cm^{-1}) are observed for all samples while others (528 cm^{-1} for Ni-Co, La ; 1128, 1382 and 1672 cm^{-1} for EA) are sample-specific. The peaks at 859 and 1585 cm^{-1} are close to those of graphite (4) but slightly downshifted. The peaks at 527 and 1189 cm^{-1} are found to be close to that of fullerene C_{60}. This is unlikely that a significant amount of fullerene exist in the samples and we rather think that these peaks are the signatures of modes involving displacements of carbon hexagons that may be similar in fullerenes and SWCN. An excellent agreement is found with zone folding calculations (4) for the modes at 859 and 1585 cm^{-1} and a correct agreement for the modes at 528,1189 and 1382 cm^{-1}. Further studies will be necessary to confirm the assignment of these modes. Unfortunately, there is no diameter-dependence of the modes in the MIR frequency range. A structural characterisation is to be undertaken in the far infrared frequency range. Such a study is in progress.

CONCLUSION

The structure of SWCN depends on the preparation technique and the catalysts mixtures used. Vibrational spectroscopy techniques are useful tools to characterize the samples in terms of diameter, polydispersity and helicoidal pitch. Correlations can be evidenced between "local" Raman characterisation and "bulk" ND measurements.

ACKNOWLEDGEMENT

One of us (SR) acknowledges the Région Languedoc-Roussillon for financial support.

REFERENCES

1. Dresselhaus, M.S., Dresselhaus G., and Eklund P.C., *Science of fullerenes and carbon nanotubes*, Academic Press ed., New-York, 1996.
2. Thess, A. *et al, Science* **273**, 483 (1996).
3. Journet, C., Maser, W.K., Bernier, P., Loiseau, A., Lamy de la Chapelle, M., Lefrant, S., Deniard, P., Lee, R., and Fischer, J.E., *Nature* **388**, 756 (1997).
4. Eklund, P.C., Holden, J.M., and Jishi, R.A., *Carbon* **33**, 959 (1995).
5. Yu, J., Kalia, R.K. and Vashista, P., *Europhys. Lett.* **32**, 43 (1995).
6. Rao, A.M *et al, Science* **275**, 187 (1997).
7. Lamy de la Chapelle, M., Lefrant, S., Journet, C., Maser, W.K., Bernier, P., and Loiseau, A., *Proc. SPIE* **2854**, 296 (1996).
8. Anglaret, E., Bendiab, N., Guillard, T., Journet, C., Flamant, G., Laplaze, D., Bernier, P. and Sauvajol, J.L., *Carbon*, in press.
9. Richter, E., and Subbaswamy, K.R., *Phys. Rev. Lett.* **79**, 2738 (1997).
10. Burger, B., Kuzmany, H., Thess, A. and Smalley, R.E., in *Proceedings of IWEPNV,* Kirchberg, Austria, 1997, World Scienctific ed.

11. Muñoz, E., Benito, A.M., Maser, W.K., Martinez, M.T., Estepa, L.C., Fernandez, J., de la Fuente, G.F. and Maniette, Y., in *Proceedings E-MRS,* Strasbourg, France, 1997, to be published in *Carbon*.
12. Laplaze, D., Bernier, P., Maser, W.K., Flamant, G., Guillard, T., and Loiseau, A., in *Proceedings E-MRS,* Strasbourg, France, 1997, to be published in *Carbon*.
13. Rols, S. *et al*, to appear elsewhere.

Mid-Infrared Investigations and Spatially Resolved Raman Spectra of Singlewalled Carbon Nanotubes

C. Thomsen[1], H. Jantoljak[1], U. Kuhlmann[1], C. Journet[2], P. Bernier[2]

[1] *Institut für Festkörperphysik, Technische Universität Berlin,
Hardenbergstraße 36, D-10623 Berlin, Germany*
[2] *Groupe Dyn. Phase Condensées, Université Montpellier II,
F-34095 Montpellier cedex 5, France*

Abstract. We performed Raman and infrared spectroscopy on singlewalled carbon nanotubes. Applying statistical methods we found two infrared-active modes at 874 ± 2 cm^{-1} and 1598 ± 3 cm^{-1}. Spatially resolved Raman measurements exhibited the strong sensitivity to symmetry and tube diameter of the low-energy mode at 165 cm^{-1} in contrast to the high-energy mode at 1985 cm^{-1}. We will discuss the results of the infrared studies as well as the Raman data in comparison with theory.

INTRODUCTION

The physical properties of singlewalled carbon nanotubes (SWNT) are based on the special closed-shell structure and were shown to be unique by experimental (e.g. [1,2]) as well as by theoretical means (e.g. [3,4]). Vibrational spectroscopy, i.e. Raman and infrared spectroscopy, yields information about the structural properties. Since only few experimental results on the infrared-active modes were published [5,6], Raman spectroscopy was proven to be a powerful tool for structural investigations [2,6-8]. In contrast to the infrared spectra, the Raman data are in good correspondence with theoretical predictions. [4]

EXPERIMENTAL

Carbon nanotubes were produced by the arc-discharge technique and a metallic catalyst (4.2% Ni and 1% Y) was used to provide the growth of singlewalled tubes. The sample we examined was taken from a deposit around the cathode and contained about 80% nanotube ropes. For details see [9].
We performed infrared reflectivity investigations using a BRUKER IFS 66v fourier spectrometer. A freestanding sample was produced by pressing the material into a

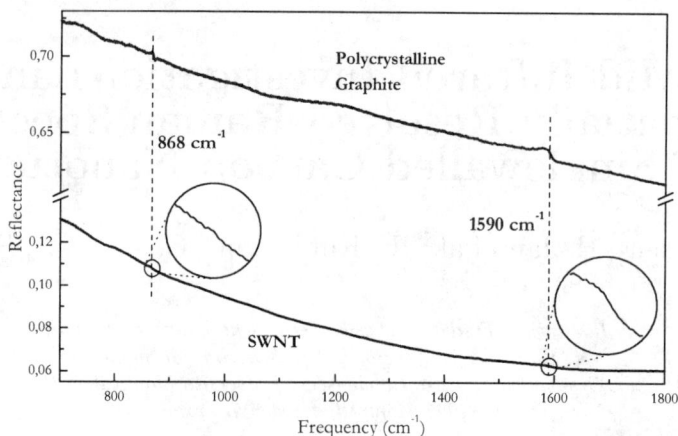

FIGURE 1. Reflectivity of SWNT in comparison with graphite. Graphite exhibits two infrared-active modes at 868 cm^{-1} and 1590 cm^{-1}. The corresponding features in the SWNT spectrum may be seen when magnified.

thin film. The spectra were collected in the spectral region of 700–7000 cm^{-1} with a resolution of 2 cm^{-1}.

Raman studies were carried out using a DILOR XY 800 triple grating spectrometer and an Ar$^+$/Kr$^+$-gas laser for excitation at 488 nm. Application of microscope optics enabled us to chose a laser power density of typically 15 kW/cm^2 and a focus diameter of about 1 µm.

RESULTS AND DISCUSSION

In Fig. 1 the infrared reflectivity of SWNT is shown. Upon strong magnification a weak feature related to lattice vibrations is seen near 1590 cm^{-1}. For comparison we performed studies on polycrystalline graphite and observed the infrared-active modes at 868 cm^{-1} and 1590 cm^{-1} in correspondence to literature. [10] To estimate the frequency of these low-intensity absorptions we analyzed the corresponding minima of the first derivatives. The measurements were repeated on different parts of the sample and on different samples to get representative results.

A statistical analysis of all results is shown in Fig. 2. The modes at 868 cm^{-1} as well as the modes at 1590 cm^{-1} are statistically distributed with respect to the frequencies. For the lower-frequency mode we find that the frequency 871±4 cm^{-1} (std. deviation) occurs significantly often, while the higher-frequency mode is found at 1597±4 cm^{-1} with statistical relevance. Both of these frequencies are higher than the respective graphite modes.

If we believe that the frequencies, which in Fig. 2 occur directly at the graphite frequencies, actually originate from graphitic parts of our sample, we may narrow down the standard deviation of our SWNT frequencies: We then find 874±2 cm^{-1}

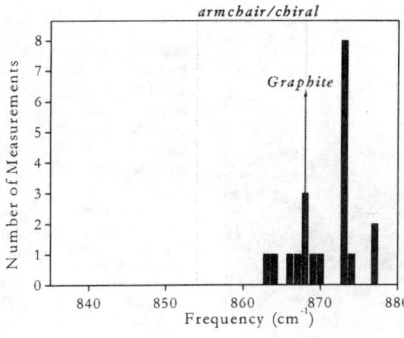 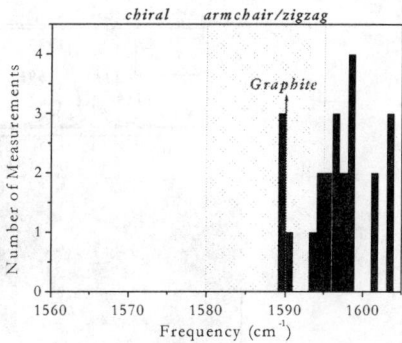

FIGURE 2. Statistical analysis of a series of reflectivity measurements like the ones shown in Fig. 1. The columns represent how often a particular frequency occured in our measurements, the graphite mode frequency is marked with an arrow, and the hatched areas indicate the theoretically predicted frequency range for various nanotube symmetries and diameters between 0.7 and 2.3 nm from Ref. [4]

and 1598±3 cm^{-1}, again significantly higher frequencies than those of graphite. The theoretical models based on force-constant calculations predict frequencies which, for the infrared-active vibrations considered here, are generally equal or lower than the graphite frequencies. [4] Although the theoretical frequencies depend on diameter and symmetry of the nanotube the results we find are outside the range for all symmetries (Fig. 2, hatched regions). We thus conclude that while the force-constant model of Ref. [4] gives a reasonable prediction for the absolute frequencies of the two high-energy infrared-active nanotube vibrations considered here, there are discrepancies with the experiment as far as the relative positions (compared to graphite) are concerned.

Using spatially resolved Raman measurements information about structural variations in the samples can be obtained. We applied microscope optics for realizing a laser focus diameter of 1 μm and a spatial resolution of 0.1 μm and performed linescans in 5 μm steps over a distance of 45 μm on the sample. On every spot a first-order Raman spectrum (100–1800 cm^{-1}) was taken.

The results on the most prominent vibrational mode bunch at about 1600 cm^{-1} is shown in a map in Fig. 3. The Raman intensities of the individual spectra were translated into a greyscale. The Raman frequencies were plotted as a function of the spot position on the linescan. From the inset the characteristic finestructure of the high-energy mode (HEM) can be seen. The main peak was detected at 1585 cm^{-1} with a shoulder to higher energies. The second prominent peak in this mode bunch was detected at 1561cm^{-1} with a shoulder to lower energies. Over the large distance of 45 μm the relative intensities of the individual phonons do not change in this mode. From the spatially resolved measurement of the HEM the examined area thus appears homogeneous.

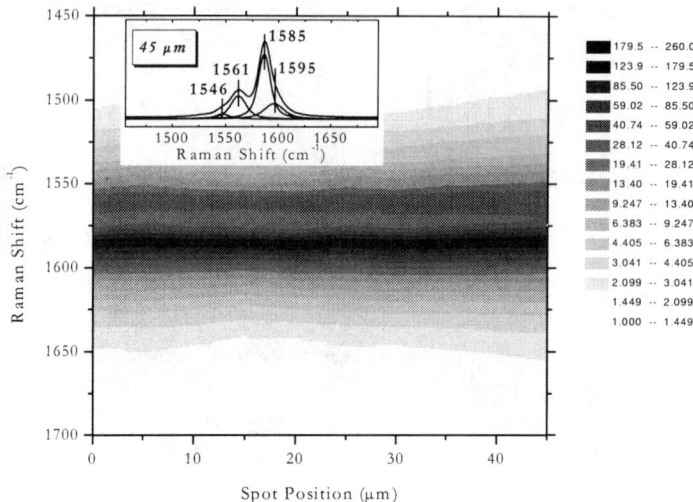

FIGURE 3. Map of the high-energy mode. The individual spectra were normalized with respect to the smaller mode at 1561 cm^{-1}.

In Fig. 4 a map of the same linescan as mentioned above is shown, but plotted for the low-frequency region. From the inset the closely arranged phonons of the low-energy mode (LEM) at 45 μm can be seen at 151 cm^{-1}, 165 cm^{-1}, 173 cm^{-1}, 178 cm^{-1}, and 184 cm^{-1}. Over the distance of 45 μm the intensities of the indiviual phonons change dramatically with respect to each other. At the spot position of 5 μm in Fig. 4 two maxima at 163 and 172 cm^{-1} were detected, but in the range of 15 and 25 μm only one maximum at 178 cm^{-1} can be seen. At 35 μm the maximum shifted to 175 cm^{-1}.

The observed changes are assumed to originate from the symmetry and diameter-dependent frequency dispersion of the nanotube vibrations. A comparison of Fig. 3 and 4 manifests the strong sensitivity of the LEM to symmetry and tube diameter. In consequence this mode yields sensitive information about the homogeneity of SWNT samples. The HEM is less suitable because of it's comparatively spot-independent signal.

Kasuya *et al.* showed that the phonon frequencies of the HEM are strongly correlated with the tube diameters [11], what is in good correspondence to our TEM and Raman data. Our samples should thus exhibit only a narrow diameter-distribution of 1.4±0.1 nm. Consequently, the structural variations responsible for the intensity changes in the LEM should be a result of different SWNT symmetries.

CONCLUSION

We have reported two infrared-active vibrations of SWNT at 874 cm^{-1} and 1598 cm^{-1}. These frequencies are higher than the corresponding ones of graphite

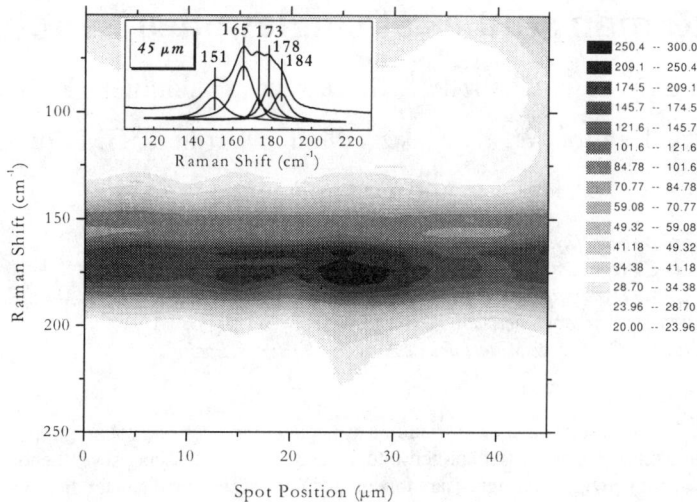

FIGURE 4. Map of the low-energy mode. The individual spectra were normalized with respect to the mode at 151 cm^{-1}.

by 6 and 8 cm^{-1}, respectively, and higher than those predicted by recent calculations.

We have presented spatially resolved Raman measurements of the LEM and HEM of SWNT and found that the LEM is much more sensitive to spatial variations than the HEM. Since the latter is known to have a diameter dependence, we presume that the observed amplitude variations of the LEM are predominantly related to variations in nanotube symmetry.

REFERENCES

1. J. W. G. Wildöer et al. Nature **391** (1998) 59.
2. A. M. Rao et al. Science **275** (1997) 187.
3. J. W. Mintmire et al. Carbon **33** (1995) 893.
4. R. A. Jishi et al. Chem. Phys. Lett. **209** (1993) 77.
5. J. Kastner et al. Chem. Phys. Lett. **221** (1994) 53.
6. B. Burger et al. in: *Molecular Nanostructures*. Proceedings of the International Winterschool on Electronic Properties of Novel Materials. Editors: H. Kuzmany, J. Fink, M. Mehring, S. Roth. World Scientific, Singapore 1997, p. 447.
7. M. Lamy de la Chapelle et al. ibid. p. 463.
8. H. Jantoljak et al. ibid. p. 459.
9. C. Journet et al. Nature **388** (1997) 756.
10. R. J. Nemanich et al. Sol. State Comm. **23** (1977) 117.
11. A. Kasuya et al. Phys. Rev. B **57** (1998) 1.

Raman studies of singlewalled nanotubes

M. Lamy de la Chapelle[a,b], S. Lefrant[a], C. Journet[c], P. Bernier[c], E. Munoz[d], A. Benito[d], W. K. Maser[d], M. T. Martinez[d], G. F. de la Fuente[e] and D. Laplaze[c].

a) Laboratoire de Physique Cristalline, IMN, Université de Nantes, France
b) Trinity College, Dublin, Ireland
c) Groupe de Dynamique des Phases Condensées, Université Montpellier 2, Montpellier, France
d) Instituto de Carboquimica CSIC, Zaragoza, Spain
e) Instituto de Ciencia de Materiales de Aragon, Zaragoza, Spain

Abstract. Raman results have been used in a systematic study of one collaret obtained by the electric arc discharge method and splitted into several parts. In studying several spots for each part, we can put in evidence a significant inhomogeneity. Indeed, the diameter distribution of the nanotubes can change drastically from one spot to the other which can be followed with the low frequency bands intensity. This could signify that the tube diameter depends on the location in the synthesis chamber. For comparison, we will present also studies of samples obtained by other methods of synthesis: laser ablation and solar energy. Information about the production process and the influence of some synthesis parameters will be presented.

Introduction

Extensive Raman studies (1, 2, 3, 4) have been performed recently on singlewalled nanotubes (SWNT's). Several features have been observed which provide a great deal of information on the tubes present in the studied samples. Then, these results allow us to use this technique as a fine characterization method at the micrometer scale. In this way, we have performed a systematic study on the synthesis products. Indeed, we present here a study of a collaret obtained by the electric arc discharge method. Also, a comparison is made with other synthesis processes : laser ablation and solar energy. In studying several spots for each case, we can put in evidence typical inhomogeneity which can be related with the influence of some synthesis parameters.

Synthesis

SWNT's were produced by creating an electric arc discharge between two graphite rods. The anode was drilled and filled with a mixture of nickel and yttrium used as catalysts. The synthesis was performed in a water-cooled chamber first evacuated and then filled with a static pressure of 660 mbar of helium. A current of 100 A was applied and a voltage of about 35 V was maintained constant by continuously translating the anode towards the cathode. Details on the synthesis can be found in Ref. 3. After the synthesis, the extracted product characterized is the "collaret" found around the deposit on the cathode, where the concentration of tubes is given to be near 70 %.

Raman experiments

All spectra have been recorded with a Jobin Yvon spectrophotometer T64000 in ambient air and at room temperature for the visible line (457.9, 514.5 and 676.4 nm) whereas the near infrared experiments (1064 nm) have been performed on a Brucker RFS100 FT Raman. In this case, we have used Surface Enhanced Raman Scattering (SERS) in depositing SWNT's, dispersed in a solvent (ethanol or CH3Cl), on a rough metallic surface (Ag, Au, Cu).

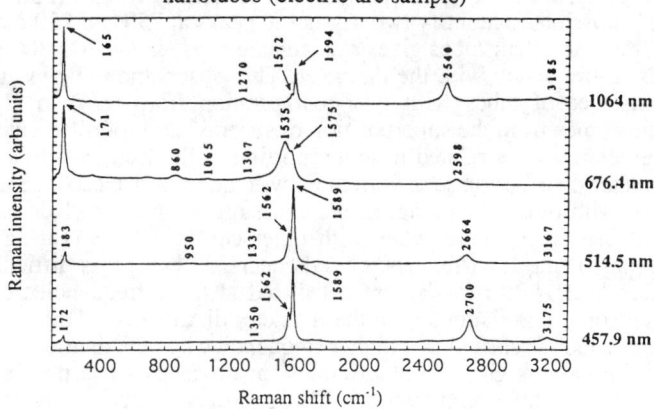

Fig 1 : Raman spectra of singlewalled carbon nanotubes (electric arc sample)

Fig 2 : Low frequency range (electric arc sample)

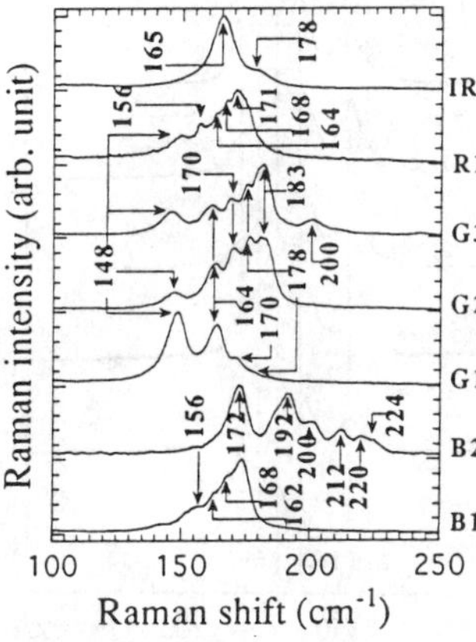

Fig 3 : Frequency range 250 - 1200 cm^{-1} (electric arc sample)

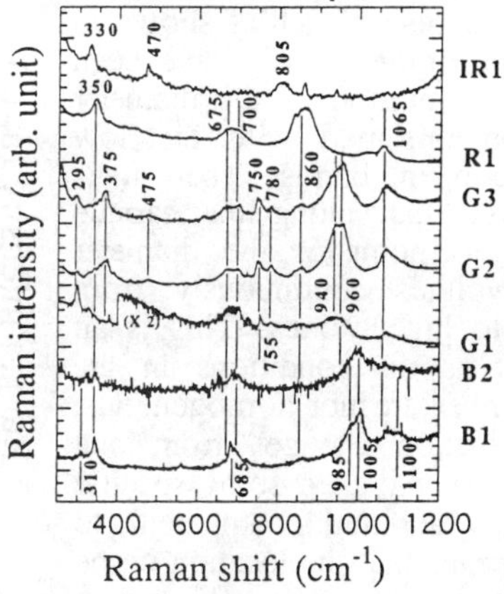

Each first order spectrum exhibits two main groups of peaks. The first one, which shows at least four components : 1589, 1562, 1550 and 1530 cm^{-1}, is the splitting of the E_{2g2} graphite mode due to the introduction of a curvature in a graphene sheet.

The second one, in the low frequency range (100-260 cm^{-1}), is related with the breathing mode whose frequency depends on the tube diameter (higher frequency, lower diameter) (1, 5). So one peak observed can be associated with one diameter. This is what is seen on the figure 2. Indeed, this figure presents different spectra recorded with different wavelengths, given by the letters on the rigth side (B : 457.9, G : 514.5, R : 676.4 and IR : 1064 nm), and different locations on the sample given by the number. It is clear that for each wavelength, the position of the peaks are identical whereas their relative intensities are changing from one location to the other. This means that each peak provides an indication of the concentration of one specific diameter relatively to the others and then the entire group gives information about the diameter distribution. In this way, we can see that the diameter distribution of the G1 spectrum is dominated by large tubes whereas this distribution is shifted to the lower diameters in the G3 spectrum.

Other features are also visible between these two frequency ranges (250-1200 cm^{-1}). In particular, it is possible to identify two asymetric peaks at 750 and 780 cm^{-1} on the G spectra. In fact, only armchair tubes gives a calculated mode in the 700-800 cm^{-1} range, which increases in frequency with the diameter. This observation allows us to put in evidence the existence of tubes with this configuration from (6/6) to (12/12). The asymetric profile comes from the superposition of several bands of different intensity as each diameter gives response related to its proportion in the location studied and then follows the diameter distribution seen with the low frequency group. Since this mode is no longer seen it with other wavelengths, we can suggest that armchair tubes are no more visible and that we observe tubes with other configurations (zigzag or chiral). Apart the overtones of the low frequency bands near 300 cm^{-1}, it is difficult to assign the other features because no modes are calculated at these frequencies even if it is possible to make some correlations with the diameter distribution. This is the case of the 950 cm^{-1} peak which is shifted to a higher frequency when the diameter decreases.

With all these informations given by the Raman spectroscopy, it is possible to make systematic studies on samples containing SWNT's. For all spectra shown below, the laser excitation wavelength is the argon line at 514.5 nm.

In a first step, we have performed experiments on one collaret. After having separated it in several parts, we have studied several spots on each as shown in the example of figure 4. We have been able to see an evolution of the diameter distribution on this part with the low frequency group of bands. Then it is clearly observed that along the sample from point 1 to point 6, the diameter distribution evolutes continuously from low diameter to higher ones. This is an indication that as the conditions in the synthesis chamber are not homogeneous, the tube production changes from one point to the other and then some specific diameters are favoured compared to others. This means that the location in the synthesis chamber has a great influence on the tube diameter.

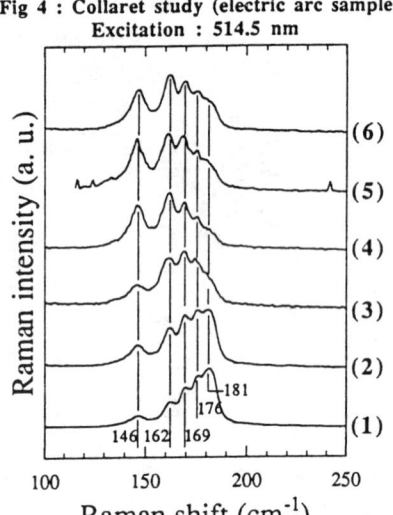

Fig 4 : Collaret study (electric arc sample)
Excitation : 514.5 nm

In a second step, we have studied samples produced by other methods, namely laser ablation and solar energy. As shown in figures 5 and 6, the two Raman spectra obtained for laser ablation and the upper one of the solar energy really look like similar to those obtained with the electric arc samples. We can always observe the E_{2g2} splitted mode, the same peaks at the same position at low frequencies or in the intermediate (250-1200 cm^{-1}) range.

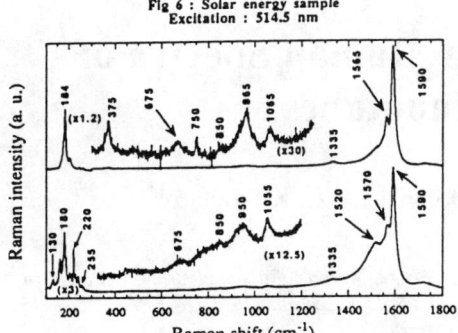

Fig 6 : Solar energy sample Excitation : 514.5 nm

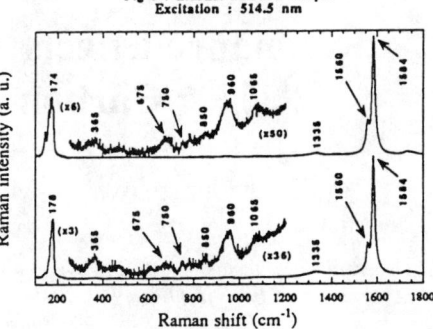

Fig 5 : LASER ablation sample Excitation : 514.5 nm

It seems that the tubes produced by the electric arc or laser ablation are identical. On the other hand, some other parts of the solar energy sample lead to the lower spectrum of fig. 6. This spectrum presents a broad peak at 1520 cm^{-1} and more than 10 peaks from 130 to 255 cm^{-1}. This corresponds to a very large diameter distribution shown by the low frequency bands and confirm by the broader splitting of the E_{2g2} graphite mode. In this case, the diameter is expected to go from a very low value near 8 A to 20 A. So, it seems that in this latter case, the synthesis conditions are not completely comparable to the two other processes, even if a large scale of SWNT's are also produced. From these onservations, it can be suggested that the sublimation parameters do not play a major role in the production of nanotubes. On the contrary, the plasma conditions after sublimation turn out to be of real importance.

Conclusion

We have presented Raman studies on SWNT's. This non destructive method is efficient to characterize these materials since we can get a diameter distribution and confirm the existence of "armchair" tubes or induced selective study by resonance effects. Therefore, it is possible to perform systematic studies. As an example, we are able to show that on the collaret, the diameter distribution changes significantly from one spot to another and then, that the location in the synthesis chamber is of a great importance in the tubes production. On the contrary, the sublimation process of graphite seem to show little influence, since similar spectra are onserved on SWNT's produced by different methods.

ACKNOLEDGEMENT

This work has been fully supported by the European Communauty through its Training and Mobility of Researcher program under network contract : NAMITECH, ERBFMRX-CT96-0067 (DG12-MIHT)

REFERENCES

1. Lamy de la Chapelle, M. et al, in *Proceedings of the SPIE conference*, 2854, 246-253, (1996)
2. Rao, A. M. et al, *Science*, 275, 187-191 (1997)
3. Journet, C., et al, *Nature*, 388, 756-758 (1997)
4. Lamy de le Chapelle, M. et al, *Carbon*, in press
5. Eklund, P. C., Holden, J. M., Jishi R. A., *Carbon*, 33/7, 959-972 (1992)

Resonance Effects in Raman Spectra of Carbon Nanotubes

E.D. Obraztsova[1], J.-M. Bonard[2], V.L. Kuznetsov[3]

[1] *General Physics Institute, RAS, 38 Vavilov street, 117942, Moscow, Russia, elobr@kapella.gpi.ru*
[2] *Departement de Physique, Ecole Polythechnique, Federale de Lausanne IPE-DP-EPFL, CH-1015 Lausanne, Switzerland*
[3] *Boreskov Institute of Catalysis, RAS, 5 Lavrentieva street, 630090, Novosibirsk, Russia*

Abstract. Simultaneous monitoring of the frequency position and the excitation-dependent behaviour of the dominating mode in a low-frequency part of the Raman spectrum of single-wall nanotubes grown by arc-discharge technique in helium atmosphere allowed to establish a presence of fractions with diameters 12.2Å, 12.5Å, 13.6Å, 13.7Å, 14.2Å in the material.

INTRODUCTION

The Raman scattering appeared to be among the techniques being able to measure the geometrical parameters of carbon nanostructures. Though this method doesn't possess a direct high spatial resolution (as HRTEM or STM), the Raman spectrum shape is strongly influenced by the atomic-scale variations of nano-geometry due to changing the selection rules for the scattering process. This allows to estimate correctly the average size of a nanostructure. Moreover in case of the single-wall carbon nanotubes (SWNT) the diameter-selective resonance enhancement of the Raman scattering intensity takes place. This is a result of a diameter-dependence of the electronic density of states (DOS) for SWNT [1-3]. An analysis of the resonances in the Raman spectra registered with a scanned excitation energy gives information about the nanotube distribution over the fractions with the different diameters.

In this work we varied the excitation energy in the range 2.4- 2.7 eV and measured the Raman spectra of SWNT produced by helium arc discharge. We tried to estimate the geometrical parameters of nanotubes and their distribution over diameter.

EXPERIMENTAL

The single-wall carbon nanotubes (SWNT) were produced by arc discharge under static pressure of helium (500 mBar) using pure graphitic electrodes: 20 mm (cathode) and 5 mm (anode) in diameter. A 3 mm diameter hole was drilled in the anode and filled with a *graphite-Ni-Y* mixture with weight proportion *2:1:1*. The

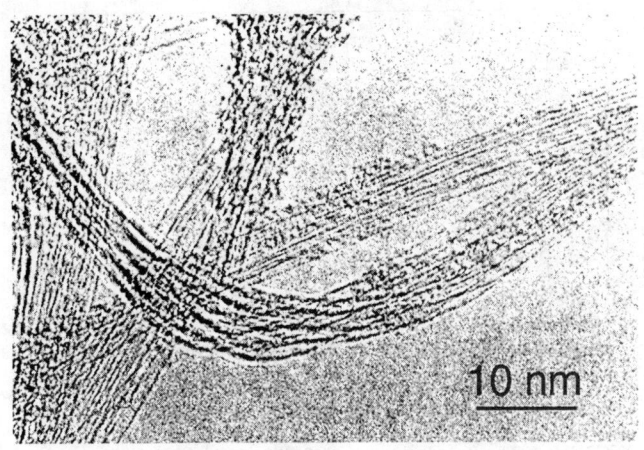

FIGURE 1. HRTEM image of SWNT grown by arc-discharge technique.

voltage and current used were 25 V and 100 A correspondingly. The nanotubes were predominantly found in the webs rather than in the cathodic deposit.

High resolution transmission electron microscopy (HRTEM) observations of the web material revealed a high density of SWNT arranged in ropes (Fig.1) which were sometimes covered by an amorphous carbon overlayer. The average tube diameter was around 13 Å.

The Raman measurements were performed for as-deposited (unpurified) SWNT material. A surface of a web-like deposit was flattened by compacting between two plane glasses. After this procedure a probe was placed under the objective of a micro-Raman spectrometer «Jobin-Yvon S-3000». The low power Ar^+-laser excitation was used to avoid SWNT modification in the laser spot. No visible changes have been observed. The excitation energy was varied in the range of the working wavelengths of Ar^+-laser (2.41-2.71 eV).

RESULTS AND DISCUSSION

The typical Raman spectrum of SWNT material is shown in Fig.2. For all excitation energies the spectra of SWNT show a splitting of the mode at 1582 cm^{-1} (being single in the spectrum of a planar graphite) into few components. Their positions and relative contributions are slightly dependent on excitation energy. The splitting is a result of the circular boundary conditions in nanotube leading to the "slicing" of the 2D-graphite Brillouin zone with a step proportional to the tube diameter [1]. As a result all "cut" points become equivalent to the Γ point ($\mathbf{k}$=0) of a Brillouin zone. The phonons with corresponding energies are allowed to participate in the Raman scattering process. The optical branch of the graphite dispersion curve is less steep than the acoustical one. Therefore the optical modes corresponding to the

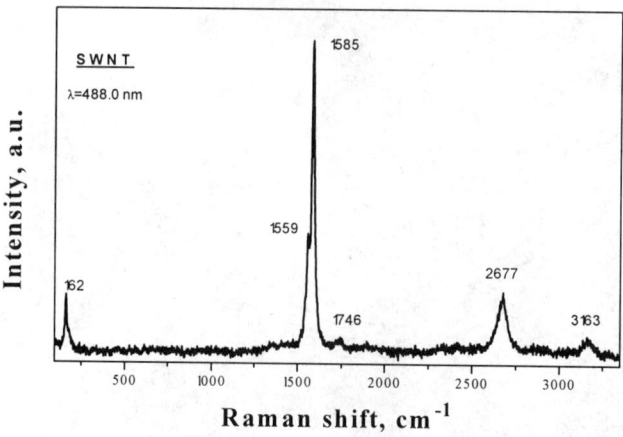

FIGURE 2. The Raman spectrum of SWNT in a wide spectral range.

nanotube fractions with different diameter easily overlap and form a broadened summary contour which is almost unchanged with the tube diameter variation. The acoustical modes appear to be more sensitive to the diameter value.

In our spectra the most dramatic excitation-induced changes occur in the spectral range 100-200 cm^{-1} (Fig.3). For each excitation energy the spectrum is dominated by the only mode: for the energy 2.7 eV - by the mode at 179 cm^{-1}, for 2.54 eV - by the mode at 162 cm^{-1}, for 2.4 eV - by the mode at 150 cm^{-1}. A frequency of the dominant (being resonant) mode decreases while the laser wavelength increases. A separation between the mirror spikes in the electronic density of states (DOS) of SWNT is smaller for the larger tube diameter. This correlates well with our observation. The laser energies used (2.4- 2.7 eV) are in the range of an optical transition between the second spikes in DOS for tubes with diameter 12-15 Å [1,2]. As the energy decreases a transition between the first spikes is also activated successively for all tube fractions beginning from the tubes of a minimal diameter.

The diameters estimated from the resonance conditions and from the mode positions coincide well. According to the calculations [4] the position of the strongest low-frequency Raman mode (so called "breathing" radial A$_g$ mode) shows no chirality dependence. Its frequency $\omega(r)$ is almost inversely proportional to nanotube radius r in the range 3Å < r < 7Å:

$$\omega(r) = \omega_{(10,10)} \cdot (r_{(10,10)} / r)^{1.0017 \pm 0.0007}, \qquad (1)$$

where $\omega_{(10,10)}$=165 cm^{-1}, $r_{(10,10)}$=6.785 Å.

In our SWNT material the fractions with diameters 12.5 Å (ω=179 cm^{-1}), 13.7 Å (ω=162 cm^{-1}) and 14.2 Å (ω=150 cm^{-1}) undergo the resonance coinciding with the "second spikes" transition. For the minimal energy used (2.41 eV) also the first-order resonance seems to be activated for the modes at 183 cm^{-1} and 165 cm^{-1}. They may be ascribed to **(9,9)** tubes (diameter 12.2 Å) and **(10,10)** tubes (diameter 13.6 Å) [2].

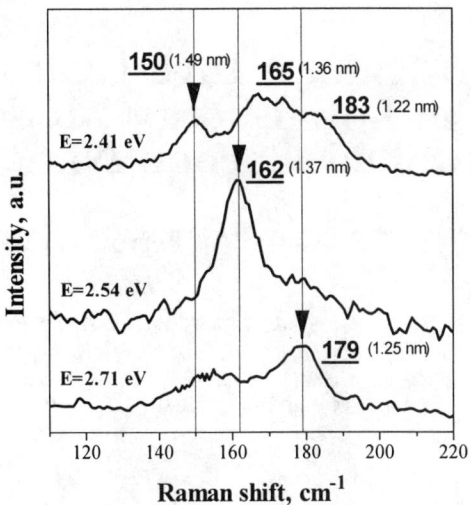

FIGURE 3. Dependence of the low-frequency part of the Raman spectra of SWNT on the laser excitation energy (E).

The resonance Raman technique seems to give more information about the diameter values of main fractions in SWNT material than about an exact tube geometry. But the diameter is the most important parameter determining the electronic DOS of a nanotube, and (as a result) a semiconducting or metallic type of its behaviour [1].

The Raman spectrum shape was the same for all probes measured confirming the homogeneity of our SWNT material. We suppose that it contains a limited number of nanotube fractions because all "breathing" modes are observed in a spectral range (150-185 cm^{-1}) which is much narrower than those (measured at the same energies) for the nanotubes produced by another techniques: 148-224 cm^{-1} [5], 142-225 cm^{-1} [6].

ACKNOWLEDGEMENTS

The work described was supported in part by Grant No. 97-02-17282 of the Russian Foundation for Fundamental Research and by CRDF grant No.RC1-251.

REFERENCES

1. Dresselhaus, M.S., Dresselhaus, G., Eklund, P.C., *Science of Fullerenes and Carbon Nanotubes*: Academic Press, New York, 1996.
2. Rao, A.M., Richter, E., Bandow, S., Chase, B., Eklund, P.C, et al., *Science* **275**, 187-191 (1997).
3. Kasuya, A., Sasaki Y., Saito, Y., Tohji, K., Nishina, Y., *Phys. Rev. Lett.*, **78**, 4434-4437 (1997).
4. Saito, R., Takeya, T., Kimura, T., Dresselhaus, G., Dresselhaus M.S., *Phys. Rev. Lett.* ,**57**, 4145-4153 (1998).
5. Lamy de la Chapelle, M., Lefrant, S., Journet, et al., in *Proc. of Euro-MRS'97, Carbon* (in press).
6. Kuzmany, H.,.Burger, B, in *Proc. of Euro-MRS'97, Carbon* (in press).

Micro-Raman Investigation of the Low-Energy Mode of Multiwalled Carbon Nanotubes

H. Jantoljak[1], J.P. Salvetat[2], L. Forró[2], C. Thomsen[1]

[1] *Institut für Festkörperphysik, Technische Universität Berlin,
Hardenbergstraße 36, D-10623 Berlin, Germany*
[2] *Institut de Génie Atomique, Ecole Polytechnique Fédérale,
CH-1015 Lausanne, Switzerland*

Abstract. We investigated multiwalled carbon nanotubes applying Micro-Raman spectroscopy. Well separated vibrational modes were found in the low-energy region at special positions of the sample; we conclude that only rare and unique structures in the sample provide this signal. We show that nanotube vibrations generate those phonons and we discuss different possibilities for their origin.

INTRODUCTION

Carbon nanotubes were established as a new allotrope form of carbon by Iijima *et al.* in 1991. [1] The strong correlation between the tubular structure and the physical properties opens up a large number of possibilities for unique applications and generated intense research activities on the structural and electronic properties of carbon nanotubes. [2]
The structural similarity between carbon nanotubes and graphite is responsible for the strong correspondence in the Raman spectra, especially in the high-energy region. Multiwalled nanotubes (MWNT) and graphite exhibit a single peak at 1580 cm^{-1}. [3] Singlewalled nanotubes (SWNT) reveal a fingerprint-like Raman mode at 1591 cm^{-1} with a typical finestructure [4] originating from symmetry and diameter-dependent dispersion as theoretical works predicted. [5] In the low-energy region the differences become more exalted. Typically, MWNT do not show any low-energy modes (LEM). The LEM in graphite is known to be at 42 cm^{-1}, [6] in SWNT the radial breathing mode is known to be at about 165 cm^{-1}, again revealing a finestructure. [4]
We found phonons in the low-energy region in MWNT, unique in shape and frequency, on special parts of the sample. The origin of these LEM will be discussed in comparison with related materials.

EXPERIMENTAL

The samples contain multiwalled carbon nanotubes produced by the arc-discharge technique. The raw soot was purified by a liquid-phase separation and a purification degree of 90 % was obtained. From transmission electron micrographs we estimated a tube length of around 200–5000 nm, a mean outer diameter of 10–20 nm, and predominantly 10–20 sheets in a tube; occasionally we observed up to 100 sheets. Details on the production and purification process were published elsewhere. [7]
For the Raman studies we employed a DILOR XY 800 triple grating spectrometer and an Ar^+/Kr^+-gas laser for excitation at 488 nm. Typically, a laser power density of 15 kW/cm^2 and a focus diameter of about 1 μm were realized applying a microscope optics. The spectra were collected in back-scattering geometry at room temperature.

RESULTS AND DISCUSSION

To obtain spatially resolved information we performed a series of Raman measurements. Using microscope optics we realized linescans with well defined steps on MWNT containing films. Examinations of the entire sample exhibited predominantly the characteristic Raman spectrum similar to graphite as mentioned above. On some special spots on the sample we observed vibrational modes in the low-energy region.
In Fig. 1 the low-energy region of seven spectra at different sample positions along a line in 1 μm steps is shown. On special points of the linescan, at 1 and 4 μm, well separated sharp peaks were observed. The signal was reproducible concerning the scattering intensity and where it occured in the linescan. Measurements over a few days revealed a stable signal. Peaks at the frequencies 166 cm^{-1}, 178 cm^{-1}, and 205 cm^{-1} were obtained with an FWHM (full width at half maximum) of 5 cm^{-1}. Additional measurements on different parts of the sample revealed LEM in the same frequency range of about 150–210 cm^{-1} in apparently random combination. Hence the peaks seem to have different origins. We assume them to originate from very specific structures. Graphitic particles were excluded since the E_{2g} phonon of graphite is known to be at 42 cm^{-1}. [6] The detected modes also do not correspond to the finestructured mode bunch at about 165 cm^{-1}, which is characteristic for SWNT. [4] We cannot, however, exclude that the LEM originate from individual SWNT or an SWNT bundle, even if it is unlikely that SWNT grew without catalysts present in the arc-discharge.
In Fig. 2 the Raman intensities of the linescan shown in Fig. 1 were translated into a greyscale. The spectra were plotted only in the frequency range of the LEM (145–250 cm^{-1}) as a function of the spot position in the linescan. From this map it can be seen that the phonons are well separated concerning the frequency and their position in the linescan. The LEM at 205 cm^{-1} reveals its maximum intensity at

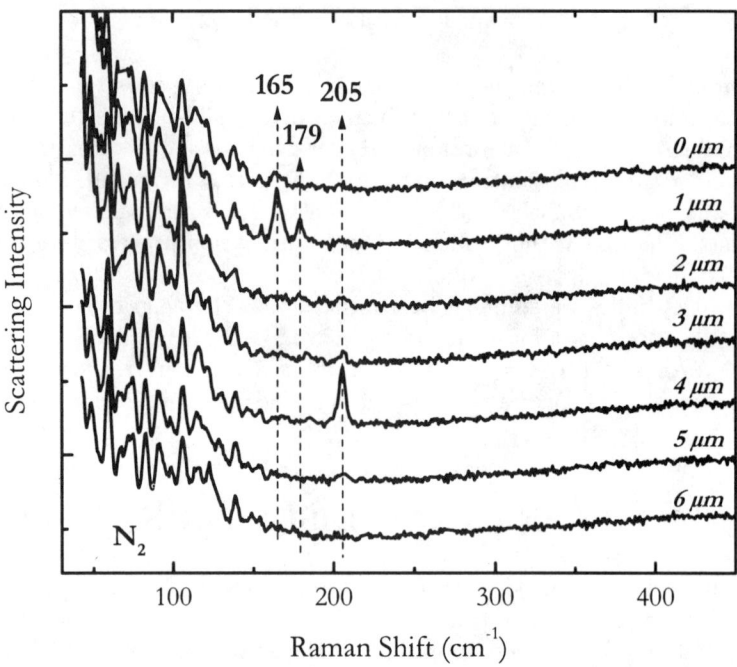

FIGURE 1. A linescan with 1 μm steps. On special spots well separated phonons of low intensity were detected. They were clearly distinguishable from the Raman signal of air (N$_2$). (The spectra are shifted for clarity.)

4 μm but remains down to 0 μm stable with low intensity. The LEM at 166 cm^{-1} and 178 cm^{-1} show their maximum intensity at 1 μm and do not appear again on other parts of the linescan.

This special arrangement in frequency and spot position, where the phonons are clearly distinguishable, is typical for MWNT. SWNT exhibit a completely different behavior in spatially resolved measurements as can be seen in Ref. [8]. The phonons are closely arranged in a finestructured mode bunch and remain present over the recorded linescan with slightly changing intensities. The frequency and spatial arrangement strengthens the assumption that the MWNT phonons in the low-energy region originate from very specific structures which are distributed over the sample. A possible explanation is that only MWNT with a core nanotube of small diameter of around 1–2 nm show a distinct LEM. According to Ref. [5] and [9] the Raman intensities should depend on diameter and symmetry; maybe only a few of the MWNT present in the sample reveal Raman intensities high enough to be detected. Furthermore we could assume a collective, radial vibration of the whole MWNT.

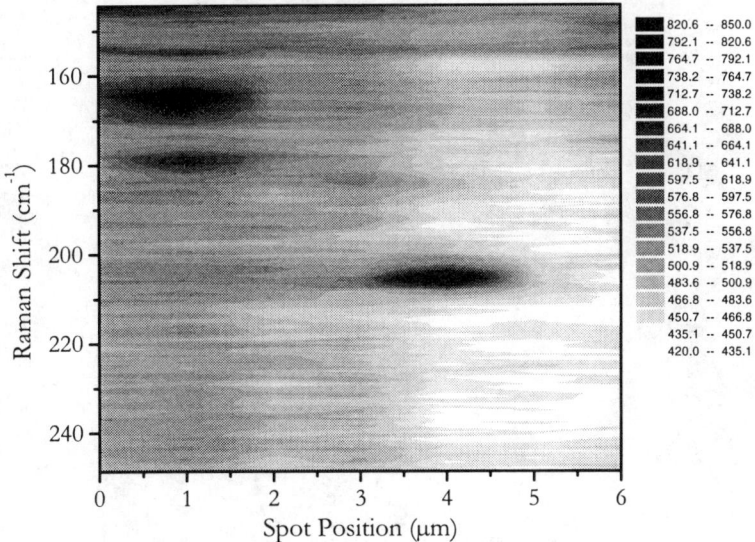

FIGURE 2. Map. The Raman intensities of the spectra in Fig. 1 are translated on a greyscale. The Raman Shift is plotted in dependence of the spot position in the linescan. The phonons (dark) are not only well separated in frequency but also in space.

CONCLUSIONS

We observed well separated vibrational modes in the low-energy region at special parts of the MWNT film. These modes seem to originate from very specific MWNT structures. Possible explanations for the origin of these phonons are vibrations of core tubes or collective vibrations. Further measurements on different MWNT samples are necessary to clarify those structures and the eigenvectors of those modes.

REFERENCES

1. S. Iijima. Nature **354** (1991) 56.
2. M. S. Dresselhaus et al. *Science of Fullerenes and Carbon Nanotubes*. (Academic Press, San Diego, 1996).
3. W. S. Bacsa et al. Phys. Rev. B **50** (1994) 15473.
4. A. M. Rao et al. Science **275** (1997) 187.
5. R. A. Jishi et al. Chem. Phys. Lett. **209** (1993) 77.
6. S.A. Solin. Physica **99** B (1980) 443.
7. J.-M. Bonard et al. Adv. Mat. **9** (1997) 827.
8. C. Thomsen et al. in: *Molecular Nanostructures*. Proc. IWEPNM. Edts: H. Kuzmany, J. Fink, M. Mehring, S. Roth. World Scientific, Singapore 1998, *this issue*.
9. R. Saito et al. Phys. Rev. B **57** (1998) 4145.

NANOTUBES: THEORY

Electronic Structure and Transport in Nanotube Ropes

C.L. Kane and E.J. Mele

Department of Physics, University of Pennsylvania Philadelphia, Pennsylvania 19104

Abstract. Electronic coupling between Carbon nanotubes gives rise to a suppression in the density of states at energies below a pseudogap energy scale, W. We consider the effect of disorder due to random wrapping vectors and orientations of nanotubes in a rope. By analyzing a model in which coupling between tubes is random but translationally invariant we find that: (1) compositional disorder (tubes with random chiralities) tends to reduce the pseudogap energy scale, and (2) due to effects of level repulsion, the dispersion of the energy bands near zero energy is nearly flat, leading to a very small velocity, dE/dk. This leads to a substantial suppression in the electrical conductivity when the Fermi energy is close to zero.

INTRODUCTION

The availability of high quality samples of single wall carbon nanotubes has stimulated great interest in their fundamental properties as well as potential practical applications [1,2]. In their native form, nanotubes are bundled together in "ropes" consisting of a regular triangular lattice of tubes. While there was initial optimism that the ropes were composed primarily of the metallic [10,10] nanotubes, it has become increasingly clear that the ropes contain tubes with a distribution of different diameters and chiralities, which can be controlled to a certain extent by varying growth conditions [3].

The electronic properties of an individual nanotube are determined by the wrapping vector characterizing the tube. In ropes, however, the electronic coupling between neighboring tubes also plays a key role [4,5]. Indeed, it has recently been pointed out that in crystalline ropes of metallic [10,10] nanotubes the intertube coupling gives rise to a suppression of the density of states near the Fermi energy. Electronic structure calculations predict a "pseudogap" of approximately $W \approx .08$ eV [5].

Though tubes pack into a triangular array, it is unlikely that the ropes are truly crystalline. In this article we argue that *disorder* in the ropes can play a key role in the rope's electronic structure. It is useful to distinguish three different categories of disorder. (1) Compositional disorder: Tubes in a rope have a distribution of chiralities and diameters. (2) Orientational disorder: Even if the rope (or a region within the rope) is composed of purely [10,10] tubes, it is unlikely that these tubes would be orientationally ordered. The nanotubes have a 10-fold rotational symmetry, which is frustrated in the 6-fold crystal environment. (3) Longitudinal disorder: Impurities and defects in lattice registry can give rise to scattering which can limit transport in a rope.

The first two classes of disorder are unique, in that the rope still possesses a translational symmetry. This is clearly the case for an orientationally disordered rope of [10,10] tubes. Moreover, in a continuum theory, a compositionally disordered rope has the same translational symmetry. Since nanotubes are believed to be quite rigid with respect to elastic deformations it is likely that registry between tubes is preserved over a substantial length. Therefore, it is sensible to explore the consequences of compositional and orientational disorder on translationally invariant ropes. The effects of longitudinal disorder, which ultimately will lead to localization, will be studied afterwards.

Translation symmetry allows the electronic eigenstates to be labeled by the conserved momentum k_z. A rope with N tubes will be characterized by N one dimensional bands. In a rope of [10,10] tubes without intertube coupling these bands are all degenerate and cross at the Fermi energy. The intertube coupling lifts this degeneracy, and leads to anticrossings in the bands near the Fermi energy, and a suppression in the density of states. The energy scale characterizing this pseudogap is related to the hopping matrix element t between neighboring tubes. In a hexagonal crystal with 6 neighbors, $W \approx 12t$. In the following, we argue (1) compositional disorder may substantially reduce the energy scale W describing the pseudogap and (2) disorder in the ropes leads level to repulsion which suppresses the *velocity* of the states at the Fermi energy which has important implications for transport.

COMPOSITIONAL DISORDER: PSEUDOGAP SUPPRESSION

Here we consider the effects of compositional disorder, in which a rope contains a distribution of tubes with different chiralities. For metallic "mod three" tubes, the electronic states at the Fermi energy are plane waves, with a wavelength determined by the underlying graphite band structure. While in a [10,10] tube, the momentum is aligned with the tube axis, in a chiral tube, the waves are tilted with respect to the axis, so that the component of the momentum along the axis is $k_F \cos\theta$, (k_F is the Fermi momentum in a [10,10] tube, and θ is the chiral angle). Since tunneling between neighboring tubes of different chirality connects states with the same momentum, this implies that for tubes of different chirality, the states coupled by tunneling are no longer degenerate. The difference in energy between the states which are coupled is given by $\Delta E = v_F k_F \Delta \cos\theta$. When $t << \Delta E$, the bandwidth $W \approx 12t$ follows from degenerate perturbation theory. However, when $\Delta E << t$, the bandwidth will be suppressed $W \approx t^2/\Delta E$ We estimate $\Delta E \approx 35\text{meV}$ for for tunneling between $\theta = 0$ and $\theta = 5°$. This energy will be larger for other combinations of tubes. This should be compared with the matrix element t connecting tubes $t \approx 10$ meV. We therefore conclude that compositional disorder leads to a suppression of the pseudogap, W.

LEVEL REPULSION

We now consider the effect of disorder on the energy levels in a rope. For simplicity we consider a rope without compositional disorder, consisting only of [10,10] tubes. Fig. 1 shows the one dimensional energy bands, obtained from a tight binding model, for (a) an orientationally ordered crystal and (b) and orientationally disordered rope. In both cases, the intertube coupling lifts the degeneracy of the bands and gives rise to anticrossings, where some of the bands do not intersect the Fermi energy, $E = 0$. This leads to the pseudogap in the density of states plotted in Fig. 2a. While the suppression of the density of states is quite similar for the orientationally ordered and disordered cases, there is an important difference in the band structures. In the ordered case, Fig. 1a, there are a number of level crossings which result from the symmetry of the crystal. The states at the Fermi energy have essentially the same velocity dE/dk as an isolated tube. By contrast, in the disordered rope, Fig. 1b, the energy levels avoid each other. This well known phenomena of level repulsion leads to nearly flat bands with small velocities due to the crowding of the energy levels near E_F.

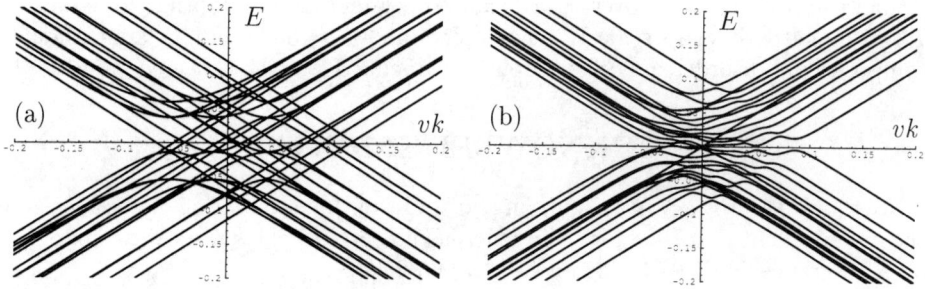

FIGURE 1. Energy levels as a function of momentum k along a rope for (a) a crystalline rope of [10,10] tubes and (b) an orientationally disordered rope of [10,10] tubes.

For a translationally invariant metallic system, conductivity is $\sigma(\omega) = A(E_F)\delta(\omega)$, where the weight of the Drude pole is ne^2/m (n and m are the electron density and mass) which is proportional to the density of states weighted by the velocity squared. In one dimension, $A(E_F) = \sum_{i,k}(e^2/\hbar)v_i(k)^2\delta(E_F - E_i(k))$, where i labels the bands bands intersecting the Fermi energy and $v_i = (1/\hbar)dE_i/dk$. In Fig. 2b we show the Drude weight A as a function of Fermi energy for the ordered and disordered cases. When the Fermi energy is near zero energy, the level repulsion and subsequent band flattening leads to a suppression in the weight.

(a) 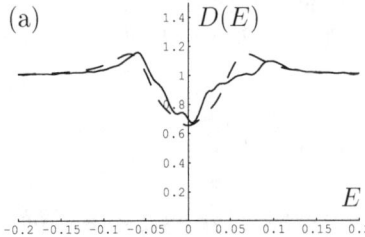 (b)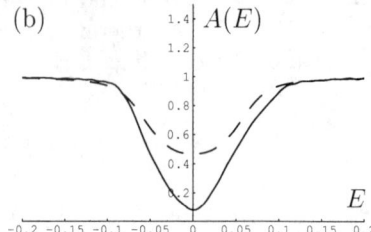

FIGURE 2. (a) Density of states and (b) weight of the Drude pole in the conductivity in a nanotube rope as a function of energy. The dashed lines are for an ordered crystal, and the solid lines is for an orientationally disordered rope. Both quantities are normalized by the energy independent D_0 and A_0 found in the absence of coupling between tubes.

Longitudinal disorder leads to a lifetime τ and broadens the delta function. Within Boltzmann transport theory, the DC conductivity is $\sigma = A(E_F)\tau$. One dimensional systems are always unstable to localization. By performing a standard weak localization analysis it is possible to estimate the localization length. We find that the localization length is strongly energy dependent for energies of order the pseudogap energy scale W. For a rope containing N tubes it varies from ℓ (the mean free path for scattering) when $E_F < W$ to $N\ell$ when $E_F > W$.

Nanotube ropes exhibit an resistivity upturn below a temperature $T^* \approx 50-200$K [6], and appear to cross over to a localized regime. It is tempting to associate T^* with the pseudogap scale W. For a crystal, $W \approx 800$K is too large, though compositional disorder may play a role in reducing T^* to the observed value.

ACKNOWLEDGMENTS

We thank J.E. Fisher, A.T. Johnson and W. Klaus for stimulating discussions and the organizers of IWEP 98 for their generous hospitality. This work was supported by NSF grants DMR 95-05425 and DMR 96-32598.

REFERENCES

1. S. Iijima and T. Ichihashi, *Nature* **363**, 603 (1993).
2. A. Thess *et al.*, *Science* **273**, 483 (1996).
3. T.W. Odom, *et al.*, *Nature* **391**, 62 (1998).
4. J.Charlier, X. Gonze and J. Michenaud, *Europhysics Letters* **29**, 43 (1995).
5. P. Delaney, *et al.*, *Nature* **391**, 466 (1998).
6. J.E. Fischer *et al.*, *Phys. Rev. B* **55**, R4921 (1997); C.L. Kane *et al. Europhysics Lett.*, in press.

Correlation effects in single-wall nanotubes

Reinhold Egger* and Alexander O. Gogolin[†]

Fakultät für Physik, Albert-Ludwigs-Universität, D-79104 Freiburg
[†]*Department of Mathematics, Imperial College, London SW7 2BZ, UK*

Abstract. The effective low-energy theory for single-wall nanotubes including the electron-electron interaction is discussed. Several experimentally relevant consequences of the correlations are described, e.g., the phase diagram, ferromagnetic tendencies, anomalous conductance laws, and a slow decay of the Friedel oscillation.

Single-wall nanotubes (SWNTs) are tubular objects which can be thought of as graphite sheets wrapped into a cylinder, with a radius R in the nm range. In a remarkable recent experiment [1], it has been demonstrated that *single* SWNTs can be manipulated and electrically contacted. Such a system is potentially very interesting because electron-electron interactions always lead to the breakdown of Fermi liquid theory in a one-dimensional (1D) metal. Interacting 1D electrons typically form a *Luttinger liquid* characterized by, e.g., the absence of Landau quasi-particles, spin-charge separation, suppression of the tunneling density of states (TDOS), and anomalous power laws [2]. Due to their structural flexibility and unique band structure, SWNTs are expected to be more stable 1D conductors than conventional systems such as long chain molecules, edge states in the fractional quantum Hall effect, or quantum wires in heterostructures. The nature of the non-Fermi-liquid state theoretically expected in a SWNT, and in particular to what extent the Luttinger liquid is realized, is the subject of our study. Here we omit detailed derivations (see Ref. [3]) and focus on experimentally relevant effects of the electron-electron interaction in metallic SWNTs.

The remarkable electronic properties of SWNTs are due to the special bandstructure of the π electrons in graphite. Wrapping a graphite sheet into a tube leads to the generic bandstructure of a metallic SWNT shown in Fig. 1. First of all, due to transverse momentum quantization, a 1D situation arises with all other bands more than 1 eV away from the Fermi surface. The typical 1D physics can thus be expected even for unusually high energy scales. Furthermore, there are only two linearly independent Fermi points $\alpha \vec{K}$ with $\alpha = \pm$ and $\vec{K} = (k_F, 0)$ [where $k_F = 4\pi/3a$ in an armchair SWNT, with a denoting the graphite lattice constant]. Since the basis of the honeycomb lattice contains two atoms, there are

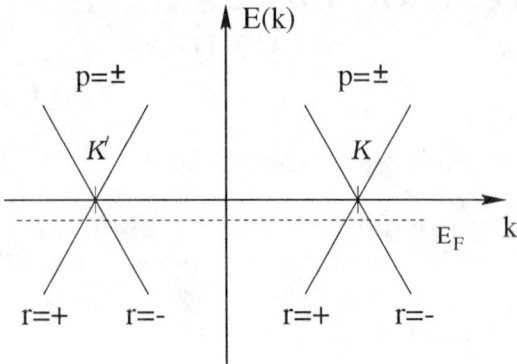

Figure 1: Bandstructure of a metallic SWNT. A right- and left-moving branch ($r = \pm$) is found near each of the two Fermi points $\alpha = \pm$ corresponding to K and K', respectively. Right- and left-movers arise as linear combinations of the sublattices $p = \pm$. The Fermi energy can be tuned by an external gate.

two sublattices $p = \pm$. As is evident from Fig. 1, the SWNT then forms a 1D quantum wire with only two bands intersecting the Fermi energy. To a very good approximation, we have a linear dispersion relation. Finally, by using a suitable gate, one can experimentally tune the average charge density on the SWNT and hence adjust the Fermi energy. In contrast to the situation in graphite, one is normally off half-filling in SWNTs, $E_F \neq 0$.

The construction of an effective low-energy theory is based on the low-energy expansion of the electron operator given in Ref. [4]. This expansion employs the Bloch functions to introduce slowly varying 1D fermion operators that depend only on the transport coordinate x. The effective low-energy Hamiltonian is then [3]

$$H = H_0 + H_{\alpha FS}^{(0)} + H_{\alpha FS}^{(1)} + H_{\alpha BS} , \qquad (1)$$

where H_0 is the kinetic energy. The interactions determine the other three terms. First, $H_{\alpha FS}^{(0)}$ is the direct forward scattering interaction. Second, $H_{\alpha FS}^{(1)}$ is due to the difference of inter- and intra-sublattice forward scattering interactions. This difference matters at short length scales due to the hard core of the Coulomb interaction. However, the corresponding coupling is very small, $f/a \approx 0.05\, e^2/R$. Finally, $H_{\alpha BS}$ is due to backscattering and described by the coupling constant $b/a \approx 0.1\, e^2/R$. The most important parameter is the *correlation strength* $K \leq 1$, which is determined solely by $H_{\alpha FS}^{(0)}$. Here a small value of K corresponds to strong correlations [2]. For a SWNT length $L = 3\mu$m, we estimate (v is the Fermi velocity)

$$K \simeq \left\{ 1 + \frac{8e^2}{\pi \hbar v} \ln(L/2\pi R) \right\}^{-1/2} \simeq 0.18 , \qquad (2)$$

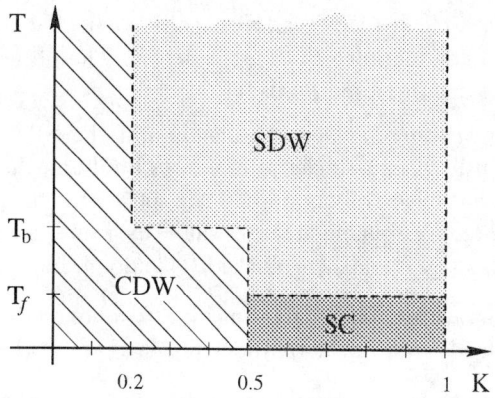

Figure 2: Phase diagram of a metallic SWNT in the $T - K$ plane.

implying very pronounced non-Fermi liquid effects.

The model (1) is equivalent to two spin-$\frac{1}{2}$ fermion chains coupled by the interactions, similar to the standard two-chain problem of coupled Luttinger liquids [5]. It is clear from the bosonized form of H [3] that for $f = b = 0$, the SWNT constitutes a realization of the Luttinger liquid. A finite b then leads to the gap $k_B T_b = D \exp[-\pi v/\sqrt{2}b]$ for charge excitations between the two chains, where $D \simeq 7.4$ eV is the bandwidth. Furthermore, a finite f leads to spin gaps characterized by $k_B T_f \simeq (f/b) k_B T_b$. As both T_b and T_f are well below the mK range, it is in practice justified to ignore these gaps. *A metallic SWNT is thus equivalent to a spin-$\frac{1}{2}$ Luttinger liquid with an additional flavor index α.*

Figure 1 shows the *phase diagram* of a SWNT in the $T - K$ plane. Since there can be no ordered state in 1D, we characterize a phase by the most slowly decaying correlations. We have studied charge-density wave (CDW), spin-density wave (SDW), and superconducting (SC) correlations. The interactions normally lead to CDW or SDW behavior. Only at extremely low temperatures ($T < T_f$) and externally screened interactions ($K > 1/2$), a singlet superconducting phase emerges. Superconductivity is therefore of little practical importance in SWNTs. We also have to distinguish the spatial oscillation period of the correlations. From Fig. 1 we see that the wavelengths $\lambda = \pi/k_F, \pi/q_F$, and $\pi/(k_F \pm q_F)$ could occur, where $q_F = E_F/v$ arises in the doped case. For instance, consider the SDW correlations which dominate for $K > 1/5$. One can now argue [3] that the slow wavelength is more important,

$$\langle s_z(x) s_z(0) \rangle \sim \cos[2q_F x] \, x^{-(3+K)/2} . \tag{3}$$

Since $\cos[2q_F x] \approx 1$ for distances x where the correlations are not already vanishingly small, Eq. (3) is of apparent *ferromagnetic* character. This reasoning might explain the observed ferromagnetic tendencies [1].

Let us now turn to *transport* through the SWNT. Since there are four transport channels, see Fig. 1, a clean SWNT adiabatically connected to leads exhibits the perfect two-terminal conductance $G = G_0 = 4e^2/h$. The observed conductance is then reduced by at least three mechanisms.

1. The *contact resistance* between the leads and the SWNT will decrease G. More importantly, it causes charging effects implying the presence of an energy scale E_c that effectively quantizes the spectrum. In Ref. [1], the charging energy $E_c \approx 2.6$ meV was much larger than the level spacing $\hbar v/L$. Transport at energy scales $\epsilon \leq E_c$ is then blocked ("Coulomb blockade").

2. At high temperatures, backscattering due to phonons or *twistons* becomes effective and leads to a linear temperature increase of the resistance [4].

3. At intermediate-to-low temperatures, *impurity backscattering* is most efficient and discussed below for the linear conductance $G(T)$. The voltage dependence of the nonlinear conductance follows the same power laws.

Disregarding resonant tunneling, the generic consequences of impurity backscattering can be studied for a single impurity of strength Λ at $x = 0$. The effective temperature scale T_i generated by the impurity is obtained from scaling arguments [6], $k_B T_i = D(\Lambda/D)^{4/(1-K)}$. For high temperatures $T \gg T_i$, the conductance corrections δG defined by $G = G_0 - \delta G$ can then be computed perturbatively. One finds $\delta G \sim (T/T_i)^{(K-1)/2}$ to lowest order in Λ^2. The full crossover to the low-temperature regime can be obtained for arbitrary K, and at temperatures $T \ll T_i$, we find $G(T) \sim (T/T_i)^{(1/K-1)/2}$. Since $K < 1$, this implies a total suppression of transport through the SWNT at $T = 0$ in the presence of a single, arbitrarily weak impurity [6]. In practice, one has to be aware that a crossover into a temperature-independent conductance occurs at $k_B T \simeq \hbar v/L$. Furthermore, if contact resistances are large, the conductance drops to zero at $k_B T \simeq E_c$. Thereby the above anomalous conductance laws are cut off. It is thus desirable to decrease the contact resistances in order to observe Luttinger liquid behavior over a wide temperature range. Finally, we have assumed here that the temperature stays large compared to the gap scales T_f and T_b. Otherwise the exponents are different [3].

We now address the TDOS which could be measured in *STM experiments*. Here electron tunneling occurs close to the impurity position, the role of the impurity being played by the STM tip itself. The presence of an impurity modifies the anomalous exponents [6], and the TDOS is $\rho_{\rm TDOS}(\epsilon) \sim \epsilon^{2\bar{\Delta}-1}$, which is different from the bulk DOS, $\rho_{\rm bulk}(\epsilon) \sim \epsilon^{2\Delta-1}$. Here the exponents are defined by the boundary and bulk scaling dimensions [2], $\bar{\Delta} = \frac{1}{8K} + \frac{3}{8}$ and $\Delta = \frac{1}{16}\left(\frac{1}{K} + K\right) + \frac{3}{8}$, respectively. Since $\bar{\Delta} > 1/2$ for $K < 1$, the TDOS always vanishes for $\epsilon \to 0$. Taking $K = 0.18$, the above exponents are 1.13 and 0.46, respectively. For a Fermi liquid ($K = 1$), we would instead have $\Delta = \bar{\Delta} = 1/2$, and both exponents are zero.

Finally, we discuss the *Friedel oscillation* building up around an impurity of strength Λ at $x = 0$, or at the end of the SWNT (which acts like a strong impurity, $\Lambda = D$). Following Ref. [7], the Friedel oscillation can be straightforwardly obtained. While for $x \gg \hbar v/k_B T$ the Friedel oscillation decays exponentially on the thermal length scale $x_T = \hbar v/k_B T$, the result for $a \ll x \ll x_T$ contains first a $2q_F$ oscillatory part [8],

$$\langle \rho(x) \rangle \sim \cos[2q_F x] (x/x_i)^{-(3+K)/4} \quad (x \gg x_i) \qquad (4)$$
$$\sim \cos[2q_F x] (x/x_i)^{-(1+K)/2} \quad (x \ll x_i) ,$$

and superimposed a rapidly oscillating contribution,

$$\langle \rho(x) \rangle \sim \cos[2(k_F \pm q_F)x] (x/x_i)^{-(3+K)/4} \quad (x \gg x_i) \qquad (5)$$
$$\sim \cos[2(k_F \pm q_F)x] (x/x_i)^{-(1+K)/2} \quad (x \ll x_i) ,$$

where $x_i = \hbar v/k_B T_i$ sets the length scale generated by the impurity. Here the important point is that the decay is always *slower* than the standard $1/x$ Fermi liquid result. Furthermore, it becomes even slower as the impurity is approached. Therefore one can expect very pronounced Friedel oscillations in a SWNT, and STM measurements of the screening cloud can potentially give access to the correlation strength K.

To conclude, we have predicted experimentally relevant correlation effects in SWNTs. Further manifestations of the electron-electron interaction can be expected in the shot noise and in frequency-dependent transport properties.

References

[1] S.J. Tans, M.H. Devoret, H. Dai, A. Thess, R.E. Smalley, L.J. Geerligs, and C. Dekker, Nature **386**, 474 (1997). See also S.J. Tans, M.H. Devoret, R.J.A. Groeneveld, and C. Dekker, preprint.

[2] A.O. Gogolin, A.A. Nersesyan, and A.M. Tsvelik, *Bosonization and Strongly Correlated Systems* (Cambridge University Press, 1998).

[3] R. Egger and A.O. Gogolin, Phys. Rev. Lett. **79**, 5082 (1997); European Phys. Journal B (in press).

[4] C.L. Kane and E.J. Mele, Phys. Rev. Lett. **78**, 1932 (1997).

[5] See, e.g., M. Fabrizio, Phys. Rev. B **48**, 15 838 (1993), and Ref. [2].

[6] C.L. Kane and M.P.A. Fisher, Phys. Rev. B **46**, 15 233 (1992).

[7] R. Egger and H. Grabert, Phys. Rev. Lett. **75**, 3505 (1995).

[8] For $K < 1/5$, the wavelength $\pi/4q_F$ is dominant [3].

NON CONVENTIONAL SCREENING OF THE COULOMB INTERACTION IN C_{60} AND IN CARBON NANOTUBES

J. van den Brink* and G.A. Sawatzky
*Laboratory of Applied and Solid State Physics, University of Groningen,
Nijenborgh 4, 9747 AG Groningen, The Netherlands*

We study the screening of the Coulomb interaction in C_{60} and carbon nanotubes. It is shown that for these systems the screening deviates strongly from the Clausius-Mossotti behavior. The short range interaction is strongly screened and the long range interaction is anti-screened, thereby strongly reducing the gradient of the Coulomb interaction. For the nanotubes a cross-over from quasi one dimensional to two dimensional screening is found with increasing radius of the nanotube. The unconventional screening explains the moderate effect of electron-electron correlations on the electronic structure of these molecules.

In model Hamiltonians set up to treat correlation effects the on-site Coulomb interaction is a very important parameter. It is usually argued that this interaction is only weakly screened relative to the longer range interactions, which justifies the neglect of the later [1]. In order to decide on the importance of the correlation effects precise knowledge of the longer range Coulomb interactions is crucial since of course it is the gradient of this interaction which really matters.

Consider a systems composed of polarizable atoms. If we assume a linear response of the atoms to an electric field, the induced dipole moment $\mathbf{p}(\mathbf{r}_i)$ on an atom on site $\mathbf{r}_i$ is proportional to the local field $\mathbf{F}(\mathbf{r}_i)$: $\mathbf{p}(\mathbf{r}_i) = \alpha_i \mathbf{F}(\mathbf{r}_i)$, where α is the atomic polarizability. The energy of the system is lowered due to the induced dipole moment by an amount of $\Delta E_i = -\frac{1}{2}\alpha_i |\mathbf{F}(\mathbf{r}_i)|^2$.

Using this, the screening energy of the Coulomb interaction between two charges can be obtained [2].

$$V(\mathbf{R}) = \frac{e^2}{R} - \frac{1}{2}\sum_i \mathbf{d}_i \cdot \mathbf{p}_i \equiv \frac{e^2}{R} - 2E_p(\mathbf{R}), \qquad (1)$$

where $\mathbf{d}_i$ is the total external monopole field, $\mathbf{p}_i$ the induced dipole moment at site i and $2E_p$ is the polarization energy. The mathematical description of the microscopic response of a system of point-dipoles to external fields was developed by Mott and Littleton [3]. The matrix representation of the equations for the induced dipoles is in Cartesian coordinates :

$$p_i^\mu = \alpha_i d_i^\mu + \sum_\gamma \sum_{j,j \neq i} M_{ij}^{\mu\gamma} p_j^\gamma, \text{ with } \mu,\gamma = x,y,z. \qquad (2)$$

The elements of the matrix that represents the dipole-dipole interactions are given by:

$$M_{ij}^{\mu\gamma} = \alpha_i(3l_{ij}^{\mu} l_{ij}^{\gamma} |l_{ij}|^{-5} - |l_{ij}|^{-3} \delta_{\mu\gamma}), \tag{3}$$

where $\mathbf{l}_{ij} = \mathbf{l}_i - \mathbf{l}_j$ is the vector connecting the two dipoles. The solution of this set of equations gives the exact effective potential for the electrons.

For a 3D system the equations above reduce to the well know Clausius-Mossotti (C.-M.) result when the summations are replaced by integrations:

$$\frac{\epsilon - 1}{\epsilon + 2} = 4\pi\alpha/3, \quad V(\mathbf{R}) = e^2/\epsilon R. \tag{4}$$

The situation is very different for a system where the dipoles and electrons are confined to a plane (2D system) or a line (1D system) [4]. As the system is an object in the 3D real space, the bare Coulomb interaction between the two electrons is inversely proportional to the distance between the charges. For a 1D system the polarization energy is calculated exactly and shown in figure 1. At very short distances (till about 2 lattice spacings) the Coulomb

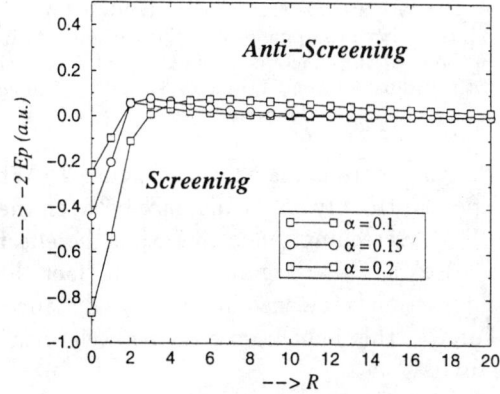

Figure 1: The polarization energy $-2E_p$ as a function of separation between the electrons in one dimension. Negative values of $-E_p$ represent screening and positive values anti-screening. Distances are in units of the lattice spacing

interaction is screened. When the separation between the charges is larger, however, the Coulomb interaction is *anti-screened*: the induced polarization results in an increased repulsion between the two charges. This behavior is markedly different from the C.-M. result.

For a C_{60} molecule we can model the screening by assuming that on each carbon site a dipole moment can be induced due to two excess charges on the molecule. The effective Coulomb repulsion between these two charges is calculated by the method described above.

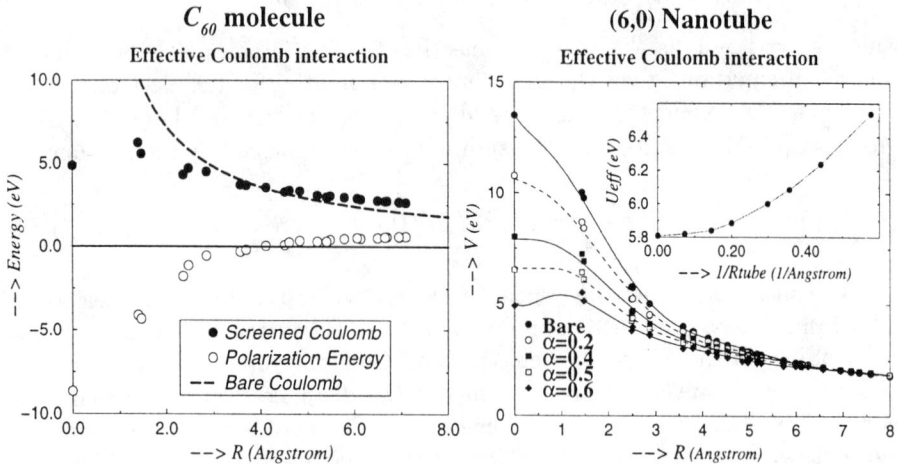

Figure 2: Left: effective Coulomb interaction as function of the separation of two electrons on a C_{60} molecule. The carbon-carbon distance is 1.391 & 1.455 Å and α=0.56 Å^3. Right: effective Coulomb interaction a (6,0) nanotube. The carbon polarizability α is varied between 0 and 0.6 Å^3. In the inset the effective on-site Coulomb interaction is plotted as a function of the inverse of the radius of the nanotube for α=0.5 Å^3. The lines are guides for the eye.

The results for the C_{60} molecule is shown in figure 2. Whereas the bare Coulomb repulsion depends strongly on the distance between the charges (ranging from ~13.5 to ~2 eV), screening tends to flatten the effective interaction (ranging from ~6 to ~3 eV). This is again due to the fact that at short distances the Coulomb interaction is screened and at large distances anti-screened, as in the 1D case. For C_{60} this behavior compares well with what has been found with other methods [5,6].

In figure 2 the effective Coulomb interaction for two electrons on a (6,0) nanotube is shown for different values of the carbon polarizability. The calculations clearly show that the long range part of the Coulomb interaction is independent of the polarizability and that only the interaction between electrons that are close by is efficiently screened. It is found that the screening depends mainly on the distance between the charges, and is independent of the position of the charges on the nanotube. In the inset is shown how the screened on-site Coulomb interaction depends on the radius of the nanotube R_{tube}. The limit $R_{tube} \to \infty$ corresponds to a (2D) honeycomb lattice. There is a smooth crossover from quasi-1D screening for nanotubes with a small diameter to 2D screening for tubes with a large radius.

Interesting also is that for realistic values of the polarization the on-site and nearest neighbor Coulomb interactions in these systems are almost equal, which really makes a Hubbard-like description very questionable.

We considered, in conclusion, a point-dipole model to account for the screening of the Coulomb repulsion in non-polar insulators. For three dimensional systems the deviations from the Clausius-Mossotti result for the intersite screened Coulomb interaction are small, but in one and two dimensional systems only local field effects contribute to the screening. At large distances the Coulomb interaction is unscreened and in 1D at intermediate distances the Coulomb interaction is even anti-screened. Applying the dipole screening model to finite size systems like the C_{60} molecule and carbon nanotubes, we showed that the effective Coulomb interaction is only weakly dependent on the distance between the electrons. So correlation effects are drastically reduced, explaining the success of (effective) one-particle theories for these large molecules.

Acknowledgments

This work was financially supported by the Nederlandse Stichting voor Fundamenteel Onderzoek der Materie (FOM) and the Stichting Scheikundig Onderzoek Nederland (SON), both financially supported by the Nederlandse Organisatie voor Wetenschappelijk Onderzoek (NWO).

References

[*] Present address: Max Planck Institut für Festkörperforschung, Heisenbergstr. 1, 70569 Stuttgart, Germany.
1. J. Hubbard, Proc. Roy. Soc. A **276**, 238 (1963)
2. J. van den Brink, M.B.J. Meinders, J. Lorenzana, R. Eder and G.A. Sawatzky, Phys. Rev. Lett., **75** 4658 (1995) and Phys. Rev. Lett., **76** 2826 (1996)
3. N.F. Mott and M.J. Littleton, Trans. Faraday Soc. 34, 485 (1938)
4. J. van den Brink and G.A. Sawatzky, preprint cond-mat/9802132
5. O. Gunnarsson and G. Zwicknagel, Phys. Rev. Lett. **69**, 957 (1992) and O. Gunnarsson, D. Rainer and G. Zwicknagel, Int. J. Mod. Phys. B **6** 3993 (1993)
6. R.W. Lof, M.A. van Veenendaal, H.T. Jonkman and G.A. Sawatzky, J. El. Spec. Rel. Phen. **72** 83 (1995)

Tight-Binding Calculation of the Elastic Properties of Single-Wall Nanotubes

E. Hernández[1], C. Goze[2], P. Bernier[2] A. Rubio[1]

(1) Departamento de Física Teórica
Universidad de Valladolid
47011 Valladolid, Spain
(2) GDPC CC026 Université Montpellier II
34095 Montpellier Cedex 05
France

Abstract. We report theoretical predictions of the energetic, structural and elastic properties of single wall nanotubes. We shall discuss estimations of the Poisson ratio and Young modulus for C, BN, BC$_3$ and BC$_2$N nanotubes. We have focused attention on both arm-chair (n,n) and zig-zag (n,0) nanotubes, with n in the range 5 to 20. For the particular case of C nanotubes we have also considered a number of chiral examples. A brief outline of the theoretical methods will be given, followed by a discussion of the results obtained, paying particular attention to the elastic properties.

INTRODUCTION

In this work we focus our attention on the elastic properties of carbon and B$_x$C$_y$N$_z$ single wall nanotubes. A more complete report of our results can be found in ref. [1]. Many potential applications of nanotubes depend critically on their mechanical properties, and therefore it is interesting to probe the stiffness of these materials along the axial direction as a function of their structure and composition. In particular, we have considered the following nanotube structures: For the calculations reported here we have used a non-orthogonal Tight-Binding [3] model derived from *first principles* calculations [4].

ELASTIC PROPERTIES OF SINGLE-WALL NANOTUBES

Although a more detailed characterisation of the elastic properties of nanotubes is possible [5], we focus our attention on the Poisson ratio (σ) and Young's modulus, especially the latter, since this is an experimentally measured quantity [6-8]. The Poisson ratio is given by

$$\frac{R - R_{eq}}{R_{eq}} = -\sigma\epsilon, \qquad (1)$$

$B_xC_yN_z$	(n,m)	D_{eq} (nm)	σ	Y_s (TPa·nm)	Y (TPa)
C	(10,0)	0.791	0.275	0.416	1.22
	(6,6)	0.820	0.247	0.415	1.22
	(10,5)	1.034	0.265	0.426	1.25
	(10,7)	1.165	0.266	0.422	1.24
	(10,10)	1.360	0.256	0.423	1.24
BN	(10,0)	0.811	0.232	0.284	0.837
	(6,6)	0.838	0.268	0.296	0.870
	(10,10)	1.390	0.263	0.306	0.901
BC_3	(10,0)	1.630	0.282	0.313	0.922
	(6,6)	1.694	0.279	0.315	0.925
BC_2N II	(7,0)	1.111	0.289	0.336	0.988
	(5,5)	1.370	0.287	0.343	1.008

Table 1: Structural and elastic properties of selected nanotubes obtained from the tight-binding calculations reported here. Young modulus values given in parenthesis were obtained from first-principles calculations. Also the value of Y with the convention $\delta R = 0.34$ nm is given for comparison.

where R is the tube radius at strain ϵ, R_{eq} is the equilibrium radius (i.e. $R(\epsilon = 0) = R_{eq}$), and σ is the Poisson ratio. The Young modulus in its conventional form is not a well defined quantity in the case of single-wall nanotubes, because it is not possible to unambiguously assign a thickness to a sheet one-atom thick [1]. For this reason we have adopted the following definition of the Young modulus, which avoids this difficulty:

$$Y_s = \frac{1}{S_{eq}} \left. \frac{\partial^2 E}{\partial \epsilon^2} \right|_{\epsilon=0}. \qquad (2)$$

Here S_{eq} is the surface area defined by the tube at equilibrium ($\epsilon = 0$) and E is the total Tigh-Binding [4] energy. In this definition Y_s has dimensions of *pressure·length*.

RESULTS AND DISCUSSION

In Table. (1) we have listed some values of Y_s obtained for seleted structures considered in this work. As can be seen from the data, there is a small dependence of Y_s on the tube diameter, the dependence becoming smaller as the tube diameter increases. In the limit of large tube diameters Y_s tends to a value close to that calculated for a flat graphitic sheet. Another conclusion that can be extracted from our results is that the carbon nanotubes are stiffer than any of the composite nanotubes. The BN nanotubes have the smallest Young modulus, but nevertheless remain considerably stiff. For comparison with the available experimental data [6-8] for multi wall nanotubes, it is useful to convert our data to the convention obtained by adopting a tube wall thickness of $\delta R = 0.34$ nm, the interlayer spacing in graphite. This was the convention adopted in a previous study of the elastic properties of carbon nanotubes [5]. When we do this, we obtain a value of Y around 1.25 TPa for carbon nanotubes with diameters in the range 1.3 nm, in good agreement with the experimental data [6,7]. For the BN nanotubes in the same range of diameters the value is around 0.9 TPa, somewhat smaller than the value measured experimentally [8], but still in reasonably good agreement.

ACKNOWLEDGEMENTS

We are grateful to G. Seifert, J.A. Alonso, M.A. López and M. Galtier for stimulating discussions. Financial support was provided by the EU through its Training and Mobility of Researchers Programme under contract ERBFMRX-CT96-0067 (D612-MITH). The use of computer facilities at C4 (Centre de Computació i Comunicacions de Catalunya) and CNUSC (Montpellier) is also acknowledged.

REFERENCES

1. E. Hernández, C. Goze, P. Bernier and A. Rubio, *Phys. Rev. Lett.* (accepted).

2. A.Y. Liu, R.M. Wentzcovitch and M.L. Cohen, *Phys. Rev. B* **39**. 1790 (1989).

3. C.M. Goringe, D.R. Bowler and E. Hernández, *Rep. Prog. Phys.* **60**, 1447 (1997).

4. D. Porezag *et al.*, *Phys. Rev. B* **51**, 12947 (1995); J. Widany *et al.* *Phys. Rev. B* **53** 4443 (1996).

5. J.P. Lu, *Phys. Rev. Lett.* **79**, 1297 (1997).

6. M.M.J. Treacy, T.W. Ebbesen and J.M. Gibson, *Nature* **381** 678 (1996).

7. E.W. Wong, P.E. Sheehan and C.M. Lieber, *Science* **277** 1971 (1997).

8. N.G. Chopra and A. Zettl, *Solid State Comm.* **105** 297 (1998).

Self-assembly and electronic structure of bundled single- and multi-wall nanotubes

David Tománek

Department of Physics and Astronomy and Center for Fundamental Materials Research, Michigan State University, East Lansing, Michigan 48824-1116, USA

Abstract. Detailed growth mechanism of single-wall nanotube bundles and of multi-wall nanotubes of carbon is investigated using *ab initio* and parametrized calculations. Our results show that single-wall tubes grow only in presence of a catalyst, whereas multi-wall tubes, stabilized at the growing edge by a covalent "lip-lip" interaction, may form in a pure carbon atmosphere. The individual tubes in these systems are likely to exhibit a low-frequency twisting motion. The weak, partly anisotropic inter-wall interaction, present in single-wall nanotube bundles and in multi-wall nanotubes, may cause significant changes in the density of states near the Fermi level.

Introduction

Carbon nanotubes, consisting of graphite layers wrapped into seamless cylinders, have been produced in the carbon arc and by laser vaporizing graphite [1–6]. Both single-wall and multi-wall systems have been observed that are up to a fraction of a millimeter long, yet only nanometers in diameter. Absence of defects and chemical inertness suggests that these molecular conductors should be ideal candidates for use as nano-wires.

The present study has been motivated by several open questions. First, the formation of single-wall nanotubes from lightly doped graphite and of multi-wall nanotubes from pure graphite is unusual in view of the fact that spherical finite-size isomers, or graphite and diamond in the bulk, are the most stable modifications of carbon. Also, the spherical counterparts of nanotubes, fullerenes such as C_{60}, have been shown to spin within the C_{60} solid, whereas spinning or twisting of individual tubes within a bundle or a multi-wall tube has not been discussed so far. Our particular interest in tube rotations results from the fact that both the spinning/twisting motion and the electronic structure of these systems depends strongly on the anisotropy of the inter-tube interaction. We suggest the onset of "orientational melting" at T^* as one possible explanation of the unusual temperature dependence of resistivity in nanotube bundles, which show a transition from non-metallic to metallic behavior when changing the sign of $d\rho/dT$ at $T^* \approx 50 - 100$ K.

These questions are addressed using a combination of *ab initio* and parametrized techniques. To obtain the total energy of nanotubes in perfect lattices or of fragments near the tube end, we use the Density Functional Formalism with a plane wave basis for the solid [7] and a numerical basis for the fragments [8]. Band structure details are studied using a parametrized Linear Combination of Atomic Orbitals (LCAO) formalism, with parameters described in Ref. [9]. The inter-tube interaction is described in analogy to the inter-ball interaction in the doped C_{60} solid [10]. Up to $102,400$ k-points in the irreducible Brillouin zone are used to determine the electronic structure of ordered nanotube lattices, the "ropes" [7].

Formation Mechanism

It is generally agreed that the most stable carbon isomers should be chains and rings for up to $N \approx 20 - 30$ atoms, and closed single-wall fullerenes for tens to hundreds of atoms [9,11–13]. Only beyond several hundred atoms should one observe a transition to spherical multi-layer "onions" [14,15]. Hence, reaction kinetics during the condensation of carbon atoms from the vapor is expected to play a crucial role in the formation of nanotubes.

Formation of Single-Wall Nanotube Ropes

The nucleus of a growing nanotube is a short cylinder that is capped on one side and open on the other side [5,8]. The system consists of carbon atoms forming hexagonal rings, six of which have been substituted by pentagonal rings. Substitution of more hexagons by pentagons (up to twelve) would gradually close the tube to a fullerene capsule; a system with less than six pentagons in the structure would resemble a cone with an increasing exposed edge. With a given number of atoms, the half-closed tubule may be either long and thin, thus minimizing the "dangling-bond" energy at the open edge, or short and fat, thus minimizing the strain energy in the cylinder. The equilibrium will lie in-between; as a matter of fact, the equilibrium diameter increases very slowly, only with the cube root of the number of atoms N. The tube diameter will eventually be frozen at a critical size estimated to be few hundred atoms under synthesis conditions; then, the rate associated with global structural rearrangements will no longer be competitive with the growth rate. For systems containing few hundred carbon atoms, the equilibrium diameter lies close to 1.4 nm, that of a (10,10) "armchair" tube [5].

Nanotube closure will necessarily be initiated by substituting pentagonal rings for hexagons at the growing edge. It has been shown in Ref. [8] that pentagon formation can be prevented catalytically by transitions metals atoms, such as Ni or Co, adsorbed at the growing edge. These metal atoms have the ability to enter a forming pentagon, with essentially no activation barrier, to form a metallacycle and thus to "masquerade" as carbon atoms. In the next step, the metal atom in the hexagonal ring is substituted by a carbon atom. This completes the formation of a hexagon at the defect-free edge of a nanotube.

Short tubes are likely to coalesce as they grow. The attractive inter-wall interaction will attempt to maximize the contact area between adjacent tubes, thus yielding a triangular lattice (a "rope") of nanotubes, observed in Ref. [5].

Formation of Multi-Wall Nanotubes

In absence of metal atoms at the growing edge, which would inhibit dome closure, a different mechanism has to be invoked to explain the growth of carbon nanotubes rather than their energetically favored spherical counterparts. In the first stage of the aggregation process, as in the case of single-wall nanotubes, we expect the formation of a graphitic hemisphere containing six pentagons. While this half-dome is being formed, a second "layer" may aggregate rapidly on this pre-formed graphitic template. This is expected especially if the carbon vapor is very dense, which is the case when focussed laser pulses are used [16].

It has been shown in Ref. [17] that the inter-wall gap at the edge of the double-dome is likely to be bridged by carbon atoms which establish a covalent "lip-lip" interaction, thus saturating eventual dangling bonds. Under this scenario, growth would occur by accretion of carbon at the stabilized edge. In presence of the strong "lip-lip" interactions, which maintain a constant inter-wall separation, eventual tube closure now would involve a concerted addition of pentagons in the inner and outer wall. This is unlikely to occur, and thus we expect the double-dome not to close, but rather to continue growing by adding hexagonal rings, as a double-wall tube. The efficiency of the "lip-lip" interaction mechanism to keep a double-wall nanotube from closing at high temperatures has been confirmed by *ab initio* Car-Parrinello calculations [18].

Electronic Structure of Single-Wall Nanotube "Ropes"

The interaction between adjacent nanotubes in a bundle or "rope" is, same as in the C_{60} solid, not completely isotropic about the tube axis. Individual (10,10) tubes are expected to librate about their axis with a relatively low frequency of $50 - 60$ cm^{-1}, close to the observed (but not identified) 41 cm^{-1} infrared-active mode [19]. Even though the activation barrier for free rotation is only $\lesssim 4$ meV *per atom*, individual tubes are not expected to rotate rigidly due to their high total mass. Since the rigidity of nanotubes is limited, finite segments are more likely to twist about the tube axis. The twisting motion is likely to be accompanied by displacement of orientational dislocations that have been frozen in during the assembly of the "ropes".

Inter-tube interactions in the ropes have been shown to modify the electronic structure of individual tubes by opening a pseudo-gap near the Fermi level [7,20,21].

With the onset of orientational melting, one would expect the pseudo-gap to smear out, thus significantly increasing the conductivity of the system.

Inter-tube coupling leads to an increase in the density of states by ≈7% outside the pseudo-gap. An even larger increase in the density of states, namely by a factor of ≈12 with respect to the pristine system, is predicted for the potassium doped system with the composition KC_8, in agreement with the observed conductivity increase by a factor of 10~20 [22].

Electronic Structure of Multi-Wall Nanotubes

Inter-wall interaction in a multi-wall nanotube may cause similar changes in the electronic structure near the Fermi level as the inter-tube interaction in "ropes" of nanotubes described above [23,24]. Our calculation for the (5,5)@(10,10) double-wall tube [25] suggests that due to inter-tube coupling, the density of states near E_F increases by ≈3%. The value of the rotational barrier *per atom* in this system is somewhat smaller than in nanotube "ropes", in agreement with results for the same system published in Ref. [26]. Off-axis displacements of ≲0.1 Å in multi-wall tubes cost essentially no energy. For the particular (5,5)@(10,10) double-tube, we expect librational modes to occur at ω_{in}≈31 cm^{-1} for the inner tube and ω_{out}≈11 cm^{-1} for the outer tube, depending on which of these tubes is pinned. As in the case of nanotube "ropes", we expect segments of individual tubes to exhibit a twisting motion rather than the entire tubes to rotate rigidly. We also expect an orientational melting transition to occur in multi-wall nanotubes, close to or below the temperature expected in single-wall nanotube "ropes".

Summary and Conclusions

Ab initio and parametrized calculations have been used to show that under specific conditions, carbon atoms condense to form nanotubes. Whereas single-wall tubes form only in presence of an atomically dispersed transition metal catalyst, it is the covalent "lip-lip" interaction at the growing edge that prevents multi-wall tubes from closure even in absence of such a catalyst. The attractive inter-wall interaction stabilizes not only multi-wall nanotubes, but causes also single-wall tubes to pack into ordered "ropes". This interaction induces additional band broadening by ≈0.2 eV, and opens up a pseudo-gap at E_F in the "ropes". Due to their large inertia, individual nanotubes do not rotate as a whole. Finite tube segments are rather expected to exert a local twisting motion. Orientational dislocations, which were frozen in during the formation of the "ropes", lower the activation barrier for tube rotations and hence the orientational melting temperature of the "ropes".

The author gratefully acknowledges financial support by the organizers of the "XIIth International Winter School on Electronic Properties of Novel Materials: Molecular Nanostructures" (IWEPNM-98) in Kirchberg (Austria), February 28 - March 3, 1998. Contributions of Young-Kyun Kwon, Young Hee Lee, Seong Gon

Kim, Philippe Jund, Richard E. Smalley, Kee Hag Lee, and Susumu Saito to this work are gratefully acknowledged.

REFERENCES

1. S. Iijima, Nature **354**, 56 (1991).
2. S. Iijima and T. Ichihashi, Nature **363**, 603 (1993).
3. D.S. Bethune, C.H. Kiang, M.S. de Vries, G. Gorman, R. Savoy, J. Vazquez, R. Beyers, Nature **363**, 605 (1993).
4. R.E. Smalley, Mat. Sci. Eng. **B 19**, 1 (1993).
5. A. Thess, R. Lee, P. Nikolaev, H. Dai, P. Petit, J. Robert, C. Xu, Y. H. Lee, S. G. Kim, D. T. Colbert, G. Scuseria, D. Tománek, J. E. Fisher, and R. E. Smalley, Science **273**, 483 (1996).
6. M.S. Dresselhaus, G. Dresselhaus, and P.C. Eklund, (Academic Press, San Diego, 1996).
7. Young-Kyun Kwon, Susumu Saito, and David Tománek (submitted for publication).
8. Young Hee Lee, Seong Gon Kim, and David Tománek, Phys. Rev. Lett. **78**, 2393 (1997).
9. D. Tománek and Michael A. Schlüter, Phys. Rev. Lett. **67**, 2331 (1991).
10. M. Schlüter, M. Lannoo, M. Needels, G.A. Baraff, and D. Tománek, Phys. Rev. Lett. **68**, 526 (1992).
11. K.S. Pitzer and E. Clementi, J. Am. Chem. Soc. **81**, 4477 (1961).
12. G. von Helden, N. Gotts, and M.T. Bowers, Nature **363**, 60 (1993).
13. J.M. Hunter, J.L. Fye, E.J. Roskamp, M.F. Jarrold, J. Phys. Chem. **98**, 1810 (1994).
14. D. Ugarte, Nature **359**, 707 (1992); Europhys. Lett. **22**, 45 (1993).
15. D. Tománek, W. Zhong, and E. Krastev, Phys. Rev. B **48**, 15461 (1993).
16. Ting Guo, Pavel Nikolaev, Andrew G. Rinzler, David Tománek, Daniel T. Colbert, and Richard E. Smalley, J. Phys. Chem. **99**, 10694 (1995).
17. Young-Kyun Kwon, Young Hee Lee, Seong-Gon Kim, Philippe Jund, David Tománek, and Richard E. Smalley, Phys. Rev. Lett. **79**, 2065 (1997).
18. J.-C. Charlier, Alessandro De Vita, Xavier Blase, and Roberto Car, Science **275**, 646 (1997).
19. W. Holmes, J. Hone, R. Mallozzi, J. Orenstein, P.L. Richards, and A. Zettl (private communication).
20. J.-C. Charlier, X. Gonze, and J.-P. Michenaud, Europhys. Lett. **29**, 43 (1995).
21. P. Delaney, H.J. Choi, J. Ihm, S.G. Louie, and M.L. Cohen, Nature **391**, 466 (1998).
22. R.S. Lee, H.J. Kim, J.E. Fischer, A. Thess, and R.E. Smalley, Nature **388**, 255 (1997).
23. Ph. Lambin, L. Philippe, J.C. Charlier, and J.P. Michenaud, Comp. Mat. Sci. **2**, 350 (1994).
24. Ph. Lambin, J.C. Charlier, and J.P. Michenaud, in *"Electronic Properties of Novel Materials"*, edited by H. Kuzmany, J. Fink, M. Mehring and S. Roth (World Scientific, Singapore, 1994), p. 130.
25. Young-Kyun Kwon and David Tománek (unpublished).
26. J.C. Charlier and J.P. Michenaud, Phys. Rev. Lett. **70**, 1858 (1993).

Simulation of STM images of 3D objects and comparison with experimental data: carbon nanotubes

Géza I. Márk, László P. Biró, and József Gyulai

KFKI-Research Institute for Technical Physics and Materials Science, H-1525 Budapest, P.O.Box 49, Hungary
E-mail: mark@sunserv.kfki.hu, URL: http://www.phy.bme.hu/mg/index.html

Abstract. Tunneling through a nanotube is a much more complex phenomenon than STM imaging of an atomically flat surface. Besides geometric convolution effects, and resonant tunneling through the two tunneling gaps: STM tip-nanotube, and nanotube-substrate, differences in electronic properties of the nanotube and of the support play a role. We used wave packet dynamical calculation of tunnel current density in the STM tip-nanotube-support system in order to separate the distortion in the STM image formation process in pure geometric and electronic effects. Simulated line cuts for the case of a nanotube on supports with similar and different electronic structures are coincident with experimental data.

INTRODUCTION

There are some differences in STM imaging of a three dimensional object "floating" over the surface of the support as compared with STM applied on flat surfaces. Convolution effects arise from the geometry at the very end of STM tip. This will produce an apparent broadening [1] of the tube. Existence of two tunneling gaps and differences in the electronic structure of the nanotube and that of the support have significant effect on tunneling current.

EXPERIMENT

Carbon nanotube STM samples were prepared by the catalytic decomposition procedure as reported earlier [2,3,1]. Constant current STM images were acquired using tunneling currents in the range of $0.2\,nA$ and bias voltages in the range of $0.4\,V$. In *Fig. 1 a, and b* the STM image and the corresponding line cut are shown for a carbon nanotube placed over other carbon nanotubes stacked in a regular way, i. e. on a "raft". An apparent broadening, (defined as $B = 2HW/h$), by a factor of 2.18 is present in the image. In *Fig. 1 c, and d* we show an STM image acquired over a nanotube on HOPG. Apparent broadening is $B = 3.24$.

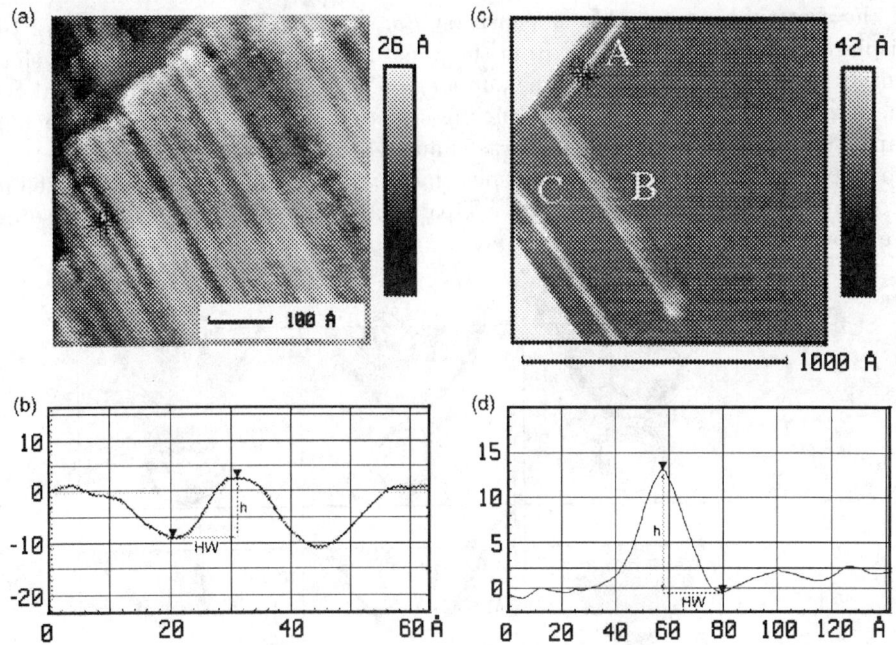

FIGURE 1. *a)* Carbon nanotube placed on a "raft"; *b)* line cut marked in the image taken through a tube well separated from other tubes. *c)* Nanotubes on graphite; *d)* line cut marked in the image taken through object A.

GEOMETRIC AND QUANTUM LINE CUT

As a first approximation, we have calculated the image distortions caused by finite tip size using a simple geometric method. The support is assumed to be a flat surface. The tube is modelled by a cylinder of $0.5\,nm$ radius floating above the support at a distance of $0.335\,nm$. STM tip is approached by a rotational hyperboloid of $0.5\,nm$ apex radius and $15\,deg$ aperture angle. Effective surface of the electrodes is $0.071\,nm$ outside the geometric surface. Geometric line cut is shown in *Fig. 3 (left)*. It is defined as the line drawn by the tip apex point, when the distance of the nearest points of the sample and tip effective surfaces remains constant.

We have performed quantum mechanical probability current calculations through a simple two dimensional potential modelling the STM tip – nanotube – support system. The potential is *zero* outside the effective surfaces of the electrodes, and

$-9.81\,eV$ inside. ($E_F = 5\,eV$, $W = 4.81\,eV$). Tunneling probability is calculated from time dependent scattering of a wave packet approaching from the bulk of the tip. Details of calculation are given elsewhere [5,6]. The probability density of the scattered wave packet is shown on *Fig. 2*. For $X_{apex} = 2\,nm$ lateral tip displacement the tip is still far from the nanotube. The wave packet is tunneling simply from the tip apex into the support. For $X_{apex} = 0\,nm$ the tip is above the nanotube. It is a resonant tunneling situation because of the two tunneling gaps. STM constant current loop was simulated by finding for each fixed lateral tip displacement that vertical tip displacement which yielded a setpoint tunneling probability value of $P_{setpoint} = 3 \cdot 10^{-3}$. Tip displacement values from this procedure are shown on *Fig. 3 (left)*.

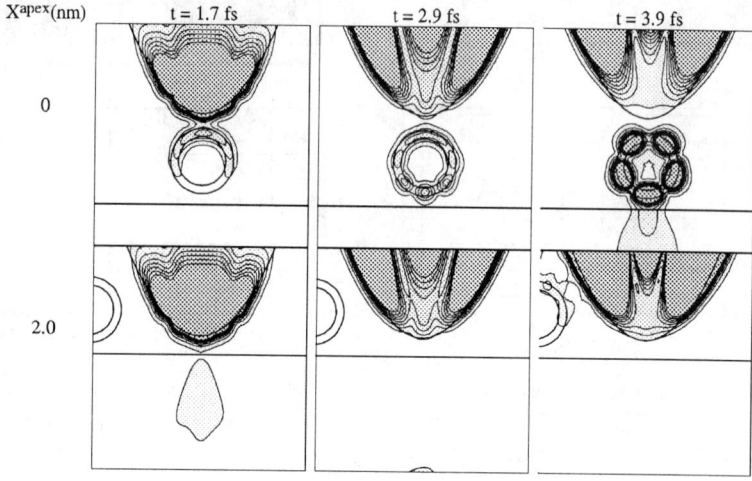

FIGURE 2. Probability density of scattered wave packet. Effective surfaces of electrodes are shown by thick lines. Size of presentation window is $3.84\,nm$. Contour lines are drawn on sqare root scale. Each frame is normalized to its maximum density.

DISCUSSION AND CONCLUSIONS

When the nanotube is placed on a support with similar electronic structure (on top of the raft, (*Fig. 1 a*), the geometric line cut does not differ significantly from the quantum line cut (Cf. Fig. 3 (left)). Major distortion that influences the apparent tube diameter is geometric convolution of the tip with the tube. When the tube is on a support with different electronic properties, the simplification used in our calculation: E_F and W in the nanotube and in the support is identical, is not valid. In case of geometric line cut this can be taken in account by increasing the value of the tunneling gap over the support as compared to the value over the nanotube. Ratio of HW to h versus increase of tunneling gap over the support is shown in *Fig. 3 (right)*. Comparing the case of the nanotube over the raft, i.e. identical electronic structure, with *Fig. 3 (right)*, one may conclude that in the experimental case the distortion agrees with the value corresponding to zero tunnel gap increase.

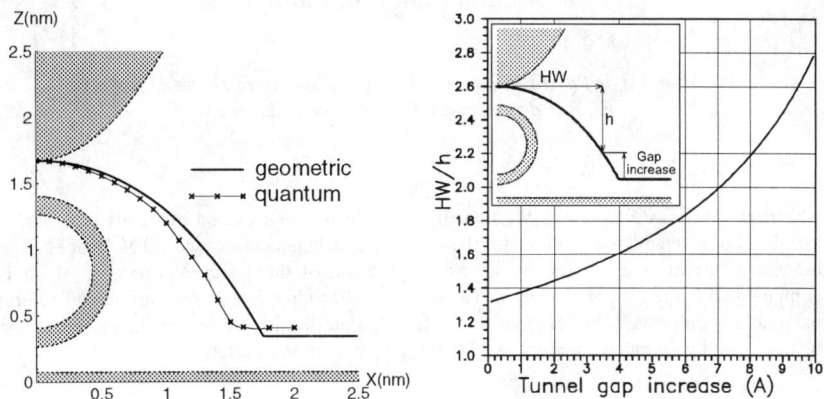

FIGURE 3. *left)* Thick solid line is geometric line cut. Crosses connected by thin solid line show calculated points of quantum line cut. *right)* HW/h versus increase of tunneling gap above the support.

Conclusions: if the electronic structure of the nanotube and support is similar the major distortion arises from geometric convolution of the shape of the tip with the tube geometry. When the electronic structure of the support is different from that of the nanotube, further distortions arise from the modification of the width of tunneling gap over the nanotube if compared to the value over the support. The second tunneling gap may introduce complications in interpretation of STS data.

Acknowledgments: helpful discussion with Prof. Ph. Lambin of FUNDP, Namur, are gratefully acknowledged. The work was partly supported by AKP grant No. 96/2-637.

REFERENCES

1. L. P. Biró et al., *Phys. Rev. B* **56**, 12490 (1997).
2. A. Fonseca et al., *J. Mol. Catal.* **107**, 159 (1996).
3. K. Hernadi et al., *Zeolites* **17**, 416 (1996).
4. L. P. Biró et al., *Scanning tunneling microscopy (STM) imaging of carbon nanotubes*, (in press Carbon). http://www.phy.bme.hu/pub/emrs97/index.html
5. G. I. Márk, L. P. Biró, and J. Gyulai, to be published.
6. http://www.phy.bme.hu/education/schrd/index.html

Electronic Processes in Scanning Tunneling Microscopy of Carbon Nanotubes

V. Meunier and Ph. Lambin

Département de Physique, Facultés Universitaires Notre-Dame de la Paix
61, Rue de Bruxelles, B-5000 Namur, Belgium

Abstract. An LCAO theory of the tunneling current between carbon materials (sp^2) and a metallic tip is developed to simulate their differential conductance and STM images. The tunneling current is expressed by a series expansion of the resolvent operator which is computed with a recursion algorithm. The differential conductance of both chiral and achiral nanotubes is calculated. We emphasize the fact that electronic and geometrical effects must be interpreted with much attention in STM imaging of carbon nanotubes.

INTRODUCTION

Since their discovery, carbon nanotubes (CN's) have attracted great attention. On the experimental point of view, much effort has been paid to synthesis[1] and characterisation. Simultaneously, theoretical works have predicted interesting properties.[2] In particular, CN's can be either metallic or semiconductor depending on their diameter and chirality. Recent experimental confirmations of these properties have been given by Scanning Tunneling Microscopy (STM) and Scanning Tunneling Spectroscopy (STS).[3] This paper aims at understanding the electronic processes involved in STM and STS experiments. The first part of this paper is a theoretical study of the tunneling current between a tip and a carbon molecule. This method is used in the second part to simulate both tunneling conductance and STM images. A discussion of our results is given in the last part. In particular, we show that STM images of CN's must be interpreted with much attention.

MODEL AND METHOD

Our method is based on a tight-binding model (TBM). The TBM has the advantage of still being useful in cases where Bloch theorem is not valid. Indeed, the description of electronic wave functions in this formalism enables us to work in the direct space. The wave functions are expressed by a Linear Combinations of Atomic Orbitals (LCAO). We have used a π-orbital approximation for the nanotube in which curvature effects have been incorporated in the hopping parameters (fitted to the band structure of graphite).

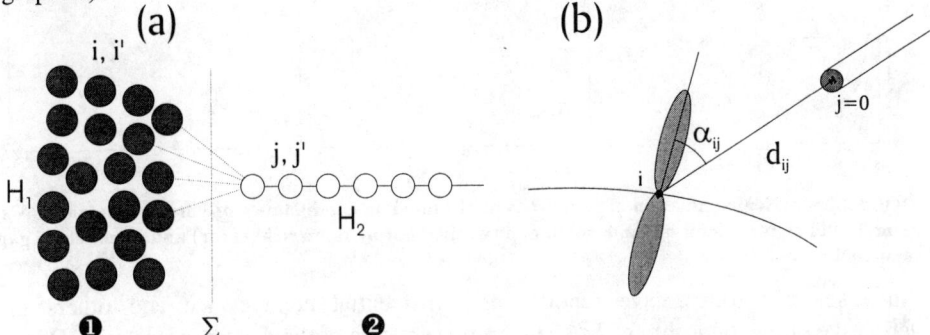

Figure 1 (a) Schematic representation of the interaction between the substrate (1) and the tip (2). H_1 and H_2 are the Hamiltonian of the unperturbed substrate and for the tip respectively. (b) Interaction between a s-type orbital of the tip and a π-orbital of the substrate

A Bardeen[4] like formulation of the tunneling current in presence of a bias voltage between a STM tip (2) and a substrate (1) can be derived as follows,

$$I = -\frac{2e}{\hbar} \int_{E_F}^{E_F+eV_{bias}} dE \sum_{i,i' \in 1} \sum_{j,j' \in 2} V_{j'i} R^{(1)}_{ii'}(E) V_{i'j} R^{(2)}_{jj'}(E - eV_{bias})$$

where $R^{(1)}_{ii'}$ and $R^{(2)}_{jj'}$ are the imaginary parts of the matrix elements of the Green function for the nanotube and for the the tip respectively.[5] The interaction V_{ij} between sites i (nanotube) and j (tip) is described in the following paragraph.

In this preliminary model (see Figure 1(b)), the tip was treated as a linear chain of atoms with s-orbitals and was assumed to be terminated with a single atom (denoted as $j=0$). Even if this approximation is inadequate in most instances, it allows us to explore geometrical and electronic effects that may give spurious effects in the imaging of the nanotube.

From this simple tip model, the parameters V_{ij} are restricted to V_{i0}. Within the Slater-Koster theory, the interaction between an s- and a π-orbital depends on the angle α_{i0} between them and on the distance d_{i0} separating the tip end and the atom i in the following way :

$$V_{i0} = V_{sp\sigma} \cos(\alpha_{i0}) e^{-\lambda d_{i0}}$$

We choose $\lambda = 1.15 \text{Å}^{-1}$ such that the current intensity is decreased by one order of magnitude when the tip moves 1Å away from the substrate.

STS SIMULATION

Experimental measurements of I-V curves of single-wall CN's have been made recently. We present our simulation for various kinds of CN's on Figure 2.

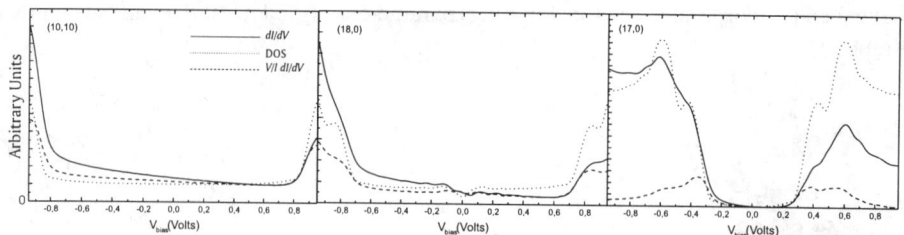

Figure 2 Numerical simulation of the differential tunneling conductance of various kinds of CN's. From left to right: a semi-metal, a small gap (washed out by numerical error) and a moderate gap semi-conductor.

Since several groups have reported the differential conductance and others the derivative of the logarithm of I as being representative of the density of states (DOS) of the nanotube, we compare these data in Figure 2. Neither dI/dV nor $V/I\, dI/dV$ is an accurate representation of the DOS. Nevertheless the essential characteristics of it (Van Hove singularities, band gap, ...) are correctly featured. The conductance curves show some asymmetry due to the off-diagonal elements $R_{ii'}^{(1)}(E)$ of the Green function of the nanotube.

STM SIMULATION

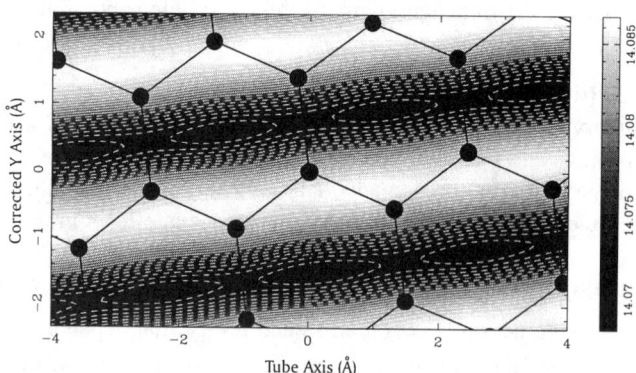

Figure 3 Computed STM image of the (11,7) carbon nanotube. The grey scale represents the distance tip-tube axis (Å). The Y axis is corrected as it is explained in the text. The atomic pattern is the projection of the atoms of the tube on the tangent plane (see Figure 4).

We have performed a constant-current STM simulation. In order to eliminate the continuous variation of the tip height due to the nanotube curvature, we have reported the distance from the tip to the tube axis. We reproduce a simulated STM image of the (11,7) nanotube on Figure 3. This choice has been guided by recent experimental measurements on this particular type of single-wall CN.[3]

One of the main goal of an STM experiment is the determination of the chiral angle of a nanotube. When atomic resolution can be achieved, the image of the hexagonal lattice at the surface of the nanotube leads to the chirality of the tube.

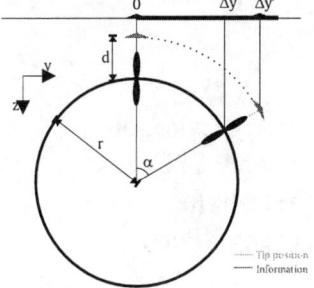

Figure 4 A schematic explanation of the origin of the deformation of an STM image due to the π nature of the sp2 orbital.

On the other hand, the tube diameter might also be determined by STM measurements. Unfortunately, the π-nature of the orbital at the surface of the nanotube affects the resulted map. At first (see Figure 4), we shall neglect the atomic corrugation. In our scan at constant current, the path of the tip is the doted line following the curved surface of the nanotube. With atomic corrugation, the tip has to move slightly upward when in the neighbourhood of an atom and downward when in the neighbourhood of the centre of an hexagon. But the height of the tip is not at a local maximum on top of an atom. Indeed, the s-π interaction is maximum when the s-orbital of the ending atom tip and the π-orbital of an atom of the nanotube are aligned. Since this effect only affects the direction perpendicular to the tube axis, it leads to a strong deformation of the image of the nanotube. From that, if no correction is taken into account, the determination of the chiral angle will be erroneous. This correction is easily performed by dividing the y coordinates by $(d+r)/r$ where d is the mean distance tip-axis and r is the nanotube radius. Because of the electronic origin of the deformation, such a correction should be still needed for more realistic models of the tip.

ACKNOWLEDGEMENTS

This work has been performed under the auspices of the inter-university research program on *Reduced Dimensionality Systems* (PAI-IUAP N. P4/10) initiated by the Belgian Federal OSTC. V.M. acknowledges a grant from the Belgian Fund for Industrial and Agricultural Research (FRIA). The authors acknowledge helpful discussions with P. Senet.

REFERENCES

1. A. Thess, et al., Science **273**, 483 (1996) ; C. Journet et al., Nature **388**, 756 (1997)
2. M.S. Dresselhaus, G. Dresselhaus, and P.C. Eklund, *Science of Fullerenes and Carbon Nanotubes* (Academic Press, San Diego, 1996)
3. J.W.G. Wildoër et al Nature **391**, 59 (1998) ; T.W. Odom et al, Nature **391**, 62 (1998) ; W. Clauss et al, *to be published*.
4. J. Bardeen, Phys. Rev. Lett. **6**, 57 (1961)
5. V. Meunier and Ph. Lambin, *to be published*

C_{60}-Based Molecular and Electronic Nanostructures

EugeniaV. Buzaneva[1], Leonid A. Bulavin[2], Valeriy E. Pogorelov[2],
Valeriy N. Yashchuk[2], Yuriy I. Prilutski[2], Sergey S. Durov[2],
Alexandr V. Nazarenko[2], Yuriy A. Astashkin[2],
Tymish Yu. Ogul'chansky[2], Grigoriy V. Andrievsky[3], Peter Scharff[4]

[1]*Department of Radiophysics and* [2]*Department of Physics, Kiev Shevchenko University, Vladimirskaya Str., 64, 252033 Kiev, Ukraine*
[3]*Institute for Therapy of the Academy of Medical Sciences of Ukraine, Postysheva Str., 2a, 310116 Kharkov, Ukraine*
[4]*Institut fur Anorganise and Analytische Chemie, TU Clausthal, Paul-Ernst-Strase 4, D-38670 Clausthal-Zellerfeld*

Abstract. Two new type of molecular/electronic nanostructures are proposed. The formation in water of highly stable fullerene aggregates with size 3.84 nm and a number of molecules C_{60} N=33 is theoretically predicted and experimentally confirmed. The modelling of the geometric structure and electronic properties of singlewalled fullerene nanotube is carried out. It is shown that the presense of heptagon-heptagon pair as a defect in the structure of this nanotube leds to the formation of the semiconductor-semiconductor heterojunction with the different value of gap.

Molecular nanostructures from fullerene C_{60} aggregates

Recently a method [1-2] has been offered for production of highly stable and finely dispersed colloidal solutions of fullerenes C_{60} in water (in the absence of any additives) and this resulted in the generation of solutions with fullerene aggregate sizes from several nanometers to 200 nm. The evaluation of the characteristics of this solution from the point of view of colloidal chemistry [3] indicated high stability with no essential changes during (12-18) months on storage at the normal conditions. Thus, the formation of clathrate-like networks of water molecules [4] around fullerene C_{60} aggregates, stabilized due to the low conformational mobility of fullerenes and geometrical matching between the structures, which may be formed by hydrogen bonding of water molecules in the clathrate and covalent bonds of the fullerene carbon atoms takes place.

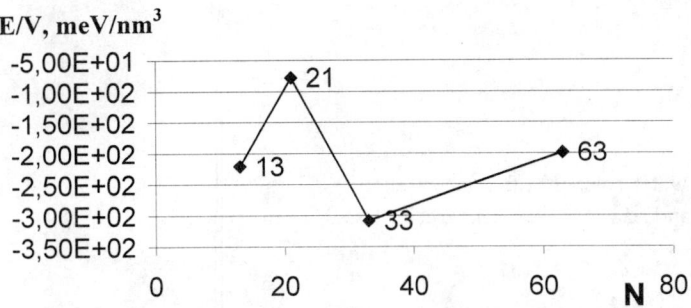

FIGURE 1. The calculated energies (normalized to the unit of volume) of possible hydrated clusters in dependence on the number of molecules C_{60} entering in them.

In our initial calculations we have assumed that the intermolecular interactions are described by a Lennard-Jones (12-6) atom-atom potential with values of parameters taken from our previous paper on simulation of solid C_{60} dynamics [5]. The calculations were carried out in the approximation of rigid molecules using the known atom-atom potential method and the dense packing principle.

The possible hydrated clusters were formed of fullerenes C_{60} by arranging the molecules along the symmetry axes C_5 (in number of 12), C_3 (in number of 20) and C_2 (in number of 30) of the single molecule C_{60} with diameter of about 0.7 nm situated in the centre.

The energy (normalized to the unit of volume) of possible hydrated clusters against a number of molecules C_{60} entering in them is given in Fig.1. It is obvious that the fullerene cluster of diameter 3.84 nm consisting of N=33 of C_{60} molecules has the most stable structure. The structure with a number of molecules C_{60} N=21 being unstable is transient between structures with N=13 and N=33.

To our opinion, the further formation of hydrated fullerene aggregates in water should be produced on the basis of highly stable cluster with a number of molecules C_{60} N=33. It keeps the icosahedral symmetry group of isolated C_{60} molecule and has to have the similar vibrational spectrum.

Raman spectrum of the fulerene water solution is given in Fig.2. This spectrum was obtained by using the double-Raman spectrometer DFS-24 with PC controling system. The spectral width of the spectrometer is 1.5 cm^{-1}. The exiting source was Cu-wapour laser with power 1 Wt. In this spectrum we can see an increasing of the frequencies of the vibrational fundamental bands for all observed modes in comparise with results for individual C_{60} molecules [6]. This increasing we connect with strengthening of the intramolecular bonds under forming nanostructural fulerene aggregates in water solution described above.

Finally, the numerical calculations show that the maximum size of fullerene aggregates produced on the basis of highly stable hydrated clusters with a number of

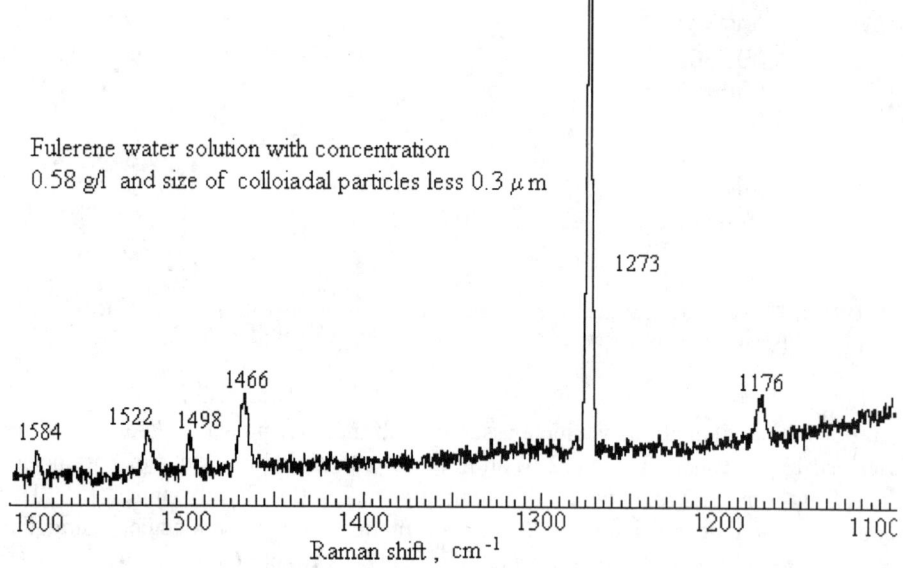

FIGURE 2. Raman spectrum of the fulerene water solution.

molecules C_{60} N=33 does not exceed 200 nm and agrees with experimental value [2].

Nanostructures on the basis of fullerene C_{60} nanotube

The discovery of singlewalled carbon nanotubes [7] have provided the opportunity to study both their mechanical and electronic properties. Theoretical calculations [8] have shown that, depending on the nanotube symmetry and diameter, these carbon nanotubes can be metallic or semiconducting.

In the present work an attempt was made to modulate the singlewalled carbon nanotube from C_{60} fullerenes. The obtained structure of singlewalled fullerene nanotube is given in Fig.3. The length of continious part of nanotube is equal to 0.43 nm.

As is known [8], the dependence of gap on the radius of singlewalled nanotube may be approximately described by the formula:

$$\varepsilon_g = \varepsilon_{\pi\pi} \frac{d_0}{R}.$$

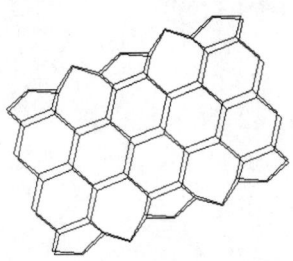 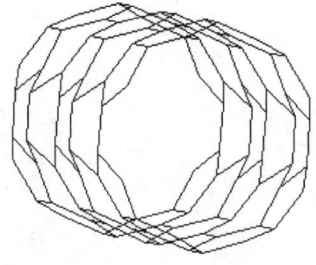

FIGURE 3. The obtained structure of singlewalled fullerene nanotube.

In our case $\varepsilon_{\pi\pi}$ - the average energy of interaction between two π-electrons, located on the single and double bonds in C_{60} molecule; d_0 - the average distance between the neighbouring carbon atoms in C_{60} molecule. The numerical calculations carried out by the use the molecular dynamics and the tight-binding models [9] have shown that $\varepsilon_{\pi\pi}$=2.35 eV and d_0=0.14 nm.

The presence of heptagon-heptagon pair as a defect in the structure of singlewalled fullerene nanotube leds to the change of its radius from 0.35 nm to 0.38 nm. As a result, the decrease of gap from 0.94 eV to 0.87 eV takes place.

Thus, the semiconductor-semiconductor heterojunction with the different value of gap is formed. It should be noted that the similar heterojunction was really observed in the experiment [10] for the singlewalled carbon nanotube.

REFERENCES

1. Andrievsky, G.V. et al, *J.Chem.Soc., Chem.Commun.*, 1281-1282 (1995).
2. Mchedlov-Petrosyan, N.O. et al, *J.Chem.Soc. Faraday Trans.* **93**, 4343-4346 (1997).
3. Heimenz, P.C., *Principles of Colloid and Surface Chemistry,* New York: Marcel Dekker, 1986.
4. Jeffrey, G.A., Saenger, W., *Hydrogen Bonding in Biological Structures,* Berlin: Springer-Verlag, 1991.
5. Prilutski, Yu.I., Shapovalov, G.G., *Phys.St.Sol. (b)* **201**, 361-370 (1997).
6. Bethune, D.S. et al, *Chem.Phys.Lett.* **179**, 181-186 (1991).
7. Bethune, D.S. et al, *Nature* **363**, 603-605 (1993).
8. Dresselhaus, M.S. et al, *Science of Fullerenes and Carbon Nanotubes,* New York: Academic Press, 1996.
9. Prilutski, Yu.I. et al, *Ukr.Fiz.Zhurn.* **42**, 1143-1145 (1997).
10. Saito, R. et al, *Phys.Rev.B.* **53**, 2044-2047 (1996).

FULLERENES

Computing Most Stable Fullerenes

Zdeněk Slanina, Xiang Zhao and Eiji Ōsawa

Laboratories of Computational Chemistry & Fullerene Science
Department of Knowledge-Based Information Engineering
Toyohashi University of Technology, Toyohashi 441-8580, Japan

Abstract. Potential energy differences alone cannot always explain why particular fullerene isomers C_n are observed. It has been demonstrated on various higher IPR isomers that entropy effects are to be included in order to get a realistic stability picture at high temperatures. Our computations along this line extend to $n=96$. However, the principle works for smaller fullerenes, too, as it is demonstrated on the case study of C_{36}.

INTRODUCTION

In a systematic large-scale computational study, we have treated all IPR isomers of fullerenes from $n=76$ till $n=96$ in a uniform way (1). A key conclusion drawn from the computations is that the computed energetics itself (i.e., potential energy changes) is not always able to produce a good agreement with observations, and that entropy terms are to be considered, too. Then, a critical comparison with observed data indicates an overall satisfactory agreement.

COMPUTATIONS

All considered cages (typically, the IPR topologies) are first optimized at the MM3 molecular mechanics level and then reoptimized with the semiempirical SAM1 method implemented in the AMPAC program package (2). In the SAM1 optimized geometries, the harmonic vibrational analysis is carried out. Rotational-vibrational partition functions are then constructed from the computed data and temperature-dependent relative concentrations evaluated. The inter-isomeric energetics is also checked at *ab initio* level, e.g., Hartree-Fock (HF) SCF computations in the 4-31G basis set (HF/4-31G) or density-functional computations at the B3LYP/6-31G* level using the G94 program package (3). The SAM1 heats of formation are reduced to the absolute zero temperature using the partition functions. A special attention is paid to the symmetry and chirality contributions into the partition functions.

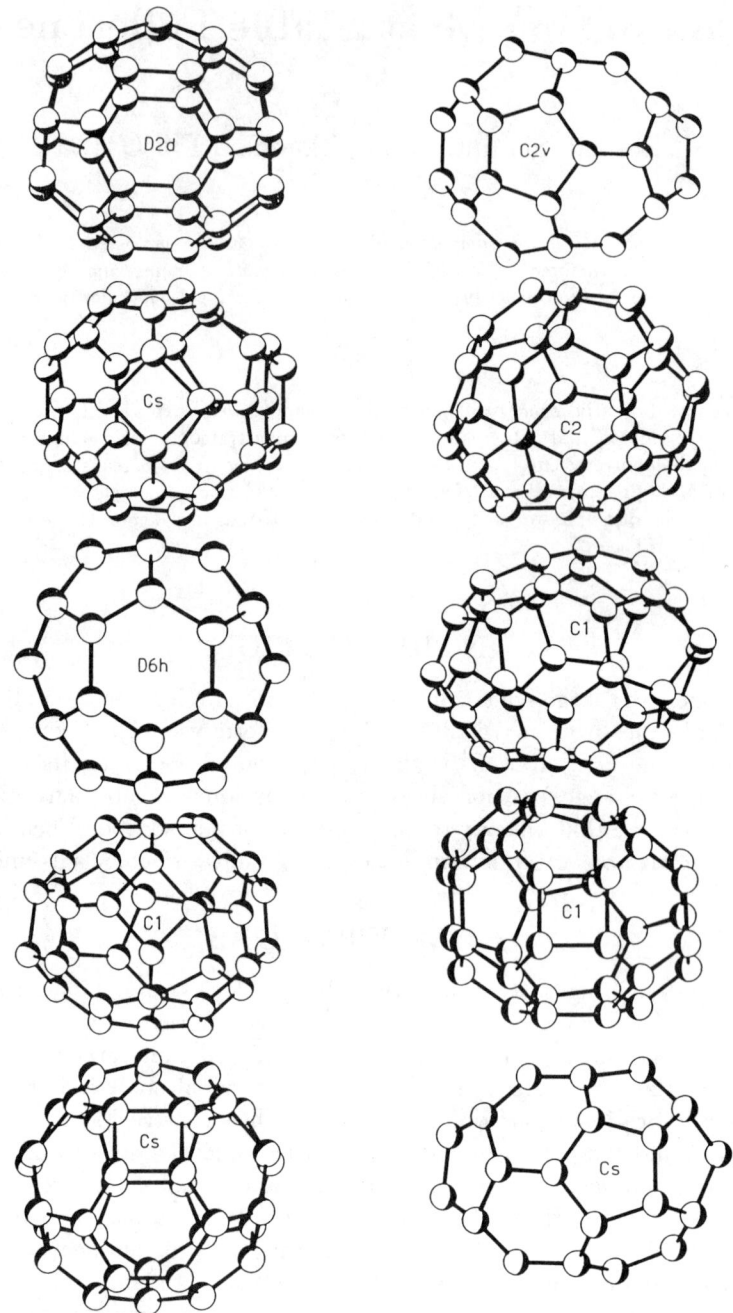

FIGURE 1. The SAM1 optimized structures of C_{36} - ten energy-lowest isomers out of a set of 566 cages (note four- and seven-membered rings in some structures).

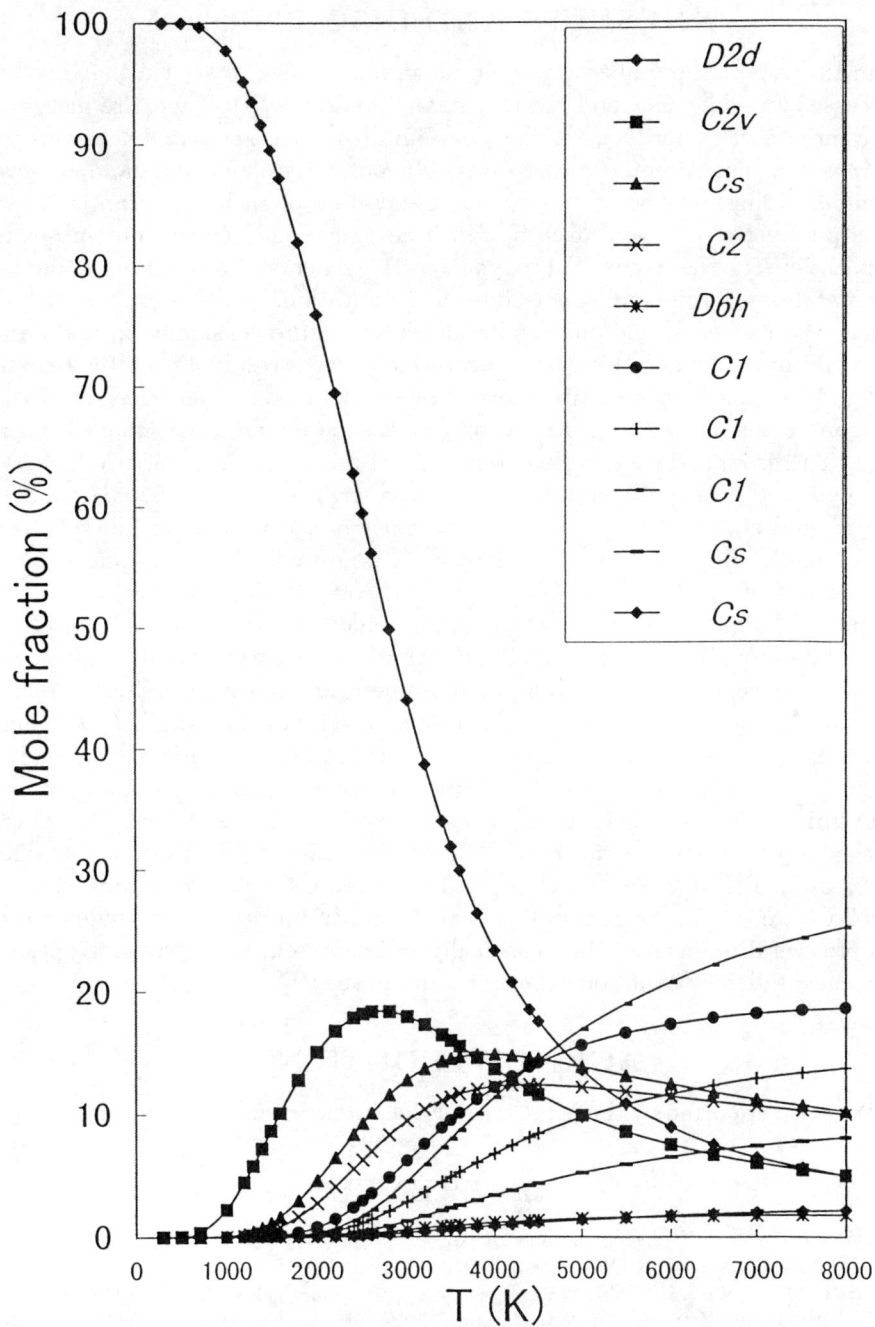

FIGURE 2. The SAM1 computed relative concentrations of the ten energy-lowest isomers of C_{36} from Fig. 1.

RESULTS AND DISCUSSION

In this report the unified treatment is applied to C_{36}, a system that has been discussed for some time and recently finally separated (4). There are just fifteen conventional cages for C_{36}, i.e. the cages built from pentagons and hexagons only. However, in our recent study (5) of C_{32} a possible role of four- and/or seven-membered rings was pointed out, too. Generation of such cages can be treated by a purely heuristic approach in which new topologies come from subsequent applications of the Stone-Wales type rearrangements. In order to reduce the computations, we introduce an additional limitation: only structures with one square, two squares, and one square and hexagon are considered in addition to the conventional C_{36} fullerenes: altogether 566 cages resulted from the search.

Fig. 1 depicts ten structures lowest in the SAM1 energy - only six of them are conventional fullerenes, three contain one square, and one a square-heptagon pair. The ground state is a D_{2d} conventional fullerene, the structure highest in energy is a C_s conventional fullerene (bottom right of Fig. 1, 162 kJ/mol above the ground state). The second lowest isomer is a conventional fullerene (C_{2v}, 36 kJ/mol). However, the third lowest structure contains one square (C_s, 77 kJ/mol). The HF/4-31G and B3LYP/6-31G* computations basically support the picture. The high symmetry (D_{6h}) conventional fullerene is located 124 kJ/mol above the ground state (it undergoes a Jahn-Teller distortion).

Fig. 2 presents the SAM1 computed temperature dependencies of the relative concentrations of the C_{36} isomers. The structure lowest in energy (D_{2d}) remains the most populated till a temperature of almost 5000 K. However, already at 2000 K or so other two or three structures become significant, too, viz. the C_{2v} conventional fullerene and the C_s *quasi*-fullerene with one four-membered ring. For example, at a temperature of 2500 K those three structures represent 59.5, 18.2, and 9.3 % of the equilibrium ten-membered C_{36} mixture. At present, only a solid-state NMR spectrum of C_{36} is available, and it does not exclude presence (4) of several isomers. If they can really be separated, the interesting option of one *quasi*-fullerene could be experimentally probed.

ACKNOWLEDGMENTS

Work is supported by the Ministry of Education, Science and Culture in Japan.

REFERENCES

1. Slanina, Z., Zhao, X., Lee, S.-L., and Ōsawa, E., *Chem. Phys.* **219**, 193-200 (1997).
2. AMPAC 6.0, Shavnee, KS: Semichem, 1997.
3. Frisch, M. J., *et al.*, GAUSSIAN 94, Revision E.2, Pittsburgh, PA: Gaussian, Inc., 1995.
4. Piskoti, C., and Zettl, A., "A New Carbon-Cage Solid: C_{36}," presented at IWEPNM, Kirchberg/Tyrol, Austria, Feb. 28 - March 8, 1998.
5. Zhao, X., Ueno, H., Slanina, Z., and Ōsawa, E., *Rec. Advan. Chem. Phys. Fullerenes Relat. Mater.* **5**, 155-166 (1997).

The First Stable Lower Fullerene : C36

C. Piskoti and A. Zettl

Department of Physics, University of California, Berkeley, and Materials Sciences Division, Lawrence Berkeley National Laboratory, Berkeley, CA 94720 USA

A new pure carbon material, presumably composed of thirty six carbon atom molecules, has been synthesized and isolated in milligram quantities[1]. It appears as though these molecules have a closed cage structure making them the smallest member of a new class of molecules known as fullerenes, most notably of which is the soccer ball shaped C_{60}[2]. However, unlike other known fullerenes, any closed, fullerene-like C_{36} cage will necessarily contain fused pentagon rings[3]. Therefore, this molecule apparently violates the isolated pentagon rule, a criterion which requires isolated pentagons for stability in fullerene molecules[4,5]. Striking parallels between this problem and the synthesis of other fused five member fused ring systems will be discussed. Also, it will be shown that certain biological structures known as clathrin behave in a manner which gives excellent predictions about fullerenes and nanotubes. These predictions help to explain the presence of abundant quantities of C_{36} in arced graphite soot.

Fused Pentagon Fullerenes

The discovery of Buckminsterfullerene (C_{60}) demonstrated the possibility of producing graphite-like molecular structures by bending the planar sp^2 carbon bonds of graphite[7]. Although such strained ring systems are common to organic chemists, "buckyballs" are unique only in that the carbon structure closes on itself requiring no passivation of bonds by hydrogen. In fact, it is useful to think of the C_{60} molecule as a continuation of the $C_{20}H_{10}$, corannulene unit (I) which has the structure of a central pentagon surrounded by fused hexagons (one third of a C_{60} molecule)[6]. Prior to the discovery of C_{60}, one could have inferred this molecule's stability based on the existence of corranulene which possesses comparable strain and curvature of the sp^2 carbon bonds.

I. II.

This method of applying information about previously synthesized fused ring systems to the problems of novel fullerenes may also be addressed to fullerenes with fused pentagons such as C_{36}. The first logical step in this progression is to pose the question of whether or not fused cyclopentano ring systems exist in organic chemistry literature. Indeed, such species have been synthesized and the first of these fused pentagon ring systems was 2,2a,3,4-tetrahyrdro-1H-cyclopent[cd]indene (II). This molecule falls into a class of 5,5,6 fused ring molecules which various groups had attempted to synthesize from the mid-1920's through the early 1950's[8]. The failure of this effort to yield any conclusive results lead some chemists to speculate that such a strained ring system may not be stable enough to exist. However, in 1956, Rapoport & Pasky developed an effective method of synthesizing and characterizing II[8]. It was concluded that the fused pentagon ring system is not as highly strained as previously believed.

Similarly, the lack of evidence of lower fullerenes in crude fullerene mass spectra has lead some to speculate that fused pentagon fullerenes cannot exist outside of vapor phase mass spectrometry experiments[4,5]. On the contrary, we have found that C_{36} is stable in solid form. Our findings, the details of which are reported elsewhere[1], indicate that the chemistry of this new fullerene is significantly different from other known fullerenes. For example, C_{36} is insoluble in the common "fullerene" solvents such as toluene or benzene. However, we have found that it is soluble in pyridine, a fact which has aided tremendously in the bulk synthesis. Also, laser desorption mass spectrometry of pristine C_{36} is difficult due to the fact that the molecule is easily destroyed by the laser pulse. One way to remedy this problem has been to use a matrix such as the fullerene soot in which C_{36} is generated. Alternatively, somewhat more robust derivatives such as $C_{36}H_6$ have been synthesized and successfully tested by mass spectrometry. We speculate that the combination of such factors may have prevented the discovery of this material until now.

Clathrin

Perhaps one of the most striking analogs to fullerenes is that of the clathrin protein which is found in various types of animal cells[6,9,10]. Clathrin is a long, rigid molecule having a molecular weight of about 180,000 a.m.u.. It is observed to link together into a hexagonal, graphite-like network on the surface of vesicles. Although this is the most preferred orientation, clathrin is known to form pentagons in order to enclose a tubular or spherical vesicle. Much can be learned from the types of structures which result, especially when applied to closed graphitic structures such as fullerenes and nanotubes. For example, clathrin coats on micro tubules possess various chiralities and diameters similar to those observed in single-walled and multi-walled nanotubes. Also, large, spherical particles were first observed to have an icosahedral, soccer ball shaped clathrin coat, like C_{60}. One might expect that any closed structure consisting of pentagons and hexagons would form but that is not the case. For example, Of the higher fullerenes analogs, only the C_{82} isomers with isolated pentagons have been observed in abundance[6].

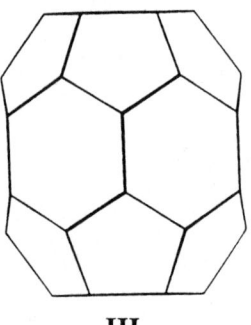
III.

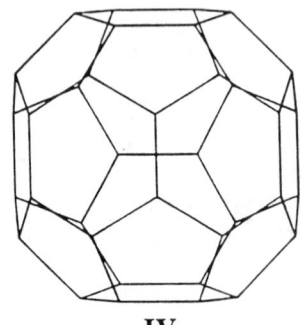
IV.

Therefore, it appears intuitive to study the structures observed for lower fullerene clathrin analogs. Given that there are over five hundred lower fullerene-like, clathrin structures possible[3], only four have been observed. Two of these structures are analogs of C_{36} isomers. One with D_{6h} symmetry (III) and one with D_{2d} symmetry (IV)[6,10]. Amazingly, recent calculations on the fifteen possible isomers of C_{36} have predicted these exact structures to be lowest in energy as well as isoenergetic[11]. Clathrin must have to obey the same rules of minimizing strain by distributing curvature evenly over the entire structure. Not surprisingly, our experimental results strongly suggest that the C_{36} samples synthesized contain the D_{6h} isomer in abundance[1].

Conclusion

To our knowledge, the isolation of C_{36} signifies the first successful bulk synthesis of a lower fullerene. We have found that although the molecule is stable, various factors such as heightened sensitivity to photochemical degradation as well as low solubility has prevented synthesis of this material in the past. The fact that isolated pentagons is not an absolute requirement for stability has been discussed. Also, parallels have been drawn to the protein clathrin which coats vesicles in a graphite-like network. Strong evidence has been presented that this material is constrained upon closure in ways which are very similar to fullerenes. This appears to make clathrin an excellent model for predicting which fullerenes should be expected.

Acknowledgments: The study of C_{36} arose out of a joint experimental/theoretical collaboration with M.L. Cohen, M. Côté, J. C. Grossman, and S.G. Louie, all of whom we thank for stimulating discussions and helpful interactions. This work was supported in part by the U. S. Department of Energy, the Office of Naval Research, and the National Science Foundation.

Correspondence and requests for materials to A. Zettl (azettl@physics.berkeley.edu).

References

1. C. Piskoti, J. Yarger, A. Zettl, Submitted to Nature
2. See, for example, *Buckminsterfullerenes*, W. Edward Billups and Marco A. Ciufolini, (VCH Publishers, New York, 1993)
3. P. W. Fowler, & D. E. Manolopoulos, *An Atlas of Fullerenes* (Clarendon Press, Oxford, 1995).
4. H. W. Kroto, Nature **329**, 529-531 (1987)
5. R. E. Smalley, Accts. Chem. Res. **25**, 98-105 (1992)
6. D. Koruga, et. al., Fullerene C_{60} (Elsevier Science Publishers, North-Holland, 1993)
7. R. C. Haddon, Accts. Chem. Res. **21**, 243-249 (1988)
8. H. Rapoport & J. Z. Pasky, J. Amer. Chem. Soc. **78**, 3788-3792 (1956)
9. T. Kanaseki & K. Kadota, J. Cell Biol. **42**, 202-220 (1969)
10. R. A. Crowther, J. T. Finch and B. M. F. Pearse, J. Mol. Biol. **103**, 785-798 (1976)
11. J. C. Grossman, M. Côté, S. G. Louie & M. L. Cohen, Chem. Phys. Lett. **284**, 344-349 (1998).

Simulated Behavior of Fullerenes at High Temperatures

Noriyuki Kurita and Eiji Osawa

*Department of Knowledge-Based Information Engineering,
Toyohashi University of Technology, Aichi, 441-8580, JAPAN*

Abstract. In order to clarify the difference in resistance to thermal decomposition between C60 and other fullerenes, we analyzed their behavior at high temperatures by using a dynamic reaction coordinate option in a semiempirical molecular orbital program. There are some differences in thermal decomposition process between C60 and C82, depending on the symmetries of both structure and lowest frequency modes. However, C58, C60, C62 and C82 behave similarly in thermal decomposition, so that the magic number of fullerenes can not be explained by only the resistance to thermal decomposition.

INTRODUCTION

One of the remaining enigmata about the mechanism of fullerene formation is the magic number problem or the very special selectivity regarding the size and structure of isolated fullerenes. Even the simplest possible question often asked among the fullerene scientists, why is C60 formed most readily than any other fullerenes, has not been answered in satisfiable detail. In order to solve the time-old problem, we start from a new hypothesis: the isolated fullerenes may be the survivors of decomposition by violent vibrations at high temperatures which the carbon clusters had to go through in the course of annealing after they are formed from vaporized carbon atoms. We expect that the isolated fullerenes are resistant to decomposition by thermal vibration. In this study, we analyzed the behavior of fullerenes at high temperatures by using a semiempirical molecular orbital program (MOPAC).[1]

METHOD OF CALCULATION

We used a dynamic reaction coordinate (DRC) option[2] in MOPAC which allows us to perform molecular dynamics at high temperatures while performing SCF calculations of electronic structure and nuclear disposition so that we can trace structural changes during the thermal decomposition of a molecule. As for a model Hamiltonian in MO method, the restricted Hartree-Fock-AM1 model[3] was used. Molecular dynamics simulations were started from the optimized structure with initial velocities of each atom assigned from the average of lowest frequency modes.

THERMAL DECOMPOSITION OF C60

Figure 1 (a) shows the change in energy during decomposition process for C60. The structure in lower energy region (time step=3820) and that in higher energy region (time step=5000) are shown in Figure 1 (b) and (c), respectively. In lower energy region, potential energy curve oscillates, and the structure of C60 is kept almost spherical. In higher energy region, C60 deforms to a flat disk-like shape (Figure 1 (c)) accumulating strain around the edge so that several bonds along the edge begin to break. We performed an energy partition analysis for this deformed structure and clarified that the bonds along the edge are more easily broken than those near the center.

By a further increase of total energy, C60 cage is broken and potential energy increases rapidly. The threshold value of potential energy for decomposition is about 40 kcal/mol/atom. In this way, C60 deforms and decomposes with keeping high symmetry, because C60 has lowest frequency modes with high symmetry as well as Ih structure.

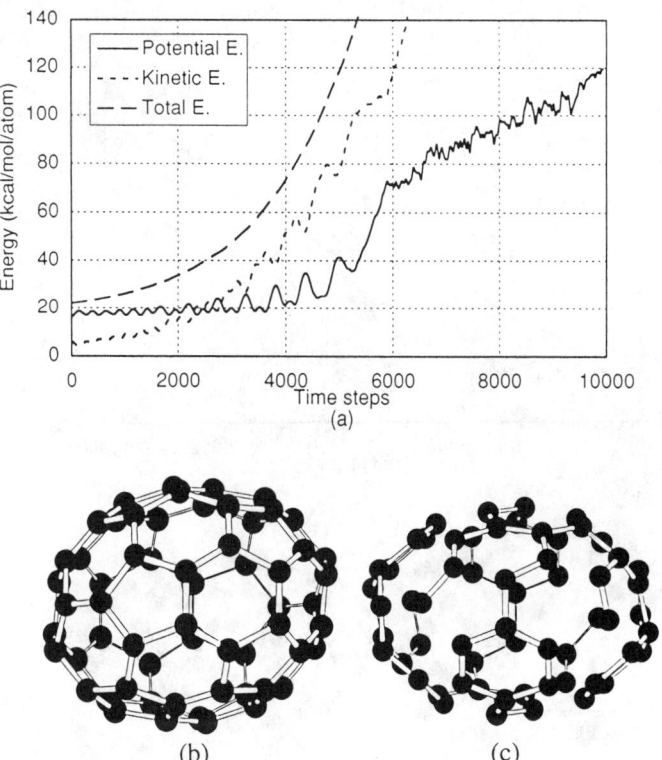

FIGURE 1. Thermal decomposition of C60: (a) change in energies during decomposition process, (b) structure at 3820 step and (c) structure at 5000 step.

THERMAL DECOMPOSITION OF C82

C82 has nine types of topologically different isomers under the isolated pentagon rule. Here we performed AM1 optimization for these isomers and for the most stable isomer the DRC analysis was performed. Figure 2 (a) shows the change in energy during decomposition process. The structure in lower energy region (time step=3560) and that in higher energy region (time step=5200) are shown in Figure 2 (b) and (c), respectively. In contrast to C60, an oblong C82 deforms to more and more oblong shapes and quickly decomposed into several pieces. In lower energy region, C82 deforms to a less symmetry structure as shown in Figure 2 (b), because of low symmetry of the lowest frequency modes. In higher energy region, C82 deforms to an oblong and flat shape (Figure 2 (c)). This deformation seems to come from the situation that C82 will lower the potential energy by making graphite-like structures on both sides of the flat structure. The threshold value of potential energy for decomposition is about 35 kcal/mol/atom, which is a little smaller than that for C60.

From the comparison between C60 and C82 results, it was clarified that not only structure but lowest frequency modes effect to the thermal decomposition process.

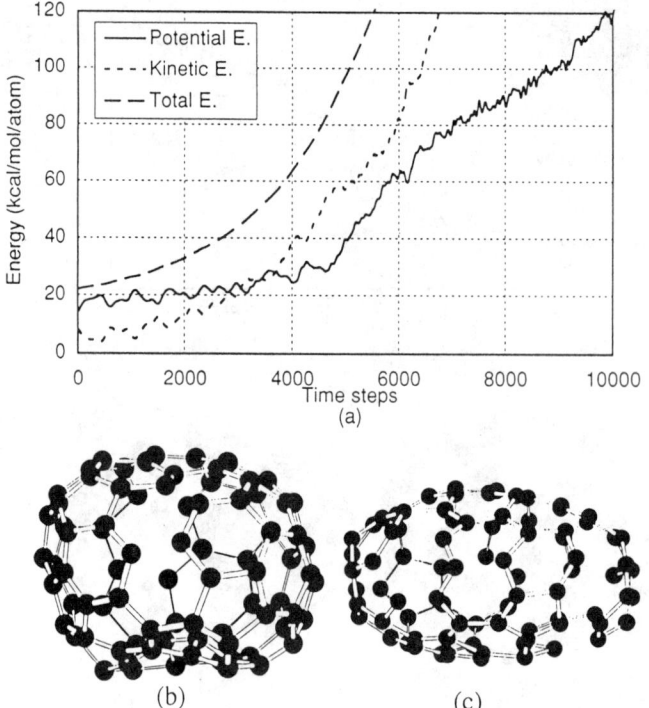

FIGURE 2. Thermal decomposition of C82: (a) change in energies during decomposition process, (b) structure at 3560 step and (c) structure at 5200 step.

SAMMARY

The thermal decomposition processes for C58 and C62 were also analyzed in the same way, the results are shown in Figures 3 and 4. From these figures, it was found that C58, C60, C62 and C82 behave similarly in thermal decomposition. Therefore, the resistance to thermal decomposition of C60 is not so large as it can explain the reason why C60 is formed most readily than any other fullerenes. In order to explain the magic number of fullerenes, other factors must be considered.

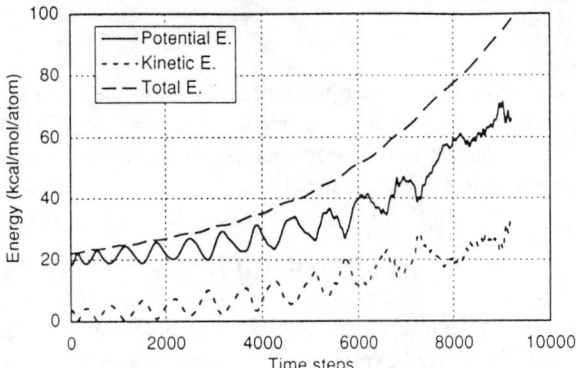

FIGURE 3. Change in energies during thermal decomposition of C58.

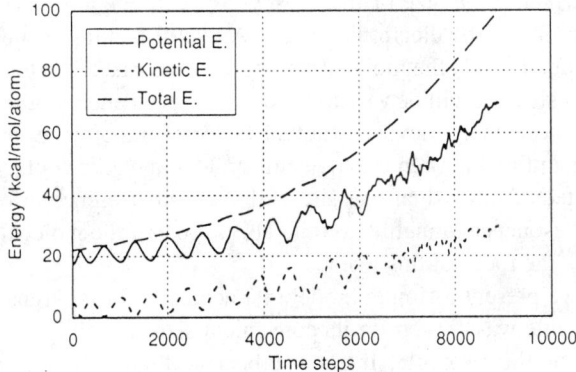

FIGURE 4. Change in energies during thermal decomposition of C62.

ACKNOWLEDGMENTS

This study is partly supported by the Nagai Science and Engineering Foundation.

REFERENCES

1. MOPAC93 version 6.01 by J. J. P. Stewart was obtained from the Japan Chemical Program Exchange, 1-7-12 Nishine-Nishi, Tsuchiura 300, Japan, program No. P049.
2. J. J. P. Stewart, L. P. Davis and L. W. Burggraf, *J. Comp. Chem.* **8**, 1117 (1987).
3. M. J. S. Dewar et al., *J. Am. Chem. Soc.* **107**, 3902 (1985).

Contact dependence of the conductance through C_{60}

Magnus Paulsson and Sven Stafström

Department of Physics and Measurement Technology, IFM, Linköping University, S-581 83, Linköping, Sweden

Abstract. The electronic transmission through a C_{60} molecule is studied with emphasis on the contact geometry. The transmission can not be understood simply by the electronic structure of the C_{60} molecule itself, interference effects related to the metal/molecule contact are in fact very important for the conductance.

INTRODUCTION

The development of molecular wires for electronic applications is an area of rapidly growing interest[1-8]. There are several aspects of the molecular electronic structure which are of importance for the conductance, for instance, the distribution of the energy levels, in particular the size of the energy gap, the length of the molecules, and localization properties of the electronic states. A useful picture of the transmission is resonance tunneling, that is if the molecule supports states that are extended between the two contacts, the resistance will be very low when the Fermi energy of the metal contacts align with one of these energy levels. Another factor of importance for the conduction is the electronic coupling between the molecule and the metallic contacts. In particular, molecules contacted at more than one site will show interference effects, making the simple picture of resonance tunneling to fail. This aspect of the molecular device is less well studied and is the focus of this work.

In this paper we present a simple model that accounts for the transmission through the C_{60}, this molecule was chosen for its convenient size as well as the different possibilities of contacting the molecule. It has also been well studied both theoretically and experimentally[9]. The model, which is summarized in Section II, involves a simple one electron Hamiltonian. By using a Green function technique we can calculate the conductance through the molecule. Full details of the model can be found elsewhere[10]. Results from the conductance calculations are presented in Section III.

METHODOLOGY

The conductance is calculated from the Landauer formula[11,12]

$$G = \frac{2e^2}{h} T(E) \quad (1)$$

where the transmission function $T(E)$ represents the transmission probability between the metal contacts for an electron with energy E. In the following we will focus on the transmission function i.e. the dimensionless conductance.

The metal and the metal/molecular contact are described in the simplest possible way. The metal is modeled by a 1D metallic wire[13] with total band width of 20 eV centered at zero energy. The metal/molecule coupling is described by hopping matrix elements $\bar{t}_l$ ($\bar{t}_r$) which couple an arbitrary number of atomic orbitals of the C_{60} to the 1D metal contacts on the left right sides of the molecule.

The Hamiltonian of the molecule is described within the tight-binding approximation. It only treats the π-electron system with a basis of $2p_z$ atomic orbitals. The interactions are restricted to electron hopping between nearest neighbors (n.n.). The values of these hopping terms depend on the C-C bond lengths. Here we adopt a simple exponential relation between the value of the hopping integral and the bond length $R_{l',l}$

$$t_{l',l} = t_0 e^{-\alpha(R_{l',l}-R_0)} \tag{2}$$

R_0 is the reference bond length, which is fixed to 1.40 Å for the C_{60}. The values of $R_{l',l}$ are obtained from optimization of the C_{60} molecule using the the semi-empirical AM1 method[14] and the values of the parameters are $t_0 = -2.5$ eV and $\alpha = 2.0 \text{Å}^{-1}$ giving a good fit to the experimental band width and band gap.

The Green function for the molecule is obtained in the usual way:

$$G(E) = (H - E)^{-1} \tag{3}$$

By connecting the molecule to the metal contacts via the hopping $\bar{t}_{l,r}$ and assuming the simple 1D metal contacts we obtain the transmission function as

$$T(E) = 4t_c^2 \sin^2 k \left| \left[\left(t_c e^{ik} I - EI - \begin{pmatrix} \bar{t}_l \\ \bar{t}_r \end{pmatrix} G(E) \begin{pmatrix} \bar{t}_l & \bar{t}_r \end{pmatrix} \right)^{-1} \right]_{(2,1)} \right|^2 \tag{4}$$

where t_c is the hopping in the one-dimensional metal contacts, and k the wave vector in the metal contacts. I is the 2×2 identity matrix and $(2,1)$ denotes one element of the inverted 2×2 matrix.

RESULTS AND DISCUSSION

The weak bond between atoms of the C_{60} molecule and the metal contacts is modeled by a constant hopping $t_{cm} = 0.5 t_0$ eV for each atomic orbital coupled to the metal contacts. This value was chosen arbitrary but with the knowledge that the qualitative results will not be changed by 'small' changes in this hopping. In the calculations below we have studied four different cases of bonding of the C_{60} to the contacts : bonding to opposite pairs of atoms along single or double bonds and bonding to opposite pentagons or hexagons. These cases were chosen as the most probable experimental orientations[15].

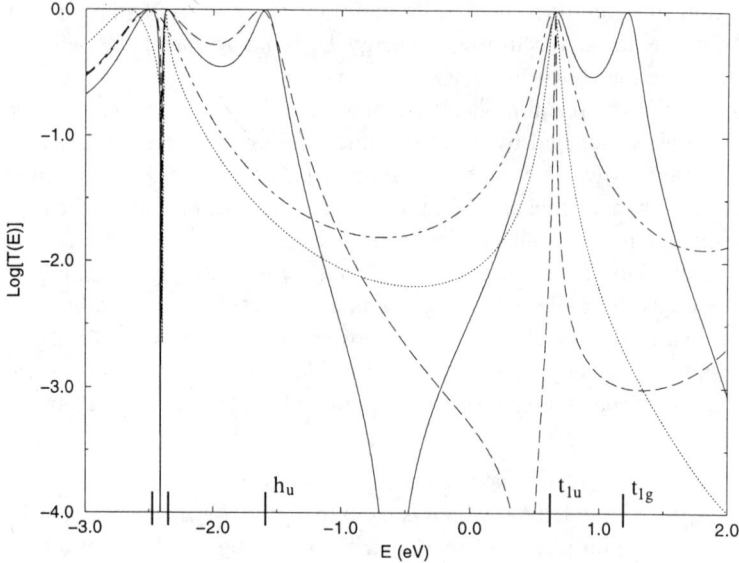

FIG. 1: Logarithm of the dimensionless conductance through C_{60} contacted through opposite pairs of atoms along single bond (solid line), along double bond (dashed line), contacted through opposite pentagons (dot dashed line) and hexagons (dotted line), also the energy eigenvalues of the C_{60} is indicated on the x-axis for reference.

The transmission function $T(E)$ for the four different cases mentioned above are shown in Fig.1, together with four of the eigen-energies of the unperturbed C_{60} molecule.

The first impression is that the conductance is strongly dependent on the contact geometries. In the case of contact through opposite single bonds (solid line) we see that the conductance peaks correspond closely to the eigen energies of the molecule. This corresponds to the simplified picture of resonant tunneling through the eigenstates of the molecule, with one exception, the destructive interference that drops the conductance to zero around mid gap. Comparing with the case of contact through double bonds (dashed line) the main differences is the disappearance of the peak corresponding to the t_{1g} level and that the destructive interference is shifted to higher energy. This large changes in conductance for the relatively small change in contact geometry is quite surprising.

The transmission in the case of contact through opposite hexagons (dotted line) and pentagons (dot dashed line) are very similar to each other with a slightly larger transmission for the pentagon case in the HOMO (h_u) - LUMO (t_{1u}) gap. In both cases, the resonant tunneling effect of the h_u and t_{1g} levels has disappeared due to interference effects caused by the symmetries of the molecular orbitals. Comparing with the single, double bond cases the destructive interference in the HOMO-LUMO gap has disappeared.

In the case of a C_{60} molecule on a metal surface contacted by a STM-tip the interference effects described above could be changed due to the over simplified model of the metal surface having only one electron-channel (the STM contact however is well described by the one electron channel model). Here we will argue that the qualitative features of the conductance will not be affected by a better model of the metal surface with more electron-channels. Several conductance calculations, not shown in the figure, have been performed for the conductance through the C_{60} contacted on one side through a single-, double-bond, pentagon or hexagon to a single atomic orbital on the opposite side. The conductance for these cases has the same qualitative behavior as the corresponding symmetric case (i.e. single-, double-bond, pentagon or hexagon contact geometry on both sides). This means that the interference effects responsible for the change of the transmission from the simplified picture of resonance tunneling can be attributed to each side of the contact, individually, and does not critically depend on the symmetry between the injection and ejection contacts. A more realistic model of a metal surface with more than one electron-channel will not change the qualitative results since the interference effects will show up on the side where the STM tip contacts the molecule.

The main conclusion of these results is that the transmission through a molecule corresponds to a simple picture of resonant tunneling when the molecule is contacted by single atomic orbitals, however when we increase the number of contact points the importance of interference effect around these contact points increases and some resonances might disappear entirely.

ACKNOWLEDGMENTS

Financial support from Swedish Research Council for Engineering Science (TFR) and Swedish Natural Science Research Council (NFR) is greatfully acknowledged.

---

[1] L. A. Bumm *et al.*, Science **271**, 1705 (1996).
[2] R. P. Andres *et al.*, Science **272**, 1323 (1996).
[3] V. Mujica, M. Kemo, and M. A. Ratner, J. Chem. Phys. **101**, 6856 (1994).
[4] M. P. Samanta *et al.*, Phys. Rev. B **53**, R7626 (1996).
[5] M. Magoga and C. Joachim, Phys. Rev. B **56**, 4722 (1997).
[6] S. Datta and W. Tian, Phys. Rev. B **55**, R1914 (1997).
[7] S. Datta *et al.*, Phys. Rev. Lett. **79**, 2530 (1997).
[8] M. A. Reed *et al.*, Science **278**, 252 (1997).
[9] C. Joachim and J. K. Gimzewski, Chem. Phys. Lett. **256**, 353 (1997).
[10] M. Paulsson *et al.*, In preparation.
[11] R. Landauer, IBM J. Res. Develop. **1**, 223 (1957).
[12] R. Landauer, Philos. Mag. **21**, 683 (1970).
[13] Z. G. Yu *et al.*, Phys. Rev. B **56**, 6494 (1997).
[14] M. J. S. Dewar *et al.*, J. Am. Chem. Soc. **107**, 3902 (1985).
[15] T. Hashizume *et al.*, Phys. Rev. Lett. **71**, 2959 (1993).

Intermolecular bond stability of C_{60} dimers and 2D pressure-polymerized C_{60}

P. Nagel[1], V. Pasler[1], S. Lebedkin[1], C. Meingast[1], B. Sundqvist[2],
T. Tanaka[3], K. Komatsu[3]

[1] *Forschungszentrum Karlsruhe - Technik und Umwelt, Institut für Nukleare Festkörperphysik, PO Box 3640, 76021 Karlsruhe, Germany.*
[2] *Department of Experimental Physics, Umeå University, S-90187 Umeå, Sweden.*
[3] *Institute for Chemical Research, Kyoto University, Uji, Kyoto 611, Japan.*

The thermal stability of C_{60} dimers and 2D pressure-polymerized C_{60} is studied using high-resolution capacitance dilatometry. The transformation of both the dimer and the polymer phases back to 'normal' C_{60} is excellently described by a simple thermally activated process, with activation energies of 1.75 ± 0.1 eV (dimer) and 1.9 ± 0.2 eV (polymer). These results are compared to previous data of 1D-polymerized C_{60} and photo-polymerized C_{60}. The thermal expansivity of the 2D-polymer phase is as much as a factor of ten smaller than that of pure C_{60} and approaches values for diamond.

INTRODUCTION

Besides the 'normal', face-centered-cubic (fcc), plastic-crystal phase of C_{60}, several other phases have been reported in which the C_{60} molecules are linked via covalent bonds to form dimers, linear chains (1D), 2- and even 3-dimensional networks [1]. These polymerized phases can be produced by different methods, such as photo-polymerization (dimers and 1D) [2], high-pressure-temperature conditions (1D, 2D and 3D) [3,4] and the recently published solid-state mechano-chemical reaction of C_{60} with KCN (dimers) [5]. The resulting polymer phases are metastable under ambient conditions and revert back to fcc C_{60} during heating to higher temperatures.

Here we study the kinetics of the polymer to fcc transformation of C_{60} dimers and 2D pressure-polymerized C_{60} using high-resolution dilatometry up to 500 K.

EXPERIMENTAL

The C_{60} dimers were synthesized by a solid-state mechano-chemical reaction of C_{60} with potassium cyanide [5]. The 2D-polymerized sample was prepared by annealing high-purity sublimed polycrystalline C_{60} at 830 K for 5 hours at a pressure of 2.0 GPa and then cooling the sample before releasing the pressure. Two high-resolution capacitance dilatometers with temperature ranges of 4-300 K [6] and 150-500 K [7], respectively, were used to measure the thermal expansion. Data were taken at constant heating (cooling) rates, and He exchange gas (10 mbar) was used to thermally couple the samples to the dilatometers. To characterize the dimer phase, Raman-spectra were taken with a FT-Raman spectrometer using a Nd:YAG-laser with a wavelength of 1064 nm.

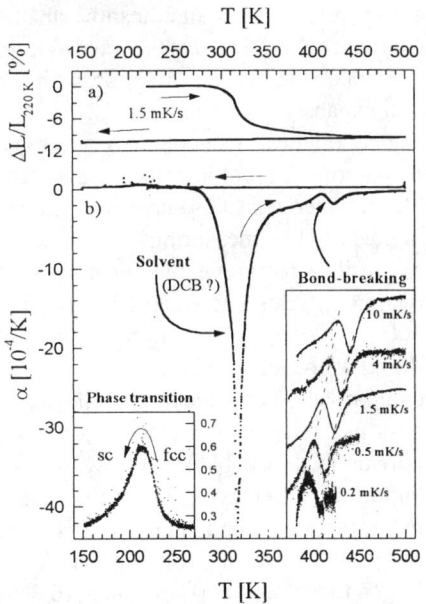

Fig. 1. Linear thermal expansion a) and expansivity b) of the C_{60} dimer.

RESULTS AND DISCUSSION

C_{60}-dimer: In Fig. 1 we present the linear thermal expansion, $\Delta L/L_0$, and the expansivity, $\alpha = 1/L \cdot dL/dT$, of the C_{60} dimer powder for the first heating cycle from 220 K to 500 K and the subsequent cooling cycle from 500 K to 150 K. Anomalies in $\alpha(T)$ are observed at about 315 K and 415 K in the heating curve and are absent in the cooling curve, indicating irreversible processes. The fcc-sc phase transition near 250 K is recovered in the cooling curve, demonstrating that the dimers have been broken apart by heating to 500 K. In order to determine which of the anomalies upon heating is due to bond breaking, Raman spectra [8] of fresh samples were taken before heating, after annealing at 385 K for three hours and after slowly heating to 500 K. These spectra clearly demonstrate that the 315 K peak is due to evaporating solvent (o-dichlorobenzene) and that bond breaking occurs around 415 K, in good agreement with differential scanning calorimetry results from Ref. 5.

The right inset of Fig. 1b shows that the bond-breaking anomaly shifts to lower temperatures with decreasing heating rates. Defining the bond-breaking temperatures $T_{bb} \equiv T_{bond-breaking}$ as either the maxima or minima of the anomalies allows one to plot the logarithm of the heating rate r versus $1/T_{bb}$, from which an activation energy is obtained. The activation energies are $E_a = 1.70 \pm 0.05$ eV (maxima) and $E_a = 1.79$ eV ± 0.05 eV (minima). To convert the heating rate to a bond breaking rate τ_{bb}^{-1}, the bond breaking was simulated with a simple model in which the fraction of dimers y obeys the differential equation

$$\frac{dy}{dT} = \frac{y}{\tau_{bb}(T) \cdot r} \qquad (1)$$

with an activated rate $\tau_{bb}^{-1}(T) = \nu_0 \cdot \exp(-E_a/k_B T)$. The only free parameter in this model is the attempt frequency, which could be determined to $\nu_0 = (2.6 \pm 3) \cdot 10^{17}$ Hz.

The resulting temperature dependent bond breaking rate of the dimer samples is in very good agreement with that of the photo-polymerized C_{60} films from Y. Wang et al. [9], which strongly suggests that these films consist primarily of dimers. (see Fig. 3)

2D pressure-polymerized C_{60}: The thermal expansivity of the 2D-polymerized sample measured along two orthogonal directions is shown in Fig. 2. A surprisingly large anisotropy is observed, which possibly could arise from an uniaxial pressure component during synthesis. The expansivity in both directions is significantly smaller than that of both 'normal' C_{60} [10] and 1D-polymerized C_{60} [11], as expected due to stronger covalent bonding between molecules. The values along the direction with the smaller expansivity approach the very small thermal expansivity of diamond [12].

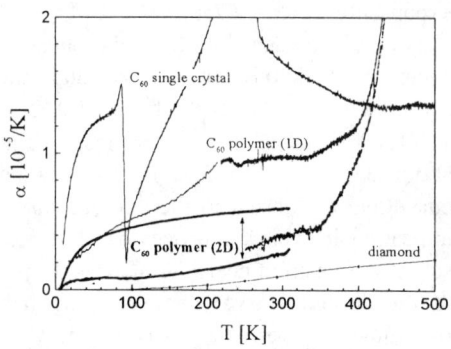

Fig. 2. Thermal expansivity of 2D-polymerized C_{60} compared to 'normal' C_{60} [10], 1D-polymerized C_{60} [11] and diamond [12].

The kinetics of depolymerization and the associated volume increase of the 2D-polymerized C_{60} sample was studied in detail by measuring the thermal expansion for repeated heating and cooling cycles between 150 K and 500 K (see Fig. 2 for the initial part of the first heating cycle). We obtained qualitatively very similar results as in our previous measurements on 1D-polymerized C_{60} [11], but with two important differences. First, the length increase upon depolymerization of the 2D polymer was almost a factor of ten larger than for the 1D-polymer (6.3 % vs. 0.7 %), implying a quite large volume increase in good agreement with the literature [3]. Second, although the bond-breaking rate τ_{bb}^{-1} exhibits an activated behavior with an activation energy $E_a = 1.9 \pm 0.2$ eV for the different heating and cooling cycles just as in the 1D-material, the attempt frequency decreases during the course of the measurement from an initial value close to the value of the 1D-polymer ($\nu_0 = 7 \cdot 10^{15}$ Hz) and ends up with $\nu_0 = 7.3 \cdot 10^{14}$ Hz (see Fig. 3). This may indicate that the sample is a mixture of 1D- and 2D-polymer phases, which are decomposing at different rates. A direct measurement of the time-dependent length-increase at 500 K near the end of depolymerization supports this interpretation; two different relaxation rates (see open-circles in Fig. 3), the values of which excellently match those of the 1D- and 2D-polymers obtained during heating and cooling cycles, are found.

Comparison: The bond-breaking rates of dimers, 1D- and 2D-polymerized materials are compared in Fig. 3. Interestingly, τ_{bb}^{-1} of the dimers is more than 3 orders of magnitude faster than that of the 1D or 2D material. This is quite surprising since experimental results [13] indicate that the dimers are energetically more stable than the

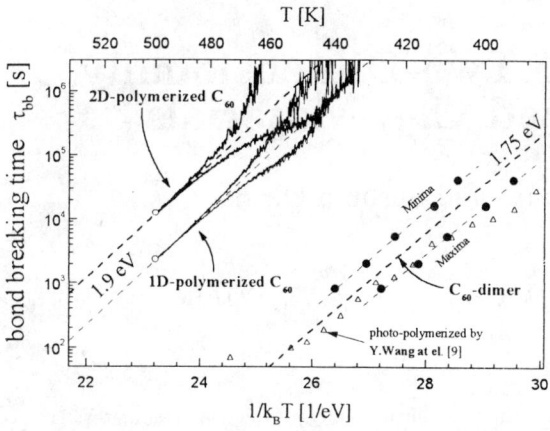

Fig. 3. Bond-breaking time τ_{bb} of the C_{60} dimer (•) and the 2D-polymerized C_{60} compared to 1D-polymerized [11] and photo-polymerized C_{60} (Δ) [9]. (see text for details)

1D or 2D materials. This difference appears to result both from different attempt frequencies and slightly different values of the activation energy. We note that the values of the activation energies found here agree very well with theoretically determined values (~ 1.7 eV) for the 2D-polymer phase [14].

CONCLUSIONS

The kinetics of bond-breaking of C_{60} dimers and 2D-polymerized C_{60} has been studied using high-resolution capacitance dilatometry up to 500 K. The bond-breaking rates are thermally activated with activation energies of 1.7 - 1.9 eV. The expansivity of the 2D-polymerized sample is much smaller than the expansivity of both 'normal' C_{60} [10] and even 1D-polymerized C_{60} [11], and approaches values of diamond [12].

REFERENCES

[1] B. Sundqvist, *Advances in Physics*, in press (1998).
[2] M. Rao et al., *Science* **259**, 955 (1993).
[3] Y. Iwasa et al., *Science* **264**, 1570 (1994).
[4] P.-A. Persson et al., *Chem. Phys. Lett.* **258**, 540 (1996).
[5] G.-W. Wang et al., *Nature* **387**, 583 (1997).
[6] C. Meingast et al., *Phys. Rev.* **B41**, 11299 (1990).
[7] P. Nagel, *Diplom thesis*, Universität Karlsruhe (1996).
[8] S. Lebedkin et al., *Chem. Phys. Lett.* **285**, 210 (1998).
[9] Y. Wang et al., *Chem. Phys. Lett.* **217**, 413 (1994).
[10] F. Gugenberger et al., *Phys. Rev. Lett.* **26**, 3774 (1992).
[11] P. Nagel et al., in *Molecular Nanostructures*, ed. H. Kuzmany, S. Fink, M. Mehring and S. Roth, World Scientific, Singapore (1998) p. 365.
[12] S.I. Novikova, *Sov. Phys. - Solid State* **2**, 1464 (1961).
[13] Y. Iwasa et al., see *these Proceedings*.
[14] S. Saito et al., see *these Proceedings*; S. Okada and S.Saito, to be published.

Energetics of Two-Dimensionally Polymerized C_{60} Materials

Susumu Saito and Susumu Okada

Department of Physics, Tokyo Institute of Technology
2-12-1 Oh-okayama, Meguro-ku, Tokyo 152-8551, JAPAN

Abstract. Energetical stabilities of the two-dimensionally polymerized C_{60} materials, the rhombohedral phase (r-C_{60}) and the tetragonal phase (t-C_{60}) are studied in detail within the framework of the density functional theory. From the study of the single polymerized layer and the isolated deformed C_{60} unit in addition to the C_{60} polymer phase of interest, the interlayer interaction is found to be sizable both in r-C_{60} and t-C_{60}. Also it is found that the bond energy of the interfullerene chemical bond in polymerized materials is less than 2 eV, which is much smaller than that of the ideal bond between sp^3 C atoms.

INTRODUCTION

The discovery of solid C_{60} in which C_{60} fullerenes form close-packed crystalline lattice [1] triggered the extensive research on this new crystalline solid carbon next to graphite and diamond. Semiconducting solid C_{60} is electronically different from insulating diamond which consists of sp^3-hybridized C atoms [2]. Although solid C_{60} consists of sp^2 C atoms as in the case of graphite, the system is a good electron acceptor while the hole injection has not been achieved. This interesing difference between two sp^2 C systems is due to the presence of five-membered rings in C_{60}, which gives rise to the electron-hole asymmetry to the system.

In addition to this third form of carbon, solid C_{60}, the polymerization of C_{60} is found to take place under the external pressure at high temperatures, realizing another new form of crystalline carbon consisting both of sp^2 and of sp^3 hybridized C atoms [3]. So far three distinct polymerized phases have been identified [4]: the rhombohedral phase, the tetragonal phase, and the orthorhombic phase. In these polymerized phases, the four-membered rings are formed between C_{60} units. Interestingly, these phases are stable at room or lower temperatures while at high temperatures the depolymerization is found to take place and they reconvert to the face-centered cubic (fcc) solid C_{60}. Therefore, the energetical stability of these new crystalline phases of solid carbon is not only of high interest but also important to address the novel polymerization and depolymerization behaviors observed.

The strength of the interfullerene bond with the four-membered ring is also an interesting issue to be studied in detail.

We have performed the density-functional total-energy calculations for these phases in order to address the above issues. In this paper, we report the details of the two-dimensionally polymerized phases, that is, the rhombohedral C_{60} and the tetragonal C_{60} with the analysis of the single polymerized sheets and the deformed isolated C_{60} units. Based on these results, we discuss the energetics of these two phases in term of the strength of the interlayer interaction as well as that of the interfullerene chemical bonds.

COMPUTATIONAL METHODS

The total energy of each system studied has been obtained by using the local-density approximation within the framework of the density-functional theory [5,6]. We use the plane-wave basis set with the norm-conserving pseudopotenail [7] to treat the valence electrons selfconsistently [8]. The cutoff energy of the plane-wave basis used is 50 Ryd. Also we use the separable approximation for the pseudopotential [9]. The present procedure has been applied to a number of carbon-based systems including diamond and graphite and has proven to work with sufficient accuracy. The interlayer interaction of graphite is found to be well described in this formalism [10]. This is an important point for the present study to discuss the interlayer interaction of the two-dimensionally polymerized C_{60} phases.

RESULTS

In Fig. 1, total energies obtained for the systems studied are summerized [11]. All the energies are measured from the energy of the graphite.

In the case of the fcc C_{60}, the isolated C_{60} is found to gain 0.018 eV per atom in forming the crystalline lattice due to the van der Waals-type interfullerene interaction. For this calculation the experimental fcc lattice constant is adopted and the internal geometry has been fully optimized. In the case of the isolated C_{60}, on the other hand, the system is placed in a large unit cell to eliminate the interfullerene interaction and its geometry has been also fully optimized.

In the case of the rhombohedral C_{60} polymer (r-C_{60}), as we have previously reported [12], the system is found to be energetically as stable as the fcc C_{60}. Interestingly, from the study of the isolated polymerized layer, the interlayer interaction of this two-dimensionally layered system is found to be sizable. The interaction energy per fullerene pair across the layer is 0.142 eV. In the study of the r-C_{60} we also used the experimental lattice constants and again the internal geometrical freedom has been optimized. This optimized geometry is used also for the hypothetical single-layer system. Furthermore in order to study the strength of the interfullerene C–C bonds, we have calculated the total energy of the deformed (D_{3d}-symmetry) isolated C_{60} unit with the same geometry as that in the unit cell of the optimized

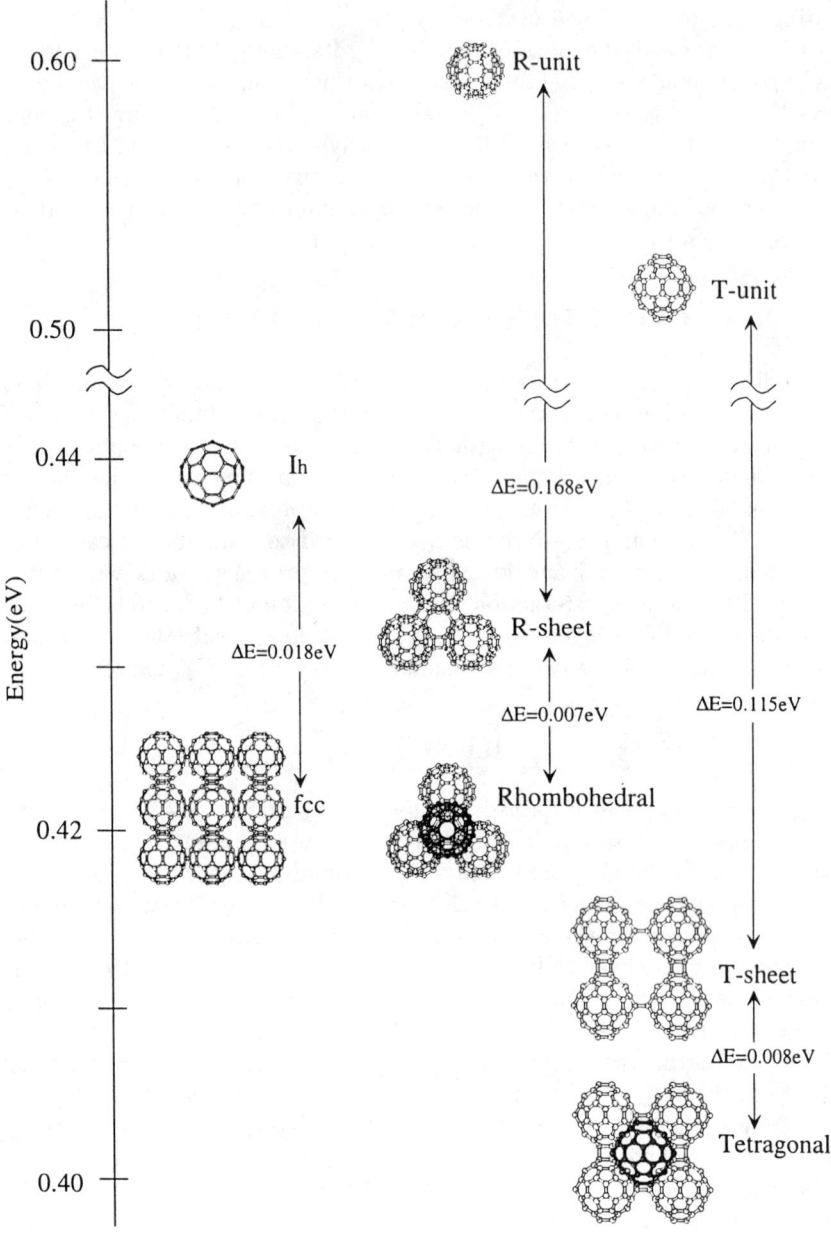

Fig. 1 Schematic energy diagram of C_{60} in various forms [11].

r-C_{60}. Comparing to the total energy of the single C_{60} layer, the energy gain due to the interfullerene interaction within the layer is found to be 3.36 eV per fullerene pair. If we attribute this value all to the interfullerene covalent bonds, its bond-energy value is to be 1.7 eV per bond. This value is in good agreement with the experimentally reported bond-breaking energy for two-dimensionally polymerized C_{60} of 1.9 eV [13].

The similar analysis has been done also for the other two-dimensionally polymerized C_{60}, the tetragonal C_{60} (t-C_{60}). The experimentally observed lattice constant values have been used. The interlayer interaction energy obtained from the total energies of the t-C_{60} and its hypothetical single-layer system is found to be 0.114 eV per fullerene pair across the layer. This energy gain per pair is found to be similar to that in r-C_{60}. We also have studied the isolated deformed (D_{2h}-symmetry) C_{60} unit with the same geometry as that in the unit cell of the optimized t-C_{60}. Surprisingly, this D_{2h} C_{60} unit is found to have much lower total energy than that of the isolated deformed D_{3d} unit of r-C_{60}. This energy difference is found to persist in the single polymerized layers of r-C_{60} and t-C_{60}, and, furthermore, in r-C_{60} and t-C_{60} themselves. The energy gain due to the interfullerene interaction witin the polymerized layer in t-C_{60} is found to be 6.90 eV per C_{60}, which corresponds to 3.45 eV per fullerene pair. If we again consider this to represent the bond energy of the interfullerene C–C bonds, it is to be 1.7 eV per bond. It is interesting to point out that the bond energy of the ideal C–C single bond between sp^3-hybridized C atoms estimated from diamond is more than 3.6 eV. Therefore, the interfullerene C–C bonding in the polymerized C_{60} materials is generally much weaker than the ideal bonds. It would be interesting to systematically study the strengths of the four-membered ring C–C bonds in other systems.

CONCLUDING REMARKS

From the present study, the energetics of the deformed C_{60} unit is found to be important in considering the energetical stability of the polymerized C_{60} systems. It is now of high interest whether the three-dimensional C_{60} polymer phases can be synthesized or not. It should be also related to the unidentified superhard carbon systems recently reported using the high-pressure method. In considering such possible new C_{60} phases, the energetics of the deformed C_{60} unit is one of the key issues to be addressed. The total-energy electronic-structure calculations will continue to play an important role in this field.

ACKNOWLEDGEMENTS

This work was supported by the Japan Society for the Promotion of Science (Research for the Future Program, No. 96P00203), The Ministry of Education, Science and Culture of Japan, its Grant-in-Aid for Creative Basic Research (No. 08NP1201), and The Nissan Science Foundation. We would like to thank

Professor A. Oshiyama for providing the program used in this work. Numerical calculations reported were partly performed at the Computer Center of the Institute for Molecular Science and the Supercomputer Center, Institute for Solid State Physics, University of Tokyo.

REFERENCES

1. W. Krätschmer, L. D. Lamb, K. Fostiropoulos, and D. R. Huffman, *Nature* **347** (1990) 354.
2. S. Saito and A. Oshiyama, *Phys. Rev. Lett.* **66** (1991) 2637.
3. Y. Iwasa *et al.*, *Science* **264** (1995) 1570.
4. M. Nunez-Regueiro, L. Marques, J-L. Hedeau, O. Bethoux, and M. Perroux, *Phys. Rev. Lett.* **74** (1995) 278.
5. P. Hohenberg and W. Kohn, *Phys. Rev.* **136** (1964) B864.
6. W. Kohn and L. J. Sham, *Phys. Rev.* **140** (1965) A1133.
7. N. Troullier and J. L. Martins, *Phys. Rev.* **B 46** (1992) 1754.
8. O. Sugino and A. Oshiyama, *Phys. Rev. Lett.* **68** (1991) 1858.
9. L. Kleinman and D. M. Bylander, *Phys. Rev. Lett.* **48** (1982) 1425.
10. Y.-K. Kwon, S. Saito, and D. Tománek, to be published in *phys. Rev. B*.
11. S. Okada and S. Saito, *to be published*.
12. S. Okada and S. Saito, *Phys. Rev.* **B 55** (1997) 4039.
13. P. Nagel, V. Pasler, S. Lebedkin, C. Meingast, B. Sundqvist, T. Tanaka, and K. Komatsu, in this proceedings volume.

Structural properties of a C_{120} crystal

Christophe Laforge, Patrick Senet, Philippe Lambin

Département de physique, Laboratoire de physique du solide,
Facultés Universitaires Notre-Dame de la Paix, 61 rue de Bruxelles, B-5000 Namur, Belgium

Abstract. With well-know models used for the study of orientational ordering transition in C_{60} crystal and with the crystallographic data from G-W.Wang and al.[Nature 387,583(1997)] , this work aimed at characterizing a possible orientational ordering transition in C_{120} crystal. The C_{120} molecular structure used is based on MNDO optimization. With the same model, we have investigated the effect of the crystal surface on the ordering phase temperature. An orientational ordering phase transition was predicted to take place at 220K in the bulk and 175K at the surface.

I. INTRODUCTION

C_{120} is formed by a [2+2] cycloaddition of two C_{60} molecules. The bonds involved in this dimerization are two double ones shared by two hexagons on adjacent C_{60} molecules. Theoretical investigations have showed that this cycloaddition, also called 66/66, was the most stable configuration[1] ,for a neutral C_{120} molecule. All four-membered ring atoms, which were sp^2 hybridized in C_{60}, become sp^3 in the dimer. The symmetry of this molecule is D_{2h}. An X-ray structural analysis[2] confirms the existence and the structure of the dumb-bell shaped C_{120}.

Crystals of C_{60} dimers have been recently synthetized[2]. Because of the proposed alignement of the C_{120} molecules in the solid[2] and due to the D_{2h} symmetry of them, a phase transition corresponding to the frustated rotation of C_{120} around the axis of the C_{60}-C_{60} covalent bond is expected. This effect is studied here for the first time in the bulk and at the surface of this molecular crystal by using the Monte Carlo algorithm.

A motivation of this work came from the problem of finding a suitable intermolecular potential describing the order-disorder phase transition in crystal C_{60}. Parameters or physical insight may be gained from the simpler case of C_{120} crystals.

The paper is organized as follows. The model of the C_{120}-C_{120} interaction and of the C_{120} crystal is presented, see II. Simulations of the order-disorder transition on the bulk as at the surface for this model are discuded in III . Prelimary conclusions end the paper,see IV.

II. MODEL

The potential between two C_{60} molecules proposed by Lu et al.[3] is used to calculate the C_{120}-C_{120} interactions. This theory has given good information on equilibrium configurations for the C_{60} below the ordering transition at 260K. In this model[3], the molecular binding energy is computed as the sum of carbon-carbon Lennard-Jones pair potential plus a Coulomb energy due to static charges distributed on the C_{60} bonds. The latter term models the electrostatic energy due to inhomogeneity of the C_{60} molecular ground-state electronic density.

The pair potential, for C_{60} is, explicitely :

$$V_{12} = \sum_{i,j=1}^{60} 4\varepsilon \left[\left(\frac{\sigma}{|r_{1i}-r_{2j}|} \right)^{12} - \left(\frac{\sigma}{|r_{1i}-r_{2j}|} \right)^{6} \right] + \sum_{m,n=1}^{90} \frac{q_m q_n}{|b_{1m}-b_{2n}|}$$

were r_{ij} is the position of jth carbon atom on the ith C_{60}, b_{ij} is the position of jth bond on the ith C_{60} and q_i is the charge affected on the ith bond.

The model makes distinction about three kinds of interacting sites on the C_{60} molecule:
* carbon atoms
* single bonds(q_s)
* double bonds($-2q_s$)

and is based on the following parameters:
$q_s=0.27$ e, $\varepsilon=2.964$ meV, $\sigma=3.407$ Å.

Due to the large size of the system, the present preliminary calculations were not carried out on the full tridimensionnal crystal but on a bidimensional model instead where the C_{120} molecules are aligned on a square lattice like in a cigar's box. The dimension of the lattice is 20*20 with fixed orientation on the boundaries.

One expects that the dimerization of the C_{60} will induce a charge redistribution on the C_{60} cages. As shown in recent first principle simulations[4], this perturbation is strongly localized in the vicinity of the four covalent C_{60}-C_{60} bonds. In agreement with this result, we modified the bonds involved in the cycloaddition : the two short bonds 66, with a positive charge $2*q_s$, became four long bonds, with a negative charge $-q_s$. The balance of the global charge on the structure followed by assuming simply that each carbon atom keeps a neutral environment.

III. RESULTS

By using the Monte-Carlo technique with decreasing temperatures, we have observed a low temperature equilibrium configuration where each molecule can be orientated along four specific angles(figure 1). The angles can be grouped in pairs corresponding nearly to the same energies. This degeneracy prevents a long range ordering. However, the lattice can be characterized by a 2x2 cell having two equiprobable configurations of the molecules belonging to it.

We characterized the order of the molecules by an order parameter η defined as a normalized sum of the scalar product of the orientation vector $\vec{S}_i$ of all the N molecules i with their low temperature orientation vector $\vec{S}_{0i}$:

$$\eta = \frac{1}{N}\sum_{i=1}^{N}\left(\vec{S}_i \cdot \vec{S}_{0i}\right)$$

$\eta=1$ corresponding there to the 2X2 low temperature phase.

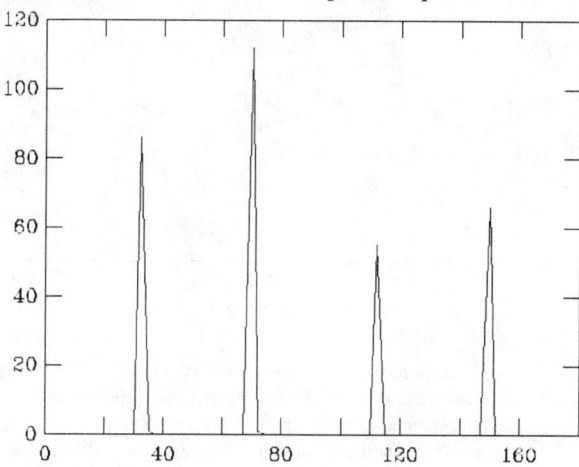

figure 1 : angular distribution of the C_{120} molecules calculated at low temperature. (the axial angles are given between 0° to 180° due to the D_{2h} symmetry of the dimer. The 0° angle corresponds to the [100] bravais vector)

A. Bulk ordering-desordering transition

As expected, an ordering-desordering phase transition is found in the Monte-Carlo simulations. It corresponds to a transition from a freely rotation around the main axis of the dimerized molecules to a frozen intermolecular configuration of the crystal characterizad by four angles(figure 1). The calculated transition temperature occurs at 220K for the present bidimensional model(figure 2). Future experiments would be able to show such a transition.

B. Surface orderering-desordering transition

The effect of a crystal surface on the molecular ordering was investigated with the same bidimensionnal model of the C_{120} crystal. A lowering of the temperature phase transition from 220K to 175K was obtained.

The surface effect observed in the present modelling of C_{120} phase transition is similar to the effect observed in C_{60}. Indeed, experimental data, from helium atom scattering and LEED pattern, have shown that the ordering phase transition temperature (fcc to sc) in C_{60} was 35K lower at the surface than in the bulk.[5]

With our computations, we were able to show up that Monte-Carlo algorithm can explain this phenomenon which is due to the number of neighbours being reduced at the surface.[6]

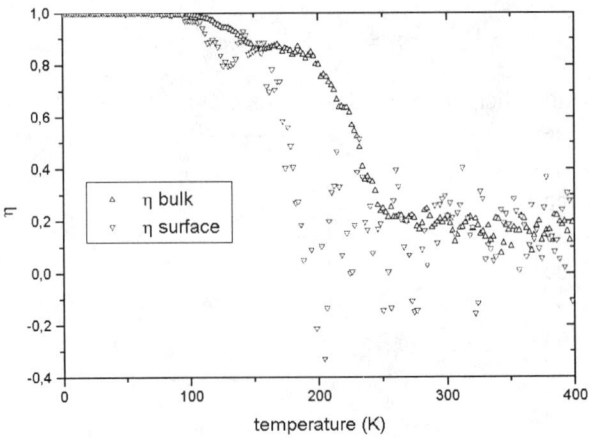

figure 2 : computed order parameter versus temperature. A phase transition corresponding to ordering-desordering transition clearly shows up. The square correspond to order parameter in the bulk, and circles to surface. A difference between the two transition temperatures is observed.

IV. CONCLUSION

A possible ordering-desordering phase transition in C_{120} molecular crystal has been investigated by using with the Monte-Carlo algorithm with a classical intermolecular potential[3]. Using simple bidimensionnal lattice model of the C_{120} crystal, one predicts an order-disorder phase transition occuring at 220K and 175K respectively for the bulk crystal and at the surface. These values are of the same order of magnitude as the phase transition temperatures measured for the C_{60} crystal at 260K for the bulk and 225K at the (111) surface.

Unfortunatly, experimental data are not yet available for C_{120}. We hope that the preliminary results would stimulate such investigations.

V. REFERENCES

1. Kürti J., Németh K., Chemical Physics Letters, 256,p.119(1996)
2. Wang G. and al, Nature 387, p.583-586 (1997)
3. Lu J.P., Li X.-P., R.M.Martin , Physical Review Letters 68(10), p.1551 (1992)
4. Esfarjani K. and al., Physical Review B, 57,1,P.223(1998)
5. Glebov A., Senz V., Toennies J.P., Gensterblum G., J.Appl.Phys., 82(5),p.2329(1997)
6. Goldoni A., Cepek C., Modesti S., Physical Review B, 54(4), p.2890(1996)
7. Savin S., Harris A.B., Yildirim T., Physical Review B, 55,21,p.14182(1997)

Electrosynthesis and Characterization of Dimers of Mono Alkyl Adducts of C_{60}

K. Noworyta,[a] P. Kuran,[b] L. Dunsch,[b] and W. Kutner[a]

[a] *Institute of Physical Chemistry, Polish Academy of Sciences, Kasprzaka 44, 01-224 Warsaw, Poland*
[b] *IFW Dresden e.V., Institut für Festkörperforschung, Helholtzstrasse 20, D-01069 Dresden, Germany*

Abstract. First results are presented on selective electrosynthesis of dimers of methyl and n-octyl adducts of C_{60}. That is, electrochemically generated C_{60}^{2-} was reacted with suitable n-alkyliodide. Cumulatively, the HPLC and MS results indicate that dimers of mono alkyl adducts were formed. Thermal decomposition of the dimers leads to respective monomers. The dimers show UV-vis spectra markedly different from each other and from that of parent C_{60}. In toluene, dimers slowly decompose to form respective monomers.

INTRODUCTION

A mono-linked dimer of *tert*-butyl adduct of C_{60} (1), bis-linked dimers $C_{120}O$ (2, 3), $C_{121}H_2$ (3) and tetra-linked dimer $C_{120}O_2$ (4) were prepared by chemical synthesis while a C_{60} [2 + 2] dimer by solid-state mechanochemical synthesis (5). Herein, we present our preliminary results on selective electrosynthesis of mono-linked dimers of methyl and n-octyl adducts of C_{60} along with their HPLC separation as well as MS and UV-vis spectroscopy characterization. By electrosynthesis in aprotic solvent solutions, alkyl adducts of fullerenes are conveniently prepared because the charge transferred and potential applied can be easily controlled (6).

EXPERIMENTAL

Instrumentation. Commercial chemicals were used for controlled-potential electrosyntheses performed by using Autolab electrochemical system (Eco Chemie, The Netherlands) with a H-type three-electrode cell. The Pt grids served as the working and auxiliary electrodes. An NaCl saturated calomel electrode, separated with a salt bridge from the working solution, was used as the reference electrode. The electrosyntheses products were separated by HPLC on Cosmosil Buckyprep columns (Nacalai Tesque, Japan). The products were identified by means of MAT 95 (Finnigan, Finland) sector field mass spectrometer using chemical ionization with methane and negative ion detection (7). UV-vis spectra were recorded on a Model UV-3100 (Shimadzu, Japan) spectrophotometer.

Quantum chemistry calculations. Semi-empirical (PM3) calculations of heat of formation for the dimers were performed by using a HyperChem 5 software (8) and a pentium 166 MHz microcomputer as well as a pentium 133 MHz two-processor computer system.

RESULTS AND DISCUSSION

Dimers of both n-octyl and methyl adduct of C_{60} were prepared by selective electrosynthesis. For that purpose, C_{60}^{2-}, electrochemically generated at constant potential in 0.1 M tetra-n-butylammonium hexafluorophosphate, (TBA)PF$_6$, in benzonitrile, was reacted with either methyl- or n-octyliodide, according to a modified procedure described earlier (6). The combined HPLC and MS results, similar for both alkyliodides used, indicate that dimers of mono alkyl adducts are formed. That is, two closely spaced peaks **1** and **2** are seen for each product mixture of electrosynthesis in the HPLC chromatogram (Fig. 1). Retention times of these peaks are much larger than that of C_{60} indicating that solubility of the products in toluene is much smaller than that of C_{60}. The m/z values determined by MS for **1** and **2** were the same and equal to 736 and 833.6 for methylated and n-octylated product, respectively. These values are equal to the m/z value for the corresponding mono alkyl adducts. Apparently, thermal decomposition of these dimer isomers results in respective monomers.

UV-vis spectra of the dimers markedly differ one from the other and from that of pristine C_{60}. (Fig. 2). That is, a broad band in a visible range, characteristic for C_{60}, is gone and UV bands are largely suppressed in the dimer spectra. In toluene solutions, dimers are unstable with respect to decomposition on a month time scale. That is, they slowly decompose to form respective monomers. This decomposition was manifested by a simultaneous slow decrease of the HPLC peaks **1** and **2** and increase of the peak corresponding to the respective monomer.

*S*emi-empirical quantum chemistry calculations of heat of formation, ΔH_f, indicate that 1,4-1,4 and 1,2-1,4 regio isomers (Scheme 1) are the most stable dimers of both methyl and n-octyl adduct of C_{60} (Table 1) and, therefore, these isomers are ascribed

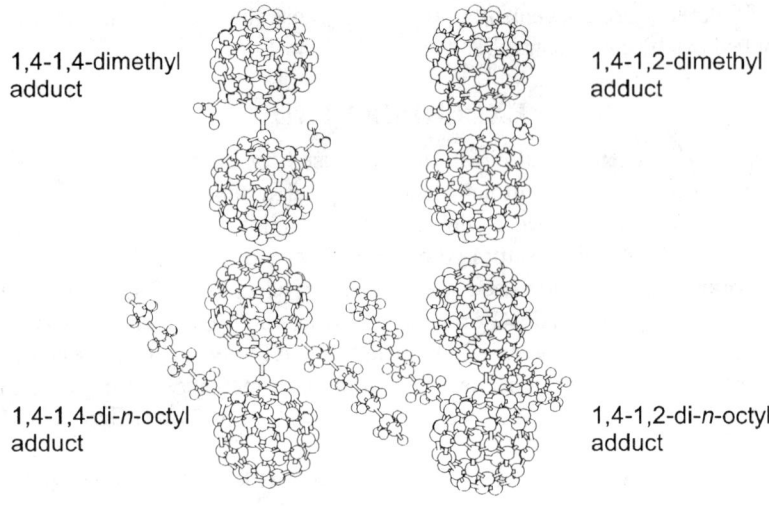

1,4-1,4-dimethyl adduct

1,4-1,2-dimethyl adduct

1,4-1,4-di-n-octyl adduct

1,4-1,2-di-n-octyl adduct

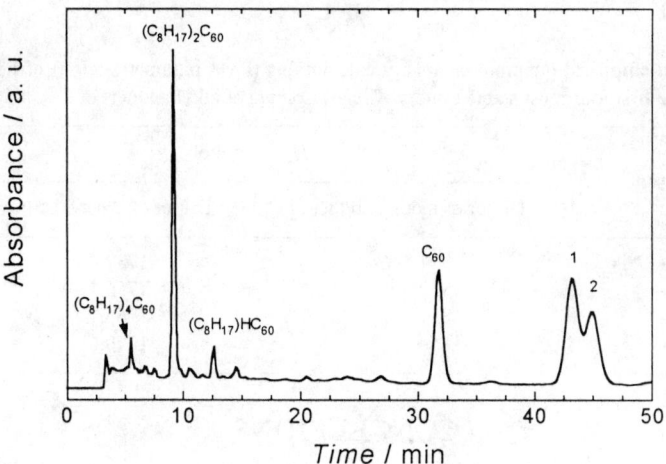

FIGURE 1. HPLC chromatogram of a product mixture of electrosynthetic n-octylation of C_{60}; Cosmosil Buckyprep column (4.6 × 250 mm); 0.95 mL min^{-1} toluene/hexane, 1 : 1 (v/v); sample volume 20 μl; λ = 340 nm.

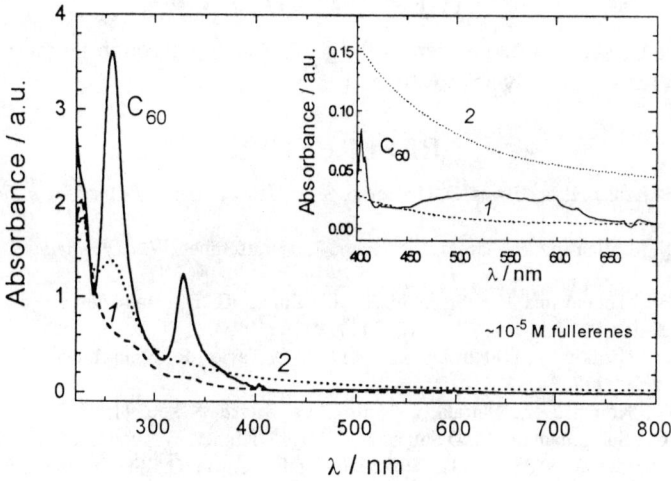

FIGURE 2. UV-vis spectra of ~10^{-5} M C_{60} and regio isomers **1** and **2** of dimer of C_{60} methyl adduct.

tentatively to **1** and **2**, respectively. Interestingly, the ΔH_f values for dimers of n-octyl adducts are smaller than those for dimers of methyl adducts. This result is in agreement with the relative product distribution determined by HPLC. Moreover, the calculated ΔH_f values for the dimers are over twice as large as those for the corresponding monomers (9).

TABLE 1. Semi-empirical quantum chemistry calculations (PM3 parametrization) of heat of formation, ΔH_f, for the first four most stable dimers of regio isomers of alkyl adducts of C_{60}.

Regio isomer	ΔH_f, kcal mol^{-1}	
	Dimer of methyl adduct of C_{60}	Dimer of n-octyl adduct of C_{60}
1,4-1,4-	1804.7	1737.4
1,2-1,4-	1819.1	1756.1
1,4-1,6-	1837.8	1766.1
1,2-1,6-	1850.0	1770.6

CONCLUSIONS

Dimers of alkyl adducts of C_{60} can be prepared selectively by controlled-potential electrosynthesis. Two different regio isomers of the dimers can be obtained. UV-vis spectra of the isomers markedly differ one from the other and from that of pristine C_{60}. Quantum chemistry calculations of heat of formation indicate that 1,4-1,4 regio isomers are the most stable dimers of both methyl and n-octyl adduct.

ACKNOWLEDGES

Financial support of Volkswagen-Stiftung (Germany), through Grant No. I/70 377 to LD and WK, is gratefully acknowledged.

REFERENCES

1. Fagan, J. P., Krusic, P. J., Evans, D. H., Lerke, S. A., Johnston, E., *J. Am. Chem. Soc.* **114**, 9697 (1992).
2. Lebedkin, S., Ballenweg, S., Gross, J., Taylor, R., Krätschmer, W., *Tetrahedr. Lett.* **36**, 4971 (1995).
3. Smith, A. B., Tokuayama, H., Strongin, R. M., Furst, G. T., Romanow, W. J., Chait, B. T., Mirza, U. A., Haller, I., W., *J. Am. Chem. Soc.* **117**, 9359 (1995).
4. Gromov, A., Lebedkin, S., Ballenweg, S., Avent, A. G, Taylor, R., Krätschmer, W., *J. Chem. Soc. Chem. Commun.* 209 (1997).
5. Wang, G-W., Komatsu, K., Murata, Y., Shiro, M., *Nature* **387**, 583 (1997).
6. (a) Caron, C., Subramanian, R., D'Souza, F., Kim, J., Kutner, W., Jones, M. T., Kadish, K. M., *J. Am. Chem. Soc.* **115**, 8505 (1993); (b) D'Souza, F., Caron, C., Subramanian, R. Kutner, W., Jones, M. T., Kadish, K. M., in *Recent Advances in the Chemistry and Physics of Fullerenes and Related Materials*, Kadish, K. M., and Ruoff, R. S., Eds., The Electrochemical Society Proceedings Series, Pennington NJ, 1994, p. 768, (c) Mangold, K.-M., Kutner, W., Dunsch, L., Fröhner, J., *Synth. Metals* **77**, 73 (1996).
7. Dunsch, L., Kirbach, U., Klostermann, K., *J. Mol. Struct.* **348**, 381 (1995).
8. HyperChem Release 5.0 for MS Windows, Hypercube, Inc. 1996. Serial No. 500-10002189
9. Noworyta, K., Kuran, P., Nantsis, E. A., Bilewicz, R., Dunsch, L., Kutner, W., in preparation.

Catalytic C_{120}: vibrational spectra and stability

B. Burger and H. Kuzmany

Inst. f. Materialphysik der Universität Wien, Austria

K. Komatsu

The Institute for Chemical Research, Kyoto University, Japan

Abstract. We present Raman and IR spectra of a chemically produced dimer of C_{60}. A detailed description of the spectra is given and a comparison to the spectra of C_{60} photopolymers and to theoretical predictions is performed. It is found that the measured spectra match the calculated ones reasonably well. Comparison to the spectrum of C_{60} photopolymerized at 380 K shows a high similarity, thus proving that this photopolymer consists mostly of dimers. Furthermore the stability of the material against laser irradiation is discussed. It is found, that all spectral features vanish upon irradiation with 514 nm. The spectrum is regained with slight shifts after 12 hours recovery.

I INTRODUCTION

Polymerization of C_{60} has been reported in several papers and has turned into an active field of fullerene research. Since the first observation of photopolymerization [1] and its explanation by *Eklund et al.* [2] many different polymeric forms of C_{60} have been discovered. In all cases the formation of the polymers is highly uncontrolled and depends on external parameters like pressure or temperature. This holds for the photopolymers as well as for the pressure polymers and the polymeric forms of AC_{60}. Thus the study of the vibrational spectra of these systems turns out to be difficult, as in most samples mixtures of different oligomers or configurations are present. A good example is the photopolymer, where one observes different spectra depending on the sample temperature during the irradiation process [5]. Even though the differences in the spectra are reasonably well defined and the assignment of the obtained spectra to different oligomer geometries by comparison to calculated model spectra is possible, a comparison with a spectrum of a pure oligomeric or polymeric phase is desirable. Since dumbell shaped $(C_{60})_2$

molecules, as produced by a pure chemical reaction, are now available, comparative measureme

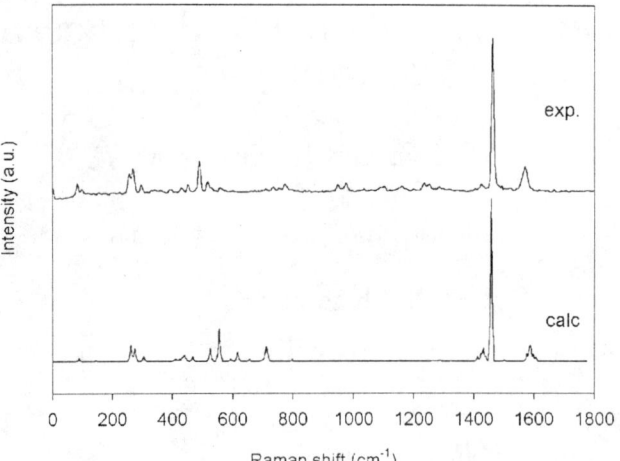

FIGURE 1. Raman spectrum of the C_{60} dimer. The line at 75 cm^{-1} is a filter artifact. The lower curve is a calculated spectrum after [4]. The part between 0 and 800 cm^{-1} is unscaled, the part above 800 cm^{-1} was downscaled by 0.887, according to a comparison of calculated and experimental Raman spectra of pure C_{60}.

II EXPERIMENTAL

The raw C_{120} material was produced using the so called vibrating mill technique. The production and purification details are described in [3]. About 1 mg of material was compacted and used for Raman measurements. Raman specra were taken under ambient conditions using an FT-Raman system equipped with a Nd:YAG laser working at 1064 nm and a liquid nitrogen cooled Ge detector. The system resolution was 4 cm^{-1}. Additional Raman measurements were made with a DILOR XY system. In that case an Argon ion laser operated at 514.5 nm was used for excitation. These measurements were made under ambient conditions as well as at a temperature of 80 K and a vacuum of 10^{-6} mbar.

For IR transmission measurements about 1 mg of the material was ground with 150 mg of KBr and pressed into pellets using a pressure of 7 Kbar. The sample chamber was evacuated to avoid signals from water vapor and CO_2. An MCT detector was employed for detection.

III EXPERIMENTAL RESULTS

In Fig. 1 the Raman spectrum of the dimer is shown. The lowest mode is observed at 99 cm^{-1}. The line at 75 cm^{-1} is an artifact due to the notch filter of the system. The lower curve is a dimer spectrum as calculated by *Porezag et al.* The agreement between the calculated spectrum and the experiment is remarkably good, particularly in the low frequency range, where the calculated spectrum is unscaled. The most intense line is observed at 1464 cm^{-1} and corresponds to the $A_g(2)$ mode of pristine C_{60}. This mode was recorded in other polymeric systems between 1457 cm^{-1} and 1462 cm^{-1}. The splitting of the fivefold degenerate H_g modes is observed as well. The splitting of the $H_g(1)$ mode is of special interest, as its components are used for the identification of different oligomers. The splitting is similar to the one measured at the material phototransformed at 380 K. This similarity of spectral features is also true for the observed stretching mode. In the dimer sample this mode is located at 99 cm^{-1}, as compared to 98 cm^{-1} in the 380 K photopol

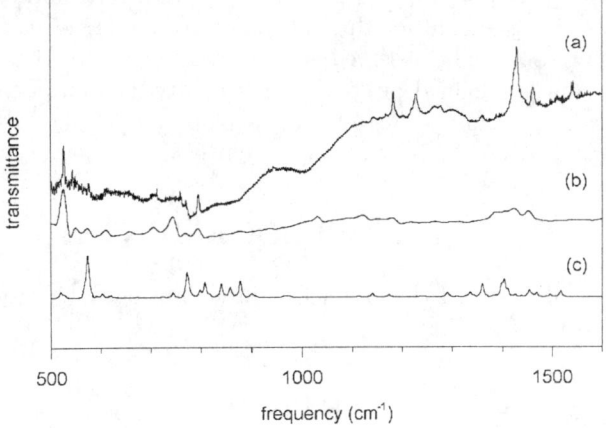

FIGURE 2. IR absorption spectrum of a photodimer (a), a chemically produced dimer (b) and calculated spectrum (c).

In Fig. 2 the IR absorption spectrum of a dimer sample is shown. The similarity to the spectrum of the photodimer is obvious, like in the case of the Raman spectrum. The most prominent mode is the $T_{1u}(1)$ derived mode at 524 cm^{-1}. The characteristic feature of polymerization, the downshift of the $T_{1u}(4)$ mode from 1428 cm^{-1} to 1424 cm^{-1} is observed as well. Additionally several new modes in the region beween 500 cm^{-1} and 800 cm^{-1} can be seen. Characteristic for the dimerization is the mode at 1461 cm^{-1}, which is seen also in the photopolymer produced at 380 K. The match with the calculated IR spectrum from [6] is good with respect to the frequencies and reasonably good with respect to the intensity.

Using 514 nm excitation for the Raman measurements leads to an overall decrease in mode intensity with irradiation time. However, unlike in the case of the polymerization of pure C_{60} no new lines or line shifts are observed. Continued irradiation results in the complete vanishing of the Raman spectrum. This observation can be made under ambient conditions as well as under vacuum and at temperatures as low as 80 K. Surprisingly the spectrum is regained if the sample is left unperturbed for a period of 12 hours. After this procedure the spectrum stays stable even under irradiation with 514 nm laser light. The only observable difference is an overall downshift of the modes of 1 - 2 cm^{-1} as compared to the untreated sample.

IV DISCUSSION

The match of the vibrational spectra with the calculated ones gives another proof for the dumbell structure of the C_{60} dimer in addition to the NMR and X-ray results presented in [3]. The analysis of the vibrational spectra of the dimer also supports the claim that the photopolymer consists mostly of dimers as formulated in [5] . The instability of the spectrum for the 514 nm excitation is not easily explained. Since no line of pure C_{60} can be observed during irradition and furthermore a nearly unchanged spectrum is regained after recovery the dissociation of the dimer into two monomers can be excluded. As a simple explanation for the bleaching of the spectra a diffusion of rest catalyst and subsequent evaporation can be assumed.

V ACKNOWLEDGEMENTS

This work was supported by the "Hochschuljubiläumsfonds der Österreichischen Nationalbank" project no. 5799.

REFERENCES

1. S.J. Duclos, R.C. Haddon, S.H. Glarum, A.F. Hebard, K.B. Lyons, *Sol. State Comm.* **80** (1991) 481
2. A.M. Rao, P. Zhou, K.A. Wang, G.T. Hager, J.M. Holden, Y. Wang, W.T. Lee, X.X. Bi, P.C. Eklund, D.S. Cornett, M.A. Duncan, I.J. Amster, *Science* **259** (1993) 955.
3. K. Komatsu et al., *Nature* **387** (1997) 583.
4. D. Porezag, M.R. Pederson, T. Frauenheim, T. Köhler, *Phys. Rev.* **B 52** (1995) 14963.
5. B. Burger, J. Winter. H. Kuzmany, *Z. Phys.* **B 101** (1996) 227.
6. J. Kürti, K. Nemeth, L. Udvardy, *Proceedings of the IWEP NM 1995* (1995) 323.

C_{120} AND $C_{120}O$: VIBRATIONAL SPECTROSCOPY AND PM3 CALCULATIONS

H.-J. EISLER, F. H. HENNRICH and M. M. KAPPES

Institut für Physikalische Chemie, Universität Karlsruhe,
76128 Karlsruhe, Germany

We present Raman and DRIFT measurements on microcrystalline C_{120} and $C_{120}O$. These are compared to semi-empirical PM3 quantum chemical calculations.

1. Introduction

Interest in covalently linked fullerene dimers ranges from photopolymerization, over RbC_{60}, to laser desorption induced coalescence. The C_{60} dimer related species C_{119} [1], $C_{120}O$ [2] and $C_{120}O_2$ [3] have been implicated as intermediates in the thermal degradation of C_{60} -in the presence of O_2. C_{120}, the C_{60} dimer has been generated by KCN catalysed mechanochemistry [4]. It is a [2+2] cycloadduct linked by two C-C bonds, which form a fourfold ring with the two cages. As part of an ongoing study of the vibrational properties of fullerene homo- and heterodimers [5], we have synthesized C_{120} and $C_{120}O$. Here, we compare their diffuse IR reflectance (DRIFT) and preresonant Raman spectra. Two points are of particular interest: (i) the nature of the three additional vibrations induced by the oxygen atom and (ii) modifications to low frequency cage-cage stretching and torsional modes due to the stronger link in going from $C_{120} \rightarrow C_{120}O$.

2. Experimental

C_{120} was synthesized in analogy to the procedure of ref. 4, by vigorously grinding a mixture of C_{60} and 20 molar equivalents of KCN in air [Specamill 6000 - Graseby Specac]. $C_{120}O$ and $C_{120}O_2$ were obtained as byproducts (probably from a $C_{60}O$ contaminant already present in the reaction mixture). C_{120} was then separated from the oxides by HPLC with a Cosmosil Buckyprep column and toluene as eluent. $C_{120}O$ was prepared and purified as previously described [5]. DRIFT spectra were recorded under ambient conditions using an appropriately equipped Bruker IFS88 FT-IR spectrometer. Dimers were dispersed as microcrystalline solids in KBr. KBr:dimer ratios were approximately 100:1 by weight. Spectra are shown for C_{60}, C_{120} and $C_{120}O$ in figure 1. Plotted is the Kubelka-Munk (KM) function $f(R)=(1-R^2)/2R$, where R is the diffuse reflectance [6]. For sufficiently thick samples, the latter is directly comparable to absorbance. The C_{60} DRIFT spectrum is in good agreement with FT-IR absorption, suggesting that DRIFT also provides a good measure of C_{120} and $C_{120}O$ absorption. PM3 vibrational frequencies/IR intensities and Raman spectra were obtained as previously described [5].

3. Results and Conclusions

The DRIFT spectrum recorded for $C_{120}O$ is close to that of C_{120} (figure 1) except for the occurrence of two strong features at 1033 and 1101 cm^{-1}. PM3 calculations confirm that these are associated with furanoid ring vibrations. Specifically, three such IR allowed modes are predicted - one near 250 cm^{-1}, below the range of our DRIFT measurement [5] and two others corresponding to those observed. Calculated IR allowed frequencies (scaled by 0.85) and intensities, are shown in figure 2 [5,7]. Frequencies are well reproduced, relative intensities are less so. In particular, predicted oscillator strengths for high frequency modes are systematically too large. This is also the case for other fullerenes and derivatives studied at the PM3 level, suggesting that a specific fullerene parameterization would be useful [7].

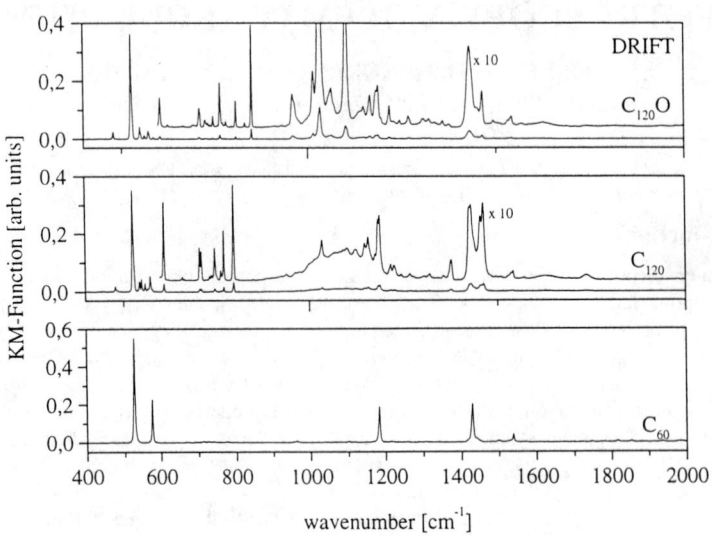

Figure 1. DRIFT measurements of C_{60}, C_{120} and $C_{120}O$.

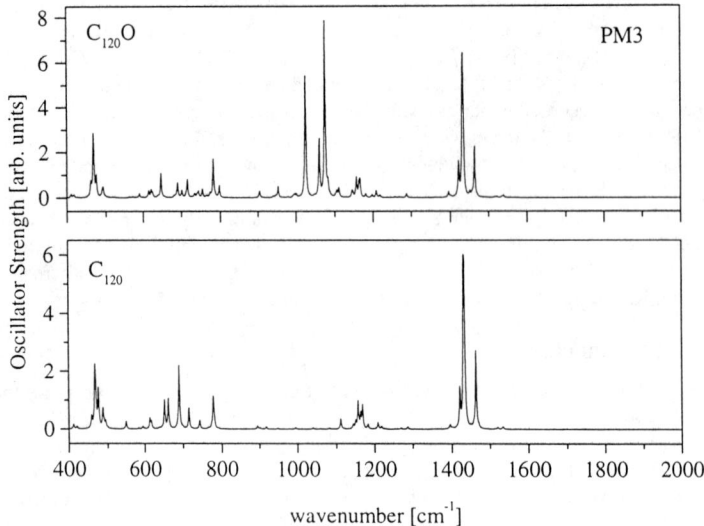

Figure 2. PM 3 calculations of IR allowed frequencies and oscillator strengths for C_{120} and $C_{120}O$ [10].

In a previous publication we have considered the systematics of low frequency cage-cage vibrations of fullerene homo- and heterodimers [5]. To this end, Raman spectra of $C_{120}O$, $C_{120}O_2$ and $C_{130}O$ (I-III) were compared with PM3 calculations of vibrational frequencies [5]. C_{120} was not then available. To first order, such molecules can be thought of as comprising two weakly coupled rigid spheres. There are then always six low frequency cage-cage vibrations, independent of linkage type: one stretching mode and five cage-cage torsions. The stretch vibration occurs near 100 cm^{-1} and is always Raman allowed.

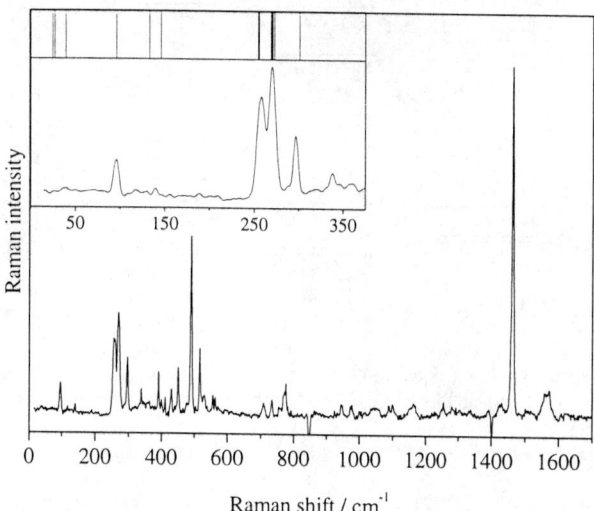

Figure 3. C_{120} Raman spectrum (794,76 nm excitation; 200 mW mm^{-2}; 2 cm^{-1} resolution) with low frequency range as insert. Superimposed are unscaled PM3 calculations.

Based on our conclusions for $C_{120}O$ and $C_{120}O_2$, we also provided a qualitative prediction of the C_{120} Raman spectrum [5]. In particular, we argued that the C_{120} Raman feature observed at 96 cm^{-1} by Lebedkin et al.[8] must by symmetry be due to the cage-cage stretch vibration. In particular for $C_{120}(D_{2h})$, the A_g symmetric stretch is expected to be the lowest frequency Raman allowed mode. To confirm this, we have obtained a C_{120} Raman spectrum to within 15 cm^{-1} of the 794.76 nm excitation line using an rubidium atomic vapour cell [9]. Figure 3 shows the full spectrum (15-1700 cm^{-1}; dips are due to caesium atomic resonances). The insert highlights the low frequency cage-cage (15-180 cm^{-1}) and squashing mode (200-400 cm^{-1}) regions. Unscaled PM3 calculations predict cage-cage vibrations at 25 (v_1; A_u), 27 (v_2; B_{1u}); 39 (v_3; B_{3u}); 96 (v_4; A_g); 133 (v_5; B_{3g}) and 146 (v_6; B_{1g}) cm^{-1}, respectively. v_4 - v_6 are Raman allowed and we therefore assign experimental features observed at 96 and 139 cm^{-1} to v_4 and v_5, respectively. Raman intensities were not calculated. Finally, in figure 4 we provide a correlation diagram for C_{120}, $C_{120}O$ and $C_{120}O_2$ cage-cage modes as calculated by PM3. Note that for all three links, there are three energetically well separated vibration types (ungerade coupled torsions, symmetric stretch and gerade coupled torsions). PM3 dimer binding energy and v_4 increase as $C_{120} \rightarrow C_{120}O \rightarrow C_{120}O_2$.

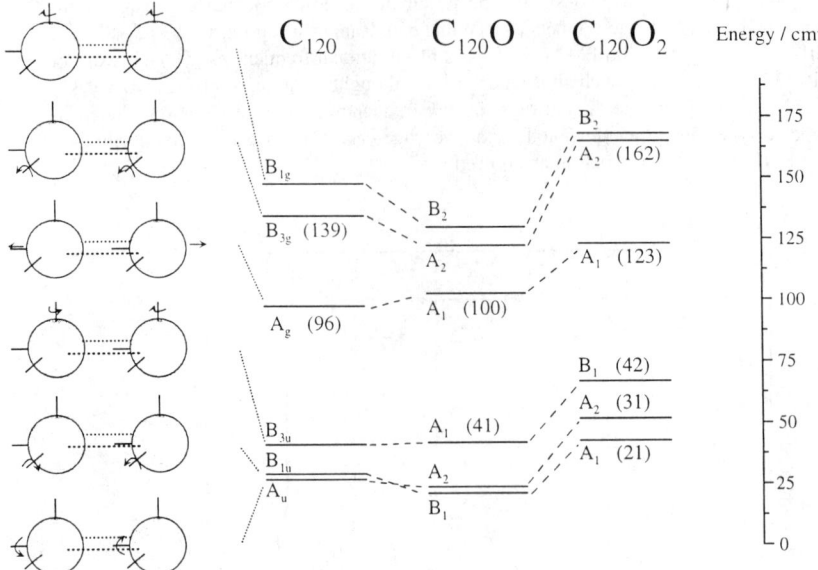

Figure 4. Correlation diagram for low frequency cage-cage vibrations of C_{120}, $C_{120}O$ and $C_{120}O_2$. Plotted are frequencies calculated at the PM3 level. Numbers in brackets are corresponding Raman spectral features.

Acknowledgements

This work was supported by the Deutsche Forschungsgemeinschaft under SFB 551 "Kohlenstoff aus der Gasphase: Elementarreaktionen, Strukturen und Werkstoffe". We thank S. Lebedkin for providing a sample of C_{120} in the early stages of our work.

References

1. A. Gromov, S. Ballenweg, S. Giesa, S. Lebedkin, W. Hull and W. Krätschmer, Chem. Phys. Lett., **267**, 460 (1997).
2. S. Lebedkin, S. Ballenweg, J. Gross, R. Taylor and W. Krätschmer; Tetrahedron Letters, **36**, 4971 (1995); A. Smith III, H. Tokuyama, R. Strongin, G. Furst and W. Romanow; J. Am. Chem. Soc., **117**, 9539 (1995).
3. A. Gromov, S. Lebedkin, S. Ballenweg, A. Avent, R. Taylor and W. Krätschmer; Chem. Commun., 209 (1997).
4. G.-W. Wang, K. Komatsu, Y. Murata and M. Shiro, Nature, **387**, 583 (1997).
5. H.-J. Eisler, F. H. Hennrich, A. Hertwig, E. Werner, C. Stoermer and M. M. Kappes, J. Phys. Chem., in press.
6. P. Kubelka and F. Munk, Z. Techn. Physik, **12**, 593 (1931).
7. H.-J. Eisler, Ph.D. Thesis, University of Karlsruhe, 1998.
8. S. Lebedkin, A. Gromov, S. Giesa, R. Gleiter, B. Renker, H. Rietschel and W. Krätschmer, Chem. Phys. Lett., in press.
9. See for example: S. Horoyski and M. Thewalt, Phys. Rev. B, **48**, 11446 (1993).
10. PM3 resonance delta functions have been folded with a constant Lorentzian to simulate experiment (2cm^{-1} halfwidth).

High Pressure Synthesis and Thermal Properties of C_{60} Dimers

Y. Iwasa, K. Tanoue, T. Mitani, A. Izuoka*,
T. Sugawara*, and T. Yagi†

Japan Advanced Institute of Science and Technology, Tatsunokuchi, Ishikawa 923-1292, Japan
*Department of Pure and Applied Sciences, University of Tokyo Meguro-ku, 153, Japan
†Institute for Solid State Physics, University of Tokyo Minato-ku, 106, Japan

Abstract. A dimeric form of C_{60} is selectively synthesized by squeezing organic molecular crystal $(ET)_2C_{60}$ at 5GPa and 200°C. Characterizations by Raman, infrared, and chromatographical studies provide unambiguous evidences for the dimeric form connected by 2+2 cycloaddition. Differential scanning calorimetry analysis showed that the heat of formation for the dimeric C_{60} is the largest among several polymeric phases of C_{60}.

Since the discovery of polymeric fullerenes in 1993, a vast variety of polymeric materials have been synthesized and characterized by various techniques.[1] In particular, the neutral polymers are attracting interest since these are new allotropes of pure carbon solids. Due to the Woodward-Hoffman's rule, polymerization in the neutral state requires some kind of perturbation, *i.e.* light-irradiation or application of pressure at high temperature. While the former is useful for thin film samples, the latter yields bulk amount of products. Another advantage of high pressure synthesis is that by tuning the temperature and pressure, one can selectively synthesize one- and two-dimensional polymers. Meanwhile, a new technique, mechanochemical method succeeded to produce bulk amount of C_{60} dimers.[2]

When pure C_{60} solids are squeezed, we always obtain polymers rather than oligomers.[3, 4] Since all C_{60} molecules are surrounded by C_{60}, application of hydrostatic pressure tends to connect all molecules. To control the degree of polymerization, we tried to squeeze a crystal of molecular compound of C_{60} with organic molecules. In such a crystal, a part of direct contact between C_{60} molecules is hindered by inserted organic molecules, providing novel arrangement of C_{60}. As a starting materials, we chose $(ET)_2C_{60}$, where ET denotes bis(ethylenedithio)-tetrathiafulvalene.[5] In this crystal, charge transfer interaction between ET and C_{60} is estimated to be significantly small, indicative of neutral states of each element. C_{60} molecules form a one-dimensional closest packing arrangement, with a regular triangle lattice framework along the c axis[5]. Interfullerene distances along the c-axis and along the diagonal direction to the c-axis are 9.923(2) and 9.919(3) Å, respectively. Since they are very close to the nearest neighbor interfullerene

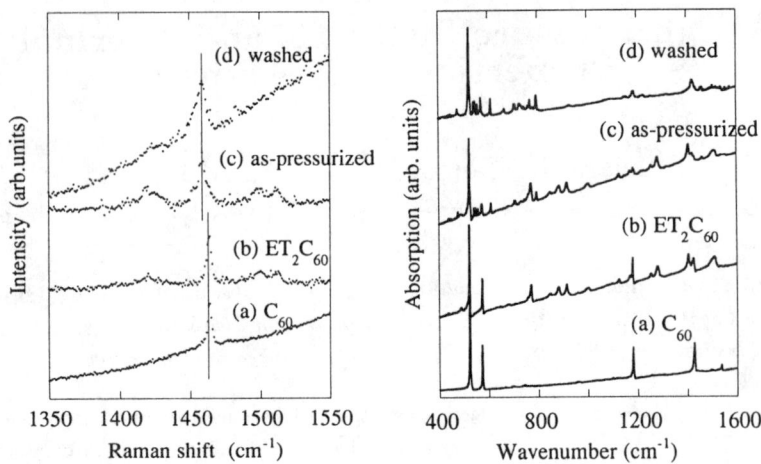

Figure 1: Left panel: Raman spectra of the pentagonal pinch mode for the I_h C_{60}. (a) pure fcc C_{60}, (b) pristine $(ET)_2C_{60}$ (c) as-pressurized $(ET)_2C_{60}$, and (d) washed by CH_2Cl_2, where ET molecules are removed. Right panel: Infrared absorption spectra for powders dispersed in KBr pellets.

distance 10.02Å in the fcc lattice of neat C_{60}, pressure-induced crosslinking in $(ET)_2C_{60}$ is highly promising.

The starting $(ET)_2C_{60}$ crystals were obtained by slow evaporation of carbon disulfide solution of stoichiometric amount of C_{60} and ET. High pressure treatment was made following the previous procedures.[3] Polycrystalline $(ET)_2C_{60}$ loaded in a gold capsule was reacted at 5GPa and and 200°C using a wedge-type cubic anvil high pressure apparatus. After squeezing, the sample was removed from the reaction capsule, followed by characterizations at ambient conditions.

The left panel of Figure 1 shows Raman spectra of the pentagonal pinch mode $A_g(2)$ for the I_h C_{60}, which is a sensitive probe for the occurrence of interfullerene bonds. In $(ET)_2C_{60}$, the $A_g(2)$ was observed at 1469.3cm^{-1}, which is identical to the position for pure C_{60}, reflecting the negligible charge transfer. In the as-pressurized $(ET)_2C_{60}$, the corresponding mode appeared at 1462.9 cm^{-1}, displaying a 6.4 cm^{-1} red-shift, indicative of occurrence of intermolecular bonds.

The right panel of Fig. 1 shows the infrared absorption spectra of several samples dispersed in KBr pellets. The spectrum of starting $(ET)_2C_{60}$ is shown by (b) in the right panel of Fig. 1. This spectrum is approximately explained as a sum of those for the element molecules, ET (not shown) and C_{60} (a), being consistent with the absence of charge transfer suggested by the structural analysis. The spectrum of as-pressurized samples is displayed by (c) in Fig. 1. Remarkable changes were observed from the spectrum of the starting $(ET)_2C_{60}$. By a careful comparison of the two spectra, we found that the peaks attributable to ET did not change by pressurization, while those from C_{60} displayed notable changes. In order to prove this more clearly, we removed the unreacted ET molecules. The pressurized $(ET)_2C_{60}$ powders were dispersed in CH_2Cl_2 followed by sonication. Only ET dissolves into CH_2Cl_2 leaving reacted C_{60} as precipitates. Figure 1 (d) shows the infrared spectrum of this precipitate obtained by filtration. Since no peaks attributable to ET were observed in the spectrum (d), we concluded that all ET molecules were successfully removed and that the obtained substance is reacted C_{60}. This substance was not soluble to toluene, suggesting an occurrence of inter-fullerene bonds. However, it is soluble to o-dichlorobenzene. Since rhombohedral and orthorhombic polymers do not dissolve even to o-dichlorobenzene, degree of polymerization is expected to be substantially smaller than those for the polymers.

In fact, the infrared spectrum (d) significantly differs from those for the 1D- and 2D-polymers reported in the previous papers. Instead, the spectrum (d) is very similar to that for the photopolymerized C_{60}. Recent studies have revealed that in the early stage of photo-induced polymerization, C_{60} dimers are dominant products.[6] We compared the spectrum (d) with that for the mechanochemically synthesized dimers.[2] The spectra for the dimer turned out to be almost identical to Fig. 1 right (d), including the peak positions and intensity distributions. These spectral evidences strongly indicate that the dominant product of high pressure treatment of $(ET)_2C_{60}$ is C_{60} dimers.

Production of C_{60} dimers from ET_2C_{60} implies that higher oligomers or ladder polymers may be synthesized by choosing appropriate starting materials and synthesis conditions. It is pointed out that even in this study we found an evidence for higher oligomers or polymers. Though most of the product of the above method is soluble to o-dichlorobenzene, there remains a slight amount of insoluble component. This insoluble part might be attributable to larger oligomers or polymers. When one considers the arrangement of C_{60} molecules in $(ET)_2C_{60}$, selective formation of C_{60} polymers is quite puzzling. One-dimensional or ladder-type polymers are likeliest products.

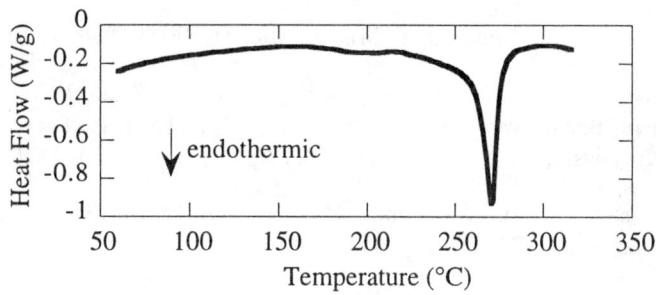

Figure 2: Differential scanning calorimetry data for C_{60} dimers synthesized by pressure. A transition to monomer phase occured in an endothermic manner with an enthalpy change of 92J/g.

Differential scanning calorimetry was made on about 10mg samples each using a Mettler-Toledo TA8000 calorimeter. Figure 2 shows DSC data for C_{60} dimer in the heating process with a heating rate of 10°C/min. A large endothermic peak was observed in the heating process, but no signal was found in the cooling scan, indicating an occurrence of irreversible transition in the heating process. The obtained substances after the DSC measurements were monomer C_{60} powders as confirmed by infrared and x-ray diffraction measurements. These data clearly showed that C_{60} dimers revert to monomer C_{60} on heating to 300°C in a similar manner to the other C_{60} polymers. This conversion takes place as an endothermic reaction, indicating that C_{60} dimer is energetically more stable than the monomeric form. The enthalpy change derived by integrating the endothermic peak was 92 ± 10 J/g, which is considerably large comparing to that for the rhombohedral phase of C_{60},[3] indicating high stability of the dimeric phase. The standard heat of formation of dimeric C_{60} has been calculated by various methods. The present experimental result is a good test for these theoretical calculations.

In summary, we have found a new route for synthesis of C_{60} dimers using a high pressure technique. Thermal analysis on the dissociation process of C_{60} dimers indicates that the dimer is more stable than monomers. Authors are grateful to K. Komatsu and Y. Murata for their help in sample characterization. They appreciate T. Uchida for his help in high pressure synthesis. This work has been supported by Grant from the Japan Society for Promotion of Science (JSPS-RFTF96P00104), from the Ministry of Education, Science, Sports, and Culture, and from Yamada Science Foundation.

References

[1] For example, *Applied Phys. A*, 1997, **64**, Special issue "Polymeric fullerenes", *Fullerene Polymers and Fullerene-Polymer Composites*, ed. P. C. Eklund and A. M. Rao, Springer, 1997.

[2] G. -W. Wang, K. Komatsu, Y. Murata, and M. Shiro, *Nature*, **387**, 583 (1997).

[3] Y. Iwasa, T. Arima, R. M. Fleming, T. Siegrist, O. Zhou, R. C. Haddon, L. J. Rothberg, K. B. Lyons, H. L. Carter Jr., A. F. Hebard, R. Tycko, G. Dabbagh, J. J. Krajewski, G. A. Thomas, and T. Yagi, *Sience*, **254**, 1570 (1994).

[4] M. Núñez-Regueiro, L. Marques, J.-L. Hodeau, O. Béthoux, and M. Perroux, *Phys. Rev. Lett.*, **74**, 278 (1995).

[5] A. Izuoka, T. Tachikawa, T. Sugawara, Y. Suzuki, M. Konno, Y. saito, and H. Shinohara, *J. Chem. Soc., Chem. Commun.*, 1472 (1992).

[6] J. Onoe and K. Takeuchi, *Phys. Rev.* **B54**, 6167 (1996).

Film Growth of $C_{59}N$ on layered materials

B. Pietzak*, C. Sommerhalter*, A. Weidinger*, B. Nuber[†], U. Reuther[†],
A. Hirsch[†]

*Hahn-Meitner-Institut Berlin, Glienickerstrasse 100, D-14109 Berlin, Germany
[†]Universität Erlangen-Nürnberg, Institut für Organische Chemie, Henkestrasse 42,
D-91054 Erlangen, Germany

Abstract. $C_{59}N$ was evaporated in an UHV system onto substrates of the layered materials HOPG, WSe_2 and mica at substrate temperatures of 25°C, 100°C and 200°C. The initial growth of the $C_{59}N$ films was investigated by atomic force microscopy (UHV-AFM) and compared to that of C_{60}. Distinct island sizes and shapes were observed for the three substrates, reflecting different mobilities and sticking factors. For $C_{59}N$ on HOPG and WSe_2, extended dendritic islands are formed indicating a large mobility of $C_{59}N$ on the substrate and a strong sticking of the molecules to the islands edges. The mobility during the growth of the second monolayer is significantly reduced therefore, nearly independent of the substrate temperature, many new nucleation centres with distances as small as 50 nm are formed on top of the 1000 - 3000 nm monolayer islands. On mica, the nucleation centres have distances of 50 - 100 nm even for the first monolayer and substrate temperatures of 200°C. This shows that the substrate interaction of $C_{59}N$ with mica is larger than that with HOPG and WSe_2.

INTRODUCTION

The electronic properties of fullerenes in the solid state can be altered by three different kinds of „doping": intercalation of metal atoms at interstitial sites of fullerene crystals, incorporation of atoms or ions inside the fullerene cage (endohedral fullerenes) and the exchange of one or more carbon atoms of the cage by hetero-atoms. Intercalation compounds exhibiting superconductivity in the A_3C_{60} phase (A = alkali metal) have been widely studied, whereas the solid state properties of endohedral fullerenes and heterofullerenes are little investigated experimentally. Recently macroscopic amounts of the stable heterofullerene $C_{59}N$ have been synthesised (1, 2) and first solid state studies of $C_{59}N$ have been reported (3, 4).

$C_{59}N$ is a structural analogue of C_{60} but the electronic properties are different. The substitution of a tetravalent carbon atom of the cage by a trivalent nitrogen atom leads to the azafullerene radical $C_{59}N^{\bullet}$ which is found to rapidly dimerise yielding the stable $(C_{59}N)_2$ dimer.

For the investigation of the solid state properties of this new material well defined films are of essential interest. Here we present first studies on the film growth of $C_{59}N$ on the layered material HOPG (highly oriented pyrolythic graphite), WSe_2 and mica and compare the growth mode with that of C_{60}.

EXPERIMENTAL

$C_{59}N$ was introduced into an effusion cell and thoroughly degassed in a vacuum of 10^{-8} mbar at a temperature of 500 K for 24 h to remove residual solvents. The thickness of the films was monitored with a quartz crystal oscillator. Films of 0.5 monolayer (ML) coverage have been grown at a rate of 0.05 ML/min on the different substrates at substrate temperatures of 293, 373 and 473 K in a vacuum of 10^{-8} mbar. The substrates were cleaved prior to loading into the deposition chamber and prebaked at 473 K for 3 h before the desired substrate temperature was set. UHV-AFM investigations were performed at 10^{-10} mbar, but the samples had to be transferred through air.

RESULTS AND DISCUSSION

Figure 2 shows an AFM image of C_{60} on HOPG. The film was grown at a substrate temperature of 373 K. Well shaped, hexagonal islands all showing the same orientations were found reflecting excellent growth conditions and a dominating substrate interaction. The second monolayer started to grow before the first monolayer was completed clearly demonstrating an island growth mechanism under these conditions. Well shaped trigonal islands were found on WSe_2 with the same deposition parameters.

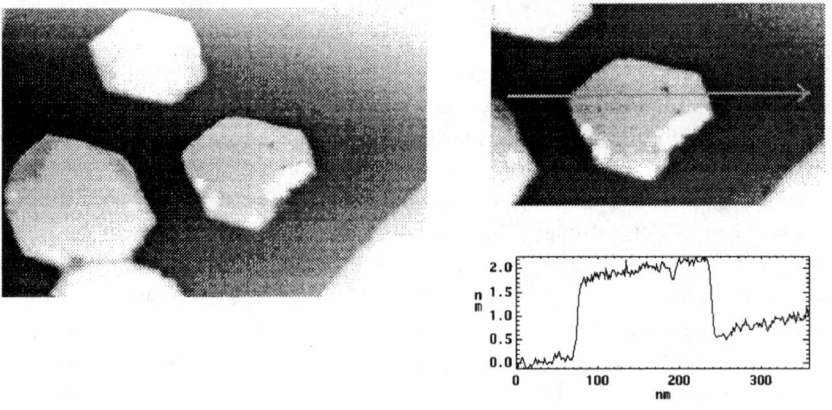

FIGURE 1: *AFM image of C_{60} on HOPG grown at a substrate temperature of 373 K. The shown area of the left picture is 500 x 400 nm in size. On the right side a section with a height scan is shown.*

In contrast to the textured growth of C_{60}, extended dendritic islands have been found for the growth of $C_{59}N$ on HOPG and WSe_2. Fig. 2 shows an AFM image of a $C_{59}N$ island on WSe_2 grown at a substrate temperature of 373 K. No preferential alignment is observed and a rather rugged island is formed. Such islands are often observed for diffusion limited aggregation. This indicates an enhanced interaction of $C_{59}N$ with the island edges prohibiting, at lower temperatures, the formation of compact islands.

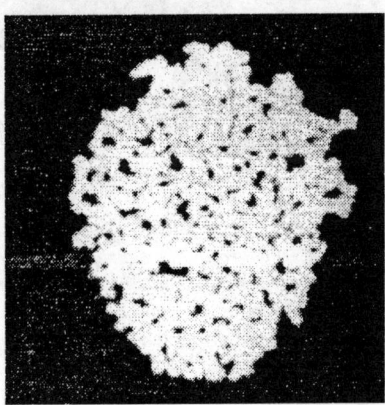

FIGURE 2: *AFM image of $C_{59}N$ on WSe_2 grown at a substrate temperature of 373 K. The area is 400 x 400 nm. In contrast to C_{60}, pronounced irregular structures are observed*

Due to the low binding energy of only 18 kcal/mol (5), the $(C_{59}N)_2$ dimer is thermally cracked during evaporation and sublimation takes place in its monomeric form. Apparently $C_{59}N$ is highly mobile on HOPG and WSe_2 and becomes immobile at the island edges.

At higher substrate temperatures, the edge diffusion is enhanced resulting in dendritic islands with less holes in it as shown in Fig.3 for a $C_{59}N$ island on HOPG grown at a substrate temperature of 473 K. Although still no preferential orientation is observed the edges are smooth. For the second monolayer the average island size is drastically reduced to only a few nanometer compared to the island size of 3 µm of the underlying first monolayer. This indicates that the mobility of $C_{59}N$ in the second and higher layers is low. The step height at the island edges was always found to be 1 nm clearly showing a layer by layer growth mechanism.

The growth behaviour of $C_{59}N$ films on mica is quite different. We find that the nucleation centres even for the first monolayer have distances of only 50 - 100 nm. This growth behaviour is observed almost independent of the substrate temperatures up to 200°C. This shows that the substrate interaction of $C_{59}N$ with mica is larger than that with HOPG and WSe_2.

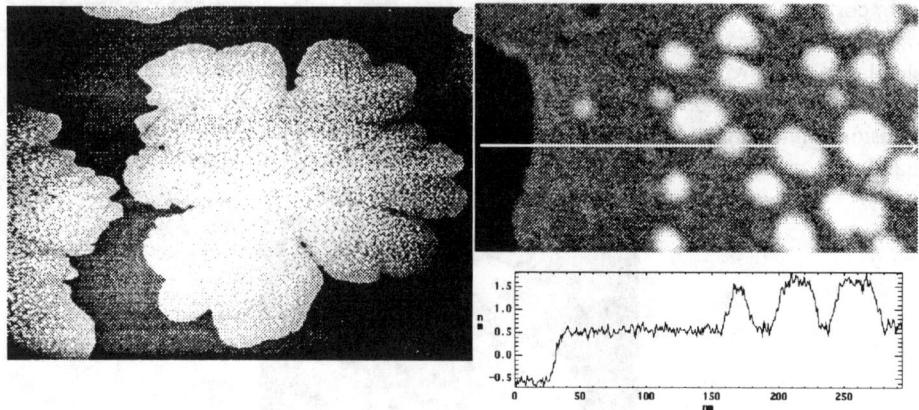

FIGURE 3: *AFM image of $C_{59}N$ on HOPG grown at a substrate temperature of 474 K. The area is 3.5 x 2.5 μm in size. The white spots are very small islands of only 100 nm in diameter and represent the second monolayer. On the right side an enlarged section of an island edge with some small second layer agglomerates is shown. A height scan is presented below.*

CONCLUSIONS

We have shown that the growth mechanism for $C_{59}N$ films on HOPG and WSe_2 is different from that of C_{60} films on the same substrates. For C_{60} well shaped islands and an island growth mechanism is found whereas for $C_{59}N$ dendritic structures and a layer by layer growth mechanism shows up. In general, for the growth of $C_{59}N$ higher temperatures are needed to obtain closed and smooth islands. Increasing the substrate temperature above 473 K may eventually lead to the formation of crystalline islands of $C_{59}N$.

REFERENCES

1. J.C. Hummelen, B. Knight, J. Pavlovich, R. Gonzalez and F. Wudl, Science, 1995, **269**, 1554
2. B. Nuber, A. Hirsch, J. Chem. Soc.; Chem. Commun., 1996, 1421
3. T. Pichler, M. Knupfer, M.S. Golden, S. Haffner, R. Friedlein, J. Fink, W. Andreoni, A. Curioni, M. Keshavarz-K., C. Bellavia-Lund, A. Sastre, J.C. Hummelen, F. Wudl, Phys. Rev. Lett., 1997, **78**, 4249
4. K. Prassides, M. Keshavarz-K., J.C. Hummelen, W. Andreoni, P. Gianozzi, E. Beer, C. Bellavia, L. Cristofolini, R. Gonzalez, A. Lappas, Y. Murata, M. Malecki, V. Srdanov, F. Wudl, Science, 1996, **271**, 1833
5. W. Andreoni, A. Curioni, K. Holczer, K. Prassides, M. Keshavarz-K., J.C. Hummelen and F. Wudl, J. Am. Chem. Soc., 1996, **118**, 11335

Investigations of Fullerenes and their Derivatives Synthesized in the Plasma Chemical Reactor

S.G. Ovchinnikov [1,2], G.N. Churilov [1,2], L.A. Solovyov [3], E.A. Petrakovskaya [1], Ya.N. Churilova [1], O.V. Chupina [2], Ya. Koretz [2], N.V. Bulina [1,4], Ya.I. Pouhova [5], A.A. Savchenko [6], V.V. Fefelova [6]

[1] *L.V.Kirensky Institute of Physics, Russian Academy of Sciences, Krasnoyarsk, 660036, Russia.*
[2] *Krasnoyask State Technical University, Krasnoyarsk, 660074, Russia.*
[3] *Institute of Chemistry and Chemical Technology, SD RAS, Krasnoyarsk, 660036, Russia.*
[4] *Krasnoyask State University, Krasnoyarsk, 660041, Russia.*
[5] *Krasnoyarsky Science Center, Russian Academy of Sciences, Krasnoyarsk, 660036, Russia.*
[6] *Institute of Medical Problems of North, Krasnoyarsk, 660022, Russia.*

Abstract. The x-ray investigations of products of synthesis and the results of investigations of iron-containing compounds of fullerene obtained in a plasma chemical reactor (PCR), and medical-biological investigations of water-soluble fullerene-iron-acetylacetone complexes are presented.

The description of axial plasma chemical reactor (PCR) that is based on the carbon plasma jet and method of synthesis of carbon cluster forms was published earlier (1). Fullerene mixture, extracted from soot, contains about 57% of C_{60} and 43% of C_{70}, which synthesized under consumption of helium 1 liter per minute.

The x-ray powder diffraction measurements were taken with DRON-4 powder diffractometer in Bragg-Brentano geometry using flat graphite monochromator and CuK_a radiation.

The main substances observed in products of synthesis are: fullerene phase (~4%), residual graphite (~7%) and amorphous phase (Figure 1a). Sample contains a distorted graphite phase which is supposed to be a product of fast condensation of carbon from plasma. The extraction from benzene solution and thermal treatment are allowed to obtain a mixture C_{60}/C_{70} with X-ray powder pattern shown on the Figure 1b.

The x-ray powder pattern analysis shown the presence of two different graphite phases in thermolys residual taken from the indestructible external ring electrode: the

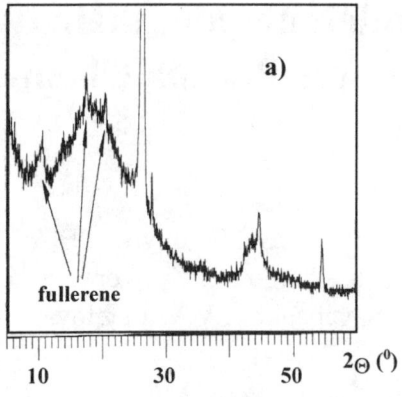 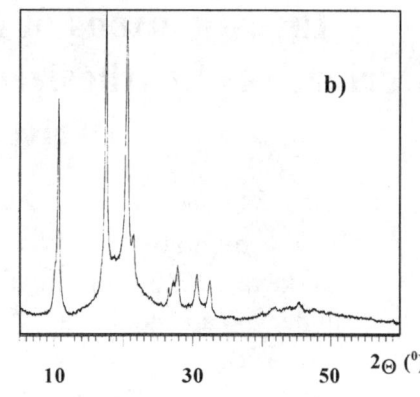

FIGURE 1. X-ray powder patterns of : (a) outcome product of the PCR taken from the walls of the reactor; (b) vacuum thermal treatment of benzene extract of C_{60} and C_{70} fullerenes mixture.

initial undistorted graphite phase with the unit cell parameters a=2.462 Å, c=6.732 Å and the distorted graphite phase.

The calculation of the unit cell parameters of distorted graphite phase based on the *00l* and *hk0* peak positions gave the following values: a=2.462 Å, c=6.852 Å. As it is seen, the unit cell parameter c of distorted graphite is greater by 0.12 Å than that of the initial graphite, while the parameter a is the same. This means that the medium distance between the graphite layers is greater by 0.06 Å (~2%) than that of conventional graphite, but the layers are undistorted. This elongation of the mean interlayer distance may be interpreted as a result of the random stacking of the graphite layers.

We synthesized fullerenes-iron compounds, using our installation, where an axial hole of the central electrode was filled by powder of carbonyl iron and investigated them by the methods IR, UV-vis and EPR.

Main characteristic absorption bands of a fullerene mixture are shown on the Figure 2a. The optical absorption spectrum of hexane solutions of fullerene extracts with iron is shown on the Figure 2b. The main changes are observed in ultra-violet bands 217 and 257 nm.: redistribution of intensity inside bands and new bands is observed. The band 328 nm and long wavelength band 480 remain constant. IR-spectrum of absorption of mixture obtained from graphite without iron have the bands with maximums on the frequencies 525, 575, 1190 and 1428 cm^{-1} (Figure 3a). The IR-spectrum of fullerene mixture obtained with iron has lines 528, 577, 1182 and 1430 cm^{-1} with initial intensities ratio. The frequencies lines 673 and 795 cm^{-1} undetected earlier (Figure 3b) are observed too. These frequencies are close to valent oscillations frequencies of iron-carbon that probably could be observed in carbides. It is well known that the carbides are not dissolved in unpolar solvents and these bands most likely correspond to coupling fullerene with iron. According to IR-spectrum we have found that carbonyls are absent in mixture. That is important because they are soluble in unpolar solvents and it would be difficult to separate them from fullerenes chemical connected with iron. Thus the researches of fullerene mixture by optical methods has shown that in fullerene mixture one fullerene complex with iron was contained at least.

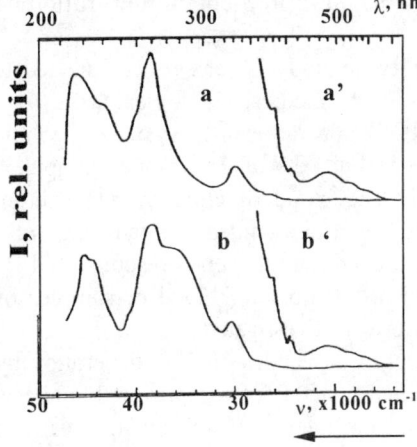

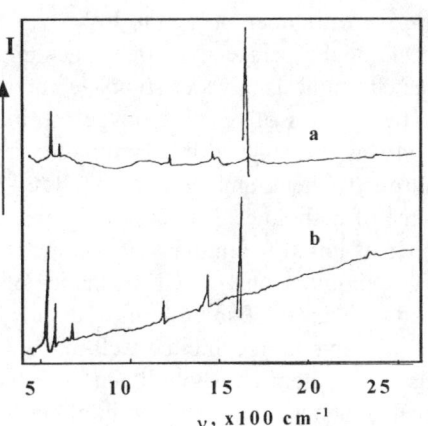

FIGURE 2. Electronic absorption spectra hexane solutions of fullerene mixture: a) obtained from pure graphite, b) from graphite with carbonyl iron. Cuvettes width: a, b - 0,1 cm; a', b' - 1 cm.

FIGURE 3. IR-spectra of an absorption of fullerene mixture: à) without iron, á) with iron.

The research of obtained fullerene mixture by ESR method was carried out with a spectrometer of X-range (SE/X-2544) under the temperatures 80-295K on hard polycrystalline samples. The ESR signal observed in fullerene that can be assigned to not coupled electrons on $C_{60} + C_{70}$ mixture (2) has line width 0,12 mT and g = 2,0023±0,0005 in our samples. This EPR signal may be explained as chemically caused defects at the fullerene molecules. Paramagnetic centers are unpaired electrons trapped in a extend π-electron system (3).

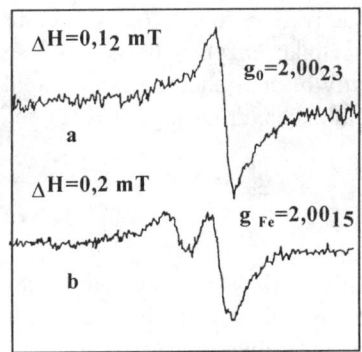

FIGURE 4. Spectra ESR a) of anion - radical C_{60}; b) of fullerene-iron complex.

The signal of synthesized iron-fullerene was shifted in the direction of field increasing on the line width (Fig. 4). The g-factor accordingly varies to g_{Fe} = 2,0015±0,0005.

Simultaneously the line intensity is reduced approximately in order of magnitude. Therefore it is possible to tell that the signal directly from iron in fullerene is not observed. However indirect proof of its presence is a noticeable drop of uncoupled electrons per 1 mole. Also it is possible to refer regularly repeated shift of a line, that may be accounted by a modification of their orbital state caused by Fe. Besides there is nominating a width line ($\Delta H \approx 50$ mT) with $g \approx 2,00$ in iron-containing fullerene samples which can belong to Fe^{3+}.

One of the perspectives of further iron-fullerene study is closely connected to directed synthesis of water-soluble compounds for the medical - biological tasks (4).

The kinetics of reactive oxygen forms (ROF) generation in a system of whole blood was investigated by chemiluminescence technique with 36-channel computer-controlled Chemiluminescence analyzer-3601. The blood of children by different degree of pathological process was investigated, that determined a broad range of a scatter of initial magnitudes of parameters of a chemiluminescence response (CLR) curve. Simultaneously a CLR kinetics of standard samples of blood or samples of blood with added 50 microlitres of distilled water was investigated.

Our preparation exhibited well-marked effect on a kinetics of ROF generation by cells in a system of whole blood. The individual analysis of samples of blood has revealed the activation of CLR in blood at 81,8 % of the patients observed in various dilutings of a preparation. In the diluting 1/500 000 the concentration of the preparation was 10^{-7} mg/ml in the calculation on fullerene, that is typical for other biologically active compounds. The influences of the fullerene complexes on the functional activity of blood leukocyte have been studied by chemiluminescence technique too. At the incubation with separate leukocytes the dose-dependent inhibit and activation effects took place. The influences of the fullerene complexes on the intracellular enzymes have been measured by cytochemical method. The increased activity of acid phosphates of granulocytes was obtained. The change of activity of succinite dehydrogenaze was not definite.

ACKNOWLEDGMENTS

This work was supported by the State committee of high education, grant on chemistry in Yaroslavl State Technical University and by the Russian Research Program "Fullerenes and Atomic clusters", grant N 97018.

REFERENCES

1. G.N. Churilov. Proceedings of International Winterschool on Electronic Properties of Novel Materials «*Progress in Fullerene Research*», Kirchberg, (1994) pp.135-138.
2. A. Bartl, L. Dunsch, J. Fröhner and U. Kirbach, *Chem. Phys. Lett.* **229** (1994) p.115.

3. I.M. Zaritskiy, S.S. Ishenko, A.A. Kontchits, S.P. Kolesnik, I.P. Vorona, S.M. Okulov, K.I. Pohodnya. *Solid State Physics,* **38** '2 (1996) pp.419-426.
4. Ya.I.Pouchova, G.N.Churilov, V.G.Isakova, A.Ya.Koretz, Ya.N.Churilova (Titarenko). *Doklady of the Russian Academy of Science,* **355** '2 (1997) pp. 269-272.

Synthesis and Properties of Novel Fullerene Derivatives

Tatiana Da Ros,[a] Maurizio Prato,[a] Dirk Guldi,[b] Enzo Alessio,[c]
Lodovico Valli,[d] Maurizio Carano,[e] Francesco Paolucci,[e]
Paola Ceroni,[e] and Sergio Roffia[e]

[a]*Dipartimento di Scienze Farmaceutiche, Università di Trieste, Piazzale Europa, 1, 34127 Trieste, Italy*
[b]*Radiation Laboratory, University of Notre Dame, Notre Dame, IN 46656, U.S.A.*
[c]*Dipartimento di Scienze Chimiche, Università di Trieste, Via Giorgieri 1, 34127 Trieste, Italy*
[d]*Dipartimento di Scienza dei Materiali, Università di Lecce, 73100 Lecce, Italy*
[e]*Dipartimento di Chimica "G. Ciamician", Università di Bologna, 40126 Bologna, Italy*

Conversion of inexpensive sources of energy, such as light, into more precious electric flow is an area of current high interest (1-5). In this field, C_{60} has generated a great deal of hope in a number of applications, such as photoconductors, photoreceptors, photoresists, and photoswitches (6). The potential use of C_{60} relies on the ability of this fullerene to accept reversibly up to six electrons in solution (7,8), which also makes it an eccellent electron acceptor in photoinduced processes (9). In addition, C_{60} has a broad range of absorptions, so that the excited states (singlet and triplet) are readily accessible (10,11). Furthermore, it has been postulated that C_{60} has a low reorganization energy (9).

This argument has been the subject of several studies, and so far, C_{60} has been used unmodified in heterogeneous mixtures with conducting polymers (12) or strong donors like phthalocyanines (13). Or, else, C_{60} has been chemically modified to improve its miscibility with conducting polymers or to attach it covalently to a donor, in order to fabricate dyads or triads of various types (6,9).

Scheme 1 shows only a few representative examples of dyads and triads which have been synthesized in recent years for exploring the use of C_{60} in photoinduced processes (14-19). The common objective of these studies has been the generation of long-lived charge-separated species, that might allow utilization of the charge separation.

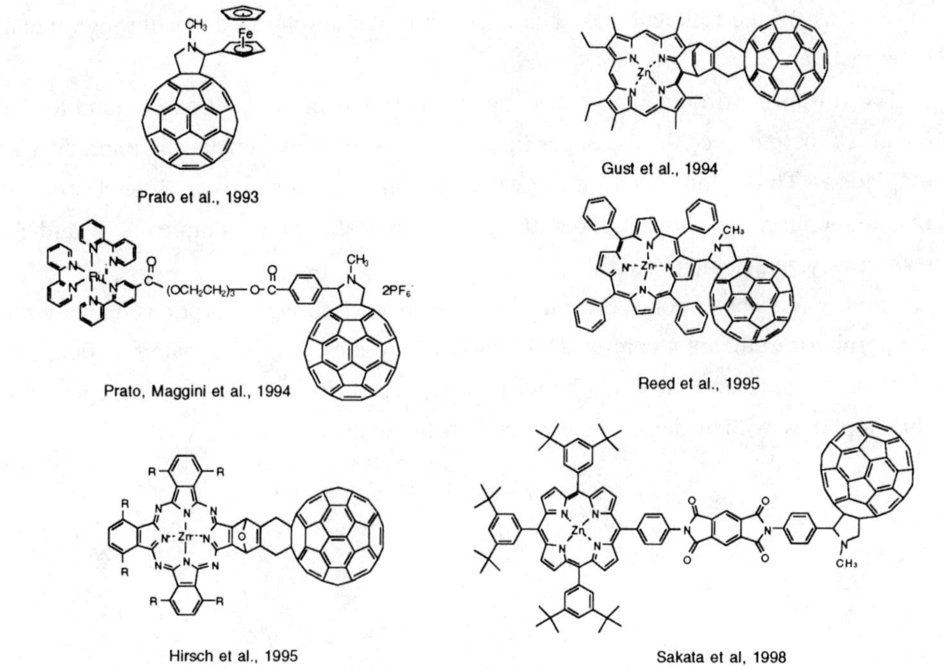

Scheme 1

The chemical modification of fullerenes, in fact, gives the opportunity to combine the uncommon properties of the fullerenes with those of other interesting materials such as electro- or photoactive species. In our group, we have developed a versatile and convenient way to functionalize C_{60}, via azomethine ylide cycloaddition (14). This reaction can be performed very easily starting from an α-amino acid and an aldehyde. In the simplest example, N-methyl glycine condenses with formaldehyde to generate the reactive species which then attacks C_{60}.

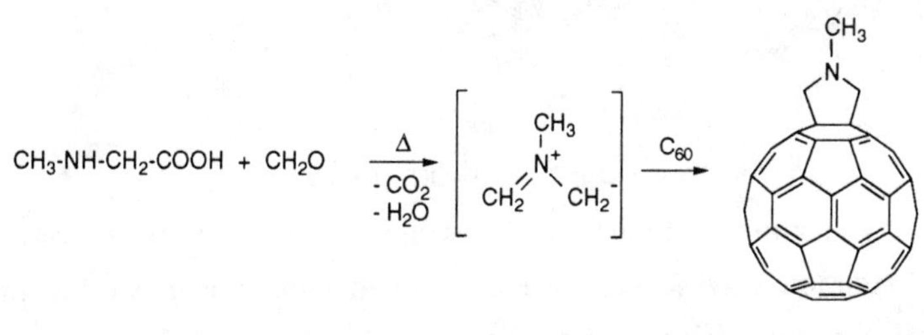

The scope of the reaction is very broad. Two different groups can be introduced simultaneously starting from N-functionalized α-amino acids and even very complex aldehydes. Using this approach, several dyads and triads have been prepared (6,9). In the present paper, we report the synthesis and properties of a novel fullerene-porphyrin dyad in which the fullerene has been attached to the porphyrin donor through metal coordination.

The Ru-porphyrin complex **1** has one ethanol molecule very weakly bound to the metal. In the presence of a stronger ligand, such as pyridine, ethanol is immediately displaced. Thus, the reaction between **1** and C_{60} derivatives **2** or **3** occurs instantaneously at room temperature, affording the new complexes **4** and **5**, respectively, in quantitative yield.

The UV-Vis absorption spectrum of dyad **4** displays strong absorptions due to both the porphyrin-ruthenium complex and the fullerene moieties. As a matter of fact, the electronic spectrum of **4** is the virtual superposition of the two independent chromophores, with no detectable interaction between the two centers.

2, R = CH_3
3, R = $CH_2CH_2OCH_2CH_2OCH_2CH_2OCH_3$

4, R = CH_3
5, R = $CH_2CH_2OCH_2CH_2OCH_2CH_2OCH_3$

The cyclic voltammetry of the dyad shows eight reversible reduction waves and two reversible oxidation waves (Table 1). It is in fact the linear combination of the electrochemical properties of porphyrin model **6** (analogous to **1**, where pyridine has substituted ethanol) and the C_{60} derivative **2**. Also this technique shows very little sign of detectable interaction between the two chromophores. The difference between the first reduction wave and the first oxidation, which gives an indication of the energy of the charge-separated state (oxidated porphyrin-reduced fullerene), is about 1.5 eV.

Table 1. $E_{1/2}$ values (V vs SCE) of the redox couples of dyad **4** and model compounds **2** and **6**, detected by cyclic voltammetry (sweep rate 0.1 V s^{-1}) in tetrahydrofuran solutions (tetrabutyl ammonium hexafluorophosphate as electrolyte), at -60°C (25°C for **6**). Working electrode = Pt.

$E_{1/2}$	+2/+1	+1/0	0/1-	1-/2-	2-/3-	3-/4-	4-/5-	5-/6-	6-/7-	7-/8-
4	+1.33	+0.98	-0.49	-1.07	-1.54	-1.67	-2.08	-2.09	-2.44	-2.90
2			-0.49	-1.06	-1.68	-2.13	-2.87			
6	+1.54	+1.14	-1.51	-2.03	-2.49					

Taken all the electrochemical and ground state features into account, it appears that dyad **4** is an excellent candidate to perform artificial photosynthesis, e.g., efficient light absorption by the antenna molecule (porphyrin moiety), followed by a rapid intramolecular electron transfer to the electron acceptor (fullerene). A first insight into intramolecular transfer dynamics was deduced from steady-state emission spectroscopy. The emission of the porphyrin core is, in fact, strongly quenched in non-polar toluene and to a similar extent in polar benzonitrile relative to the porphyrin model. In this light, further experiments were carried to probe dyad **4** under radiative conditions, e.g. time-resolved pico- and nanosecond flash photolysis. They show that the nature of the processes involved and the resulting products therefrom depend strongly on the solvent polarity. In toluene solution, energy transfer from the photoexcited metalloporphyrine to the fullerene predominates, with generation of the long lived triplet of the fullerene derivative, as detected by transient visible spectroscopy (centered at $\lambda_{max} \approx 700$ nm). In order to promote the driving force (ΔG) for a possible electron transfer and, in turn, to favor this intramolecular process over an energy transfer, the solvent polarity was changed. As a consequence of increasing ΔG, for example, in benzonitrile solution, electron-transfer occurs, affording the formation of the charge-separated species ($C_{60}^{\bullet -}$)-Ru(TPP)$^{\bullet +}$ with typical absorption of the C_{60} radical anion around 1040 nm and of the metalloporphyrin radical cation between 600 and 800 nm.

To obtain ordered films, dyad **5** was synthesized. In fact, we have already observed that for a successful LB deposition, a hydrophilic chain like in **5** is necessary (20). Compound **5** alone did not give monomolecular layers at the air/water interface when compressed, but an excellent deposition on a glass substrate could be achieved by mixing **5** with arachidic acid in a 1:4 ratio. Studies of energy/electron transfer occurring on solid substrates are currently underway.

ACKNOWLEDGEMENTS

This work was supported in part by the Office of Basic Energy Sciences of the U.S. Dept. of Energy and by C.N.R. through the program "Materiali Innovativi (legge 95/95)". Financial support from the European Community (TMR progam USEFULL, contract N° ERB FMRX-CT97-0126) is gratefully acknowledged. This is contribution No. NDRL- 4053 from the Notre Dame Radiation Laboratory.

REFERENCES

1. Wasielewski, M. R. *Chem. Rev.* **92**, 435-461 (1992).
2. Gust, D., Moore, T. A. and Moore, A. L. *Acc. Chem. Res.* **26**, 198-205 (1993).
3. Kurreck, H. and Huber, M. *Angew. Chem. Int. Ed. Engl.* **34**, 849-866 (1995).
4. Sakata, Y., et al. *Pure Appl. Chem.* **69**, 1951-56 (1997).
5. Fox, M. A. and Chanon, M. *Photoinduced Electron Transfer* (Elsevier, Amsterdam, 1989).
6. Prato, M. *J. Mater. Chem.* **7**, 1097-1109 (1997).
7. Xie, Q., Pérez-Cordero, E. and Echegoyen, L. *J. Am. Chem. Soc.* **114**, 3978-3980 (1992).
8. Ohsawa, Y. and Saji, T. *J. Chem. Soc, Chem. Commun.* 781-782 (1992).
9. Imahori, H. and Sakata, Y. *Adv. Mater.* **9**, 537-46 (1997).
10. Ebbesen, T. W., Tanigaki, K. and Kuroshima, S. *Chem. Phys. Lett.* **181**, 501-504 (1991).
11. Arbogast, J. W., et al. *J. Phys. Chem.* **95**, 11-12 (1991).
12. Sariciftci, N. S., Smilowitz, L., Heeger, A. J. and Wudl, F. *Science* **258**, 1474-1476 (1992).
13. Schlebusch, C., Kessler, B., Cramm, S. and Eberhardt, W. *Synth. Met.* **77**, 151-152 (1996).
14. Maggini, M., Scorrano, G. and Prato, M. *J. Am. Chem. Soc.* **115**, 9798-9799 (1993).
15. Maggini, M., Donò, A., Scorrano, G. and Prato, M. *J. Chem. Soc., Chem. Commun.* 845-846 (1995).
16. Linssen, T. G., Dürr, K., Hirsch, A. and Hanack, M. *J. Chem. Soc., Chem. Commun.* 103-104 (1995).
17. Liddell, P. A., et al. *Photochem. Photobiol.* **60**, 537-541 (1994).
18. Drovetskaya, T., Reed, C. A. and Boyd, P. *Tetrahedron Lett.* **36**, 7971-7974 (1995).
19. Imahori, H., Yamada, K., Hasegawa, M., Taniguchi, S., Okada, T. and Sakata, Y. *Angew. Chem., Int. Ed. Engl.* **36**, 2626-2629 (1998).
20. Wang, P., Chen, B., Metzger, R. M., Da Ros, T. and Prato, M. *J. Mater. Chem.* **7**, 2397-2400 (1997).

Phase diagram of the C_{60}/C_{70} system

D. Havlik[1], M. Steinmetz[2], P. Huber[2], W. Schranz[1], M. Enderle[2], K. Knorr[2]

[1] Institute for Experimental Physics, University of Vienna, Strudlhofgasse 4, A - 1090 Vienna, Austria

[2] Technical Physics, University of Saarland, Bau 38, 3. OG., D-66123 Saarland, Germany

Abstract. We have performed powder X-ray measurements on C_{60}/C_{70} mixed crystals with 10-30% C_{60}, at several temperatures ranging from 20 to 300 K. These measurements confirm our assumption (based on thermal expansion and elastic measurements) that $(C_{60})_x(C_{70})_{1-x}$ alloys form orientational glass for $0.1 < x < 0.3$. Furthermore, we were able to isolate several small single crystals of C_{60}/C_{70}, and analyze them using the rotating crystal method, which clearly shows that our samples are indeed well-defined mixtures, and rules out the possibility of phase separation.

INTRODUCTION

In our previous work [1,2], we have shown that mixed crystals of $(C_{60})_x(C_{70})_{1-x}$ can be grown by controlled sublimation for $0 < x < 0.3$ and $0.7 < x < 1$. The combination of elastic and thermal expansion measurements showed fast decline of the phase transition temperature with growing concentration of "guest" molecules, especially on the C_{70}-rich side of the phase diagram. Elastic measurements also showed the dynamic anomaly in the elastic measurements, that was not seen in thermal expansion. By combining the results of all the measurements, we came to the conclusion that C_{70}-rich crystals with $\approx 10\%$ to 30% C_{60} stay disordered on cooling, and form orientational glass at the temperature $T_g \approx 130$ K However, there were two important questions left:

1. It was not clear whether our samples are really well-defined crystals, or just (semi-)amorphous mixtures.

2. Since all of the measurements have been done above 80 K, there was a possibility that C_{60}-rich samples could have a structural phase transition at lower temperatures, instead of forming an orientational glass.

In order to answer these questions, we performed room temperature single-crystal X-ray measurements and temperature-dependent powder X-ray measurements on

FIGURE 1. Typical shape of the sublimated C_{60}/C_{70} mixed crystals with 10-30% C_{60}. Squares in the background are 1x1 mm^2

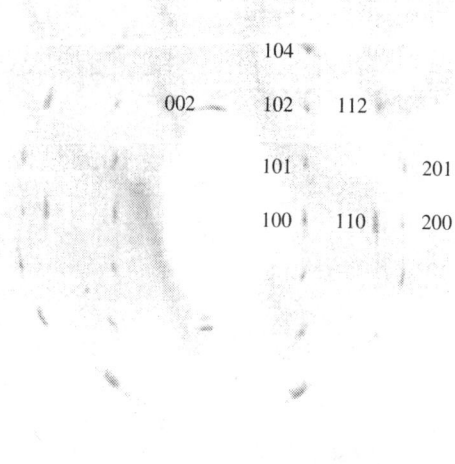

FIGURE 2. Rotating-crystal image of a $(C_{60})_x(C_{70})_{1-x}$ crystal with 25% C_{60}.

C_{60}/C_{70} mixed crystals. Thereby, we have concentrated on the samples with 10-30% C_{60}, for which we expected to find the orientational glass.

RESULTS

Figure 1 shows a photography of the sublimated $(C_{60})_x(C_{70})_{1-x}$ crystals, in the "interesting" ($0.1 < x < 0.3$) region. Size of the pen-shaped crystallites varied with the concentration, the largest crystals (2-3 mm long, with a diameter of d≈0.3mm) were obtained for x≈ 0.11. Some of them were found to be single crystals, and we were able to analyze them with the rotating-camera method. A typical result is shown on figure 2. All the crystallites had the hcp structure. One curious result was that the quality of the crystals improved with addition of C_{60} (contrary to what we expected!). One possible explanation would be that C_{60}/C_{70} crystals obtain some kind of superstructure ordering at special C_{60}:C_{70} ratios (1:4?). However, we have not found any evidence of the different crystal structure between the high-temperature phase of C_{70} and C_{70}-rich mixtures.

Simultaneously with the single-crystal measurements, we have made several temperature-dependent powder diffraction measurements. The powder measurements reveal that C_{70}-rich crystals with more than 11.5% C_{60} show only thermal expansion, but no structural phase transitions down to 20 K (Figure 3).

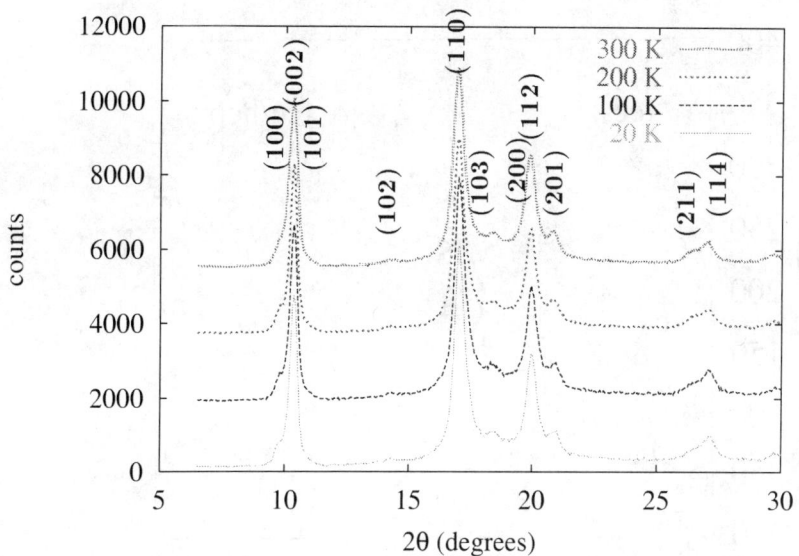

FIGURE 3. Intensity as a function of 2θ for the C_{60}/C_{70} crystal with 11.5% C_{60}, at 20, 100, 200 and 300 K. Curves were shifted for clarity.

SUMMARY

We have performed both room temperature single crystal and powder X-ray measurements at several temperatures of $(C_{60})_x(C_{70})_{1-x}$ alloys for $(0.1 < x < 0.3)$. Powder spectra of this mixtures are very similar to C_{70} spectra published by M.C. Valsakumar at. al. in [3], which were shown to belong to close packed hcp. with additional disorder due to stacking faults and amorphous background. However, unlike the pure C_{70}, this mixtures don't undergo any phase transitions on cooling (see Fig. 3) down to T=20 K. Combined with the results of our previous (thermal expansion and low frequency elastic) measurements, which show a dynamic anomaly at T≈130K, this directly proves that C_{60}/C_{70} mixed crystals with 11.5 to 30% C_{60} form orientational glass. The phase diagram of C_{60}/C_{70} mixtures, obtained with a combination of the thermal-expansion, elastic and X-ray measurements is shown in figure 4. There is still one open question, concerning the second (dhcp-monoclinic) phase transition in C_{70}-rich crystals. It seams that this phase transition merges with the first (hcp-dhcp) at less than 5% C_{60}, which would explain the existence of both the structural phase transition and dynamic anomaly in elastic moduli at this concentration. However, we dont have any measurements to directly prove this assumption.

Acknowledgments: The present work was supported by the Österreichischen Fonds zur Förderung der wissenschaftlichen Forschung under project Nr. +P10924-PHY

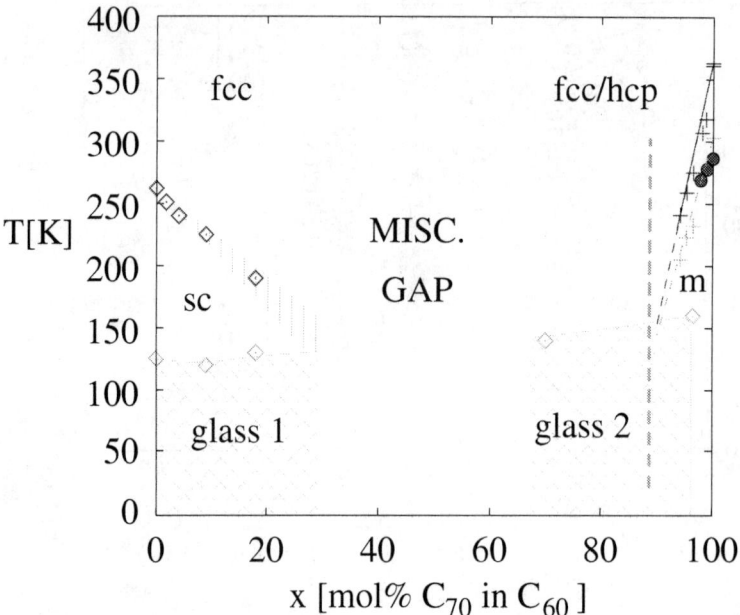

FIGURE 4. $(C_{60})_x(C_{70})_{1-x}$ phase diagram. The phase transition temperatures were taken from [1]. The vertical dotted line corresponds to the powder X-ray measurements shown in Fig. 3.

REFERENCES

1. D. Havlik, W. Schranz, M. Haluška, H. Kuzmany, and P.Rogl. Thermal expansion measurements of c_{60}/c_{70} mixed crystals. *Solid State Comm*, 104(12):775–779, 1997.
2. D. Havlik, W. Schranz, M. Haluška, J. Winter, and H. Kuzmany. C_{60}/C_{70} alloys - a model system for the orientational glass? In *Proceedings of IWEPNM-11*, Kirchberg, 1997.
3. M.C. Valsakumar, N.Subramainan, M. Yosuf, P. Ch. Sahu, Y. Hariharan, A. Bharathi, V. Sankara Sastry, J. Janaki, G. V. N. Rao, T. S. Radhakrishnan, and C. S. Sundar. Crystal structure and disorder in solid c_{70}. *Phys. Rev. B*, 48(12):9080–9085, 1993.

Catalytic Reduction of 1,2-Dihaloethanes by Electrochemically Generated C_{60}^{n-} ($n = 2$ and 3)

F. D'Souza,[a] J-P. Choi,[a] Y. Y. Hsieh,[a] K. Shriner,[a] and W. Kutner[a,b]

[a] *Department of Chemistry, Wichita State University, Wichita, KS 6720-0051, USA*
[b] *Institute of Physical Chemistry, Polish Academy of Sciences, Kasprzaka 44, 01-224 Warsaw, Poland*

Abstract. 1,2-Dihaloethanes were catalytically reduced by electrochemically generated C_{60}^{n-} ($n = 2$ or 3) anions in 0.1 M tetrabutlammonium hexafluorophosphate in benzonitrile under cyclic voltammetry and rotating disk electrode (RDE) voltammetry conditions at platinum electrodes. The second-order catalytic rate constants, determined by the RDE voltammetry under pseudo-first-order conditions against the 1,2-dihaloethanes, largely increase in the order: 1,2-dichloroethane < 1,2-dibromoethane < 1,2-diiodoethane. Alkanes, alkenes and monohalogenated alkanes were products of the electrocatalyses.

INTRODUCTION

Redox potentials of six one-electron reversible $C_{60}^{n-/(n+1)-}$ couples are largely separated (1). Therefore, stable with respect to disproportion C_{60}^{n-} anions can be easily prepared by means of selective chemical or electrochemical reduction of C_{60}. These anions react readily with halogenated organic compounds according to the two different limiting reaction schemes. On one hand, C_{60}^{n-} ($n \geq 2$) anions can react with *n*-iodoalkanes and *ipso*-diiodoalkyl derivatives yielding a range of fullerene alkyl adducts, R_nC_{60} ($n = 2, 4,..$) (2, 3), as well as methanofullerene derivatives, RC_{61} (4), respectively, and on the other, C_{60}^{n-} anions can reduce electrocatalytically *vic*-dibrominated and monobrominated alkyl derivatives (5, 6). In our study of the effect of halogenated alkanes on electrochemical behavior of C_{60}, i.e., the preference of either a C_{60} adduct formation or electrocatalytic reduction of halogenated organic compounds, we observed that electrocatalytic reduction, resulting in dehalogenation, is preferred over the formation of organofullerenes for α,ω-diiodoalkanes (7). Herein, we present cyclic voltammetry and rotating disk electrode voltammetry results of electrocatalytic reduction of 1,2-dihaloethanes by C_{60}^{n-} ($n = 2$ or 3) along with the constant potential bulk electrolysis results followed by the GC-MS product analysis.

EXPERIMENTAL

Chemicals. C_{60} (+99.95%) was from BuckyUSA (Bellaire, TX). All other chemicals were procured from Aldrich Chemicals (Milwaukee, WI) and used as received.

Electrochemical Measurements. Cyclic (CV) and differential pulse (DPV) voltammetry experiments were performed by using a Model 263A potentiostat/galvanostat of EG & G (Princeton, NJ). The rotating disk electrode (RDE) voltammetry was carried out at a 0.17 cm² platinum disk electrode by using a Model AFCB1 bipotentiostat, MSRX Speed Control Unit and AFMSRX Modulated Speed Rotator of Pine Instrument Co. (Grove City, PA). A Pt wire was used as the counter electrode and an Ag/AgCl electrode as the reference one.

GC-MS Measurements. The GC-MS analysis was performed by using a Model Q-Mass 910 of Perkin Elmer Co. (Norwalk, CT) on a DB-1 non-polar phase fused silica capillary column (30 m × 0.253 mm) operating in an electron ionization mode.

RESULTS AND DISCUSSION

Qualitative results were obtained by CV while quantitative ones by the RDE voltammetry because the latter is superior in discriminating background currents of irreversible electroreduction of 1,2-dihaloethanes recorded in the absence of C_{60}.

In benzonitrile solution of C_{60} containing 1,2-dihaloethane, both the second and third electroreduction CV peak, corresponding to the $C_{60}^{-/2-}$ and $C_{60}^{2-/3-}$ electroreduction, respectively, is increased, as compared to the peaks obtained in the absence of the 1,2-dihaloalkane (Fig. 1, Curve *1*). Moreover, the relevant electro-oxidation peaks of C_{60}^{n-} (n = 2 or 3) are gone indicating electrocatalytic reduction of 1,2-dihaloethanes:

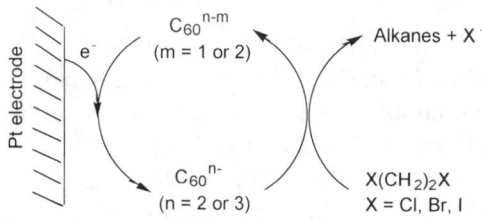

In the presence of 1,2-$C_2H_4I_2$, two new pairs of CV electro-oxidation and electroreduction peaks are seen during cycling over the positive potential range, followed the initial electroreduction scan (Fig. 1, Curve *4*). The formal redox potentials of these pairs are centered at ca. 0.19 V and 0.64 V and correspond to the irreversible I^-/I_3^- and quasi-reversible I_3^-/I_2 redox couples, respectively (8). Similarly, new electro-oxidation peaks at E_{pa} = 0.72 V and E_{pa} = 0.83 V are observed (Fig. 1, Curves *2* and *3*) in the presence of 1,2-$C_2H_4Br_2$ and 1,2-$C_2H_4Cl_2$, respectively. They correspond to the irreversible Br^-/Br_2 and Cl^-/Cl_2 electro-oxidation, respectively (8). These results indicate that halogens are released in solution during the catalytic electroreduction. By holding for ca. 1 min the potential at -1.2 V or at -1.6 V, i.e., at the values more negative than those for the second or third electroreduction peak of C_{60}, respectively, and subsequent potential cycling over the examined redox processes, we found no new peaks in the voltammograms. Hence, no C_{60} adducts are formed on a CV time scale.

In excess of 1,2-dihaloethanes with respect to C_{60}, that is, under pseudo-first-order conditions against the 1,2-dihaloethanes, the measured limiting currents by the RDE voltammetry are independent of the rotation rate, for small rotation rates. Then, the second-order electrocatalytic rate constant, k, can be calculated (9)

$$k = i_{cat}^2 / (n F A C_R)^2 D_R C_S \qquad (1)$$

where i_{cat} is the catalytic limiting current, D_R and C_R are the diffusion coefficient and concentration of C_{60}, respectively, C_S is the concentration of the substrate, i.e., 1,2-dihaloalkane, F is the Faraday constant, n is the number of electrons transferred and A is the electrode surface area. The values of i_{cat} were calculated by extrapolating the curves of limiting currents, i_{lim}, vs. square root of rotation rate to the zero abscissa. These curves deviate positively from the linear Levich plots of i_{lim} vs. square root of rotation rate, observed in the absence of 1,2-dihaloethanes. The obtained dependence of i_{cat} vs. square root of C_S is linear both for $C_{60}^{-/2-}$ and the $C_{60}^{2-/3-}$ for all 1,2-dihaloethanes (Fig. 2). The calculated k values (Table 1) for 1,2-$C_2H_4I_2$ are larger by three and four orders of magnitude than those for 1,2-$C_2H_4Br_2$ and 1,2-$C_2H_4Cl_2$, respectively.

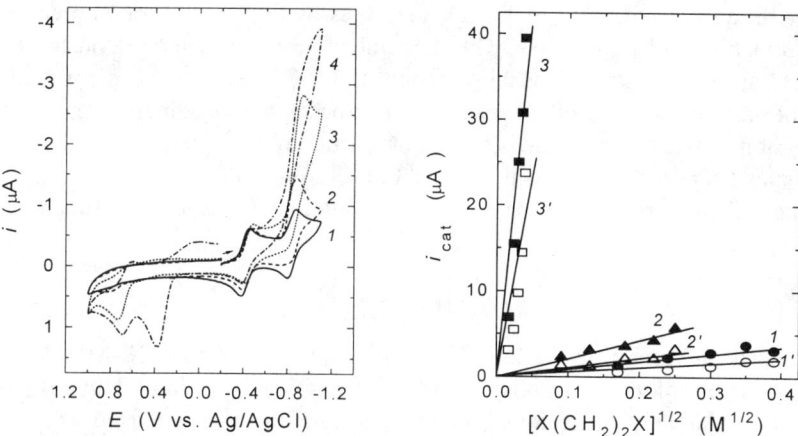

FIGURE 1. Cyclic voltammograms for 0.15 mM C_{60} - curve *1*; 0.15 mM C_{60}, 2.5 M 1,2-$C_2H_4Cl_2$ - curve *2*; 0.15 mM C_{60}, 1.5 M 1,2-$C_2H_4Br_2$ - curve *3*; and 0.15 mM C_{60}, 0.3 mM 1,2-$C_2H_4I_2$ - curve *4* in 0.1 M (TBA)PF$_6$ in benzonitrile. Scan rate 0.1 Vs^{-1}.

FIGURE 2. Dependence of the RDE voltammetry catalytic currents, i_{cat}, on the square root of the 1,2-dihaloethane concentration. 1,2-$C_2H_4Cl_2$ - curves *1* and *1'*, 1,2-$C_2H_4Br_2$ - curves *2* and *2'*, and 1,2-$C_2H_4I_2$ - curves *3* and *3'*, for the second, $C_{60}^{-/2-}$, - curves *1*, *2* and *3* as well as third, $C_{60}^{2-/3-}$, - curves *1'*, *2'* and *3'* electroreduction of C_{60}. 0.1 mM C_{60}, in 0.1 M (TBA)PF$_6$ in benzonitrile.

Products of the electrocatalytic reductions were determined by using GC-MS after constant potential bulk electrolysis of 1,2-dihaloethane solutions of C_{60} at a Pt grid electrode. In the presence of 1,2-$C_2H_4I_2$, ethane and ethene, with trace monoiodoethane, is the major product of the electrolysis performed either at -1.2 V or -1.6 V, i.e., at the values at which C_{60}^{2-} or C_{60}^{3-} is generated, respectively. Both for 1,2-$C_2H_4Br_2$ and 1,2-$C_2H_4Cl_2$, monohaloethane is the major electrolysis product. The amount of ethane and ethene generated increases with the increase of the electrolysis time. However, prolonged electrolysis also results in small amounts of the C_{60} alkyl adducts. This is manifested by new one-electron reversible DPV peaks of potentials more negative by 115 and 90 mV than those for $C_{60}^{0/-}$ and $C_{60}^{-/2-}$, respectively (3, 4).

TABLE 1. Electrocatalytic rate constants for reduction of 1,2-dihaloethanes by C_{60}^{n-} in 0.1 M (TBA)PF$_6$, benzonitrile, determined by voltammetry at the rotating disk electrode.

1,2-Dihaloethane	k, M^{-1} s^{-1}		
	$C_{60}^{0/-}$	$C_{60}^{-/2-}$	$C_{60}^{2-/3-}$
1,2-Dichloroethane	-	$(3.5 \pm 0.2) \times 10$	$(5.5 \pm 0.2) \times 10$
1,2-Dibromoethane	$(1.6 \pm 0.1) \times 10^2$	$(3.2 \pm 0.2) \times 10^2$	
1,2-Diiodoethane	-	$(2.8 \pm 0.1) \times 10^5$	$(9.4 \pm 0.2) \times 10^5$

CONCLUSIONS

Electrocatalytic reduction of 1,2-dihaloethanes by C_{60}^{n-} ($n = 2$, or 3) predominates over formation of C_{60} adducts. The determined second-order rate constants of the electrocatalysis indicate that dehalogenation of 1,2-$C_2H_4I_2$ is much more facile than those of the two other 1,2-dihaloethanes. Electrocatalytic deiodination leading to formation of the ethane and ethene is the major reaction route in case of 1,2-$C_2H_4I_2$ while monohaloethanes are generated in case of 1,2-$C_2H_4Br_2$ and 1,2-$C_2H_4Cl_2$. Apparently, 1,2-dihaloethanes are dehalogenated stepwise yielding ethane as the ultimate reaction product.

ACKNOWLEDGES

We thank BG Products, Wichita KS, for making accessible its GC-MS facility. Financial support of the US-Polish Maria Sklodowska-Curie Joint Fund II, through Grant No. PAN/NSF-96-275 to WK and FD, is gratefully acknowledged.

REFERENCES

1. For a recent review, see: Chlistunoff, J., Cliffel, D. and Bard, A. J. in *Handbook of Conductive Molecules and Polymers.* Volume 1 *Charge-transfer Salts, Fullerenes and Photoconductors.* Ed. Nalwa, H. S., Chichester, 1997, John Wiley & Sons, Chapter 7.
2. Caron, C., Subramanian, R., D'Souza, F., Kim, J., Kutner, W., Jones, M. T., and Kadish, K. M., *J. Am. Chem. Soc.*, **115**, 8505 (1993).
3. D'Souza, F., Caron, C., Subramaniam, R., Kutner, W., Jones, M. T., and Kadish, K. M. in *Recent Advances in the Chemistry and Physics of Fullerenes and Related Materials*, Eds. Kadish, K. M. and Ruoff, R. S., The Electrochem. Society Proceedings Series, Pennington, NJ, **94**, 768 (1994).
4. Boulas, P. L., Zuo, Y., and Echegoyen, L., *Chem. Commun.*, 1547 (1996).
5. Huang, Y. and Wayner, D. M., *J. Am. Chem. Soc.*, **115**, 367 (1993).
6. Fuchigami, T., Kasuga, M. and Konno, A., *J. Electroanal. Chem.*, **411**, 115 (1996).
7. D'Souza, F., Choi, J-P., Hsieh, Y.-Y. and Kutner, W. *J. Phys. Chem. B*, **102**, 212 (1998).
8. Mann, C. K., and Barnes, K. K., *Electrochemical Reactions in Nonaqueous Systems,* Marcel Dekker, New York, 1970, Chapter 7.
9. Bard, A. J. and Faulkner, L. R., *Electrochemical Methods: Fundamentals and Applications.* John Wiley, NewYork, 1980, Chapter 11.4.4.

Fullerene Radicals, Electrochemistry and Electron Spin Resonance:
Part A: Anomalous rotational dependence of the ESR signals of single crystals of $[C_{70}][I][(C_6H_5)_4P]_2$
Part B: Electrochemical and ESR characterization of mono-anionic radicals of four minor isomers of C_{84}: $[84]C_1$, $[84]C_s(V)$, $[84]D_{2d}(I)$ and $[84]D_2(III)$

José-Antonio Azamar,[1][†] Rodolphe Clérac,[1] Claude Coulon,[1]
John Dennis,[2] Hisanori Shinohara [2] and Alain Pénicaud [1]

[1] *Centre de Recherche Paul Pascal, CNRS, UPR 8641, Avenue A. Schweitzer, 33600 Pessac, France*
[2] *Department of Chemistry, Nagoya University, Nagoya 464-8602, Japan*

Abstract. In part A, the angular dependence of ESR spectra of single crystals of $[C_{70}][I][(C_6H_5)_4P]_2$ is presented showing evidence for an anomalous behaviour qualitatively explained in terms of a weak inter-radicals interaction. In part B, preliminary results on the electrochemical and ESR characterization of four recently isolated C_{84} minor isomers are presented.

PART A: ANOMALOUS ROTATIONAL DEPENDENCE OF THE ESR LINE INTENSITY IN SINGLE CRYSTALS OF $[C_{70}][I][(C_6H_5)_4P]_2$

ESR spectra on single crystals of the above mentionned C_{70} salt, grown as described in (1) have been performed. From room temperature down to 14 K, only one, symmetrical, line is observed. Below 14 K, the signal splits into four lines. An analysis of the angular dependence of the g-factor for the four lines has been performed at He

---

† permanent address: Centro de Investigación y de Estudios Avanzados, IPN, Unidad Mérida, Departamento de Física aplicada, Apartado postal 73 "Cordemex", 97310 Mérida, Yuc. México.

temperature in three mutually perpendicular planes. The resulting rotation figures actually allow a "structure determination" by ESR : Indeed, knowing the room temperature unit-cell (tetragonal, a = 12.682(4) Å, c = 21.660(3) Å) and Laue symmetry (4/mmm),(1) it is found that the 4-fold symmetry axis is retained at low temperatures but that the $C_{70}^{\bullet-}$ radicals long molecular axis are tilted 30° away from the 4-fold axis. The angular dependence of the ESR signals can then be accounted for with an uniaxial tensor for the g factor of the $C_{70}^{\bullet-}$ radicals with principal values $g_{\parallel}$ = 2.016 and $g_{\perp}$ = 2.000 (Figure 1). However, at certain positions,(145° in Fig.1) the signals, instead of crossing each other - as the normal theory (straight and dotted lines), where the radicals are considered non-interacting, predicts - repell each other. This is qualitatively interpreted as a simple two levels interaction. From the difference in g-factor, one can estimate the energy of the interaction to be 0.2 mK. Such an anomalous behaviour has actually already been observed, albeit with no interpretation for it, in DEM $(TCNQ)_2$. (2) A quantitative analysis of this phenomenon is underway in our laboratory.

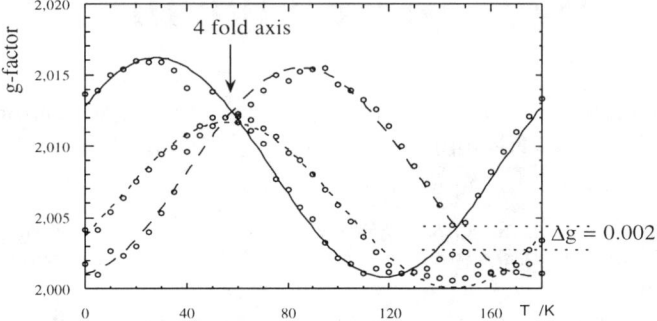

Figure 1. Rotation dependence of the g-values of the four signals in $[C_{70}][I][(C_6H_5)_4P]_2$

PART B: ELECTROCHEMICAL AND ESR CHARACTERIZATION OF MONO-ANIONIC RADICALS OF FOUR MINOR ISOMERS OF C_{84}: $[84]C_1$, $[84]C_S(V)$, $[84]D_{2d}(I)$ AND $[84]D_2(III)$

[84]fullerene is known to have 24 isolated pentagons isomers (3) of which $[84]D_{2d}(II)$ and $[84]D_2(IV)$ are the two major isomers. (4,5) Recently five minor isomers ($[84]C_1$, $[84]C_s(V)$, $[84]D_{2d}(I)$, $[84]D_2(III)$ ans $[84]C_2(IV)$) have been separated by HPLC and identified by ^{13}C NMR.(6) Each of these being a unique molecule in its own right, it is interesting to characterize it spectroscopically per se. Furthermore, by performing an electrochemical/ESR characterization of all seven isomers in different solvents, one can hope to get insight about the electronic structure of these new species and identify general trends as well as peculiarities between [84]fullerene isomers. Recently some of us developped an electrochemical cell to perform cyclic voltammetry and bulk electrolysis of minute quantities (< 100 µg) of compounds.(7) Additionnally, the electrolyzed solution can be vacuum-transferred to a quartz tube for ESR characterization of the formed radical. We report here on preliminary experiments on the four title [84]fullerene isomers in benzonitrile, 0.05 M PPN+Cl- , (PPN+ = $[(C_6H_5)_3P]_2N^+$).

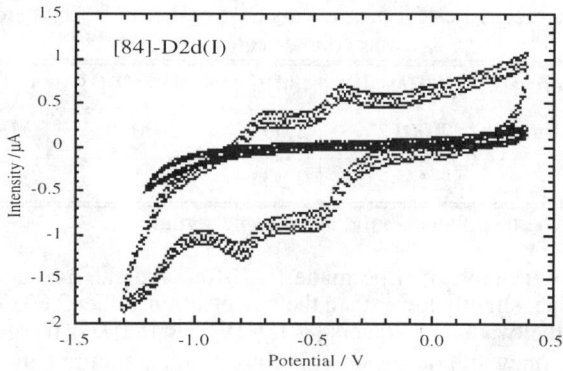

Figure 2. Cyclic voltammogram of [84]D_{2d}(I). 100 mV/S, benzonitrile, 0.05 M PPN+Cl-, ref: Ag/Ag+, underlying curve is the baseline.

Figure 2 shows a typical cyclic voltammogram for [84]D_{2d}(I). Up to four reversible waves can be seen. Table 1 lists the reduction potentials for the four isomer studied to date. Already, one can see a trend in two of those minor isomers: their first reduction potential is significantly lower than the two major isomers,(E= -0.59 V,(7)) a fact which should be related to a lower energy LUMO and hence an easier reduction. [84]C_1 cyclic voltammogram in the same conditions as the other isomers presents a well defined but anomalously wide reduction wave (ca 500 mV). This fact will require confirmation by repeating the experiment in benzonitrile/PPN.Cl and other solvent/electrolyte systems.

Table 1. Half-wave potentials for the reactions: $C_{84}{}^{n-}$ + 1e- $C_{84}{}^{(n+1)-}$; potentials are referred to the Fc/Fc+ couple, measured under identical conditions (E (Fc/Fc+) = 0.22 V vs Ag/Ag+)

isomer :	[84]D_{2d}(I)	[84]C_s(V)	[84]D_2(III)	[84]C_1
0 / 1-	-0.64	-0.42	-0.47	-1.2*
1- / 2-	-0.99	-0.79	-0.80	
2- / 3-	-1.36		-1.16	
3- / 4-			-1.5	

* the peak to peak separation was ca. 500 mV (see text)

Bulk electrolysis of the solutions were performed at a potential slightly more negative than the first reduction potential for each isomer. ESR spectra for the frozen electrolyzed solutions were performed from 230 K down to 4 K (benzonitrile being a polar solvent, the ESR cavity cannot be tuned for higher temperatures which are too close to the melting point of the solvent (260 K)). From 200 to 4 K, [84]C_1, [84]C_s(V) and [84]D_2(III) present an isotropic signal whereas [84]D_{2d}(I) exhibits an anisotropic signal. The ESR characteristics of the spectra recorded for the four radicals are listed in Table 2.

Table 2. ESR characteristics for the radicals $C_{84}{}^{\bullet-}$ at 4 K in frozen benzonitrile (PPN+ as counter-cation)

isomer:	$[84]D_{2d}$(I)	$[84]C_s$(V)	$[84]D_2$(III)	$[84]C_1$
g-factor	2.0012*	2.0012	2.0016	2.0006
ΔH (Gauss)	2.16*	0.33	0.91	0.47

* approximate values only; anisotropic signal.

The following observations can be made : All four radicals present a temperature-independent g-value, slightly lower than the free electron value (2.0023), as has already been reported for the two major isomers $[84]D_2$(IV) and $[84]D_{2d}$(II).(g = 2.0008,(7)) In all four cases, the linewidth decreases slightly with temperature before rising again at low temperature. We assign the similar behaviour to a solvent related relaxation mechanism.

Between 200 and 230 K, additionnal sharp lines appear with increasing intensity when increasing temperature. When actually removing the tube from the ESR spectrometer, we noticed that part of the sample was in the liquid state; thus, we assign those sharp spectra to solution spectra. Similar measurements in different solvents will allow to clear that point.

In conclusion, the feasibility of microscale electrochemical/ESR spectroscopy, demonstrated earlier (7) is now being applied to a "real" system of samples of limited quantities. Their full study and its analysis as a function of the molecular orbitals energy levels for the different C_{84} isomers is underway in our laboratory.

ACKNOLEDGEMENTS

The French-Mexican cooperation program (action ECOS/ANUIES M96E01) is gratefully acknowledged. AP thanks the IWEPNM'98 for partial support.

REFERENCES

1. A. Pénicaud, A. Pérez-Benítez, R. Escudero, C. Coulon, *Solid State Commun.*, **96**, 147-150 (1995).
2. C.F. Schwerdtfeger, S. Oostra, G.A. Sawatzky, *Phys. Rev. B*, **25**, 1786-1790 (1982).
3. D. E. Manolopoulos, P. W. Fowler, *J. Chem. Phys.*, **96**, 7603-7614 (1992). P.W. Fowler and D.E. Manolopoulos, An atlas of fullerenes, Clarendon Press, Oxford (1995).
4. K. Kikuchi, N. Nakahara, T. Wakabayashi, S. Suzuki, H. Shiromaru, Y. Miyake, K. Saito, I. Ikemoto, M. Kainosho, Y. Achiba, *Nature*, **357**, 142-145 (1992).
5. R. Taylor, G. J. Langley, A. G. Avent, J. S. Dennis, H. W. Kroto, D. R. M. Walton, *J. Chem. Soc. Perkin Trans. 2*, 1029-1036 (1993).
6. T.J.S.Dennis et al., submitted to *J.Phys.Chem.*; T.J.S.Dennis et al. IWEPNM '98 Kirchberg Proceedings.
7. J. Antonio Azamar-Barrios, Eduardo Muñoz P. and Alain Pénicaud, *J. Chem. Soc. Faraday Transactions*, 3119-3123 (1997).

In Situ ATR-FTIR Spectroscopic Investigations during Electrochemical Reduction of Fullerene Thin Films

Helmut Neugebauer, Carita Kvarnström[1] and N. Serdar Sariciftci

Physical Chemistry, Johannes Kepler University of Linz, A-4040 Linz, Austria

Abstract. Spectroelectrochemical investigations of the redox processes of C_{60} fullerene films in Li^+ and TBA^+ containing electrolytes using *in situ* FTIR-ATR spectroscopy are shown. Using Li^+, the reduction to C_{60}^{2-}-salt and the reoxidation occur in two processes, whereas with TBA^+ only one reduction process and two less resolved reoxidation processes are observed.

INTRODUCTION

Electrochemistry provides controlled conditions (defined redox potential, controllable amount of charge) for the redox processes of solid state fullerene films (1). Fourier transform infrared (FTIR) spectroscopy is a powerful tool to get information on structural and molecular properties of fullerenes. By combining the two methods in a special setup using cyclovoltammetry and attenuated total reflection (ATR) spectroscopy with germanium reflection elements as working electrodes (2), spectral information *in situ* during electrochemical redox processes of C_{60} films in organic electrolytes is obtained.

EXPERIMENTAL

The FTIR-ATR spectroelectrochemical setup and the method are described in (2). Working electrode: Pt covered Ge reflection element with a C_{60} film, electrode area: 0.63 cm^2, reference electrode: Ag/AgCl, counter electrode: Pt foil, electrolyte solution: 0.1 M lithiumperchlorate ($LiClO_4$) or tetrabutylammoniumperchlorate ($TBAClO_4$) in acetonitrile, temperature: -4 °C, scan rate: 5 mV/s, C_{60} films (0.6 – 1 μm) prepared with hot wall beam epitaxy (HWBE), FTIR spectrometer: Bruker IFS66S, MCT-Detector, coadding of 32 interferograms for each spectrum, resolution: 4 cm^{-1}. Each spectrum covers a range of about 90 mV in the cyclovoltammogram.

---
[1] On leave: Analytical Chemistry, Åbo Akademi University, FIN-20500 Åbo, Finland

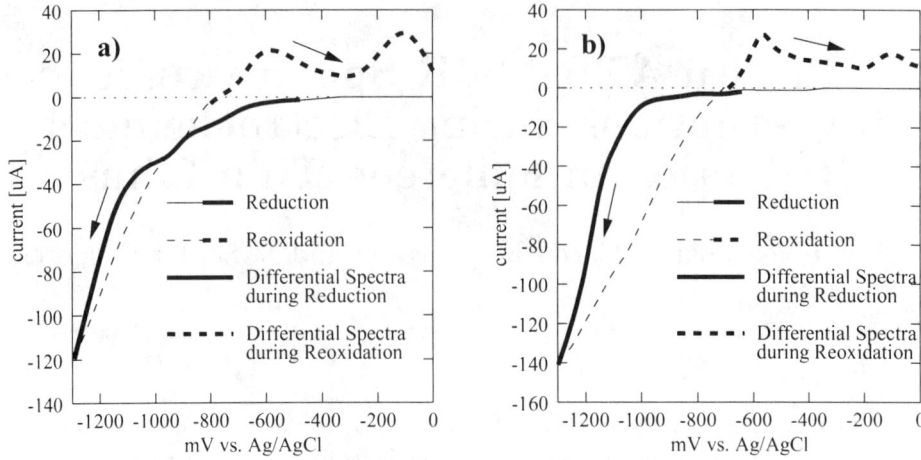

FIGURE 1: Cyclovoltammograms in a) Li^+ and b) TBA^+ containing electrolytes.

RESULTS AND DISCUSSION

Figure 1 shows cyclovoltammograms of C_{60} films with Li^+ and TBA^+ containing electrolytes, which indicate a retarded reduction process, especially in presence of TBA^+. Similar behaviour has been observed by other authors during doping with TBA^+ and has there been explained by film resistivity effects (1,3).

During reduction and reoxidation IR spectra were recorded consecutively. Differential spectra obtained by calculating each difference between two consecutive spectra are shown in Figure 2 for the reduction processes and in Figure 3 for the reoxidation processes in Li^+ and TBA^+ containing electrolytes. With differential spectra, the dynamic formation and consumption of substances can be seen by upwards and downwards pointing IR bands, respectively.

During reduction (Figure 2), a negative IR band at 1428 cm^{-1}, attributed to a decreasing amount of C_{60} (4,5) on the electrode surface, is found in both electrolytes. Using Li^+, an upwards pointing band appears at 1393 cm^{-1}, which upon further reduction shifts to 1375 cm^{-1}. With TBA^+, the process is retarded and only at a later state of the reduction a band at 1375 cm^{-1} can be seen. According to the literature (4,5), a band at 1393 cm^{-1} can be assigned to the formation of C_{60}^--salts. Since C_{60}^{3-} has been reported to show a band at 1363 cm^{-1} (4,5), we attribute the band at 1375 cm^{-1} to C_{60}^{2-}-salts. From the spectral data, the reduction of a C_{60} film in Li^+ containing electrolyte shows two processes $C_{60} \to C_{60}^- \to C_{60}^{2-}$ (partially overlapped), whereas with TBA^+ the formation of the C_{60}^{2-}-salt occurs in one process $C_{60} \to C_{60}^{2-}$, due to the more retarded reduction reaction.

In the 1440-1470 cm^{-1} wavenumber range additional electrode potential dependent IR bands can be seen. A band at 1467 cm^{-1} occurs predominately in the reaction $C_{60} \to C_{60}^-$ (lower part in Figure 2a), and a band at 1442 cm^{-1} in the

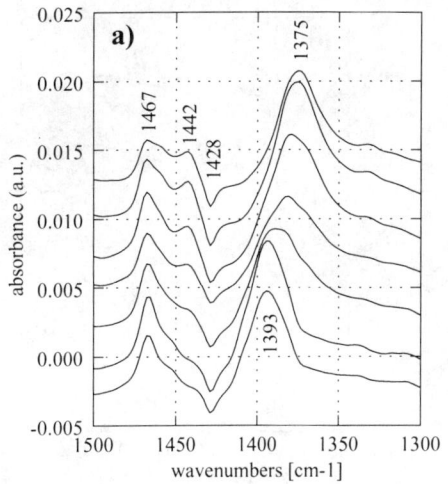 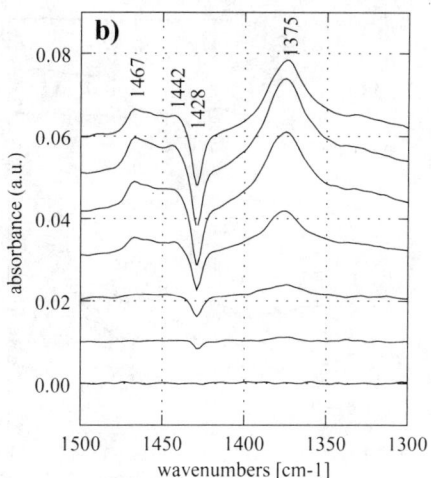

FIGURE 2: Differential spectra during reduction in a) Li$^+$ and b) TBA$^+$ containing electrolyte. The spectra are separated. Sequence: bottom to top.

reaction $C_{60}^- \rightarrow C_{60}^{2-}$ (upper part in Figure 2a). The additional spectral features seen during reduction may be due to structural changes taking place in the film during doping (6) or the formation of dimers of $C_{60}^{-\bullet}$ as proposed in (7,8).

In both electrolytes, the reoxidation processes occur in two steps around -600 and -100 mV/SCE (Figure 1). Comparing the differential spectra (Figure 3), the processes are better resolved in the case of Li$^+$. At the more negative potential, a decreasing IR band at 1375 cm^{-1} and an increasing band at 1393 cm^{-1} (lower part in Figure 3a) indicate the oxidation of C_{60}^{2-}-salt to C_{60}^--salt. Upon further reoxidation of C_{60}^- to C_{60} (upper part in Figure 3a) a decreasing IR band at 1393 cm^{-1} and an increasing band at 1428 cm^{-1} are obtained. Using TBA$^+$, the processes are less resolved (Figure 3b). The intermediate C_{60}^--salt can be seen by an increasing absorption at 1410 cm^{-1} (lower part in Figure 3b) and a decreasing absorption around 1400 cm^{-1} (upper part in Figure 3b). Morphological effects due to the larger size of the TBA$^+$ cation are probably responsible for the differences to the results using Li$^+$.

The spectral features in the 1440-1470 cm^{-1} wavenumber range during reoxidation occur in the reverse sequence compared to the reduction reaction. The band at 1442 cm^{-1} vanishes in the reaction $C_{60}^{2-} \rightarrow C_{60}^-$, and the band at 1467 cm^{-1} in the reaction $C_{60}^- \rightarrow C_{60}$ (Figure 3a). Again, the effects are less resolved with TBA$^+$ (Figure 3b).

To summarize, FTIR ATR spectroelectrochemistry provides valuable information on the redox processes of C_{60} films. Comparing the results in Li$^+$ and TBA$^+$ containing electrolytes, Li$^+$ shows two reduction processes $C_{60} \rightarrow C_{60}^-$-salt $\rightarrow C_{60}^{2-}$-salt and two clearly separated reoxidation processes, whereas in the case of TBA$^+$ a

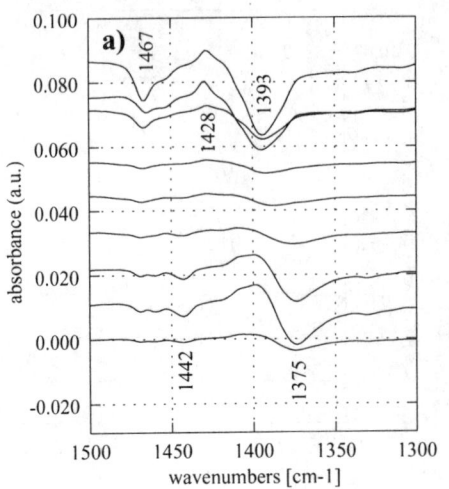 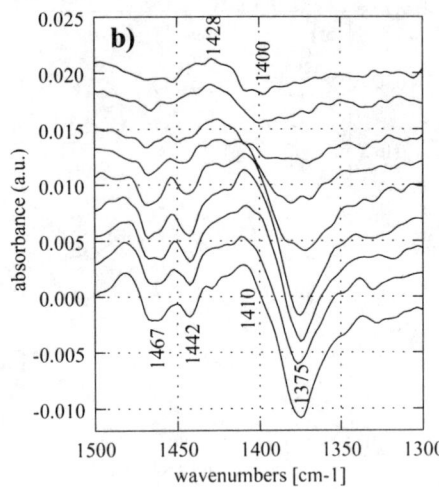

FIGURE 3: Differential spectra during reoxidation in a) Li$^+$ and b) TBA$^+$ containing electrolyte. The spectra are separated. Sequence: bottom to top.

more retarded reduction with a single process $C_{60} \rightarrow C_{60}^{2-}$-salt and poorly resolved reoxidation processes are observed.

ACKNOWLEDGEMENTS

The authors acknowledge the "Fonds zur Förderung der wissenschaftlichen Forschung" of Austria (Project No. P11457-CHE) for financial support; Prof. Dr. H. Sitter, Institute of Semiconductor Physics, Johannes Kepler University Linz, for preparing the C_{60} films by the HWBE method; the Research Institute of the Foundation of Åbo Akademi University for grant (C. Kvarnström).

REFERENCES

1. Chlistunoff, J., Cliffel, D., and Bard, A. J., *Handbook of Organic Conductive Molecules and Polymers*, Vol. *1*, Nalwa, H. S. (ed), West Sussex: J. Wiley & Sons Ltd., 1997, ch. 7
2. Neugebauer, H., and Zhao Ping, *Microchim. Acta [Suppl.]* **14**, 125 (1997)
3. Koh, W., Dubois, D., Kutner, W., Jones, T. M., and Kadish, K. M., *J. Phys. Chem.* **97**, 6871 (1993)
4. Dresselhaus, M. S., Dresselhaus, G., and Eklund, P. C., *Science of Fullerenes and Carbon Nanotubes*, San Diego: Academic Press, Inc., 1996, ch. 4
5. Pichler, T., Winkler, R., and Kuzmany, H., *Phys. Rev. B* **49**, 15879 (1994)
6 Tomura, K., Nishizawa, N., Takemura, D., Matsue, T., and Uchida, I., *Chem. Lett.* 1365 (1994)
7. Heinze, J., and Smie, A., *Proc. Electrochem. Soc.* **94-24**, 1117 (1994)
8. Dunsch, L., *Proc. Electrochem. Soc.* **94-24**, 1068 (1994)

High-Resolution Vibronic Spectroscopy of Fullerenes in Shpolskii Systems

A. N. Starukhin[1], B. S. Razbirin[1], A. V. Chugreev[1], Yu. S. Grushko[2],
V. N. Zgonnik[3], E. Yu. Melenevskaya[3],
M. Happ[4], F. Henneberger[4]

(1) A.F.Ioffe Physical-Technical Institute of RAS, 194021, St. Petersburg, Russia
(2) St. Petersburg Nuclear Physics Institute of RAS, 188350, Gatchina, Russia
(3) Institute of Macromolecular Compounds of RAS, 199004, St. Petersburg, Russia
(4) Humboldt-University, 10115, Berlin, Germany

Abstract. The new results on high-resolution spectroscopy of C_{70} molecules in crystalline matrices are presented. By using luminescence excitation spectroscopy the vibronic structure of the first electronically excited singlet state was recorded separately for different non equivalently sited groups of C_{70} molecules in Shpolskii matrix. The drastic increase of the triplet emission intensity in matrix-isolated fullerene derivative C_{60}-polypyrrole as compared with the intensity of the triplet emission of pristine C_{60} was observed.

It was shown that fullerene molecules embedded in some crystalline matrices reveal the Shpolskii effect and are characterised by narrow-line emission and absorption spectra corresponding HOMO-LUMO optical electronic transitions [1,2]. At T=2K tens of sharp lines with half-width of about 3 cm^{-1} were observed in the vibronic spectra of C_{70} and C_{60} molecules as well as some C_{60} hydrocarbon and metalloorganic derivatives in these Shpolskii systems [3]. The method opens up opportunities for utilising the advantages of high-resolution spectroscopy. In this way the Zeeman effect on fullerenes (C_{70}) has been observed for the first time [4]. The polarisation effect in the optical spectra of non-spherical fullerene molecules (C_{70}) in monocrystalline toluene matrix was firstly revealed [5].

Optical electronic spectroscopy is, in fact, the only method to get information about the energies of molecular vibrational modes in electronically excited states of molecules under study. However, vibronic absorption spectra of fullerene molecules, even in Shpol'skii matrices, are characterised by the presence of number of overlapping bands, which makes the interpretation of the spectra difficult. As an example in Fig. 1 is depicted the absorption spectrum of C_{70} molecules in crystalline toluene, a group of lines in the region 1.86-1.89 eV being due to purely electronic optical S_0-S_1 transitions in the C_{70} molecules. One of the reasons for the complicated structure of the absorption spectra is the presence of several kinds of fullerene centres in host matrices, so that the spectrum observed is factually a superposition of the absorption spectra of the

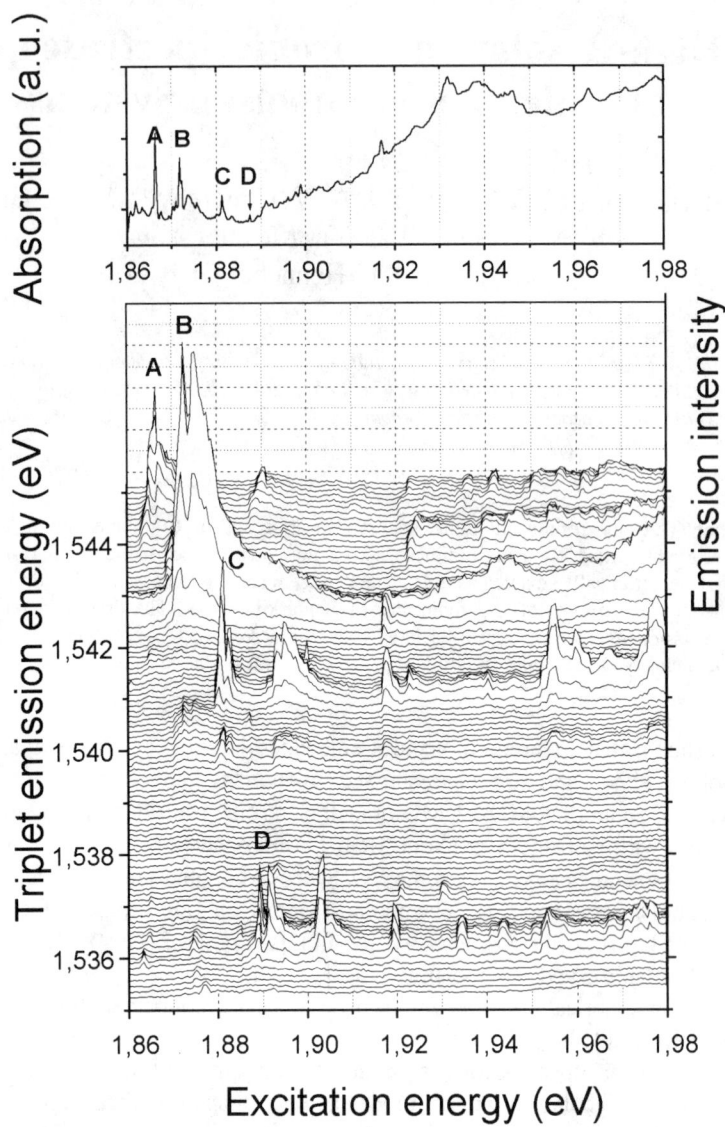

FIGURE 1. Absorption spectrum (upper curve) and luminescence excitation spectra of C_{70} molecules in Shpol'skii matrix. T = 5 K. The luminescence excitation was performed with the use of a cw tuneable dye laser.

different centres that are characterised by slightly different energies of the S_0-S_1 gaps. Specifically, the lines labelled *A, B, C, D* answer to four such non-equivalently sited groups of the C_{70} molecules in toluene matrix. In order to avoid this effect we studied excitation spectra of fullerene luminescence in Shpol'skii systems in conditions of detecting the intensity of emission lines which relate to one of the groups only. Like the absorption spectra, the excitation spectra give information on the structure of the electronically excited states of the molecules. Fig. 1 represents the luminescence excitation spectra of C_{70} molecules embedded in crystalline toluene matrix. Different curves in Fig. 1 correspond to the excitation spectra of different emission lines in the spectrum that is due to optical vibronic transitions from the lowest electronically excited triplet state T_1 to the ground state S_0 of C_{70} molecules. Well-structured excitation spectra related to the excitation of triplet emission in four groups A, B, C, D of C_{70} molecules are clearly seen. (Note that some lines in the excitation spectra are complex in structure.) The analysis of the excitation spectra shows that they are characterised by a number of common features, however some distinctions between the excitation spectra are also marked. The analysis yields the energies of the vibrational modes of C_{70} molecules in the S_1 state. On the average, these energies are about 5% less than those in S_0 state. A detailed discussion of the excitation spectra and of the electronic structure of the lowest electronically excited state of C_{70} molecules will be published elsewhere.

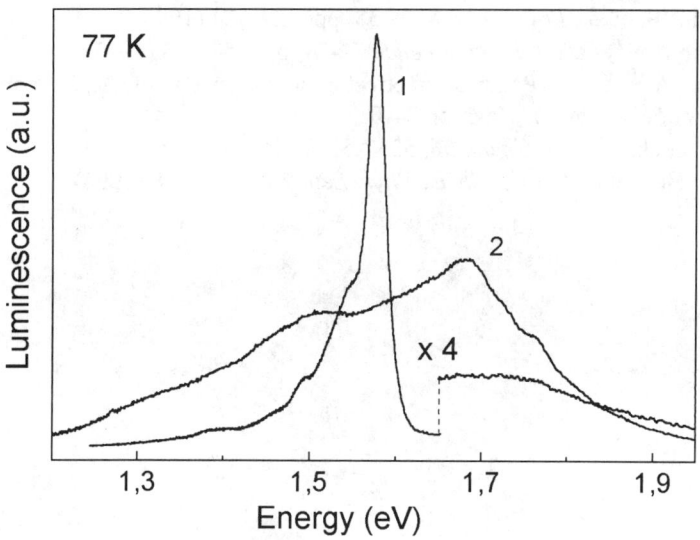

FIGURE 2. Luminescence spectra of C_{60}-polypyrrole (1) and of pristine C_{60} (2) in toluene matrices. Excitation by Ar^+-laser ($h\nu_{exc} = 2.54$ eV)

Besides pure fullerenes we have made studies into spectroscopy of some fullerene derivative among them C_{60} with added polypyrrole chains. The derivative C_{60}-polypyrrole (C_{60}-PP) was synthesised as a product of pyrrole polymerisation in the presence of fullerene C_{60} under influence of free radical initiator and UV radiation.

When embedded in crystalline toluene C_{60}-PP does not show any Shpolskii effect. On the other hand, we observed very intensive triplet emission of matrix-isolated C_{60}-PP at low temperatures (see Fig. 2). This situation is just opposite to the case of C_{60} for which triplet emission intensity is very weak [6]. According to our measurements emission bands detected in the spectral region 1.40-1.62 eV are characterised by their lifetimes about 150 µs, which is a strong evidence in favour of their triplet origin. The high intensity of the triplet emission can be explained by the relative decrease of the radiative lifetime of the triplet state in C_{60}-PP as compared with C_{60}. Probably, the effect observed can be considered as an analogue of the "heavy atom effect" that leads to the increase of the oscillator strength of the triplet-singlet transition in molecules [7].

This work was supported by the RFBR (grant No. 97-03-33634a) and by the Ministry of Sciences (grants 020/2, 94039 and 98060).

REFERENCES

[1] Razbirin B.S. et al., *JETP Lett.* **60**, 451-455 (1994).
[2] Razbirin B.S. et al., "A new approach to the optical study of fullerenes: fullerene molecule in crystalline matrix" in *Proceedings of the IWEPNM*, 1995, pp. 261-265.
[3] Razbirin B.S. et al., *Phys. Solid State* **38**, pp. 522-524 (1996)
[4] Starukhin A.N. et al., *J. Luminescence* **72-74**, pp. 457-458 (1997)
[5] Chugreev A.V. et al., "Polarisation effect in optical spectra of C_{70} fullerene" in *Proceedings of EXCON'96*, 1996, pp.7-10.
[6] Zeng Y. et al., *J. Phys. Chem.* **96**, 5237-5239 (1992).
[7] Van den Heuvel D.J. et al., *Chem. Phys. Lett.* **231**, 111-118 (1994).

Studies of Photoinduced Charge Transfer in Conjugated Polymer-Fullerene Composites by Light-Induced ESR

V. Dyakonov[+], G. Zoriniants, M. Scharber, C. J. Brabec,
R. A. J. Janssen*, J. C. Hummelen**, and
N. S. Sariciftci

Physical Chemistry, Johannes Kepler University of Linz, A-4040 Linz, Austria
*Laboratory of Organic Chemistry, TU Eindhoven, P.O.Box 513, 5600 MB Eindhoven, The Netherlands
**Organic and Molecular Inorganic Chemistry, University of Groningen, Nijenborg 4, 9747 AG Groningen, The Netherlands

Abstract. In this work we present comparative studies of the photoinduced electron transfer (PIT) in a number of conjugated polymer/fullerene composites and in pure components by using light-induced electron spin resonance. PIT from the polymer onto fullerene in the composites results in the appearance of two LESR lines: i) g=2.0025, attributed to the positive polaron on the polymer chain and ii) g=1.9995 originating from the fullerene anion-radical. These signals have different spin-lattice relaxation times with different temperature dependencies, the low field line possessing the longer relaxation time. The amount of light induced charges is proportional to the square root of the excitation light intensity, which indicates bimolecular recombination of polarons. Both lines have reversible light-induced ESR components and persistent light-induced contribution attributed to trapped polarons.

INTRODUCTION

Photoinduced electron transfer (PIT) in composites of conjugated polymers (CP) and fullerenes is a recently discovered phenomenon enabling the production of photovoltaic (PV) devices. Upon addition of few percent of a strong electron acceptor, such as fullerene into conjugated polymer, the following effects have been found: photoluminescence quenching in the subpicosecond range (1), dramatic increase of photo-conductivity (2), and improved photovoltaic effect (3). These effects are interpreted within a model of ultrafast electron transfer from polymer onto fullerene molecules with the formation of pairs of $(D^+...A^-)$ - type which have a high rate of dissociation and small rate of recombination. To demonstrate the electron transfer in

---

[+] New address: Energy and Semiconductor Physics, Carl v. Ossietzky University of Oldenburg, 26111 Oldenburg, Germany.

alkoxy-PPV/C_{60} and alkyl-PT/C_{60} composites, light-induced ESR (LESR) has been applied in (4) and revealed the appearance of two signals. These signals were attributed to radical anions of the fullerene and positive polarons (P^+) on the polymer. LESR spectra of composites made from fullerenes and thiophene oligomers consist also of two signals (5). However, the LESR spectrum of CP/fullerene composites is not clearly understood. First, the lineshape is not very well described. Therefore g-factors cannot be precisely determined. Second, the origin and the fate of polarons in this charge separation process is not clear. In undoped PPV film, a dark ESR was observed additional to a LESR signal and attributed to positive polarons (6). The CT in pristine PPV films was observed by using sensitive PL-detected magnetic resonance (PLDMR) techniques (7). Therefore, one would expect in the ESR spectrum of a pristine CP a correlated amount of photoinduced negative and positive charged radicals. In contrast to (6), the PPV used in (7) did not show any dark or light induced ESR signal. Further, the amount of photoinduced spins of each type is not well determined. Integration of the LESR curve (4) implies an equal amount of C_{60}^- and P^+ photoinduced paramagnetic species in MEH-PPV/C_{60}. The presence of P^- on the chain is therefore ruled out.

EXPERIMENTAL

We used a commercial Bruker EMX (X-band) ESR spectrometer with nitrogen-flow cryostat, allowing measurements between 77K to 300K. Excitation light was provided from an Ar^+-ion laser. Different CPs and their composites with pure and functionalized fullerenes were investigated. Polymers studied were: 3,7-dimethyl-octoxy-methyloxy-poly-p-phenylene-vinylene (PPV) and poly-(3,4-di-[(R,S)-2-methylbuthoxy]thiophene (PMBT). PPV is soluble in xylene at elevated temperatures. The fullerenes were: 1-(3-cholestanoxycarbonyl)-propyl-1-phenyl-[6,6]C_{61} (denoted as PCBM) and C_{60} (99.5%) purchased from MER Corp. Films were produced from solutions with the following ratios: PPV/PCBM (1:3 weight ratio), PPV/C_{60} (3:1) and PMBT/PCBM (1:1). Further experimental details are described elsewhere (8). We distinguish between a prompt LESR signal ("light-on" signal corrected on "light-off" and "dark" signals), and a persistent one ("light-off" corrected on the "dark", and/or "annealed" signals). Correction on the "annealed" ESR is necessary to ensure that no irreversible photochemical process occurred in the sample. Note that the composites we studied here were also used for the fabrication of the solar cells (9).

RESULTS AND DISCUSSION

Pristine PPV films do not show any dark or light-induced ESR signal. This indicates a high degree of purity of the polymer and a low concentration of paramagnetic defects. Films of PCBM do not show any dark ESR signal either, but exhibit a very weak LESR signal. The nature of this signal is not yet clear. A possible explanation could be a PIT of intramolecular type. Addition of PCBM to PPV gives rise to a dramatic increase of the LESR signal (Fig. 1). Since the particular lineshapes cannot be deduced from the spectrum, we can only speculate that the LESR spectrum may correspond to two paramagnetic species with different g-factors.

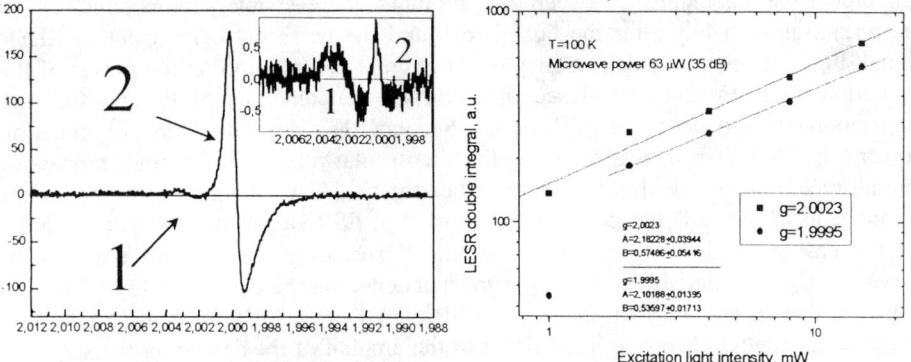

FIG. 1. LESR of Alkoxy-PPV/PCBM composite. 1 - Polymer$^+$ line, 2 - Fullerene$^-$ line. (inset - 20μW)

FIG. 2. Light intensity dependence of prompt component of LESR in Alkoxy-PPV/PCBM.

Spectra with similar parameters were also found for the other composites, PMBT/PCBM and PPV/C$_{60}$.

The lineshape distortion, in general, can be due to different mechanisms. Explanations for this lineshapes are: g-factor anisotropy in randomly oriented samples or the Overhauser effect. These assumptions are not yet ruled out, and will be a subject of future work. However, the very existence of two distinct light-induced species is proved by the following set of experiments.

The inset to Fig. 1 shows the LESR spectrum at $P_{\mu\omega}$= 20μW. A different microwave power saturation behaviour of the two LESR signals is clearly observed. The saturation of the low-field line occurs at much lower microwave power, which indicates a slower relaxation. Increasing the temperature from 90 to 200 K shifts the saturation towards higher microwave powers, which means shorter spin relaxation times and is an indicative for a phonon-assistant relaxation. Therefore, the difference in the microwave saturation behaviour allows to distinguish between the two overlapping lines: at high power of 200 mW we only see the high-field line, at power lower than 20 μW the low-field signal is observable only. Detailed studies of the saturation behaviour of this spectrum (8) revealed g-factors for particular lines as: 2.0023 (low-field line) and 1.9995 (high-field line). The high-field line is attributed to the fullerene anion-radical, and the low-field line is attributed to P$^+$ on the polymer chain.

A further consequence of the described difference in saturation behaviour is that the amounts of light-induced spins responsible for the LESR, as deduced by double integration of LESR spectrum, will only be valid at lowest microwave powers, when both lines are not saturated, i.e. below 50 μW.

After illumination, the LESR signals do not disappear completely. At T<200 K some of the paramagnetic species remain detectable for many hours. Only heating up (annealing) the sample to room temperature in the dark eliminates these ESR signals. Therefore, we distinguish between the prompt, or reversible, component, which disappears immediately after the light is off, and the persistent component of LESR signal. Fig. 2 presents the dependency of the prompt contribution on the power of the excitation light. First, one can see that nearly equal amounts of P^+ and C_{60}^- are contributing to the prompt LESR signal. Second, they depend on the illumination power I as $I^{1/2}$. This is a clear evidence for bimolecular relaxation, i.e. for a process of mutual recombination within the photoinduced radical pairs. Contrary, the persistent components ("light-off" corrected on the "dark") of ESR signals do not depend on the light intensities, and the amount of persistent P^+ species is much higher than that of persistent C_{60}^- (for details see (8)). The independence on the excitation light intensity indicates that these contributions to the ESR signal originate from spins trapped by some defect states. Therefore it is related to the amount of the defects in the samples. Thermally assistant escape from the traps is responsible for the annealing of persistent LESR signals. Non-equal amount of persistent spins clearly points to the presence of some other kind of charge carriers in the system, for instance, negative polymer polarons.

ACKNOWLEDGEMENTS

We thank Dr. H. Neugebauer for valuable discussions. Contribution of G. Zorinyants was supported by J. Kepler University Fellowship and by Russian Foundation for Basic Research, Grant 97-03-32164a. V. Dyakonov thanks FWF for financial support by a Lise Meitner fellowship under M00419-PHY.

REFERENCES

1. N.S. Sariciftci, L. Smilowitz, A.J. Heeger, and F. Wudl, Science **258**, 1474 (1992).
2. C.H. Lee et.al., Phys. Rev. B **48**, 15425 (1993).
3. G. Yu, J. Gao, J.C. Hummelen, F. Wudl, A.J. Heeger, Science **270**, 1789 (1995).
4. N. S. Sariciftci, L. Smilowitz, A. J. Heeger and F. Wudl, Science **258**, 1447 (1992).
5. R. A. J. Janssen, M. P. Christiaans, K. Pakbaz, D. Moses, J. C. Hummelen, N. S. Sariciftci, J. Chem. Phys. **102**, 2628 (1995).
6. K. Murata, Y. Shimoi, S. Abe, S. Kuroda, T. Noguchi, T. Ohnishi, Chem. Phys. **227**, 191 (1998).
7. V. Dyakonov, in *Primary Photoexcitations in Conjugated Polymers*, N. S. Sariciftci, ed. (World Scientific, Singapur, 1998).
8. V. Dyakonov, G. Zoriniants, M. Scharber, C.J. Brabec, R.A.J. Janssen, J.C. Hummelen and N. S. Sariciftci, Phys. Rev. B (1998), submitted.
9. C. Brabec et al., *ibid*.

Optical Properties and Heat of Solution of Fullerenes Taking Account the Aggregation of Fullerenes In Solutions

Bezmelnitsyn V.N., Eletskii A.V., Okun M.V

Russian Research Center *"Kurchatov Institute"*, 123182, Kurchatov sq. 1, Moscow, Russia. E-mail: eletskii@imp.kiae.ru

Abstract. The droplet model of fullerene clusters in solutions is used for analysis of experimental data on the concentration dependence of the third order nonlinear optical susceptibility of fullerene C_{60} benzene solution and also for determination of heat of solution of fullerenes versus concentration and temperature. The absorption frequency of fullerene molecules involved into clusters is removed out the four-wave mixing resonance which causes the decrease in a number of fullerene molecules interacting with radiation. Therefore the third order nonlinear optical susceptibility is defined only by separate C_{60} molecules which are not involved into clusters. The use of cluster size distribution function of aggregated fullerenes in solutions found earlier provides quite good coincigence between the measured dependencies $\chi^{(3)}(C)$ and concentration dependence of separate fullerene molecules. The same function is used for determination of the concentration and temperature dependencies of the heat of solution of aggregated fullerenes. The calculated results shed a light on the reason of apparent contradiction between measured data of various authors.

INTRODUCTION

Many extraordinary features of fullerenes in solutions are caused by the aggregation phenomenon. This phenomenon predicted theoretically [1] and confirmed experimentally by the light scattering methods [2, 3] and by the photoluminescence spectroscopy [4] appears as a trend of fullerenes in solutions to the formation of clusters consisting of a number of fullerene molecules. The cluster origin of solubility of fullerenes determines the nonmonotone temperature dependence of solubility of fullerenes in some organic solvents [1, 5]. The aggregation phenomena affect not only thermodynamic but also kinetic properties of fullerenes in solutions and cause specifically the temperature and concentration dependencies of diffusion [6] and thermodiffusion coefficients [7].

The fullerene aggregation phenomenon is reflected on the optical characteristics of fullerene solutions. Specifically, this phenomenon causes solvatochromic behaviour of fullerene solutions which are exhibiting in sharp changes in optical or raman spectra of C_{60} and C_{70} fullerene solutions as a result of rather small changes in solvent composition [3,8,9,10]. Besides of that, the fullerene aggregation results in an abnormal concentration dependence of nonlinear third order optical susceptibility of C_{60} dissolved solution in benzene [11]. In present work using the cluster size distribution function calculated previously [1,6,7], the quantitative description for this dependence is developed. It is shown that the phenomenon of aggregation of fullerenes in solutions results in

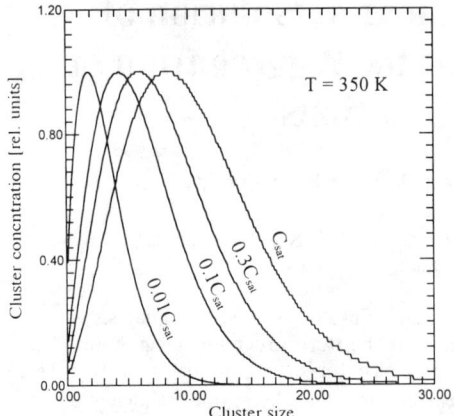 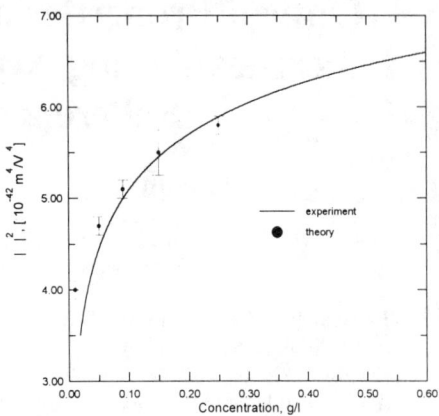

Fig. 1. The cluster size distribution function of C$_{60}$ fullerene solutions in benzene calculated on the basis of relationship (2) - (3).

Fig. 2 The concentration dependence of the third order responce [11] (+) and the number density of separate C$_{60}$ molecules (line)

the shift of the resonance frequency of four-wave mixing so that only separate C$_{60}$ fullerene molecules, which are not involved into clusters, interact effectively with laser radiation.

CLUSTER SIZE DISTRIBUTION FUNCTION

The thermodynamic approach to the detailed description of the fullerene aggregation phenomenon developed in [1] on the basis of a droplet model of clusters provides th folowing expression for the cluster size distribution function (CSDF) in saturated solution:

$$f_n = g_n \exp\left(\frac{-An + Bn^{2/3}}{T}\right) \qquad (1)$$

Here the parameter A refers to the equilibrium difference in the thermodynamic specific energies of the interaction of a fullerene molecule with its surrounding in the solid phase and in the body of cluster, and the parameter B corresponds to the similar difference for fullerene molecules placed on the surface of a cluster in solution; g_n is a statistical weight of the cluster of size n supposed to be temperature independent; T is the solution temperature (we express the parameters A and B in temperature units). The parameters A and B are determined through fitting the temperature dependence of fullerene solubility calculated on the basis of the droplet model of clusters [1] with the experimental one [5]. According to the droplet model of cluster valuable at condition $n \gg 1$ the solubility of fullerene C is proportional to the integral

$$C = g \int_{n=1}^{\infty} n \exp\left(\frac{-An + Bn^{2/3}}{T}\right) dn \qquad (2)$$

Here the parameter g is an averaged statistical weight of a cluster. The fitting procedure [1] results in $A = 320$ K, $B = 970$, which are weakly depending on the sort of solvent (benzene, toluene and CS_2) [5]. At the temperature range $T < 260$ K the parameter A increases by the value of the enthalpy of phase transition $h = 850$ K, which according to relationship (1) causes the essentially complete disappearance of aggregation phenomenon in this temperature range.

In the case of unsaturated solution the CSDF can be found from a condition of a balance between clusters of any size in solution that is expressed by requiring of equality of chemical potentials of clusters of any size [6]:

$$C_n = \lambda^n \exp\left(\frac{-An + Bn^{2/}}{T}\right), \quad (3)$$

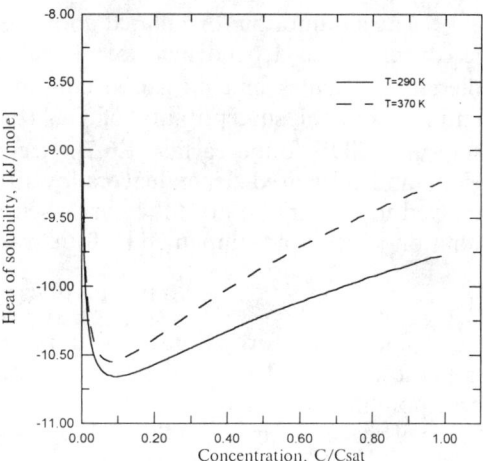

Fig. 3. The concentration dependencies of the heat of solubility of C_{60} fullerene in toluene, benzene and CS_2 calculated on the basis of droplet model (7).

Here parameters A and B are defined above and the parameter λ depends on the total solution concentration [1,6,7] and is determined from the normalization condition:

$$C(T) = C_0 \int_{n=1}^{\infty} n\lambda^n \exp\left(\frac{-An + Bn^{2/3}}{T}\right) . dn \quad (4)$$

The concentration depending multiplier C_0 can be found equating the concentration defined from (5) to the saturated value [6,7]. The CSDF in C_{60} fullerene solution calculated from (2) - (3) at various concentrations and temperature of solution are presented on fig. 1.

NONLINEAR OPTICAL SUSCEPTIBILITY

The concentration dependence of the third order nonlinear optical susceptibility measured in [11] for C_{60} fullerene solution in benzene using the Nd-YAG laser ($\lambda = 1.064$ μm) emitting 50 psec pulse of up to 5 mJ energy is presented on fig. 2. The large magnitude of nonlinear susceptibility (per one C_{60} molecule) shows the resonant mechanism of four wave mixing. Another feature of experimental data represented on fig. 2 is the nonlinear shape of concentration dependence $\chi^{(3)}(C)$. This reveals the decrease in a portion of fullerene molecules interacting with laser radiation in resonant mode. One can conclude that aggregation phenomenon results in the shift of resonance frequency of four-wave mixing process. Fullerene molecules involved in clusters are excluded from the resonant interaction with laser radiation that in turn results in a saturation of concentration dependence of nonlinear response.

The assumption formulated above is confirmed through the comparison of the concentration dependencies of the calculated number density of separate C_{60} fullerene molecules and measured concentration dependence of the third order nonlinear optical susceptibility of C_{60} (fig. 2). The calculation was performed using the CSDF found earlier [1,6,7]. A good agreement between the calculation results and measured dependence allows to believe that the fullerene molecules involved in clusters do not take part essentially in resonant interaction with laser radiation which is accompanied by four-wave mixing.

HEAT OF SOLUTION

The cluster formation in fullerene solutions determines concentration and temperature dependencies of heat of solution of fullerenes. It is caused by the corresponding dependencies of the CSDF determined above. The free energy of a cluster of size n is expressed through the equation

$$E_n = n(An - Bn^{2/3}) \tag{5}$$

where the constants A and B were defined above. The concentration and temperature dependencies of the heat of solution calculated using the expression (5) and CSDF determined above are given on fig. 3. As is seen the cluster formation changes considerably the thermal effect of solution of fullerenes. Thus, at zero concentration of solution, containing only separate fullerene molecules but not clusters (the concentration of solution is less than 0.1% of saturation) the heat of solution is accounted as -5.6 kJ/mol, whereas for solution of concentration as high as 15% of saturation (averaged size of a cluster is about 7) that is accounted as -10.6 kJ/mol. Seemly, such a strong dependence of the thermal effect of solution is a main reason of the apparent contradiction of measured data on the heat of solution of fullerenes, obtained by various authors: ΔH= -II kJ/mol [5], ΔH= -8.6 kJ/mol [12], ΔH= -8 kJ/mol [13]; ΔH= -13.6 kJ/mol [14 J.

This work was supported by the Russian Foundation for Basic Research.

REFERENCES

1. Bezmelnitsyn V.N., Eletskii A.V., Stepanov E.V., *J. Phys. Chem.* **98** 6665 (1994).
2. Ying Q., Marecek J., Chu B., *Chem. Phys. Lett.* **219** 214 (1994).
3. Ghosh H.N., Sapre A.V., Mittal J.P. J.Phys. Chem. **100** 9439 (1996).
4. Suzuki K., Ahn J.S., Iwasa Y., He A.Q., Outuka N. Mitani T. In: Molecular Nanostructures" ed. H.Kuzmany, J.Fink, M.Mehring, S.Roth. World Scientific. 1998. P.524.
5. Ruoff R.S., Malhotra R., Huestis D.L., Tse D.X., Lorents D.C. *Nature* **362** 140 (1993).
6. Bezmelnitsyn V.N., et al., *Sov. J. Chem. Phys.*, **13** №12, c. 156 (1994).
7. V.N. Bezmelnitsyn et al., *Physica Scripta,* **53**, 368-370, 1996.
8. Sun J.P., Bunker C.E., *Nature* **365** 398 (993) 557.
9. Gallagher S.H. et al., *J. Phys. Chem.* **99** 5817 (1995).
10. Gallagher S.H. et al., *J. Am. Chem. Soc.,* **116** 2091 (1994); *Chem. Phys. Let.,* **248** 353 (1996).
11. Blau W. J., et all. *Phys. Rev. Lett.* **67** 11 (1991).
12. Smith A.L., Walter E. In: "*Recent Advances in the Chemistry and Physic of Fullerenes and Related Materials*". Ed. K. Kadish, R.Ruoff. The Electrochemical Soc., Pennington, 1995.
13. Sivaraman N., et al. Ibidem, 1994, vol 94-24, pp. 443-458.
14. Smith A.L., Li D., King B., Zimmerman G., ibidem, 1994, vol.94-24, pp. 156-165.

A spectroscopic study of the fullerene/metal interface in C_{60}-Al multilayers

T. Pichler, T. Böske, M. S. Golden, M. Knupfer, M. Sing, J. Fink, Ch. Jung[1], C. Hellwig[1] and W. Frentrup[2]

Institut für Festkörper- und Werkstofforschung Dresden, Postfach 270016, D-01171 Dresden.
[1] *BESSY GmbH, Lentzeallee 100, D-14195 Berlin.*
[2] *Humboldt-Universität zu Berlin, Institut für Physik, Invalidenstr. 110, D-10115 Berlin.*

Abstract. From x-ray absorption spectroscopy at the C K edge we show that interface effects in C_{60}-Al multilayers give rise to a predominantly covalent interaction between the fullerene molecules and the metal, with minimal ionic charge transfer. In addition, an anomalous q-dependence of the spectral weight of the Al charge carrier plasmon is observed in electron energy-loss measurements, which could be related to the confinement of the Al layers between the fullerene dielectric.

INTRODUCTION

The electronic properties of the interface between metals and conjugated carbon systems are crucial in the understanding of the physics of charge injection and transport in devices such as organic-based light emitting diodes. C_{60} is by far the most intensively studied conjugated carbon system, and is thus an ideal model substance for the investigation of the fundamental interactions occuring at the carbon/metal interface. One strategy regarding the investigation of such interfaces is the study of ultrathin fullerene films on metal surfaces using surface sensitive methods [1]. An alternative strategy involves the investigation of multilayers in which the interface dominates the properties of the sample as a whole. This route has hardly been studied up to now, even though it offers the advantage of the applicability of standard volume sensitive methods such as infrared spectroscopy, transport measurements [2] or electron energy-loss spectroscopy (EELS) in transmission.

The Al-C_{60} multilayers studied here consisted of 17 bilayers of nominally 30 Å C_{60} and 30 Å Al, and were grown in UHV by successive deposition of C_{60} and Al onto KBr single crystal substrates. For the x-ray absorption spectroscopic (XAS) studies, the multilayers were grown and measured in UHV, whilst for the EELS measurements they were first floated off the KBr substrates and then transferred into the spectrometer. For all multilayers a C_{60} termination was chosen in order to help prevent oxidation of the Al surface. EELS experiments were performed in transmission in a purpose built spectrometer [3], with an energy and momentum resolution of 160 meV and 0.06 Å^{-1}. The XAS experiments at the C K edge were performed using linearly polarised synchrotron radiation from the PM5 monochromator [4] at BESSY I (Berlin), with an energy resolution of 140 meV. For the multilayers the absorption was monitored via the volume sensitive fluorescence yield (FY) mode, whereas the corresponding spectrum of C_{60} was recorded from the fullerene cap layer using total electron yield (TEY) detection.

RESULTS AND DISCUSSION

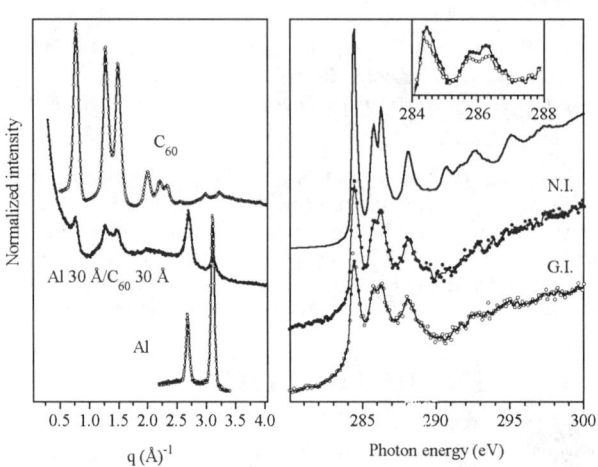

FIGURE 1. Left panel: electron diffraction profiles of C_{60}, a 30 Å/30 Å Al/C_{60} multilayer and polycrystalline Al metal. Right panel: C1s excitation spectra for C_{60} (solid line) and a 30 Å/30 Å Al/C_{60} multilayer, the latter recorded with the polarisation vector of the synchrotron radiation lying either in the multilayer plane (full circles), or at an angle of 20° with respect to the surface normal (open circles).

The left panel of Fig. 1 shows the electron diffraction profile of a 30 Å/30 Å Al/C_{60} multilayer. The appearance of the Bragg peaks of the individual constituents points to the crystalline nature of the component layers and to a relatively well-ordered interface region in the multilayer. In the right-hand panel of Fig. 1 the C1s excitation spectra (XAS) of bulk C_{60} and the 30 Å/30 Å Al/C_{60} multilayer are compared. The C1s excitation spectrum contains information related to the C2p-derived unoccupied electronic density of states of the fullerene molecules in the multilayer, which below 290 eV correspond to transitions into the unoccupied π^* molecular orbitals. Above 290 eV the features are related to transitions into the unoccupied σ^* electronic states. These σ^* features can also be discussed in terms of so-called shape resonances [5], whereby their position and intensity can be understood within a multiple scattering description involving the core-ionised

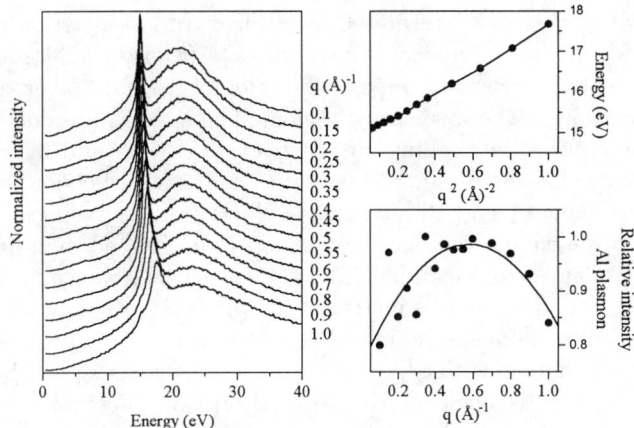

FIGURE 2. Left panel: momentum dependence of the loss function of a 30 Å/30 Å Al/C_{60} multilayer. The pronounced peak at around 15 eV for low **q** is the Al volume plasmon. Right panels: q-dependence of the dispersion relation (upper panel) and the spectral weight (lower panel) of the Al volume plasmon relative to the intensity of the C_{60} $\pi+\sigma$ plasmon in the multilayer.

atom and, in the case of C_{60}, the surrounding atoms of the same molecule [6]. In this way the σ^* features encode information regarding the shape of the molecule.

It is immediately clear that the unoccupied electronic levels of the C_{60} molecules are significantly perturbed in the multilayer, as expressed by the increased width of the π^* features. This broadening can be taken as a signal either of a symmetry-induced lifting of the degeneracy of the t_{1u}-derived states (although the core hole in the final state has already broken the I_h symmetry [7]) or of the reduced lifetime of the final state due to interaction of the fullerene MO's with the Al.

Importantly, there is little sign of a simple ionic charge transfer from Al to C_{60}, which would result in a strong reduction in intensity of the first, LUMO-derived feature and a shift of the other π^* features to lower energy [8]. As C1s excitation spectroscopy is a powerful, direct probe of charge transfer in doped fullerene systems, the clear contrast with the charge transfer of up to six electrons postulated from transport measurements of C_{60}-Al multilayers [2] suggests a reappraisal of the transport data is called for. In addition, the shape resonances above 290 eV have practically disappeared in the C1s excitation spectra of the multilayer, indicating a marked deviation of the shape of the C_{60} molecules at the buried interface compared to pristine C_{60}. The suppression of the shape resonances in the multilayer spectra is as strong as that observed in studies of ordered C_{60} monolayers on Al (111), in which a strong covalent interaction was concluded [6].

The XAS data of Fig. 1 also show a small polarisation dependence. The inset shows the low lying electron addition states on an expanded scale where it can be seen that in the grazing incidence geometry, the LUMO-derived feature is both narrower and less intense (as are also the next two transitions) than in normal incidence. This narrowing parallels the behaviour observed in single C_{60} monolayers on Al(110), where this was interpreted in terms of the localisation of the excited state on the side of the C_{60} molecule furthest away from the substrate [9].

In Fig. 2 we show the density response function of the multilayer, as represented by the loss function in the low energy region. The spectra are dominated by the volume plasmon of the aluminium charge carriers ($\sim$ 15 eV) and the $\pi + \sigma$ plasmon of C_{60} ($\sim$ 22 eV). The Al plasmon exhibits a quadratic dispersion in $\mathbf{q}$ (shown in the upper right panel of Fig. 2), as expected for a free electron gas [3]. However, the momentum dependence of the Al plasmon intensity (normalised to the non-dispersive $\pi + \sigma$ plasmon of the fullerene) is rather unusual (shown in the lower right panel of Fig. 2). At first, as $\mathbf{q}$ is increased the plasmon intensity increases, before going through a maximum and tailing off at higher momentum transfer. In general, plasmon intensities fall off with $\mathbf{q}^2$, resulting from the increasing number of channels available for decay into single particle excitations [3]. Further experiments are currently underway in which the thickness of the Al and C_{60} layers is systematically varied in order to clarify whether this behaviour is a result of confinement effects.

In summary, we have shown that the interface interaction between C_{60} and Al in bulk, multilayer samples is dominantly of a covalent nature as expressed by a minimal ionic charge transfer and a significant perturbation of the unoccupied electronic levels of the fullerene. These results are in keeping with those from ultrathin C_{60} films on metal surfaces. In addition, an unusual momentum-dependent damping of the Al charge carrier plasmon is observed.

We acknowledge the BMBF for financial support (05-625 BDA). T.P. thanks the European Union for funding under the 'Training and Mobility of Researchers' programme.

REFERENCES

1. See for example: S. J. Chase et al., Phys. Rev. **B46**, 7873 (1992); S. Modesti et al., Phys. Rev. Lett. **71**, 2469 (1993); A. J. Maxwell et al., Phys. Rev. **B49**, 10717 (1994); M. R. C. Hunt et al., Phys. Rev. **B51**, 10039 (1995).
2. A. F. Hebard et al., Phys. Rev. **B50**, 17740 (1994).
3. J. Fink, Adv. Electron. Electron Phys. **75**, 121 (1989).
4. H. Petersen et al., Rev. Sci. Instrum. **66**, 1 (1995).
5. J. Stöhr, 'NEXAFS Spectroscopy', Springer Verlag, Berlin (1992).
6. A. J. Maxwell et al. Phys. Rev. **B 52**, R5546 (1995).
7. B. Wästberg et al., Phys. Rev. **B 50**, 13031 (1994).
8. M. S. Golden et al., J. Phys.: Condens. Matt. **7**, 8219 (1995).
9. A. J. Maxwell et al., Phys. Rev. Lett. **79**, 1567 (1997).

FULLERIDES

The electronic structure of doped fullerenes studied using high energy spectroscopy

T. Pichler, M. S. Golden, M. Knupfer and J. Fink

Institut für Festkörper- und Werkstofforschung Dresden, Postfach 270016, D-01171 Dresden.

Abstract.
The electronic properties of fullerenes can be engineered by three types of doping: intercalation, on-ball doping and endohedral doping. Here we present an overview of the investigation of the electronic structure of intercalated, hetero and endohedral fullerenes using high energy spectroscopic techniques. In particular, we focus on information obtained regarding the electronic density of states and the ground state of the A_xC_{60} (A = alkali metal), $(C_{59}N)_2$ and Tm@C_{82} materials.

INTRODUCTION

In principal, doping of fullerenes can be done in three ways: doping from outside the molecules (intercalation), doping from inside (endohedral doping) or changing the molecular structure itself by a substitution of C atoms with heteroatoms like N or B (heterofullerenes), as it is depicted schematically in Fig. 1. All these routes have been successfully followed, resulting in a large number of fullerene intercalation compounds (fullerene salts), endohedrally doped fullerenes (metallofullerenes) or heterofullerenes. These developments were also successful in their aim of property engineering whereby fullerene-based materials now support a wide range of properties including metallic conductivity, superconductivity or ferromagnetism.

In this paper, we illustrate some of the issues involved in the electronic structure and physical properties of doped fullerenes by treating one example of each of the doping methods mentioned above in some detail. This overview underscores not only the richness of the physics encountered in the doped fullerenes, but also the important role played by high energy spectroscopy in their investigation.

C_{60} and the endohedral fullerenes studied here were produced using the Krätschmer/Huffman carbon arc method [1]. $C_{59}N$ was produced via an organic synthetic route [2]. For the high energy spectroscopic studies thin films were prepared by sublimation in ultrahigh vacuum. For details of the sample preparation, see Ref. [3]. The EELS measurements in transmission were carried out with a spectrometer described in detail elsewhere [4]. For the valence level (core level) excitations and elastic scattering data, the momentum resolution of the EELS

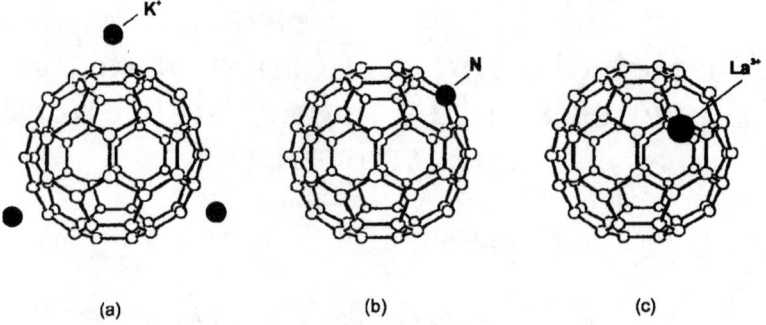

FIGURE 1. A schematic representation of the three different ways to dope fullerenes: (a) intercalation, (b) on-ball doping, (c) endohedral ('in-ball') doping.

spectrometer was set to 0.05 (0.15) Å^{-1} with an energy resolution of about 120 (160) meV. The PES experiments were carried out using a hemispherical electron analyser, together with a noble gas discharge lamp providing He I radiation (21.22 eV, overall resolution 100 meV) or an x-ray source providing monochromatised Al Kα radiation (1486.6 eV, overall resolution 350 meV).

INTERCALATED FULLERENES

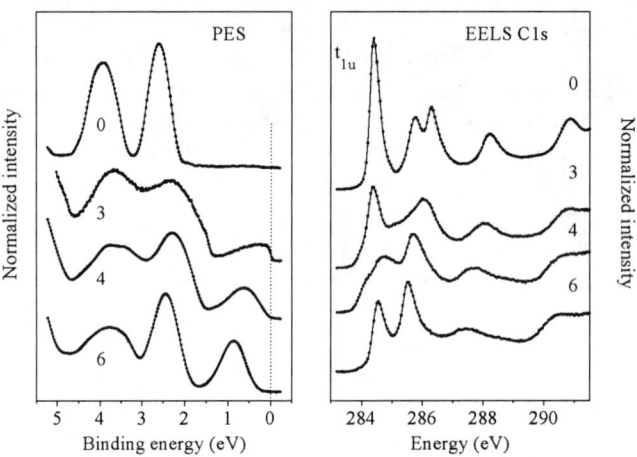

FIGURE 2. Left panel: photoemission spectra of potassium intercalated K$_x$C$_{60}$ for $x = 0, 3, 4, 6$. The dotted line marks the position of E_F. Right panel: C 1s excitation spectra for the same compounds.

Figure 2 shows the photoemission and C1s excitation spectra for the K$_x$C$_{60}$ system with x = 0, 3, 4 and 6. Upon intercalation the incremental occupation of

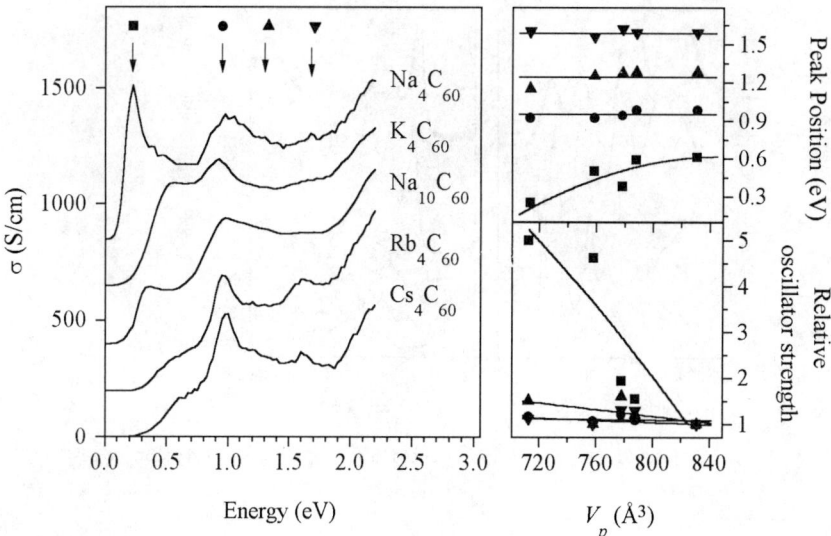

FIGURE 3. Left panel: optical conductivity of A_4C_{60} and $Na_{10}C_{60}$. Right panel: the experimentally determined position and relative oscillator strength of the four lowest lying transitions as a function of the volume of the primitve unit cell, V_p.

the LUMO can be clearly seen in the spectra by the growth of a new feature located in the former gap (PES) and reduction in the LUMO spectral weight in EELS. In K_6C_{60}, the LUMO-derived feature is no longer visible in the C1s excitation spectra, as the t_{1u} band is completely filled and thus lies completely below the Fermi level (E_F) in the PES spectrum. An understanding of the spectra for C_{60} and K_6C_{60} presented in Fig. 2 is relatively straightforward: both materials have completely filled electronic levels which results in their observed insulating nature. For K_3C_{60} a considerable density of states at the Fermi level and a clear Fermi cutoff are visible in the PES indicating the metallic ground state for this compound. The anomalous width of the PES profile for the half-filled t_{1u}-derived states can be accounted for by the joint effects of coupling to both molecular vibrations and to the charge carrier plasmon during the photoemission process [5]. An alternative interpretation in terms of a shift of spectral weight due to the effects of correlation [6] cannot explain the magnitude of the spectral weight observed at 600 meV in high resolution PES measurements [5,7]. On further intercalation to $x \sim 4$ the LUMO-derived peak has grown further and shifted away from E_F, resulting in small or zero intensity at the chemical potential. The insulating, non-magnetic behaviour of the A_4C_{60} compounds despite the formal partial (2/3) filling of their t_{1u} bands cannot be explained within a one-electron picture. In the literature, these sytems have been proposed to be either Mott-Hubbard [6] or Jahn-Teller insulators [8].

To clarify this question, some of us have undertaken a systematic study of the

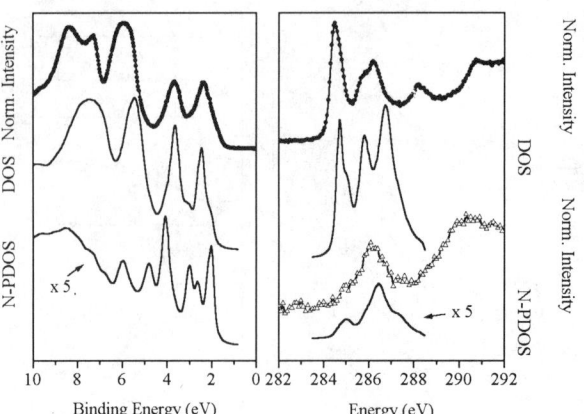

FIGURE 4. Valence band photoemission spectra and core level absorption edges of $(C_{59}N)_2$ compared with theory.

optical conductivities of A_4C_{60} and $Na_{10}C_{60}$ (the latter has a 1/3-filled conduction band and a loss function very similar to that of the A_4C_{60} sytems), which concluded that these compounds possess a Mott-Hubbard ground state [9]. The results on which this conclusion is based are shown in Fig. 3, in which the optical conductivities are fitted using a Lorentz model with four oscillators. This analysis reveals that the energy position and the spectral weight of the energetically lowest lying transition is strongly dependent on the volume of the primitive unit cell, whereas the three higher lying transitions do not show such behaviour. This fact enables the assignment of the first structure to transitions between adjacent molecules (the higher lying ones being due to intra-molecular excitations). The behavior of the first transition, which represents the energy gap, can in fact be understood within a simple Mott-Hubbard description which takes into account the bare on-site correlation energy, its screening in the solid, as well as the band width. As the latter two are functions of the lattice constants and symmetry [9], this explains the observed V_p-dependence seen in Fig. 3. Thus, the success of this simple model in the description of the optical properties of the A_4C_{60} compounds (as well as $Na_{10}C_{60}$) is strong evidence for a Mott-Hubbard ground state in these systems.

HETEROFULLERENES

$C_{59}N$ is one example of how to achieve a formal n-type doping by substituting C-atoms with heteroatoms (in this case N). Fig. 4 shows a comparison of the valence band PES and the C1s and N1s (Ref. [10]) excitation spectra of $(C_{59}N)_2$ [11]. Also included are the calculated N-derived partial DOS (PDOS) and the total DOS, which have been obtained from DFT-calculations and broadened to facilitate the comparison with the experiment [11,12]. The agreement between the calculation

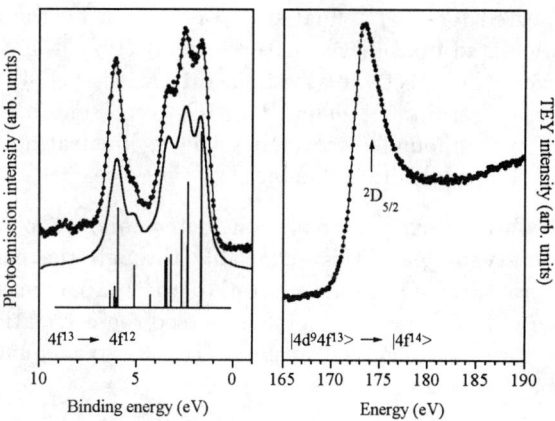

FIGURE 5. Left panel: Photoemission spectrum of the Tm $4f$ levels of Tm@C$_{82}$ together with the calculated multiplets for $4f$ photoemission from a divalent Tm ion. Right panel: Tm $4d$ excitation spectrum of Tm@C$_{82}$ measured using x-ray absorption spectroscopy

and experiment is remarkable and leads to the following picture for the electronic structure of the heterofullerene [11]:

- the HOMO of $(C_{59}N)_2$ is shifted to lower binding energy than that of C_{60} and is spatially located mainly on the N atoms and the intermolecular bond.

- the energy distribution of the lowest lying unoccupied states of the heterofullerene resembles that in C_{60}. These states in the heterofullerene are almost of pure carbon character, being spatially located at the opposite ends of the $C_{59}N$ molecules to the dimer bond.

This demonstrates that heterofullerene formation, while far from representing a simple doping in the sense of traditional semiconductor physics, results in a novel and interesting reconstruction of the electronic states of C_{60}.

ENDOHEDRAL FULLERENES

The last of the three doping methods to be considered is endohedral doping, which is unique to closed cage-like structures such as the fullerenes [13]. One of the key parameters which controls the electronic properties of endohedrally doped fullerenes is the amount of charge that is transferred from the encaged atom to its fullerene host. Until recently, the charge distribution in monometallofullerenes of the lanthanide ions has been discussed almost exclusively in terms of $M^{3+}@C_{82}^{3-}$ [14]. The first and only example of a purely divalent lanthanide monometallofullerene discovered up until now is Tm@C$_{82}$. This is illustrated in Fig. 5, where the Tm $4f$ photoemission and Tm $4d$ excitation spectrum of Tm@C$_{82}$ are shown. The former (measured with monochromatised Al:K$_\alpha$ radiation) exhibits the characteristic multiplet splitting expected for a $4f^{12}$ final state, as is seen by the quantitative agreement with the calculated multiplet structure. In addition, the single-peaked nature of the Tm $4d$ excitation spectrum attests to the closed-shell $4f^{14}$ final state in the electron addition spectrum [15]. Thus, Tm@C_{82} represents the purest and most stable divalent Tm compound discovered to date, illustrating the unusual properties accessible in the endohedral fullerenes.

These three case studies covering intercalation, hetero and endohedral doping exemplify the wide variety of properties achievable through the engineering of fullerenes' electronic structure. Of great interest to the fundamental researcher is also the large number of basic phenomena (such as electron correlation, electron-phonon coupling, dimersation and charge transfer) necessary to adequately describe the electronic structure of doped fullerenes.

We acknowledge the BMBF for financial support under contract number 13N6676/7. T.P. thanks the European Union for funding under the 'Training and Mobility of Researchers' program.

REFERENCES

1. W. Krätschmer et al., Nature **347**, 354 (1990).
2. J.C. Hummelen et al., Science **269**, 1554 (1995).
3. M. S. Golden et al., J. Phys.: Cond. Mat., **7**, 8219 (1995).
4. J. Fink, Adv. Electron. Electron Phys. **75**, 121 (1989).
5. M. Knupfer et al., Phys. Rev. **B47**, 13944 (1993).
6. G. A. Sawatzky in *Physics and Chemistry of Fullerenes and Derivatives*, Eds. H. Kuzmany, J. Fink, M. Mehring, and S. Roth, World Scientific, Singapore, pp 373-378 (1995).
7. O. Gunnarsson, Z. Phys. B **104**, 279 (1997).
8. Y. Iwasa et al., Synth. Met. **70**, 1361 (1995); R. Kerkoud et al., Synth. Met. **77**, 205 (1996).
9. M. Knupfer and J. Fink, Phys. Rev. Lett. **79**, 2714 (1997).
10. The N1s excitation spectrum is downshifted in energy by the difference between the N1s and C1s core level binding energies in XPS: 400.7 - 285.2 = 115.5 eV.
11. T. Pichler et al., Phys. Rev. Lett. **78**, 4249 (1997).
12. W. Andreoni et al., J. Amer. Chem. Soc. **118**, 11 335 (1996).
13. for an overview of the current status of research into endohedral fullerenes see the special issue 'Endohedral Fullerenes', Ed. E. E. B. Campbell, Appl. Phys. A **66**, 241-319 (1998).
14. The only exception so far is the work of Kessler et al., Phys. Rev. Lett. **79**, 2289 (1997) who propose an oxidation state of less than three for the La in La@C_{82}.
15. T. Pichler et al., Phys. Rev. Lett. **79**, 3026 (1997).

Rb_1C_{60} as a 3D electronic system

Martin Fally[*][1] and Hans Kuzmany[†]

Institut für Experimentalphysik
†*Institut für Materialphysik*
Universiät Wien, Strudlhofgasse 4, A-1090 Wien, Austria

Abstract. We use a random-phase-approximation expression of an interacting electron model with a tight-binding energy dispersion to describe the spin susceptibility $\chi(\mathbf{Q})$ of the A_1C_{60} systems. By slightly varying the band-filling k_F, the anisotropy of the electron overlaps $t_\parallel/t_\perp$ and/or the interaction strength λ we are able to describe the electronic and magnetic properties of Rb_1C_{60} as well as K_1C_{60} by a 3D model. The results are compared with measurements from the literature.

The A_1C_{60} compounds in their polymeric low temperature phase have gained a lot of interest. Their properties have been investigated by different experimental techniques: X-ray [1], NMR [2-4], ESR [5,6], resistivity and thermopower measurements [7]. In several of these works the properties of the Rb_1C_{60} were claimed to be typical for a 1D conductor. In contrast, the electronic bandstructure has been determined by LDA calculations for Rb_1C_{60} [8] as well as for K_1C_{60} [9] to be essentially 3D. Only recently have we shown that the crucial point - the existence of a metal-insulator transition in Rb_1C_{60} and the absence of such an instability in K_1C_{60} despite their similar structures - can be explained by assuming that the two 3D electronic systems differ only slightly in their bandfillings or anisotropy [10].

In this contribution we present the results of a refined calculation, i.e. the phase diagram in the $k_F - t_\parallel/t_\perp$-plane and weak interactions $1.1 < \lambda/t < 1.6$ considering the full expression for the tight binding energy dispersion. Moreover the $\mathbf{Q}$-dependence of the phase transition temperature for a particular set of parameters appropriate to describe the properties of Rb_1C_{60} is examined. Finally we evaluate the spin lattice relaxation time by applying our 3D model. A good qualitative agreement between the experimental data and the calculations was achieved.

Within the RPA the dynamic susceptibility of a correlated system is:

$$\chi(\mathbf{Q},\omega,T) = \frac{\chi_0(\mathbf{Q},\omega,T)}{1 - \lambda\chi_0(\mathbf{Q},\omega,T)} \quad (1)$$

---

[1]) Fally@pap.univie.ac.at

where λ is a phenomenological constant which describes the mean-field. The dynamic response function of the uncoupled system reads:

$$\chi_0(\boldsymbol{Q},\omega,T) = \sum_k \frac{f(\epsilon_k,T) - f(\epsilon_{k-Q},T)}{\hbar\omega + (\epsilon_k - \epsilon_{k-Q}) - i0^+} \qquad (2)$$

where f denotes the Fermi distribution and

$$\epsilon(\boldsymbol{k}) = -2t_\parallel \left[\cos(ak_x) - \cos(ak_F)\right] - 2t_\perp \left[\cos(bk_y) + \cos(ck_z)\right] \qquad (3)$$

is the nearest neighbor tight-binding energy dispersion with k_F the bandfilling and $t_{\parallel,\perp}$ the transfer integrals parallel and perpendicular to the polymeric chain, respectively.

As the A_1C_{60} systems are known to be nearly isotropic ($t_\parallel/t_\perp \approx 1$) with a more or less half filled conduction band ($k_F \approx \pi/(2a)$) we investigated this region in the k_F vs. $t_\parallel/t_\perp$ plane in more detail. We use the set of parameters for Rb_1C_{60} from [10]: $k_F = 0.95\pi/(2a)$, $t_\parallel/t_\perp = 0.955$. These values and a T_c of 50 K are consistent with an interaction constant of λ =186 meV. By a small variation of these parameters the properties of the system changes dramatically: The phase transition can disappear, the NMR-relaxation rates can decrease and become temperature independent, the critical $\boldsymbol{Q}_c$-vector can change, etc. A good example for such a behavior are the systems Rb_1C_{60} and K_1C_{60}.

As a first step we calculate the stability of the metallic phase as a function of the bandfilling and the electron overlap ratio for three values of λ. The stability of the metallic phase in the RPA is determined by Stoner's criterion:

$$\chi_0(\boldsymbol{Q}_c, 0, T_c) = \lambda^{-1} \qquad (4)$$

The determination of the phase diagram was performed by an analytical integration of Eq. (1) along k_x, numerical integrations along the other directions for $T_c = 0$ K and $\omega = 0$ and then solving Eq. (4). The result for three different interaction constants is shown in Fig. 1. The bold lines (phase boundaries) in Fig.1 separate the regions of stability for the metallic and the insulating (spin density wave) phase. The critical wave vector was assumed to be $\boldsymbol{Q}_c = (2k_F, \pi/b, \pi/c)$. This is valid for $T_c = 0$ in the upper right region as long as the Fermi surface is closed. For $T_c > 0$ the critical Q-vector is different as will be shown below. By inspecting the upper right of Fig. 1 (enlarged) it is evident that several possibilities exist how to cross the phase boundary by slightly changing one of the parameters.

As a next step we focus on the problem of the critical vector $\boldsymbol{Q}_c$ for a particular value of the electron overlap ratio $t_\parallel/t_\perp$ =0.955. For quasi 1D systems like the well known charge transfer salts it is known that for $T \equiv 0$ the best nesting is achieved for the vector that connects the inflection points of the Fermi surface [11]. However, in our case the Fermi surface is closed and therefore this will not apply. Moreover we have a system with a phase transition temperature around 30–50 K for Rb_1C_{60} and so we have to find the critical vector $\boldsymbol{Q}_c$ at these temperatures. The phase transition

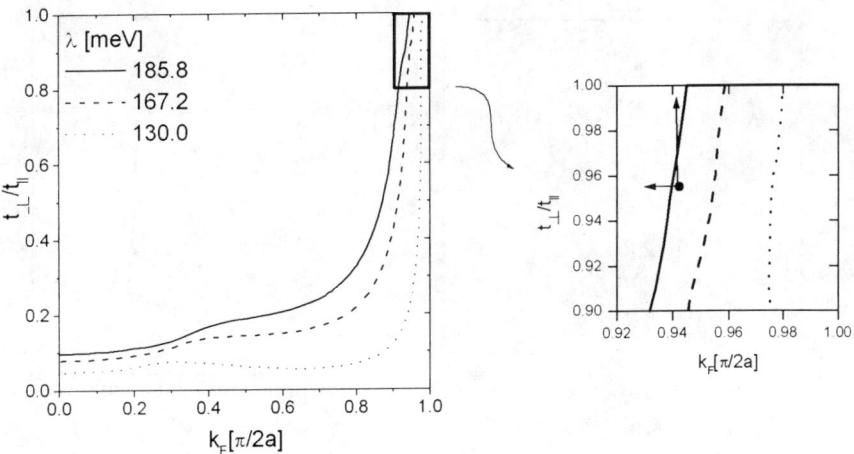

FIGURE 1. Phase diagram as a function of the bandfilling and the electron overlap ratio for three different electron-electron couplings: bold line $\lambda =186$ meV, dashed line $\lambda =167$ meV and dotted line $\lambda =130$ meV, resp. The right figure zooms in the region of interest.

temperature for a given set of parameters $k_F, \lambda = 186$ meV, $t_\parallel/t_\perp = 0.955$ is obtained by solving Eq. (4) so that for the particular $\mathbf{Q} = \mathbf{Q}_c$ the transition temperature T_c is maximized. By varying k_F, the transition temperature **and** the critical wave vector $\mathbf{Q}_c$ change. Fig. 2 shows the $\mathbf{Q}$-dependence of the transition temperature as a function of the bandfilling. If the deviation from half filling is not too large, the best nesting occurs at $\mathbf{Q}_c = (\pi/a, \pi/b, \pi/c)$. For a transition temperature of $T_c \equiv 0$ K the best nesting vector is $\mathbf{Q}_c = (2k_F, \pi/b, \pi/c)$. Within these limits $\mathbf{Q}_c$ follows a complicated law for Q_{cx} whereas $Q_{cy} = \pi/b$ and $Q_{cz} = \pi/c$, respectively. This may be important for finding the critical vector in neutron scattering.

Measurements of the spin lattice relaxation time T_1 by means of NMR on the A_1C_{60} systems were reported by [2-4]. Both, the ^{13}C and the ^{87}Rb, ^{133}Cs relaxation rates were determined. $1/TT_1$ is enhanced compared to an uncorrelated system and increases strongly when approaching the transition temperature for the Rb_1C_{60}. The relaxation rate of K_1C_{60}, on the other hand, obeys the temperature independent Korringa law and is less enhanced. The explanation given by several authors for this difference is that the Rb_1C_{60} is a quasi one-dimensional system whereas K_1C_{60} is three dimensional.

The spin lattice relaxation rate can be evaluated immediately for the above model by applying the fluctuation-dissipation theorem. The crucial parameter is the electron-electron interaction constant λ. The role of latter in the enhancement of the nuclear spin-lattice relaxation with a conduction-electron-nucleus interaction of the form $A(r)\mathbf{s} \cdot \mathbf{I}$ can be written as [12]:

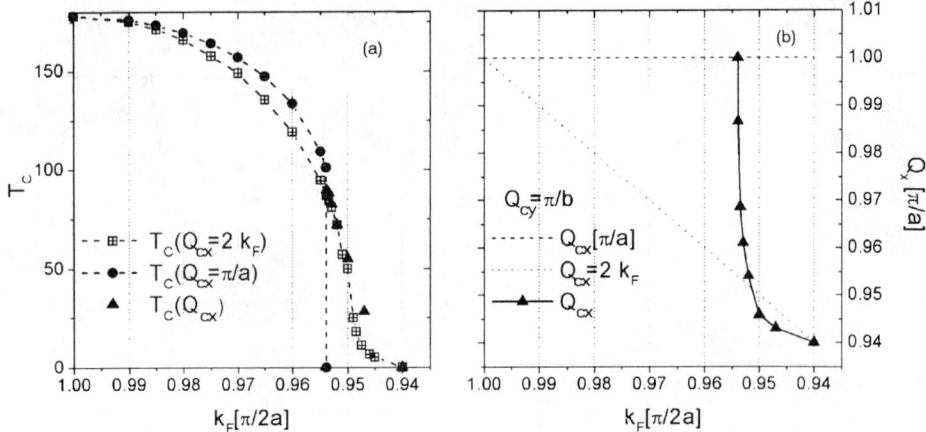

FIGURE 2. (a)$T_c(k_F)$ for some "special" Q_{cx}-vectors. (b) The Q_c-dependence of T_c as a function of the bandfilling; $\lambda = 186$ meV, $t_{\parallel}/t_{\perp} = 0.955$.

$$\frac{1}{T_1 T} \propto \sum_Q |A_Q|^2 \chi''(Q, \omega_0, T)/\omega_0 \qquad (5)$$

where A_Q is the Q-dependent matrix element between two Bloch states. For the ^{13}C nucleus the hyperfine constant A_Q exhibits no Q-dependence [4]. We restrict our calculations to this case. However, it is easy to verify that the influence of the Q-dependence of the hyperfine constant for the Rb and the K nuclei on the relaxation rates is small as well (cf. inset of Fig.2 in [4]). Within the RPA in the fast motion regime (i.e. $\omega\tau \ll 1$) the nuclear relaxation rate is obtained from Eq. (1) and Eq.(5) as:

$$\frac{1}{T_1 T} \propto \sum_Q \frac{1}{(1 - \lambda \chi_0'(Q, 0, T))^2} \qquad (6)$$

As shown above $\chi'(Q, 0, T)$ does not necessarily peak at $Q_x = 2k_F$. In that case the main contributions to the sum in Eq.(5) cannot be estimated in contrast to what was done done in Ref. [4]. Evaluating Eq. (6) by integration over Q for three different coupling constants we obtained the enhanced relaxation (for K_1C_{60}) and its increase with temperature (for Rb_1C_{60})as shown in Fig. 3.

In conclusion we have shown that a simple 3D model is consistent with drastic changes of the properties as is the case for Rb_1C_{60} and K_1C_{60} for only small changes in the bandfilling and/or the electron overlaps and/or the electron-electron coupling strength. The NMR relaxation rate and its temperature dependence has been evaluated. From Eq. (6) it is clear that the strongly temperature dependent susceptibility gives the main contribution to the spin-lattice relaxation rate. The

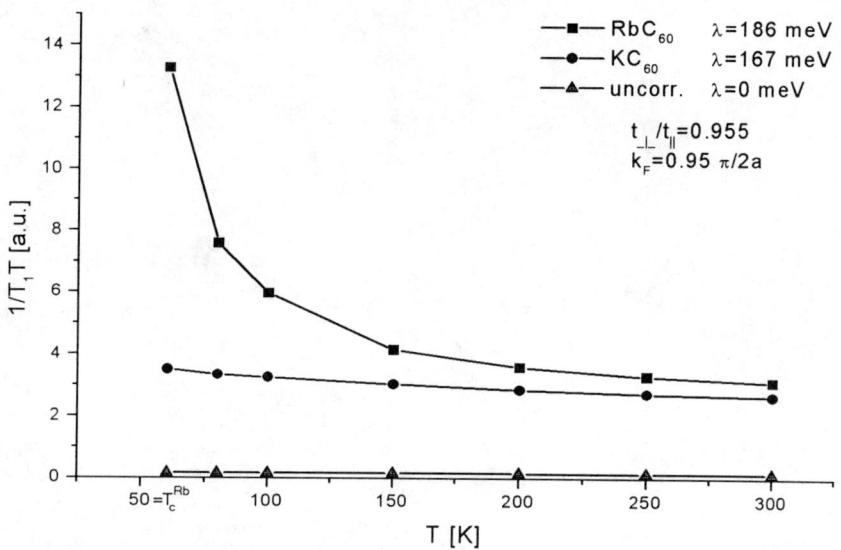

FIGURE 3. $1/T_1T$ for an uncorrelated system, Rb_1C_{60} and K_1C_{60}, resp., as calculated using Eq. (6).

enhancement reflects the electron-electron coupling. The agreement between the theoretical results and experimental data from [4] is excellent. The analysis given here is similar to the one performed in cuprate superconductors [13].

Acknowledgement This work was supported by the OeNB, Jubiläumsfonds, project N° 6318 and by the TMR project, contract N° ERBFMRX-CT97-0155.

REFERENCES

1. Stephens,P.W. et al., *Nature* **370**, 636 (1994).
2. Thier, K.-F. et al., *Phys. Rev. B.* **53**, R496 (1996).
3. Alloul, H. et al., *Phys. Rev. Lett.* **76**, 2922 (1996).
4. Brouet, V. et al., *Phys. Rev. Lett.* **76**, 3638 (1996).
5. Chauvet,O. et al., *Phys. Rev. Lett.* **72**, 2721 (1994).
6. Bommeli, F. et al., *Phys. Rev. B.* **51**, 14 794 (1995).
7. Hone, J. et al., *Phys. Rev. B.* **52**, R8700 (1995).
8. Erwin, S.C. et al., *Phys. Rev. B.* **51**, 7345 (1995).
9. Schulte, J., and Böhm, M.C., *Chem.Phys.Lett.*,**252**,367-374 (1996).
10. Fally, M., and Kuzmany, H., *Phys. Rev. B.* **56**, 13 861 (1997).
11. Jafarey, S., *Phys. Rev. B.* **16**, 2584 (1977).
12. Ehrenfreund, E. et al., *Phys. Rev. Lett.* **28**, 873 (1972).
13. Bulut, N. et al., *Phys. Rev. Lett.* **64**, 2723 (1990).

On a description of the normal-state resistivity in alkali-metal doped C_{60}

Yuanhe Huang[*], Akihiro Ito, and Kazuyoshi Tanaka[†]

Department of Molecular Engineering, Graduate School of Engineering, Kyoto University, Sakyo-ku, Kyoto 606-8501, Japan

Abstract. Two kinds of theoretical approaches, Ziman's resistivity formula ρ^Z and a usual approximate solution of Boltzmann equation for a small applied electric field ρ^B, to the resistivity expression for the alkali-metal doped C_{60} have been examined and compared. The condition under which they lead to the identical result of the normal-state resistivity of A_3C_{60} has also been afforded. It has been found that the both expressions give almost identical result on the intermolecular-mode scattering above the superconducting T_C but quite different one for the intramolecular-mode (ω_n) scattering except under such an extreme condition $\hbar\omega_n \ll E_F$ and $k_B T \gg \hbar\omega_n$. It has been concluded that it is suitable to employ the Ziman's formula for description of the normal-state resistivity of A_3C_{60}.

I. Introduction

Discovery of superconductivity in alkali-metal doped C_{60} (A_3C_{60}; A=K or Rb) has stimulated a great number of experimental and theoretical works on the electronic structures and the physical properties, including the normal-state transport properties above the superconducting temperature T_C.[1] The experimental works have shown that the temperature dependence of resistivity $\rho(T)$ in the both K_3C_{60} and Rb_3C_{60} is metallic-like above T_C even up to *ca.* 520 K without any evidence of saturation.[1] In general, there are at least two kinds of theoretical approaches to the resistivity expression.[2-6] One is Ziman's resistivity formula[7] $\rho^Z(T)$ and the other a usual approximate solution $\rho^B(T)$ of Boltzmann equation. The former formula gives an isotropic and energy-independent resistivity. The latter conventional solution of Boltzmann equation is good for weak scattering case but its breakdown in the strong scattering has not been well understood. Although it has become clear in our previous study[6] that the experimentally obtained intramolecular and intermolecular parts of $\rho(T)$ of A_3C_{60} can be fitted quite well to $\rho^Z(T)$ and to $\rho^B(T)$, respectively, its rationalization has not been well discussed hitherto. Here we attempt to clearly differentiate these two approaches and examine the condition under which they lead to the identical result of the normal-state resistivity of A_3C_{60}.

II. Expressions of $\rho(T)$

In this section, two kinds of expressions $\rho^Z(T)$ and $\rho^B(T)$ are summarized. For scattering of the intermolecular modes, we consider the contribution only from the acoustic phonons[6] corresponding to the relative movement of C_{60} molecules against each other ($20 < \omega < 70$ cm^{-1}).[8] Based on the treatment for the conventional molecular crystals[9] appropriately modified into 3D-C_{60} and for one-phonon process, the electron-phonon (e-p) interaction matrix elements are obtained.[6]

For the intermolecular mode scattering, under assumption of spherical Fermi surface, Ziman's expression of resistivity finally leads to,

$$\rho^Z_{\text{inter}}(T) = \frac{32\pi^2 k_B T^3}{\hbar \omega_p^2 \Theta_m^2} \lambda_{\text{inter}} \int_0^{\frac{\Theta_m}{T}} \frac{x^3}{(e^x - 1)(1 - e^{-x})} dx , \qquad (1)$$

where $\lambda_{\text{inter}} = N(E_F) V_{\text{inter}}$ (V_{inter}: the intermolecular coupling strength; $N(E_F)$: the density of states at the Fermi level per spin of a C_{60} molecule), ω_P is the plasmon frequency ($= \sqrt{4\pi n e^2/m}$), and $x \equiv \hbar \omega_q / k_B T$ and $\Theta_m = \hbar \omega_m / k_B$. The upper limit of acoustic phonon frequency are expressed by ω_m.

For the intramolecular mode scattering, we get the following formula:

$$\rho^Z_{\text{intra}} = \frac{8\pi^2 N(E_F)}{\omega_p^2 k_B T} \sum_n \frac{\omega_n \sum_{p,p'}^3 (\partial \varepsilon^{pp'}/\partial Q_n)^2}{(e^{\hbar \omega_n/k_B T} - 1)(1 - e^{-\hbar \omega_n/k_B T})} . \qquad (2)$$

Here p designates the triply degenerate LUMO $\{\phi_p\}$ ($p = 1, 2,$ and 3)), Q_n the normal coordinate of the n-th intramolecular phonon mode, and $\varepsilon^{pp'} \equiv \langle \phi_p | \hat{H} | \phi_{p'} \rangle$.

On the other hand, considering only the scattering on the Fermi surface and employing the relaxation-time approximation, the resistivity formulae due to the intermolecular and the intramolecular mode scatterings are obtained by solving the Boltzmann equation perturbatively:

$$\rho^B_{\text{inter}}(E_F) = \frac{32\pi^2 k_B T^3}{\hbar \omega_p^2 \Theta_m^2} \lambda_{\text{inter}} \int_0^{\frac{\Theta_m}{T}} \left(\frac{1}{e^x - 1} + \frac{1}{2}\right) x^2 dx , \qquad (3)$$

$$\rho^B_{\text{intra}}(E_F) = \frac{8\pi^2 N(E_F)}{\omega_p^2 \hbar} \sum_n \frac{\sum_{p,p'}^3 (\partial \varepsilon^{pp'}/\partial Q_n)^2}{(1 - e^{-\hbar \omega_n/k_B T})(1 + e^{\hbar \omega_n/k_B T})} \left[\left(1 - \frac{\hbar \omega_n}{E_F}\right)^{1/2} + \left(1 + \frac{\hbar \omega_n}{E_F}\right)^{1/2}\right] . (4)$$

III. Discussions

Let us first consider the scattering by the intermolecular modes. From a simple consideration, it is found that $\rho^Z_{\text{inter}}(T) = \rho^B_{\text{inter}}(T) \propto T$ at high temperature. For comparison, these resistivities calculated are shown in Fig.1 for the parameters $\lambda_{\text{inter}} = 0.085$, $\Theta_m = 100.7$ K, and $\hbar \omega_p = 1.2$ eV originally employed in Ref. 13. It is shown that $\rho^Z_{\text{inter}}(T)$ and $\rho^B_{\text{inter}}(T)$ vary linearly with temperature as expected for low-frequency intermolecular-mode scattering and that $\rho^Z_{\text{inter}}(T) < \rho^B_{\text{inter}}(T)$ in the range 20-520 K. This inequality becomes rather obvious at low temperatures (20-50 K), but the difference is still only 0.013 mΩ cm at 20 K. Therefore, $\rho^Z_{\text{inter}}(T)$ and $\rho^B_{\text{inter}}(T)$ give almost the same results over the whole temperature range.

The expressions of intramolecular resistivity ρ^Z_{intra} and ρ^B_{intra} (Eqs.(2) and (4)), on

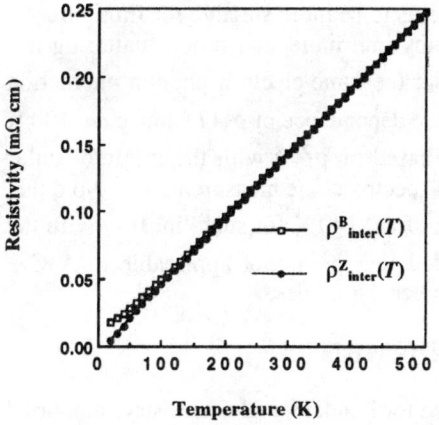

FIG. 1. The temperature dependence of resistivity of A_3C_{60} due to the intermolecular-mode scattering. Note this result is the same for the both K_3C_{60} and Rb_3C_{60}.

FIG. 2. Temperature dependence of the ratio $\rho^B_{intra}/\rho^Z_{intra}$ for A_3C_{60}. The references denote the sources of the set of the intrtamolecular coupling strengths $\{v_n; n=1\text{-}8\}$ used for the calculation.

the other hand, are quite different. Under the conditions of $\hbar\omega_n \ll E_F$ and $k_BT \gg \hbar\omega_n$, it is found that the two kinds of expressions of ρ_{intra} give the identical results. However, these conditions would never be satisfied in actual A_3C_{60} due to the following reasons: (i) The highest frequency of the intramolecular H_g mode that can couple to t_{1u} electrons is equal to 1575 cm^{-1} corresponding to 0.195 eV≈2263 K[10] and E_F's estimated are 0.3-0.35 and 0.19-0.20 eV for K_3C_{60} and Rb_3C_{60}, respectively,[11] and, therefore, the relationship $\hbar\omega_n \ll E_F$ is unrealistic, and (ii) $k_BT \gg \hbar\omega_n$ cannot be realized since A_3C_{60} will be decomposed at 2000 K.

Using the published data of the intramolecular-mode coupling strengths[3,14-17] and frequencies[10] that can fit the temperature dependence of resistivity at constant pressure,[1-3,6] we find $\rho^Z_{intra} > \rho^B_{intra}$ above the superconducting transition temperature ($T_C \approx 18$ and 30 K for K_3C_{60} and Rb_3C_{60}, respectively) for E_F=0.3 eV(K_3C_{60}) and 0.2 eV(Rb_3C_{60}). Temperature dependence of the ratio $\rho^B_{intra}/\rho^Z_{intra}$ is shown in Fig. 2. The data set of $\{v_n; n=1\text{-}8\}$ giving the largest coupling to the two highest modes, $H_g(7)$ and $H_g(8)$, results in a maximum (≤0.67) of the $\rho^B_{intra}/\rho^Z_{intra}$ curve at ca. 200 K. Other data sets give monotonous increase in $\rho^B_{intra}/\rho^Z_{intra}$ with the temperature increase, but less than 0.78 even at 520 K. Hence, compared with the Ziman's resistivity formula, usual solution of the Boltzmann equation based on the perturbation treatment gives smaller contribution to the normal-state resistivity under the same electron-intramolecular phonon interaction.

Most of the theoretical works[2-6] have used Ziman's formula for study of the normal-state resistivity. Moreover, although we made efforts to fit the normal-state resistivity at constant pressure, $\rho_P(T)$, by using ρ^B, it was not successful. $\rho_P(T)$ shows a linear temperature dependence above the room temperature for Rb_3C_{60}, whereas $\rho_V(T)$ shows lower onset temperature such as ca. 100 K for such linearity. Generally speaking, such linear temperature-dependence starting from lower temperature requires more contribution from the coupling of low-frequency phonons such as those of intermolecular modes. In this sense, ρ^B seems to be more suitable for fitting $\rho_V(T)$, since the contribution from the low-frequency intermolecular mode scattering has relatively larger weight in ρ^B than in ρ^Z under the same electron-phonon interaction strength. Actually, however, linear temperature dependence of $\rho_V(T)$ above *ca.* 100 K could be explained in a successful manner based on ρ^Z [18] with the intramolecular coupling strengths obtained from the Raman spectroscopic measurement.[19] Note that ρ^B can give only a higher onset temperature of *ca.* 140 K for such linearity with the same parameters. Therefore, it is concluded that ρ^B is not applicable to A_3C_{60} characterized by strong coupling to the intramolecular modes.

ACKNOWLEDGMENTS

This work is a part of the project of Institute for Fundamental Chemistry, supported by Japan Society for the Promotion of Science-Research for the Future Program (JSPS-RFTF96P00206). One of the authors (YH) is grateful to the financial support from the said project.

REFERENCES

*Permanent address: Department of Chemistry, Beijing Normal University, Beijing 100875, China.
†Author to whom correspondence should be addressed.

1. See, for instance, O. Gunnarsson, Rev. Mod. Phys. **69**, 575 (1997) and references therein.
2. V. H. Crespi, J. G. Hou, X. -D. Xiang, M. L. Cohen, and A. Zettl, Phys. Rev. B **46**, 12064 (1992).
3. V. P. Antropov, O. Gunnarsson, and A. I. Liechtenstein, Phys. Rev. B **48**, 7651 (1993).
4. J. S. Lannin and M. Mitch, Phys. Rev. B **50**, 6497 (1994).
5. L. Degiorgi, E. J. Nicol, O. Klein, G. Grüner, P. Wachter, S. -M. Huang, J. Wiley, Phys. Rev. B **49**, 7012 (1994).
6. K. Tanaka, Y. Huang, and T. Yamabe, Phys. Rev. B **51**, 12715 (1995).
7. See, for instance, G. Grimvall, *The Electron-Phonon Interaction in Metals* (North-Holland, Amsterdam, 1981), p. 212.
8. P. Zhou, K. -A. Wang, P. C. Eklund, G. Dresselhaus, and M. S. Dresselhaus, Phys. Rev. B **48**, 8412 (1993).
9. E. M. Conwell, Phys. Rev. B **22**, 1761 (1980).
10. D. S. Bethune, G. Meijer, W. C. Tang, J. H. Rosen, W. G. Golden, H. Seki, C. A. Brown, and M. S. deVries, Chem. Phys. Lett. **179**, 181 (1991).
11. K. Sugihara, T. Inabe, Y. Maruyama, and Y. Achiba, J. Phys. Soc.Jpn. **62**, 2757 (1993).
12. V. Korenivski and K.V. Rao, J. Supercond. **8**, 67 (1995).
13. Y. Takada, J. Phys. Soc. Jpn. **65**, 3134 (1996).
14. C. M. Varma, J. Zaanen, and K. Raghavachari, Science **254**, 989 (1991).
15. M. Schluter, M. Lannoo, M. Needels, G. A. Baraff, and D. Tomanek, J. Phys. Chem. Solids **53**, 1473 (1992).
16. J. C. R. Faulhaber, D. Y. K. Ko, and P. R. Briddon, Phys.Rev. B **48**, 661 (1993).
17. G. Gunnarsson, H. Handschuh, P. S. Bechthold, B. Kessler, G. Gantefor, and W. Eberhardt, Phys. Rev. Lett. **74**, 1875 (1995).
18. Y. Huang and K. Tanaka, Phys. Rev. B **57**, (1998), in press.
19. J. Winter and H. Kuzmany, Phys. Rev. B **53**, 655 (1996).

Mott transition and superconductivity in alkali-doped fullerides

O. Gunnarsson and E. Koch

Max-Planck-Institut für Festkörperforschung, D-70506 Stuttgart, Germany

R.M. Martin

Department of Physics, University of Illinois, Urbana, Illinois 61801

Abstract. We study the effects of deviations from cubic symmetry on superconductivity and on Mott transitions by comparing A_3C_{60} and $NH_3K_3C_{60}$. The reduced frustration in $NH_3K_3C_{60}$ favours a Mott transition at a lower ratio U/W, where U is the Coulomb repulsion of two electrons on the same molecule and W is the width of the t_{1u} band. The closeness of $NH_3K_3C_{60}$ to a Mott transition may influence the superconductivity negatively, due to an inefficient screening of the Coulomb repulsion.

INTRODUCTION

The cubic alkali-doped fullerides A_3C_{60} (A= K, Rb) are superconductors with T_c growing with the lattice parameter a [1]. This is at least qualitatively well understood in terms of the electron-phonon coupling λ growing with a [1]. In contrast, many noncubic fullerides are not superconducting, or have a lower T_c than cubic fullerides with the same molecular volume. For instance, the orthorhombic $NH_3K_3C_{60}$ is believed to be an antiferromagnetic insulator at normal pressure and small T [2]. Under pressure the deviation from cubic symmetry is reduced, and the system becomes superconducting [2]. T_c is lower than for the cubic fullerides with the same molecular volume, but the difference is reduced as $NH_3K_3C_{60}$ becomes more cubic. Results like this have lead to the suggestion that deviations from cubic symmetry are bad for superconductivity. This is discussed below.

ELIASHBERG THEORY

To study the effects of noncubic symmetry, we have solved the Eliashberg equation [3] for systems with cubic and orthorhombic symmetries. We first calculate the hopping matrix elements of cubic A_3C_{60} and orthorhombic $NH_3K_3C_{60}$, using a tight-binding formalism [4–6]. To make the two calculations as comparable as possible, we rescale the lattice parameters uniformly to obtain the band width $W = 0.62$ eV for both systems. Orientational disorder is introduced for both systems, although it is found experimentally only for A_3C_{60}. The orientational disorder smoothens the density of states and avoids the possibility that one system may have a large density of states $N(0)$ at the Fermi energy and therefore a large electron-phonon coupling λ. In this way we isolate the effects of the symmetry alone. The substantial difference in the lattice parameters ($a = 14.97$, $b = 14.90$ and $c = 13.69$ Å) in the orthorhombic system, leads to quite different hopping in the (110), (101) and (011) directions, with the largest hopping matrix element in the (011) direction about twice as large as the largest (110) element. A five-fold degenerate H_g Einstein phonon with the full Jahn-Teller character is included on each site. The coupling $\lambda/N(0) = 0.1$ (giving $\lambda \sim 0.6$) and the phonon frequency $\omega_{ph} = 0.2$ eV are used in both cases. The solution of the Eliashberg equation gives $T_c = 120 \pm 5$ K for both compounds. The large value of T_c is due to the neglect of the Coulomb pseudopotential in this calculation. The important point is that there is no difference in T_c between the two systems. We therefore conclude that at least within the Eliashberg framework and with everything else equal, the deviation from cubic symmetry does not hurt superconductivity.

MOTT TRANSITION AND FRUSTRATION

Several noncubic fullerides are insulators at normal pressure, for instance $NH_3K_3C_{60}$ and A_4C_{60}. We therefore study the effects of deviations from cubic symmetry on Mott transitions. In the cubic A_3C_{60} (A= K, Rb) the molecules sit on a fcc lattice. It is then possible to hop in a closed loop with an odd (three) number of hops. In analogy with the treatment of antiferromagnetism, we refer to this as frustration. The effects on the Mott transition are studied below. The orthorhombic lattice of $NH_3K_3C_{60}$ has the same topology as the fcc lattice, but the hopping in a triangular loop involves one hop with a matrix element which is only half as large as the other two. The frustration is therefore substantially reduced.

On dimensional grounds, it is expected that the Mott transition takes place when $U/W \sim 1$, where U is the Coulomb interaction between two electrons on the same molecule and W is the t_{1u} band width. In a system with an orbital degeneracy N this ratio tends to be increased by a factor of $\sqrt{N}$ [7]. Here we consider the effects of frustration [8] for a system with $N = 1$. In our earlier studies we found it useful to consider the limit when U is large. To find out about the Mott transition, this requires an uncontrolled extrapolation to intermediate values of U, but it nevertheless gives a quite useful indication of the ratio U/W at the transition.

The simplest system with frustration consists of three s-like orbitals on a triangle. We therefore study this system in more detail. The orbitals are connected by hopping matrix elements t, where t is assumed to be negative. A perfectly bonding state can then be formed by using the coefficient $1/\sqrt{3}$ for all orbitals. The

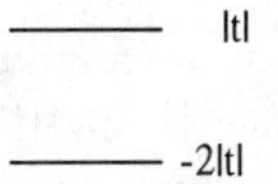

———— |t|

———— -2|t|

FIGURE 1. One-particle spectrum for three s-orbitals on a triangle and connected by hopping integrals t ($t < 0$).

corresponding energy is $-2|t|$. Due to the frustration, however, it is not possible to form a completely antibonding state. Therefore, the antibonding state only has the energy $|t|$ instead of $2|t|$ and the one-particle band width is only $W = 3|t|$ (see Fig. 1). This can be compared with a non-frustrated system, such as a square, which has $W = 4|t|$.

We consider the many-body case, assuming that there is no orbital degeneracy. To determine whether the system is an insulator, we calculate the band gap

$$E_g = A - I = E(M+1) + E(M-1) - 2E(M). \tag{1}$$

Here A is the affinity energy, I is the ionization energy and $E(K)$ is the energy of a system with K electrons. $M = 3$ is the number of electrons at half-filling. We now focus on the case of a large value of U. It is then very easy to calculate $E(3) = 0 + O(t^2/U)$, since hopping is suppressed for $K = 3$ and U large. We next consider the case of one extra electron, $K = 4$. We form the basis state number 1 in Fig. 2. In this state there is an extra electron on the upper left atom. This extra occupancy can hop clockwise around the triangle as illustrated by the states in the upper row. In state 4, the extra occupancy has returned to the original site. State 4 is, however, different from state 1, since the spins on the other two sites have been flipped. However, if the extra occupancy hops another lap around the triangle, the spins are flipped once more, and the resulting state is identical to state 1. This state was reached after an even number (six) hops. In the large U limit we then obtain a 6×6, problem with the extreme eigenvalues $\pm 2t$. The same result is obtained for the case of an extra hole ($K = 2$). Thus in the many-body case the effects of frustration are not noticed for this model. The gap then becomes

$$E_g = U - 4|t| = U - \frac{4}{3}W. \tag{2}$$

The prefactor 4/3 results from the frustration, e.g., that W has been reduced by frustration while in this case the reduction of band gap due to hopping is completely

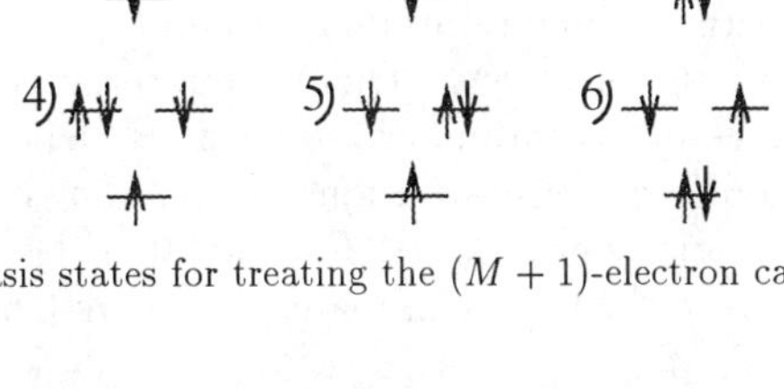

FIGURE 2. Basis states for treating the $(M+1)$-electron case for a triangle.

unaffected by frustration. If the same calculation is performed for a square, we obtain $E_g = U - 4|t|$ again, but since $W = 4|t|$ in this case, we find $E_g = U - W$, e.g., with the prefactor 1. In more general cases, the effects of frustration are more complicated [8], but the prefactor in front of W in Eq. (2) is nevertheless larger than one in all cases studied [8].

To test these ideas for intermediate values of U, we have performed $T = 0$ fixed node projection quantum Monte Carlo calculations [9,7] for models of A_3C_{60} and $NH_3K_3C_{60}$. These models describe the three-fold degenerate, partly occupied t_{1u} level. We include the on-site Coulomb interaction U and the hopping integrals $t_{im,jm'}$ between the molecules. This leads to the Hubbard-like model

$$H = \sum_{i\sigma} \sum_{m=1}^{3} \varepsilon_{t_{1u}} n_{i\sigma m} + \sum_{<ij>\sigma mm'} t_{ijmm'} \psi^{\dagger}_{i\sigma m} \psi_{j\sigma m'} + U \sum_{i} \sum_{\sigma m < \sigma' m'} n_{i\sigma m} n_{i\sigma' m'}, \quad (3)$$

where the sum $<ij>$ is over nearest neighbor sites. The hopping integrals $t_{imjm'}$ have been obtained from a tight-binding parametrization [4,5], taking the appropriate lattice structure into account. The orientational disorder is included for A_3C_{60} but not for $NH_3K_3C_{60}$ in accordance with experiment, and the hopping integrals are chosen accordingly [6]. We have performed calculations for clusters with up to 64 molecules and extrapolated the calculated gap E_g to infinite cluster size. We then find that the Mott-Hubbard transition takes place for

$$\frac{U}{W} \sim \begin{cases} 2 & \text{for} \quad NH_3K_3C_{60} \\ 2.5 & \text{for} \quad A_3C_{60} \end{cases} \quad (4)$$

$NH_3K_3C_{60}$ has the same volume per molecule as the superconducting Rb_2CsC_{60}, which has the A_3C_{60} lattice. Since superconductivity has only been observed for A_3C_{60} systems with a slightly larger lattice parameter than Rb_2CsC_{60}, this system is probably close to a Mott transition. According to tight-binding calculations, the band widths for $NH_3K_3C_{60}$ and Rb_2CsC_{60} are similar, with the width of $NH_3K_3C_{60}$ slightly larger ($\sim 10\%$). In view of the different ratios U/W for a Mott transition in these two types of systems, it is then not surprising that $NH_3K_3C_{60}$ is on the insulating and Rb_2CsC_{60} on the metallic side of a Mott transition.

SUPERCONDUCTIVITY

We now address the superconductivity and the unusual T_c versus a dependence in $NH_3K_3C_{60}$ [2]. The electron-phonon interaction induces a weak (~ 0.1 eV) attractive interaction between the electrons. This is counteracted by a strong Coulomb repulsion (~ 1.5 eV). For conventional systems it is believed that the Coulomb repulsion is strongly reduced by retardation effects, allowing the electron-phonon interaction to drive the superconductivity. For the fullerides it has been argued that the retardation effects from the higher subbands are weak [10], but that within RPA the screening is so efficient that superconductivity nevertheless becomes possible [10]. This, however, raises questions about the accuracy of the RPA approximation. Recently, this accuracy has been studied within a quantum Monte Carlo approach for a model of the type described above (Eq. (3)). It was found that RPA is surprisingly accurate until the system gets close to the Mott transition, where the screening becomes inefficient in the more accurate QMC calculation [11].

This suggests the following picture. At normal pressure $NH_3K_3C_{60}$ is a Mott insulator. As pressure is applied the system becomes a metal, but at first the screening is poor and T_c is reduced or superconductivity may be completely suppressed. As the pressure is increased the system moves away from the Mott transition and the screening becomes more efficient. The corresponding reduction of the Coulomb interaction leads to an increased T_c. This is contrary to the behavior for the A_3C_{60}, where the screening probably already is efficient, and the dominating effect is the reduction of the electron-phonon coupling as the lattice parameter is reduced.

SUMMARY

We have studied the effects of deviations from cubic symmetry on superconductivity and on Mott transitions by comparing A_3C_{60} and $NH_3K_3C_{60}$. We find that the reduced frustration in $NH_3K_3C_{60}$ leads to a Mott transition for a lower ratio of U/W than for the cubic system. It is then not surprising that $NH_3K_3C_{60}$ is a Mott insulator, while Rb_2CsC_{60} is a metal although the two systems have similar band widths. By solving the Eliashberg equation, we find that deviations from cubic symmetry is not by itself bad for superconductivity. It may, however, harm superconductivity indirectly by putting the system closer to a Mott transition, which may lead to a less efficient screening of the Coulomb repulsion.

REFERENCES

1. Fleming, R.M. et al., Nature **352**, 787 (1991).
2. Rosseinsky, M.J. et al., Nature **364**, 425 (1993); Zhou, O. et al., Phys. Rev. B **52**, 483 (1995); Iwasa, Y.et al., Phys. Rev. B **53**, R8836 (1996).
3. Eliashberg, G.M., Zh. Eksp. Teor. Fiz. **38**, 966 [Sov. Phys. JETP **11**, 696 (1960)].
4. Gunnarsson, O. et al., Phys. Rev. Lett. **67**, 3002 (1991).
5. Satpathy, S. et al., Phys. Rev. B **46**, 1773 (1992).
6. Mazin, I.I. et al., Phys. Rev. Lett. **26**, 4142 (1993).
7. Gunnarsson, O., Koch, E., and Martin, R.M., Phys. Rev. B **54**, R11026 (1996).
8. Gunnarsson, O., Koch, E., and Martin, R.M., Phys. Rev. B **56**, 1146 (1997).
9. ten Haaf, D.F.B. et al., Phys. Rev. B **51**, 353 (1995); van Bemmel, H.J.M. et al., Phys. Rev. Lett. **72**, 2442 (1994).
10. Gunnarsson, O., and Zwicknagl, G., Phys. Rev. Lett. **69**, 957 (1992); Gunnarsson, O., Rainer, D., and Zwicknagl, G., Int. J. Mod. Phys. B **6**, 3993 (1992).
11. Koch, E., Gunnarsson, O., and Martin, R.M., (to be publ.).

Dielectric screening in doped Fullerides

Erik Koch[a,b], Olle Gunnarsson[a], and Richard M. Martin[b]

[a] *Max-Planck-Institut für Festkörperforschung, 70569 Stuttgart, Germany*
[b] *Department of Physics and Materials Research Laboratory,
University of Illinois, Urbana, IL 61801, USA*

Abstract. For conventional superconductors the electron-electron interaction is strongly reduced by retardation effects, making the formation of Cooper pairs possible. In the alkali-doped Fullerides, however, there are no strong retardation effects. But dielectric screening can reduce the electron-electron interaction sufficiently, *if* we assume that the random-phase approximation (RPA) is valid. It is not clear, however, if this assumption holds, since the alkali-doped Fullerides are strongly correlated systems close to a Mott transition. To test the validity of the RPA for these systems we have calculated the screening of a test charge using quantum Monte Carlo.

INTRODUCTION

In order to lead to an effective attraction between electrons, the electron-phonon interaction has to overcome the strong electron-electron repulsion. For conventional superconductors retardation effects strongly reduce the Coulomb repulsion [1]. The resulting effective interaction is described by the dimensionless Coulomb pseudopotential μ^*. In the above sense, the alkali-doped Fullerides are non-conventional superconductors, since for them retardation effects are very inefficient in reducing the Coulomb repulsion [2]. One can, however, conceive an alternative mechanism for reducing μ^*: Within the random phase approximation (RPA) the Coulomb interaction is strongly reduced by the very efficient dielectric screening [2].

It is not clear, however, how well the random phase approximation describes the screening in the alkali-doped Fullerides. RPA works well in the limit of weak interaction, while in the opposite limit it is qualitatively wrong. Very little is known about the screening in the intermediate region, where interaction and kinetic energy are comparable. This is the region where the superconducting Fullerides, being close to a Mott transition, lie.

To find out how well RPA works for the alkali-doped Fullerides, we have studied the static screening of a test charge in a Hubbard-like model using quantum Monte Carlo methods. Our results indicate that RPA gives a surprisingly accurate description of the static screening on the metallic side of a Mott transition.

QUANTUM MONTE CARLO CALCULATIONS

To describe the conduction electrons in A_3C_{60} we use the multi-band Hubbard-like Hamiltonian

$$H_0 = \sum_{\langle i,j \rangle \, m,m',\sigma} t_{ij\,mm'} c^\dagger_{im\sigma} c_{jm'\sigma} + U \sum_{i,\, m\sigma < m'\sigma'} n_{im\sigma} n_{im'\sigma'}.$$

The indices i, j label the sites of the fcc lattice, $\langle i,j \rangle$ denoting nearest neighbors. Each C_{60}-molecule randomly takes one of two orientations [3]. The hopping matrix elements $t_{ij\,mm'}$ between the three-fold degenerate t_{1u} orbitals on neighboring sites are obtained from a tight-binding parametrization [4]. The Hubbard interaction U is varied to study the effect of correlations. Experimental estimates give $U \approx 1.5\,eV$ [5].

To study the static screening, we introduce a test charge q on a site i_q. The corresponding term in the Hamiltonian is

$$H_1(q) = qU \sum_{m\sigma} n_{i_q m\sigma}.$$

We can then find the response of the system upon introduction of the test charge by comparing the electron density n_q on site i_q for the system with the test charge, to the electron density n_0 for the unperturbed system: $\Delta n = n_0 - n_q$.

We determine the ground state properties of the Hamiltonians H_0 and $H_0 + H_1$ (for finite fcc clusters) using diffusion Monte Carlo with the fixed-node approximation [6]. In this method we construct a trial wave function $|\Psi_T\rangle$ and allow it to diffuse towards the exact (within the fixed node approximation) solution $|\psi_0\rangle$. For the trial function, we make a generalized Gutzwiller ansatz: $\langle R|\Psi_T\rangle = g^{d(R)} g_q^{n_q(R)} \langle R|\Phi\rangle$. $|\Phi\rangle$ is the Slater determinant calculated using the random phase approximation. The Gutzwiller factors reflect the two interaction terms, where for a given configuration R of the electrons $d(R) = \langle R| \sum_{i,\,m\sigma<m'\sigma'} n_{im\sigma} n_{im'\sigma'} |R\rangle$ and $n_q(R) = \langle R| \sum_{m\sigma} n_{i_q m\sigma} |R\rangle$. We optimize the free parameters g and g_q using a correlated sampling technique.

Diffusion Monte Carlo only provides us with the mixed estimator $\langle \Psi_T|\mathcal{O}|\Psi_0\rangle/\langle \Psi_T|\Psi_0\rangle$ of an observable $\mathcal{O}$. If, however, the trial function $|\Psi_T\rangle$ is close to the ground state, we can estimate the ground state expectation value using the extrapolated estimator

$$\frac{\langle \Psi_0|\mathcal{O}|\Psi_0\rangle}{\langle \Psi_0|\Psi_0\rangle} \approx 2\, \frac{\langle \Psi_T|\mathcal{O}|\Psi_0\rangle}{\langle \Psi_T|\Psi_0\rangle} - \frac{\langle \Psi_T|\mathcal{O}|\Psi_T\rangle}{\langle \Psi_T|\Psi_T\rangle}.$$

The last term, the expectation value of the observable for the trial function, is easily calculated using variational Monte Carlo. To make sure that the extrapolated estimator gives reliable results, we have compared our Monte Carlo calculations with the results from the exact diagonalization for small systems. For larger systems, where exact diagonalization is not possible, we have checked that the trial function is close to the ground state wave function.

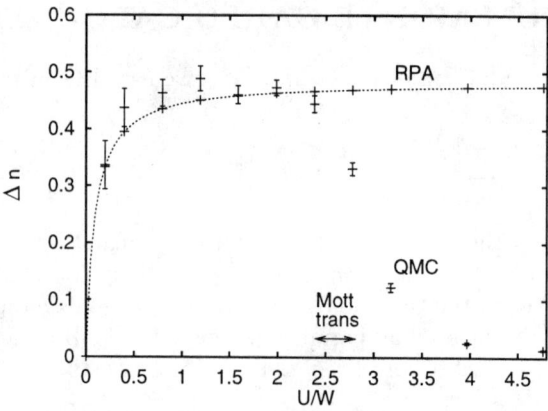

FIGURE 1. Screening of a test charge $q = 0.5\,e$ as a function of U for a cluster of 32 molecules (band width $W = 0.63\,eV$). The crosses (interpolated by the dotted curve) show the results of a calculation using the random phase approximation (RPA). The results of the quantum Monte Carlo (QMC) calculations are given with their respective error bars. The region where the system undergoes a Mott transition is indicated. The RPA screening remains rather accurate almost all the way up to the Mott transition, but fails badly in the insulating region.

RESULTS AND DISCUSSION

We have calculated the screening charge Δn for finite clusters of different size. For a given test charge, all calculations give similar results. As an example, Fig. 1 shows the screening of a test charge $q = 0.5\,e$ as a function of the Hubbard interaction U for a cluster of 32 molecules.

Looking at the results from the quantum Monte Carlo calculations, we can distinguish two qualitatively different regions. For weakly interacting systems (U/W small) the screening is very good, while it breaks down completely for large U/W, where the system is a Mott insulator. RPA somewhat underestimates the screening for $U/W \lesssim 1.0$ [7]. For intermediate values of U/W it works surprisingly well and actually remains rather accurate almost all the way up to the Mott transition. Beyond this point RPA becomes qualitatively wrong.

The failure of the RPA in the strongly interaction regime can be easily understood. In RPA it only costs kinetic energy to screen the test charge. Since in the strongly interacting region W/U is small, the cost in kinetic energy is small compared to the interaction with the test charge. Therefore, when a test charge is put on some site, RPA predicts that almost the same amount of electronic charge leaves that site in order to minimize the interaction with the test charge. Such a calculation neglects, however, the fact that in a strongly correlated system the electrons that leave the site with the test charge have to pay a large correlation energy. In the QMC calculations this is taken into account properly, leading to the observed breakdown in the screening for large U/W.

The gradual decrease in the efficiency of the screening just before the Mott transition that can be seen in Fig. 1 suggests a mechanism for understanding the anomalous behavior of $NH_3K_3C_{60}$. At normal pressure this system is an insulator. Under pressure it becomes superconducting, with T_c increasing with pressure [8]. This is in contrast to the other alkali-doped Fullerides, where T_c drops with pressure. As pressure is applied to the Mott insulator, U/W decreases and the screening becomes more and more efficient, reducing the effective Coulomb interaction μ^*, i.e. increasing T_c. As U/W further decreases, the screening saturates. Now the reduction of the electron-phonon interaction (because of the decreasing density of states) dominates, and the standard behavior of the alkali-doped Fullerides is recovered.

To summarize, our results indicate that RPA is surprisingly accurate allmost all the way up to the Mott transition. Since retardation effects are inefficient, this has important implications for the superconductivity in the alkali-doped Fullerides.

ACKNOWLEDGMENTS

This work has in part been supported by the Department of Energy (grant DEFG 02-91ER45539). E.K. thanks the Alexander-von-Humboldt-Stiftung for financial support under the Feodor-Lynen-Program.

REFERENCES

1. see e.g. J.R. Schrieffer, *Theory of Superconductivity*, Benjamin, New York, 1964
2. O. Gunnarsson and G. Zwicknagl, *Phys. Rev. Lett.* **69**, 957 (1992);
   O. Gunnarsson, D. Rainer, and G. Zwicknagl, *Int. J. Mod. Phys.* B **6**, 3993 (1992);
   O. Gunnarsson, *Rev. Mod. Phys.* **69**, 575 (1997)
3. P.W. Stephens, L. Mihaly, P.L. Lee, R.L. Whetten, S.M. Huang, R. Kaner, F. Diederich, and K. Holczer Nature **351**, 632 (1992)
4. O. Gunnarsson, S. Satpathy, O. Jepsen, and O.K. Andersen, *Phys. Rev. Lett.* **67**, 3002 (1991); S. Satpathy, V.P. Antropov, O.K. Andersen, O. Jepsen, O. Gunnarsson, and A.I. Liechtenstein, *Phys. Rev.* B **46**, 1773 (1992)
5. R.W. Lof, M.A. van Veenendaal, B. Koopmans, H.T. Jonkman, and G.A. Sawatzky, *Phys. Rev. Lett.* **68**, 3924 (1992); P.A. Brühwiler, A.J. Maxwell, A. Nilsson, N. Mårtensson, and O. Gunnarsson, *Phys. Rev.* B **48**, 18296 (1993)
6. D.F.B. ten Haaf, H.J.M. van Bemmel, J.M.J. van Leeuwen, W. van Saarloos, and D.M. Ceperley, *Phys. Rev.* B **51**, 13039 (1995);
   H.J.M. van Bemmel, D.F.B. ten Haaf, W. van Saarloos, J.M.J. van Leeuwen, and G. An, *Phys. Rev. Lett.* **72**, 2442 (1994)
7. Such an underestimation is also found in the electron gas. See, e.g. L. Hedin and S. Lundqvist, Solid State Physics **23**, 1 (1969)
8. O. Zhou, T.T.M. Palstra, Y. Iwasa, R.M. Fleming, A.F. Hebard, P.E. Sulewski, D.W. Murphy, and B.R.Zegarski, *Phys. Rev.* B **52**, 483 (1995); Y. Iwasa, H. Shimoda, T.T.M. Palstra, Y. Maniwa, O. Zhou, T.Mitani, *Phys. Rev.* B **53**, R8836 (1996)

Frequency dependent ESR study of the magnetic phase transition in $NH_3K_3C_{60}$

F. Simon[1], A. Jánossy[1], Y. Iwasa[2], H. Shimoda[2], G. Baumgartner[3] and L. Forró[3]

[1]*Institute of Physics, Technical University Budapest, H-1521 Budapest, Hungary*
[2]*Japan Advanced Institut of Science and Technology, Tatsunokuchi, Ishikawa 923 12, Japan*
[3]*Laboratoire de Physique des Solides Semicristallins, IGA, Departement de Physique, Ecole Polytechnique Federale de Lausanne 1015 Lausanne, Switzerland*

Abstract. $NH_3K_3C_{60}$ undergoes a magnetic ordering phase transition below 40K which is probably the reason that this system is not a superconductor at ambient pressure. We measured the magnetic field dependence (H = 0.3 , 2.7, 5.4 and 8.1 T) of the electron spin resonance of $NH_3K_3C_{60}$. Below 40 K the resonance broadens and shifts as the magnetic order develops. The broadening and shift are much larger at 0.3 T than at higher fields, but do not follow the 1/H dependence expected for an antiferromagnetic order. The magnetic spin susceptibility measured by the ESR intensity changes only little, if at all, through the transition. At 9 GHz (0.3 T) results depend on the cooling rate from ambient temperatures to temperatures below 50K.

$NH_3K_3C_{60}$, a fulleride first synthetized by Rosseinsky et al.[1] is at the borderline of a magnetic insulator and a superconducting metal. The crystal structure at ambient pressure has an orthorhombic distortion and a 6% volume expansion with respect to the fcc K_3C_{60} parent compound. Lattice expansion of A_3C_{60} superconductors with the same crystal structure increases the density of states and the superconducting transition temperature[2]. $NH_3K_3C_{60}$ has a unit cell volume of 3054 A^3 and a T_c higher than 30 K is expected from an extrapolation of the T_c vs lattice parameter diagram of fcc compounds. Instead, at ambient pressures it has a magnetic ground state as evidenced from low field ESR[3] and μSR[4]. The lack of superconductivity at ambient pressures is not due to the non cubic structure since under pressures larger than 10 kbar it is a superconductor[5] with T_c=28 K.

In this paper we present the magnetic field dependence of the electron spin resonance spectrum at 9, 75, 150 and 225 GHz (resonance fields of about H_r = 0.3, 2.7, 5.4 and 8.1 T) of $NH_3K_3C_{60}$. The aim was to find the antiferromagnetic resonance (AFMR) in the ground state. In a previous study of the 9 GHz ESR of $NH_3K_3C_{60}$ the onset of a magnetic order appeared as a rapid decrease of the resonance intensity. We were motivated by the recent observation[6] of the AFMR of orthorhombic RbC_{60} in which the width and shift of the resonance follows a $1/H_r$ dependence characteristic of the antiferromagnetic powder spectrum.

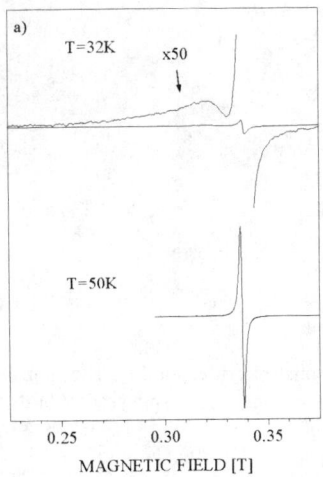
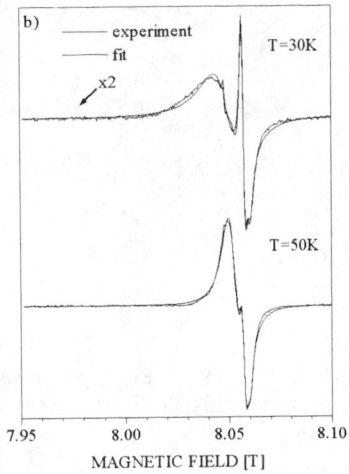

FIGURE 1. Broadening of the electron spin resonance spectra of $NH_3K_3C_{60}$ due to static internal fields below the transition at 40 K. Resonance frequency a) 9 GHz, b.) 225 GHz. The broadening is more important at low applied fields. The residual low temperature narrow lines are from a small amount of minority phase.

Samples were prepared following standard methods. X-ray diffraction showed a high phase purity. Purity was confirmed by the low temperature ESR spectra in which residual unbroadened lines had low intensities. Sample A was sealed under helium atmosphere into a quartz tube and ESR was measured at all fields. Sample B (synthetized independently from A) was mixed with vacuum grease in the dry-box under argon atmosphere and measured at the higher fields (2.7 - 8.1 T). Results for Samples A and B were similar although Sample A showed broader lines and somewhat larger shifts at low T. X-band ESR was recorded on a Bruker spectrometer in Lausanne, high frequency ESR was recorded at Budapest.

Typical ESR spectra are shown In Figure 1. The 9 GHz results are in good agreement with Ref. 2. The ESR intensity is proportional to the magnetic spin susceptibility and increases by a factor of 2 as the temperature decreases from 300 K to 50 K (Figure 2). This is a large variation for a simple metal, but is far from the Curie dependence of an insulator with localized moments. The linewidth increases linearly with field as $\Delta H = \Delta H_0(T) + a(T)H_r$. The X band linewidth (the half width at half height of the best fit Lorentzian absorption line) has a peak at 125 K where the intensity has also a small anomaly (Figure 2). $a(T)$ increases as the transition is approached, $a(200K) = 1 \times 10^{-4}$ while $a(60K) = 5 \times 10^{-4}$. On the other hand the resonance field, H_r, changes less than 100 ppm from 50 to 300 K. Thus the field does not smear the transition since static fields not only broaden but usually also shift the resonance.

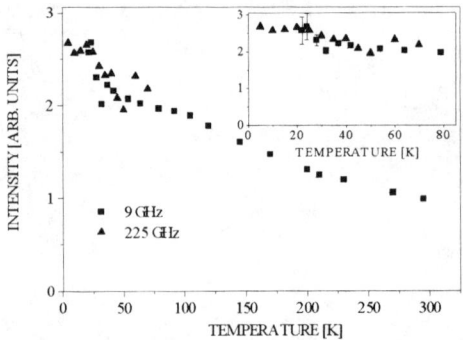

FIGURE 2. The ESR intensity is proportional to the magnetic spin susceptibility. The contribution of impurity phases is negligible. Insert: surprisingly, there is little change of the spin susceptibility through the transition.

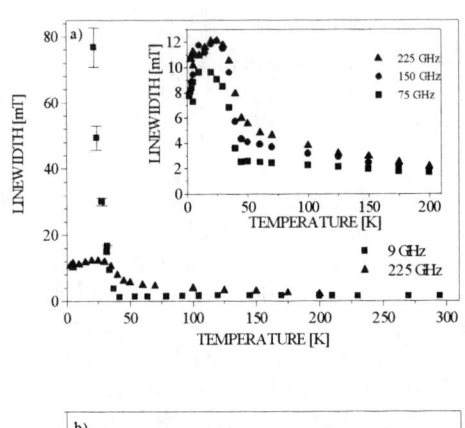

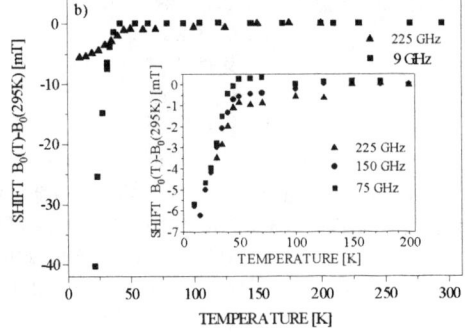

FIGURE 3. a) ESR linewidth and b) resonance field shift versus temperature at various frequencies. Above the transition the line broadens linearly with applied field. Internal static fields shift and broaden the resonance below the transition. The line is much more shifted and broadened at 9 GHz than at higher frequencies.

Below 40 K the resonance broadens and shifts to lower fields rapidly for all frequencies. These effects are quite different at 9 GHz than at higher frequencies (Figure 3). The linewidth and shift are field independent from 75 to 225 GHz and saturate as a function of temperature at 20 - 30 K. At 9 GHz the broadening and shift are much larger than at higher frequencies and depend on sample cooling rate. In samples cooled within a few minutes from ambient T to below 50 K the resonance broadened so rapidly that it could not be followed. For these samples only a decrease in the intensity of the narrow line was observed, like in previous report[3]. In slowly cooled samples a broad component carrying most of the intensity appears below 40 K. Below 22 K and at 9 GHz the line is too broad to be observed. At higher frequencies we did not detect any difference between slowly and rapidly cooled samples. The resonance intensity is unchanged both at 9 and 225 GHz within experimental accuracy through the transition. We searched in vain for magnetic field and thermal hysteresis effects between 0 and 9 T and 2.5 to 55 K.

The low temperature state is magnetically ordered as evidenced by the large field dependent broadening and shift. The order is most probably antiferromagnetic. We may rule out ferromagnetism or ferrimagnetism as there is no increase of the resonance intensity through the transition. Although the magnetic order is certainly not perfect since μSR revealed a large distribution of fields[4], the system is not a spin glass since in that case we would observe a magnetic and thermal history dependence.

In powders, the antiferromagnetic resonance (AFMR) width is of the order of $\Delta H = (H_E H_a)/H_r$ if H_r is larger than the spin flop field, $(H_E H_a)^{1/2}$ but smaller than the exchange field H_E. The broadening of the AFMR is proportional to the square of the sublattice magnetization. A shift of similar magnitude as the broadening is expected. The anisotropy field, H_a, in fullerenes is due to dipolar fields and is of the order of $H_a = 10$ mT for a the magnetic moment of 1 μ_B per fullerene molecule. At $H_r = 0.3$ T and 22 K we measure $\Delta H = 0.1$ T and with the above values the exchange field is of the order of $H_E = 3$ T. The assignment of the observed resonance to an AFMR is uncertain since we do not observe the $1/H_r$ narrowing with applied field expected for the AFMR of a powder. Instead of an $1/H_r$ dependence, a residual shift and width persists up to high fields after an initial narrowing between 0.3 and 2.7 T. We have no consistent explanation for the observed behavior.

REFERENCES

[1] M. J. Rosseinsky, D. W. Murphy, R. M. Fleming, O. Zhou Nature (London) **364**, 425 (1993).
[2] R. M. Fleming, A. P. Ramirez, M. J. Rosseinsky, D. W. Murphy, R. C. Haddon, S. M. Zahurak, A. V. Makhija Nature (London), **352**, 787 (1991).
[3] Y. Iwasa, H. Shimoda, T. T. M. Palstra, Y. Maniwa, O. Zhou, T. Mitani Phys. Rev. B **53**, R8836 (1996).
[4] K. Prassides, K. Tanigaki, Y. Iwasa J. Phys. Chem.Solids **58**, 1697 (1997).
[5] O. Zhou, T. T. M. Palstra, Y. Iwasa, R. M. Fleming, A. F. Hebard, P. E. Sulewski, D. W. Murphy, B. R. Zegarski, Phys. Rev. B **52**, 483 (1995).
[6] A. Jánossy, N. Nemes, T. Fehér, G. Oszlányi, G. Baumgartner, L. Forró Phys. Rev. Lett. **79**, 2718, (1997).

Antiferromagnetic Resonance in Rb_1C_{60}

M. Bennati*, R. G. Griffin*, S. Knorr[†], A. Grupp[†], and M. Mehring[†]

*Francis Bitter Magnet Laboratory and Department of Chemistry,
Massachusetts Institute of Technology, 02139 Cambridge MA, USA
[†] 2. Physikalisches Institut, Universität Stuttgart,
Pfaffenwaldring 57, 70550 Stuttgart, Germany

Abstract. High-frequency (94 and 140 GHz) ESR was used to investigate the magnetic properties of the low-dimensional conductor Rb_1C_{60}. Below 35 K new features of the electron spin resonance are distinguished from the CESR signal of the conducting phase. The analysis of the resonance linewidth and line shift allows a clear identification of a frequency-dependent antiferromagnetic resonance line (AFMR) below 25 K. The characteristic temperature T_N for the ordering transition is 25 K. Between 25 K < T < 35 K precursor effects are observed. The temperature dependence of the order parameter, i.e., the sublattice magnetization, was determined from first and second moment of the AFMR. The results are compatible with a 3D magnetic ordering.

INTRODUCTION

The magnetic properties of the organic conductor Rb_1C_{60} have been recently subject of several studies in order to clarify the nature of the insulating low-temperature phase [1–7]. Static magnetic moments were found by μSR [2,3] and NMR [4,5], but the interpretation of the results in term of an ordered magnetic state was rather controversial. A clear indication for an antiferromagnetically ordered ground state arised from antiferromagnetic resonance (AFMR), first performed by Jánossy et al. at 75, 150, and 225 GHz microwave frequencies [6] and confirmed by our work [7] at 94 and 140 GHz. This kind of resonance consists in a collective excitation of the two antiferromagnetic sublattices, which in case of a disordered material (powder sample) leads to a characteristic frequency dependence of the resonance broadening and resonance shift in the ESR absorption. In fact, at magnetic fields much higher than the characteristic spin-flop field, AFMR line broadening and line shift scale with the inverse of the external field [8]. In this contribution we discuss the analysis of the powder AFMR spectra and how we determined the Néel transition temperature as well as the temperature dependence of the order parameter.

EXPERIMENTAL

Experiments at 140 GHz frequencies were performed with a home-built spectrometer in a TE_{011} cylindrical cavity. The 94 GHz data were recorded with a Bruker spectrometer *Elexsys E 680*. Both spectrometers are equiped for phase sensitive detection which allows careful separation of absorptive and dispersive signal components. Rb_1C_{60} powder sample was prepared in our laboratory (Stuttgart) following standard routines.

RESULTS

Figure 1 displays the temperature dependence of the ESR absorption at 140 GHz below the metal-to-insulator phase transition (≈ 50 K). Very similar spectra were recorded also at 94 GHz. The conduction electron spin resonance starts to broaden below 35–40 K and gradually results in the strong asymmetric pattern of Fig. 1 (bottom). The shoulder in the center ($B = 4978$ mT, $g = 2$) is due to a small amount of residual paramagnetism.

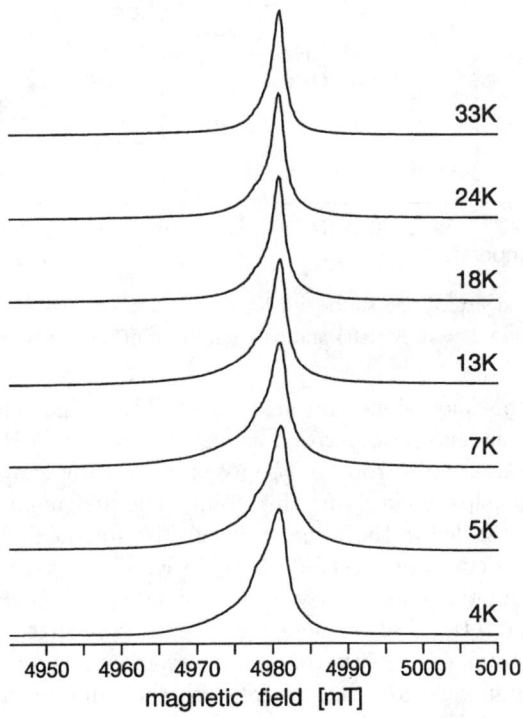

FIGURE 1. T dependence of AFMR line shape at 140 GHz.

Since width and center of the powder patterns cannot be extracted easily from the experimental data we apply a moment analysis of the line shape. The first moment and the square root of the second moment are related to line center and width, respectively. It is expected [7] that their frequency dependence scales with the inverse of the external field; at our microwave frequencies (94 and 140 GHz) this results in a ratio of 1.49. M_1 and $\sqrt{M_2}$ were extracted from the ESR-absorption line and plotted against the temperature in Fig. 2.

DISCUSSION

The T dependence of $\sqrt{M_2}$ is represented in Fig. 2 (right). Below ≈ 35 K two regions are distinguished. For 25 K $< T <$ 35 K, $\sqrt{M_2}$ increases but no frequency dependence is observed. Below $T = 25$ K the values at different frequencies deviate and the ratio at the low-temperature end is approximately 1.4, as expected for AFMR.

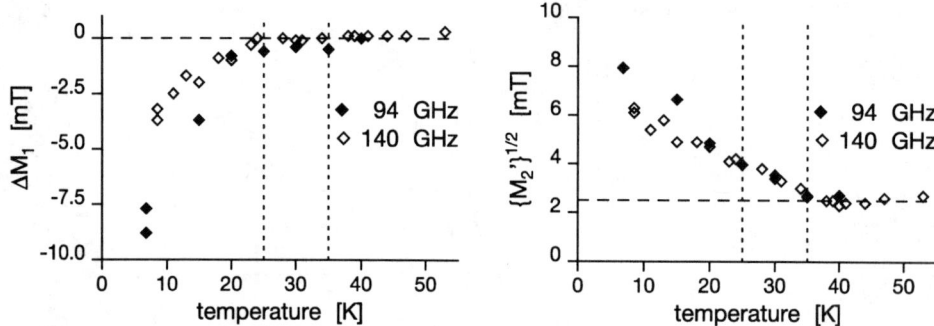

FIGURE 2. T dependence of line shifts and linewidths, represented as deviation of the first moment from its value at $T = 40$ K (left) and as square root of the second moment (right).

However, from this plot alone the onset of AFMR is not clearly defined. A weak frequency dependence above 25 K, as expected for AFMR, could indeed be compensated by g-factor anisotropy, as proposed by Jánossy et al. [6]. The analysis of the first moment helps in clarifying this point. The first moment is displayed in Fig. 2 (left) and it is scaled to the value at 40 K. Also for the first moment no shift and frequency dependence are observed above 25 K. Therefore, the analysis of two "independent" parameters consistently indicates the onset of AFMR below 25 K. We assign this value to the Néel temperature. We propose that the line broadening above 25 K arises from spin fluctuations in a precursor regime. The determined Néel temperature is in very good agreement with the values estimated from NMR (25 K) [4] and μSR (20 K) experiments [2,3]. From the $\sqrt{M_2}$ we also determined a spin-flop field value of $H_c \approx 350$ mT.

ORDER PARAMETER

The order parameter, i.e., the sublattice magnetization, is expected to be proportional to $\sqrt{H_A H_E}$ if the major contribution to the anisotropy energy arises from dipolar interaction and where H_A is the anisotropy field and H_E is the exchange field [8]. This condition might be well fulfilled in this sample since other anisotropic contributions, like spin-orbit coupling or superexchange, are expected to be very small. As M_1 and $\sqrt{M_2}$ are both proportional to the product $H_A H_E$ [7] the temperature dependence of M_0 can be extracted from two "independent" plots, $\sqrt{M_{1,\text{AFMR}}}$ versus T and $\sqrt[4]{M_{2,\text{AFMR}}}$ versus T. From the total moments only the AFMR contribution has to be considered, which is vanishing above 25 K. Therefore, M_1 and M_2 were first scaled to their values at 25 K, then plotted against T. In order to obtain a common representation for two frequencies the curves were first fitted by the same power law $M_0 \propto (T_N - T)^\beta$ and then normalized to unity at $T = 0$ (Fig. 3).

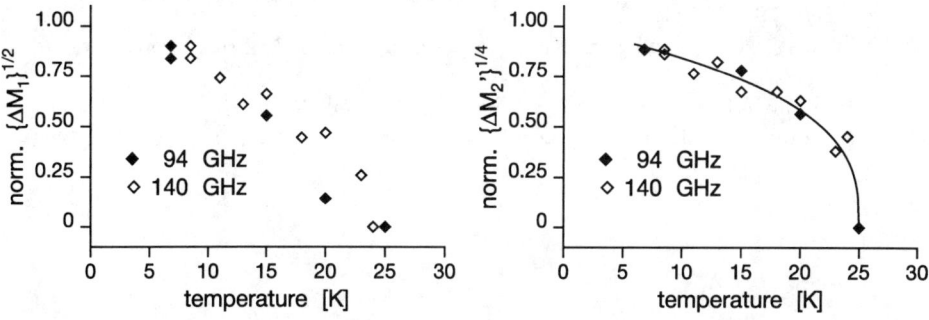

FIGURE 3. Left: Normalized $\{\Delta M_1 / \Delta M_1 (T=0)\}^{1/2}$ derived from the data shown in Fig. 2 (left). Right: Normalized $\{\Delta M_2' / \Delta M_2' (T=0)\}^{1/4}$ derived from the data shown in Fig. 2 (right). The solid line represents the theoretical curve $(T_N - T)^\beta$ with $T_N = 25$ K and $\beta = 1/3$.

We note that the T behaviour is very similar at both frequencies. In Fig. 3 (right) we show that the curve can be described by the standard fit for a 3D antiferromagnet with a critical exponent $\beta = 1/3$ in the high-T limit. It is interesting that we do not observe saturating behaviour of M_0 at the low-temperature end. This is consistent with a very small energy gap, i.e., a critical field $H_c \approx 350$ mT, so that spin waves can be thermally excited until $T < 1$ K. A similar T behaviour was also observed in $(\text{TMTSF})_2\text{PF}_6$, a low-dimensional organic conductor with a spin-density-wave ground state and a very small critical field [9].

ACKNOWLEDGEMENT

We are grateful to K.-F. Thier for providing the sample and we would like to acknowledge S. Foner for helpful discussions. M.B. was supported by a postdoctoral grant given by the Deutsche Forschungsgemeinschaft (DFG). M.M. acknowledges financial support by the Fonds der Chemischen Industrie and the recent installation of a 94 GHz ESR spectrometer provided by the DFG.

REFERENCES

1. O. Chauvet, G. Oszlányi, L. Forró, P. W. Stephens, M. Tegze, G. Faigel, and A. Jánossy, *Phys. Rev. Lett.* **72**, 2721 (1994)
2. W. A. MacFarlane, R. F. Kiefl, S. Dunsiger, J. E. Sonier, and J. E. Fischer, *Phys. Rev. B* **52**, 6995 (1995)
3. Y. J. Uemura, K. Kojima, G. M. Luke, W. D. Wu, G. Oszlányi, O. Chauvet, and L. Forró, *Phys. Rev. B* **52**, 6991 (1995)
4. V. Brouet, H. Alloul, Y. Yoshinari, and L. Forró, *Phys. Rev. Lett.* **76**, 3638 (1996)
5. K.-F. Thier, M. Mehring, and F. Rachdi, in: *Fullerene and Fullerene Nanostructures*, Proceedings of the IWEP NM96, World Scientific, p. 93 (1996)
6. A. Jánossy, N. Nemes, T. Fehér, G. Oszlányi, G. Baumgartner, and L. Forró, *Phys. Rev. Lett.* **79**, 2718 (1997)
7. M. Bennati, R. G. Griffin, S. Knorr, A. Grupp, and M. Mehring, *Phys. Rev. B, in press*
8. S. Foner, in: *Magnetism*, edited by Rado and Suhl, Academic Press, New York (1963)
9. J. B. Torrance, H. J. Pedersen, and K. Bechgaard, *Phys. Rev. Lett.* **49**, 881 (1982)

Electronic Correlations And Magnetic Ordering In CsC$_{60}$

K.-F. Thier*, M. Mehring*, F. Rachdi[†]

*2. Phys. Institut, Universität Stuttgart
Pfaffenwaldring 57, 70550 Stuttgart, Germany
† USTL Montpellier, France

Abstract. We investigate the spin arrangement in the magnetically ordered orthorhombic phase of CsC$_{60}$ by a comparison of ^{133}Cs NMR data and simulations of a 3D ordered anisotropic antiferromagnet. Consistent results are obtained for two different magnetic ordering vectors. The agreement between simulation and experiment can be further improved by including a substantial amount of magnetic disorder. In addition, the metastable low temperature cubic phase is investigated using ^{13}C NMR. We find a strongly correlated metallic system that transforms to the semiconducting dimer phase at $T = 140$K.

I MAGNETIC ORDERING IN CsC$_{60}$

Meanwhile it is well established that in the polymeric phase of RbC$_{60}$ and CsC$_{60}$ at about 30K a magnetic transition occurs. The physical nature of this transition, however, is still not completeley revealed. It is especially interesting if this transition is connected to low dimensional electronic properties of the polymer chains as was suspected by Chauvet et al. [1]. Band structure calculations, however, predict a more 3D electronic behaviour due to the reduced π-electron density in the [2+2] cycloaddition region and due to the large interchain overlap [2]. μSR studies showed the existence of a spontanous magnetization but could not detect signs of long range order [3–5]. Whereas the antiferromagnetic character of the spin order could be confirmed by antiferromagnetic resonance experiments [6,7], its detailed magnetic structure is still unresolved.

NMR can determine such magnetic ordering by using the nuclear spins as a sensitive probe for the local field at a specific lattice site. In powdered samples instead of a unique line shift a characteristic distribution of line shifts is expected resulting from the random orientation of the crystallites with respect to the external field. ^{133}Cs is a suitable nucleus for such a study, since in CsC$_{60}$ all Cs ions are magnetically equivalent above the phase transition and broadening effects due to quadrupolar interactions are neglible. Experiments on CsC$_{60}$ also show a significant magnetic broadening of the ^{133}Cs spectra below 30K [8]. From the triangular

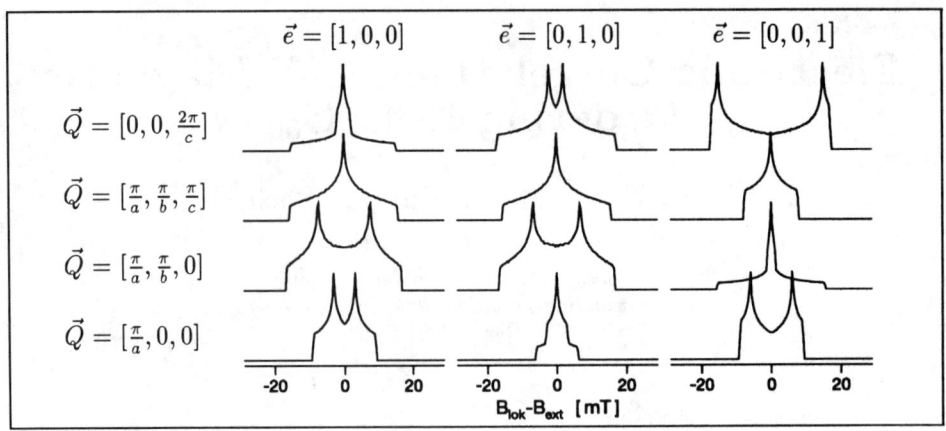

FIGURE 1. Simulated local field distributions at the Cs site in antiferromagnetically ordered CsC$_{60}$ assuming different order vectors $\vec{Q}$ and easy directions $\vec{e}$. All simulations have been performed for fields above the spin-flop transition.

shape of the observed field distribution Brouet et al. concluded a one dimensional antiferromagnetic ordering of the polymer chains. We have performed a similar analysis, but contrary to previous interpretations, where CsC$_{60}$ has been treated as an isotropic antiferromagnet, we interpret our data within the model of 3D *anisotropic* magnetic ordering. The consideration of magnetic anisotropy is necessary due to the orthorhombic crystal structure and turns out be very important for the interpretation of the experimental data.

Different types of magnetic ordering can be distinguished by their ordering vector $\vec{Q}$ and their easy axis orientation $\vec{e}$. The vector $\vec{Q}$ determines the spatial spin distribution according to

$$\vec{S}_i(\vec{r}_i) \propto \vec{S}(0) \exp(i\vec{Q}\vec{r}_i + i\phi) \qquad (1)$$

The orientation of the spin vectors $\vec{S}$ with respect to the crystal frame depends on the direction of the easy axis $\vec{e}$ which corresponds to a minimum of the magnetic anisotropic energy. If no external field is applied the sublattice magnetization $\vec{S}_i$ will always align parallel to $\vec{e}$. Upon applying an additional external field $\vec{B}$, the magnetization will eventually rotate out of the easy direction, but will always be coplanar to a plane spanned by the vectors $\vec{B}$ and $\vec{e}$. This is still true in the spin-flop phase, where the sublattice magnetization and magnetic field are always perpendicular. The huge effect of the easy direction on the NMR spectra can be seen in fig. 1 where local field distributions at the Cs site have been calculated for different vectors $\vec{Q}$ and $\vec{e}$. In the simulations, a powder average over 400000 orientations has been performed and the dipolar fields of all C$_{60}$ molecules within a sphere of radius $r = 35$Å have been included. Besides a point dipole model, were

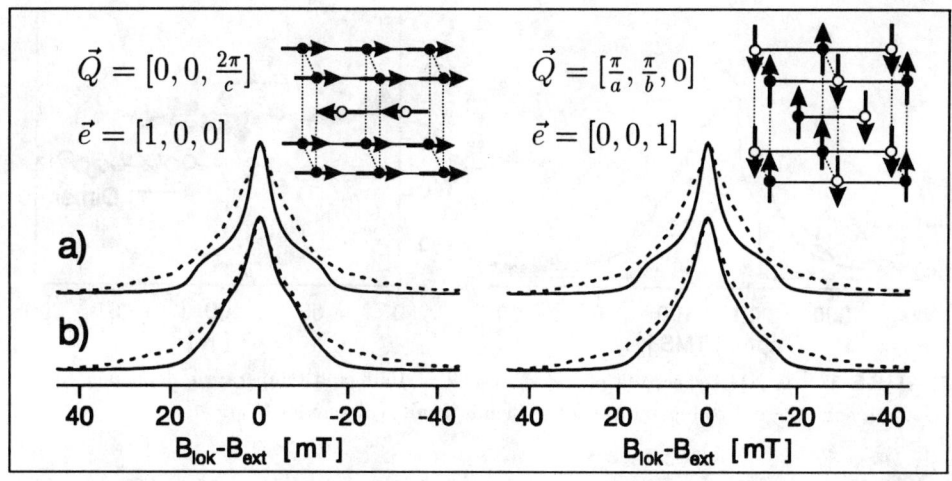

FIGURE 2. a) Comparison of experiment and simulation for two different types of antiferromagnetic order were the easy direction is determined by dipole-dipole couplings. The dotted line shows the experimentally observed field distributions. The calculated filed distribution has been convoluted with a lorentzian in order to take into account the finite intrinsic linewidth of the ^{133}Cs Spectrum. b) The same simulations assuming a magnetic disorder of about 10%.

the electron spin is located at the center of the C_{60} molecule also different inhomogeneous distributions on the C_{60} molecule have been considered. We find that the shape of the spectra shows no significant influence from a spin redistributions on the C_{60} as long as the density function is consistent with the crystal symmetry.

There are at least three different interactions, which can contribute to the crystalline ansiotropy energy: the crystal field splitting, anisotropic exchange couplings and dipole-dipole-interaction. For CsC_{60}, however, important contributions are expected only from the dipolar coupling between different electronic spins. In this case an anisotropy field of about 2mT is expected, which is consistent with the results from antiferromagnetic resonance studies [7]. Assuming that the observed anisotropy is due to the dipole-dipole-interaction, the easy axis is no more a free parameter and can be calculated for each $\vec{Q}$.

Fig 2 shows two specific cases where we find the best qualitative agreement between experiment and theory. The corresponding magnetic structures are also indicated in Fig. 2. The structure with $\vec{Q} = [0, 0, \frac{2\pi}{c}]$ corresponds to a magnetic ordering within the chemical unit cell and ferromagnetically ordered chains as proposed by Mele et al. . Such order should occur if the interchain exchange couplings are of the same order of magnitude as the couplings along the chain. The second possibility $\vec{Q} = [\frac{\pi}{a}, \frac{\pi}{b}, 0]$ involves antiferromagnetically ordered chains and a doubling of the unit cell in the a and b direction. This type of ordering is expected if the exchange couplings along the polymer chains dominate. The agreement between

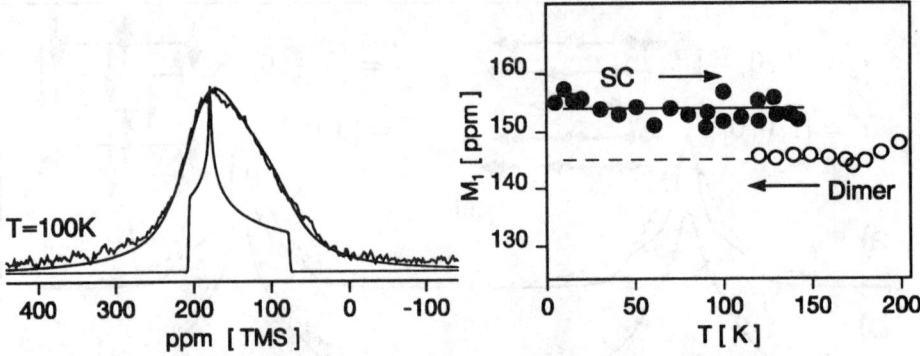

FIGURE 3. ^{13}C NMR spectrum of sc-CsC$_{60}$ at $T = 100$K and temperature dependenc of its first moment. The dottet line indicates the chemical shift referenc for C$_{60}^-$.

simulation and experiment can be considerably improved by taking into account the existence of magnetic disorder. Fig. 2 b) shows a simulation where we assumed a magnetic misalignement of about 10% of the electron spins. Such magnetic disorder could be caused by chain-ends and phase boundaries.

We note that we could describe our experimental data also with an order vector $\vec{Q} = [\frac{\pi}{a}, -\frac{\pi}{b}, \frac{\pi}{b}]$ and an easy axis within the $a - b$ plane. This easy axis is, however, inconsistent with the predictions from dipolar anisotropy.

II ELECTRONIC CORRELATIONS IN THE SIMPLE CUBIC PHASE

If CsC$_{60}$ is cooled rapidly from the high temperature fcc phase to temperatures below 100K another cubic phase is obtained instead of the orthorhombic polymer phase. In order to determine the electronic properties of this metastable phase we performed ^{13}C measurements on quenched CsC$_{60}$. Fig 3 shows a ^{13}C spectrum taken at 100K. Despite the spectrum is quite featureless we can fit an average tensor powder pattern with the principal axis values of (76, 175, 207)ppm convoluted with a 66ppm broad Lorentzian distribution. These principal axis values show a small paramagnetic shift with respect to the chemical shift tensor of C$_{60}$. We interpret this shift as a Knight shift which is a typical sign for metallic systems. Subtracting the chemical shift values of pristine C$_{60}$ with a 2ppm correction due to the charge on the C$_{60}$ molecule, an isotropic Knight shift of $K = 8$ppm and a dipolar Knight-Shift tensor with the principal axis values of $K_{d,ii} = (-19, -19, 38)$ppm is obtained. The temperature dependence of the first moment of the spectrum is also shown in fig. 3. For the simple cubic phase we find a temperature independent paramagnetic shift, as expected for a Pauli-like susceptibility. Above 140K the Knight shift disappears as the sample transforms into the semiconducting dimer phase.

Longitudinal and transverse relaxation measurements also have been performed

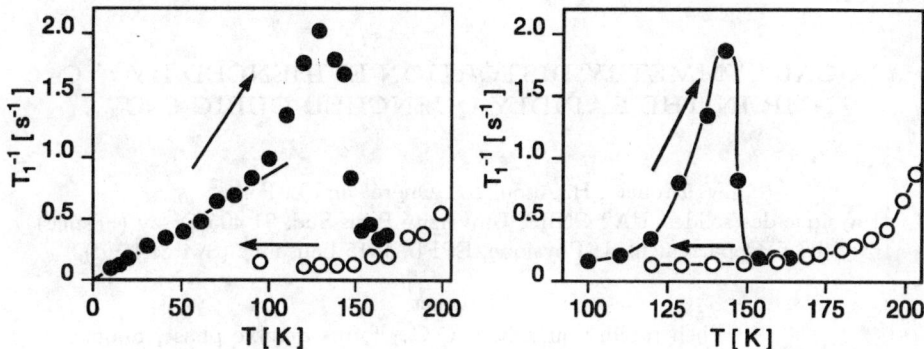

FIGURE 4. Longitudinal ($1/T_1$) and transversal relaxation rates ($1/T_2$) of the ^{13}C magnetization in the simple cubic phase of CsC$_{60}$.

and the results are shown in Fig 4. Above 100K fluctuations are present that lead to a strong increase of the T_1 and T_2 relaxation rates. These fluctuations may be related to the irreversible transition into the dimer phase. Below 100K, the longitudinal relaxation rates show a linear increase in temperature and can be well described by a Korringa formalism according to [9]

$$\frac{1}{T_1} = C T K_{iso}^2 (1 + \frac{1}{2}\epsilon) S_K \qquad (2)$$

with $\epsilon = (K_{33} - K_{iso})^2 / K_{iso}^2$ and $C = 4\pi k_B/\hbar \, (\gamma_n/\gamma_e)^2$. The fact that the experimentally determined Knight shift tensor is only the average value of a distribution, makes a precise determination of S_K difficult. Taking into account a distribution of the principal axis values with a width of 66ppm, we obtain values of $S_K \gtrsim 6$. This value is large compared to other alkali fullerides where $S_K = 2.5..3.5$ [9] and points to strong electronic correlations. The temperature independent enhancement factor shows, however, that these correlations are not connected to critical fluctuations which could be interpreted as a precursor of another phase transition.

REFERENCES

1. O. Chauvet et al. Phys. Rev. Lett. **72**, 2721 (1994).
2. S.C. Erwin et al. Phys. Rev. B **51**, 7345 (1995)
3. L. Cristofolini et al., J. Phys.: Condensed Matter **7**, L567 (1995).
4. W.A. MacFarlane et al., Phys. Rev. B **52**, 6995 (1995).
5. Y.J. Uemura et al., Phys. Rev. B **52**, 6991 (1995).
6. A. Janossy et al., Phys. Rev. Lett. **79**, 2718 (1997).
7. M. Benatti et al., this proceedings and Phys. Rev. B **57** July (1998).
8. V. Brouet et al., Phys. Rev. Lett. **76**, 3638 (1996).
9. M. Mehring et al., Phil. Mag. B **70**, 787 (1994)

LOCAL SYMMETRY DISTORTION EVIDENCED BY ^{133}Cs NMR IN THE RAPIDLY QUENCHED CUBIC CsC$_{60}$.

V. Brouet[1], H. Alloul[1], F. Quéré[1] and L. Forro[2]

[1] Physique des solides, UA2 CNRS, Université Paris-Sud, 91 405 Orsay (France)
[2] IGA, Département de Physique, EPFL, 1015 Lausanne (Switzerland)

Abstract

When rapidly quenched, CsC$_{60}$ forms a cubic phase, about which very little information is available at present. The alkali NMR reveals the existence of two different ^{133}Cs sites, which implies a symmetry distortion with respect to the orientationally ordered cubic structure, proposed by structural measurements. We will show that the two lines are intrinsic to the cubic quenched phase and will raise the question of the role of the associated symmetry distortion in the electronic properties of this phase.

1 Introduction

A systematic investigation of the electronic properties of fullerides as a function of band filling is desirable, but it turns out to be rather complex and only a few stoichiometries can be formed. Among these compounds, AC$_{60}$ is the only known odd electron system other than A$_3$C$_{60}$ and is therefore particularly interesting to study. Below room temperature, a polymerized phase is formed which exhibits new and exciting properties, and much work has been devoted to understanding them in the past years. On the other hand, there is little information on the high temperature cubic (HTC) phase - which is directly comparable to A$_3$C$_{60}$ - mainly because of its restricted range of temperature. This phase has a Curie-like susceptibility [1,2] which suggests localization of the electrons, but there is no direct evidence that it is insulating [3].

To get more information, it is natural to try to prevent polymerization by quenching this phase. For KC$_{60}$ and RbC$_{60}$, a completely distinct phase is obtained, where C$_{60}$ molecules form dimers [4]. More recently, it has been shown that in CsC$_{60}$, the cubic structure can be preserved and is orientationally ordered (Pa$\bar{3}$) [6]. This cubic quenched (CQ) phase transforms irreversibly into the dimer phase above 130 K. Surprisingly, the first study of this phase [5] revealed an ESR Pauli-like susceptibility, at variance with the HTC Curie-like susceptibility, and motivates further investigation.

We will focus in this paper on an unexpected aspect revealed by ^{133}Cs NMR in this CQ phase, that is the existence of two ^{133}Cs sites, even though only the octahedral site of the fcc structure is occupied by the alkali. The situation is somehow reminiscent of the A_3C_{60} case, where three alkali NMR lines are observed, whereas only two sites are expected. In CsC_{60}, it is even more dramatic since the two lines have similar intensities.

2 Alkali spectrum

The splitting of the ^{133}Cs spectrum into two well-defined lines can be clearly seen on Figure 1, in the 120 K spectrum. One appears around -100 ppm, which is an usual shift range for alkali fullerides (we will refer to it as Non-Shifted (NS) line), the other is strongly shifted at 800 ppm (shifted (S) line). When the temperature is decreased to 10 K (see spectrum at 30 K in Figure 1), there are huge variations of shifts, but the two lines can always be clearly distinguished and conserve similar relative intensities within 15 %[a].

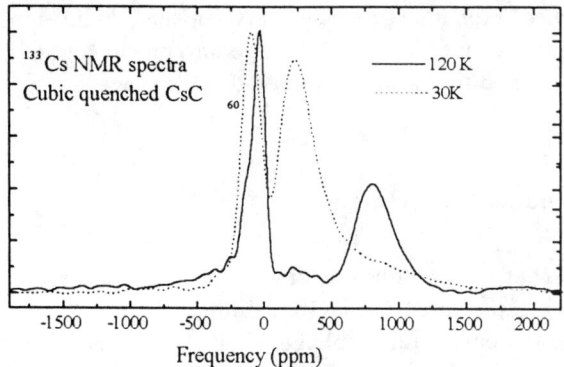

Figure 1: ^{133}Cs NMR spectra at 120 K and 30 K in the cubic quenched phase.

The quenched phase was obtained by immersing the sample into liquid nitrogen, after one hour thermalization at 530 K. ^{13}C NMR clearly shows the transition from the CQ phase to the dimer at around 130 K. At the same temperature, the two lines disappear irreversibly, implying that both are related to the CQ phase. Although the NS line appears at roughly the same position as the dimer line, it is narrower and its relaxation time (T_1) is two orders of magnitude smaller at 120 K, so that they can be unambiguously separated. In addition, we have not detected any anomaly in the NMR behavior of our sample in the HTC, polymer or dimer phases with respect to any data previously

---

[a]We have checked that quadrupole effects are sufficiently similar for the two lines to be observed under the same conditions, without intensity distortion.

published. Besides NMR, the sample was also investigated by X-ray diffraction and ESR in the CQ phase. No spurious phase could be singled out by X-ray diffraction, which further confirmed the orientationally ordered structure [8]. The ESR susceptibility is almost temperature independent, in agreement with reference 5. Three other samples were investigated and all display these two ^{133}Cs lines with similar intensity ratios. Their behavior was completely similar, although in one sample, the transition to the dimer phase was found to be partially reversible.

In addition to the difference in shifts, the two lines have also very different relaxation rates. Below 80 K, they differ by two orders of magnitude, which indicates that the hyperfine coupling and/or electronic properties are also radically different between the two sites. Interestingly, a huge variation of the relaxation rate of the NS line is observed in the vicinity of the transition to the dimer, and at 130 K, the two lines have nearly equal relaxation rate. This is likely due to chemical exchange processes between the two lines just before the transition and suggests that they would merge into one at higher temperature, if the transition to the dimer state were not to take place.

3 Double resonance experiment

A definitive proof that the two lines belong to the same phase was provided by a double resonance experiment (SEDOR), similar to the one used in the case of A_3C_{60} to probe the environment of the T' site [7]. In this experiment, one of the two ^{133}Cs lines is selectively irradiated, Figure 2 (left) shows the resulting spectrum for the NS line. This spectrum is then observed again, when the magnetization of the other line is flipped by another RF pulse at a time τ. No modification would result if the two sites were not coupled. On the contrary, if they are sufficiently close from each other, their dipolar coupling yields an intensity loss. This is clearly observed in Figure 2 and proves that the sites are intimately mixed in a single phase. The same kind of results (not shown) have been obtained for the S line.

Figure 2 (right) shows the dependence of this effect as a function of τ. The relative intensity reduction - the so called sedor fraction (SF) - behaves in the expected way for small τ, i.e. SF$\propto 2\Delta^2\tau^2$ as shown in figure 2. In principle, Δ^2 should allow us to extract the mean distance between the two types of sites, but knowledge of their distribution in the structure is required, and is missing in our case. Nevertheless, we have checked that, assuming six nearest neighboring sites, leads to a reasonable site-site distance of a few angstroms.

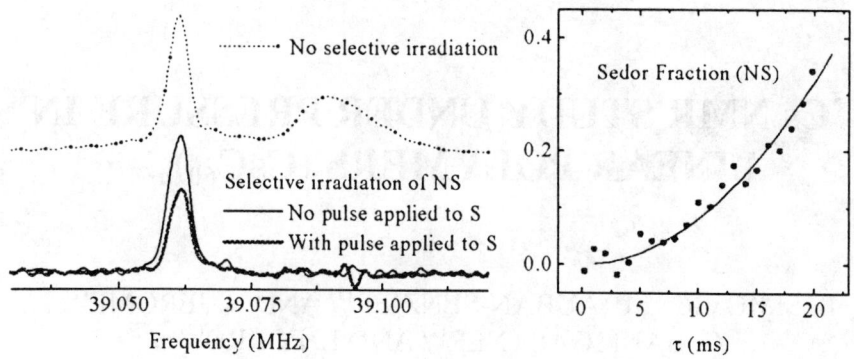

Figure 2: SEDOR double resonance experiment on the NS line. Left : Spectra under the indicated conditions. Right : Sedor fraction as a function of τ.

4 Conclusion

We have shown that the two ^{133}Cs lines observed in the CQ phase are intrinsic. Presently, there is no clear understanding of the origin of these two sites. In the cubic Pa$\bar{3}$ structure, the only expected structural difference between alkali sites would come from the existence of different orientations at the neighboring C_{60} balls (there are here a major and a minor orientation [6]). It seems surprising that Cs could be so sensitive to these effects, unless they are associated with electronic distortions. The combination of fine details of the structure and electronic properties will likely be necessary to provide a complete understanding of this phase.

1. O. Chauvet et al., Phys. Rev. Lett. 72, 2721 (1994).
2. R. Tycko et al., Phys. Rev. B 48, 9097 (1993).
3. M.C. Martin et al., Phys. Rev. B 49, 10818 (1994).
4. G. Oszlanyi et al., Phys. Rev. B 51, 12228 (1995), Q. Zhu et al., Phys. Rev. B 51, 3966 (1995).
5. M. Kosaka et al., Phys. Rev. B 51 17, 12018 (1995).
6. A. Lappas, M. Kosaka, K. Tanigaki and K. Prassides, J. Am. Chem. Soc. 117, 7560-7561 (1995).
7. C. H. Pennington et al., Phys. Rev. B 54, R6853 (1996).
8. G. Bendele, P.W. Stephens et al, in preparation.

^{133}Cs NMR STUDY UNDER PRESSURE IN LINEAR POLYMERS (CsC$_{60}$)$_n$.

B. SIMOVIC[1], P. AUBAN-SENZIER[1] AND D. JEROME[1]
G. BAUMGARTNER[2] AND L. FORRO[2]

[1] *LABORATOIRE DE PHYSIQUE DES SOLIDES (ASSOCIE AU CNRS), UNIVERSITE PARIS SUD, 91405 ORSAY-CEDEX FRANCE*
[2] *DEPARTEMENT DE PHYSIQUE ECOLE POLYTECHNIQUE FEDERALE DE LAUSANNE, 1015 LAUSANNE, SWITZERLAND*

Abstract. Alkali metal doped fullerides reveal a polymerized crystal structure with parallel chains of covalently bounded C$_{60}$ molecules. The electronic structure shows a narrow p-electron band. The existence of strong electron-electron correlations in this fulleride is supported by the establishment of an antiferromagnetic ground state at low temperature. We have performed ^{133}Cs-NMR relaxation time T$_1$ measurements on (CsC$_{60}$)$_n$ versus pressure and temperature. The salient and most unexpected result of our work is the fact that T$_1$ is not sensitive to pressure (up to 9kbar) in the temperature range 100 and 300K unlike T$_1$ measurements performed under pressure on ^{13}C in the parent coumpound (RbC$_{60}$)$_n$. This finding allows to rule out a purely magnetic relaxation mechanism for the alkali atom. However, the relaxation becomes strongly pressure dependent at low temperature (T<100K). In particular a new ground state (found to be non magnetic and possibly spin-Peierls) establishes above 5kbar with the concomitant activation of 1/T$_1$ below 30K.

Introduction

Recently, ESR measurements [1] have shown that in spite of similar structures [2], the three polymerized compounds (AC$_{60}$)$_n$ (A=K, Rb and Cs) exhibit a strongly different behaviour in their electronic properties. (KC$_{60}$)$_n$ is metallic at any temperature unlike (RbC$_{60}$)$_n$ and (CsC$_{60}$)$_n$ which both undergo a three dimensional magnetic transition [1,3] below 50K. The narrow ESR linewidth observed in (RbC$_{60}$)$_n$ and (CsC$_{60}$)$_n$ suggests low dimensional properties which is in contradiction with theoretical band calculations [4]. Moreover, the NMR study [5] of (CsC$_{60}$)$_n$ at ambient pressure has shown a different behaviour of the spin-lattice relaxation rate T_1 between ^{133}Cs and ^{13}C. This experimental fact is consistent with the presence of one dimensional antiferromagnetic fluctuations up to room temperature. In this context, NMR under pressure is particulary relevant to study a possible crossover between one dimensional regime and a three dimensional one in the electronic properties.

Experimental

NMR measurements were performed on ^{133}Cs in a powdered sample using a CXP200 spectrometer. The spin-lattice relaxation rates were determined by a conventional technique of saturation recovery experiment at the frequency irradiation 43 MHz in 7.8 Tesla.
Our pressure system is a home-made double-stage CuBe pressure cell using inert fluor as hydrostatic pressure medium. This sytem allows us to correct for losses of pressure during cooling.

Results and discussion.

Figure1 displays the linewidth behavior of ^{133}Cs versus temperature and pressure. The inhomogeneous linewidth of the spectrum is correlated to the local distribution of electronic magnetic moments around the nucleus. When the correlation time τ of the three dimensional antiferromagnetic fluctuations is large enough to have $\gamma\Delta H\tau > 1$, the local distribution of magnetic moments appears to be static. Hence the important broadening observed at 1 bar and 3 kbar is due to the onset of a magnetic ground state at low temperature. So at 5kbar, the disappearance of this magnetic broadening indicates the suppression of the magnetic ground state as it has been reported on ^{13}C NMR under pressure in $(RbC_{60})_n$ [6].

Figure 1

Figure 2

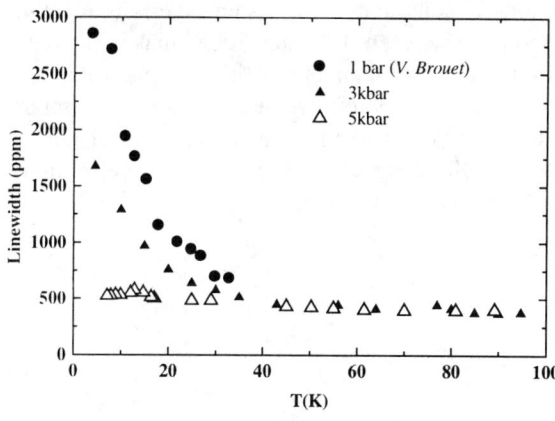

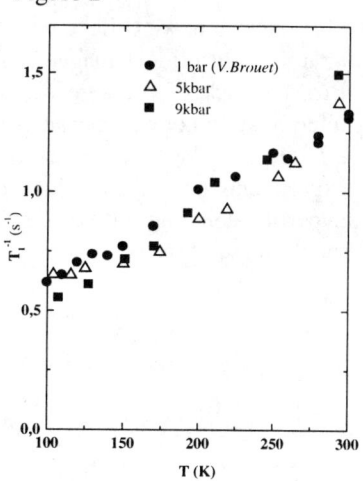

Figure 2 shows the behavior of the spin-lattice relaxation rate between 100 and 300K at various hydrostatic pressures. ESR measurements [7] under pressure on CsC_{60} have shown a decreasing of the uniform susceptibility at a rate of 10%/kbar. This strong effect of pressure has also been observed in $(RbC_{60})_n$ on ^{13}C NMR [6]. The fact that T_1 is pressure-independent rules out for ^{133}Cs any relaxation mechanism based on the spin susceptibility. But, it is important to notice the fact that ^{133}Cs nucleus has one of the smallest quadrupolar moment in nature. So, it is difficult to think that the main contribution to the relaxation is due to electric field gradient fluctuations.

At low temperature (<100K), the spin-lattice relaxation rate becomes strongly dependent on pressure. The sharp peak observed at 5kbar on $(T_1T)^{-1}$ (figure 3a) indicates clearly a non metallic behaviour (usually characterized by a Korringa law : $(T_1T)^{-1}$=Cst). The activation of $(T_1)^{-1}$ below 30K (figure 3b) is the signature of gapped excitations which rapidly disappear with pressure.

Figure 3a   Figure 3b

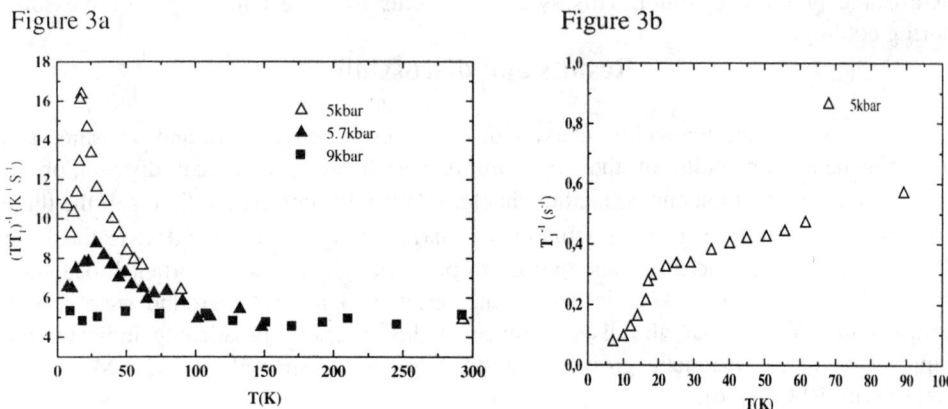

In conclusion, we believe that the activated behaviour of spin-lattice relaxation rate at 5 kbar without critical magnetic fluctuations is consistent with a spin-Peierls transition. But, the lack of pressure dependence on T_1 between 100 and 300K makes T_1 data interpretation not easy because we still do not know the nature of fluctuations which are effective in the ^{133}Cs relaxation. To clarify this point, we are performing similar experiments on ^{13}C for which the main contribution to the relaxation is correlated to hyperfine coupling with electronic spins. It will be easier to observe, on ^{13}C relaxation, the decrease of the susceptibility due to the formation of a singlet state.

Anyway, this unusual pressure behaviour supports the idea of low dimensional electronic properties in $(CsC_{60})_n$.

Acknowledgment

We wish to acknowledge V. Brouet for the communication of her NMR data at ambient pressure.

References

1. F. Bommeli *et al.*, Phys.Rev B **51**, 3966 (95)
2. P.W. Stephens *et al.*, Nature **351**, 632 (91)
3. A. Janossy *et al.*, Phys. Rev. Lett **79**, 14, 2718 (97)
4. S. C. Erwin, G. V. Krishna and E. J. Mele., Phys.Rev B **51**, 7345 (95)
5. V. Brouet, H. Alloul, Y. Yoshinari, and L. Forro, Phys. Rev. Lett **76**, 19, 3638 (96)
6. P. Auban-Senzier, D. Jérome, L. Forro and F. Rachdi, J. Phys. I France **6**, 2181 (96)
7. L. Forro et al., Proceedings **IWEPNM 96**, 102

Low-frequency modes of CsC_{60} phases

J.L. Sauvajol and E. Anglaret

Groupe de Dynamique des Phases Condensées (UMR 5581)
Université Montpellier II
34095 Montpellier Cedex 5, France

Abstract. In this communication we review Raman and inelastic neutron results concerning the low-frequency vibrational modes for CsC_{60} phases obtained upon slow and rapid cooling from the high temperature disordered phase (T> 370 K).

INTRODUCTION

The alkali-doped AC_{60} (A= K, Rb, Cs) compounds exhibit rich and original phase diagrams, including a polymer phase obtained upon slow cooling from the high temperature fcc phase (1-3) and several metastable phases obtained upon rapid cooling (3-5). Kosaka et al studied the phase diagram of slowly-cooled and 77 K-quenched CsC_{60} by means of calorimetry and X-ray diffraction (3). Upon slow cooling a stable-in-air orthorhombic phase which made of covalently bonded polymer chains is obtained below 370 K: the CsC_{60} polymer phase. Upon rapid cooling at least five different metatastable phases are evidenced between 20 K and 370 K, namely a 3D conducting orientationally ordered phase below 160 K, a dimer phase in the temperature range 160 K- 220 K, and two other 3D conducting phases above 220 K before the recovering of the polymer phase above 270 K (3).

In this communication we review Raman scattering and inelastic neutron scattering studies of the low-frequency vibrational modes in the different CsC_{60} phases.

RESULTS AND DISCUSSION

Raman results

In a previous work we have shown that the room-temperature low frequency Raman spectra of polymer RbC_{60} and CsC_{60} excited at 776 nm exhibit a well-defined peak around 30 cm^{-1} (6). In this early work the assignment of this excitation was discussed and its attribution to the interball strectching mode in long oligomers was suggested. In the following, in agreement with theroretical predictions and neutron results (7), this mode was definitely assigned to a libration of the polymer chain. Recently the temperature dependence of the position and width of this low-frequency excitation in the polymer phase of RbC_{60} was reported (8). In the whole polymer-phase temperature range, the behavior of this mode is that expected for regular anharmonicity: quasi linear softening and broadening of this mode with the temperature. In addition no significant change of this mode was observed at the metal-insulator transition temperature ($T_{M/I}$ =50 K). As expected for a chain libration, this mode vanishes at the polymer/disordered phase transition ($T_{P/D}$=370 K). In this communication we report a Raman investigation of the temperature dependence of this excitation in the polymer phase of CsC_{60}. The spectra (excited at 776 nm) measured below and above the polymer/disordered phase transition are displayed on figure 1.A. We used the signal recorded in the $H_g(1)$ mode range as a probe of the structure of the sample at the vicinity of the phase transition (figure 1.B). Indeed it is known that the $H_g(1)$ mode is splitted in at least three components in the polymer phase, by contrast a single line was observed in the disordered phase (6,9). From this, we unambiguously state that the chain libration vanihes in the CsC_{60} disordered phase. The comparable dependence of this low-frequency excitation in RbC_{60} and CsC_{60} confirms the analoguous behaviour of both compounds.

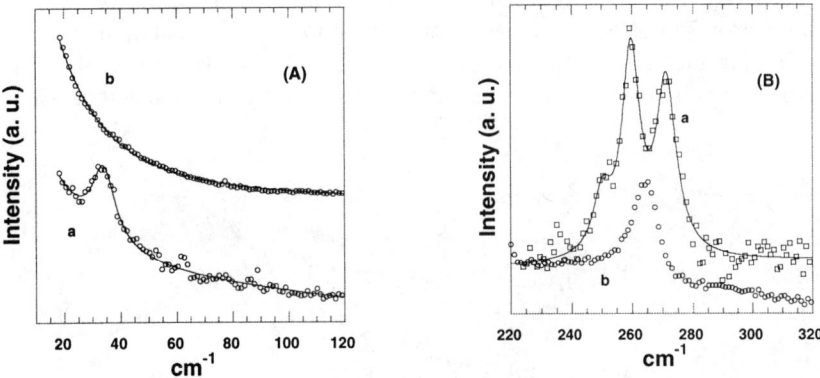

Figure 1. Low-frequency Raman spectrum (A) and Raman spectrum in the $H_g(1)$ range modes (B) . a) T< $T_{P/D}$, b) T>$T_{P/D}$. Solid lines: fit with Lorentzian line shapes. Excitation wavelength= 776 nm.

Neutron results

The vibrational low-frequency excitations of CsC_{60} in its polymer phase was investigated in a large temperature range (20 K- 320 K) by inelastic neutron scattering (10). In figure 2.A are displayed the signal measured at three temperatures. In the whole temperature range the profile of the signal does not change with the temperature.

The signal essentially displays two contributions around 0.65 THz (22 cm^{-1}) and 1.15 THz (38 cm^{-1}). The frequencies of these peaks are close to those observed in TOF experiments on RbC$_{60}$ (7). The Q-dependence of these two peaks are similar to that of the libration of C$_{60}$ confirming the assignment of these modes to libration-like excitations: chain libration and twist of the chain (7,10).

The inelastic neutron spectra of quenched CsC$_{60}$ at different temperatures (the same that for the polymer neutron spectra) are displayed in figure 2.B. By contrast with the behaviour in the polymer phase the signal significantly changes with the temperature. At low temperature a single component is observed around 0.6 THz (20 cm^{-1}), at 210 K the signal displays a double-peak structure with components at 0.4 THz (13 cm^{-1}) and 0.7 THz (25 cm^{-1}), at 270 K a quasi-elastic component dominates the spectrum. These different vibrational features can be assigned on the basis of the sequence of phase transitions stated by Kosaka et al (3) and recalled in the introduction. In the range of the ordered monomer phase (below 160 K) the single peak continuously softens with heating, follows a temperature dependence close to that of pristine C$_{60}$ libration and the Q-dependence of its intensity is similar to that of both the librations of ordered C$_{60}$ phase and polymer phase (7,10,11). This clearly indicates that this mode can be attributed to (C$_{60}$)$^{-}$ libration. It is to our knowledge the first measurement of a librational excitation in the low temperature phase of quenched CsC$_{60}$. The two-peak structure observed in the range of the dimer phase:160 K-220 K (3) is assigned as the intrinsic vibrational signature of this phase. We propose to assign the low-frequency component of this structure to the libration of the (C$_{60}$)$_2^{2-}$ dimer around its axis and the high frequency component to a twist of the dimer. The quasi-elastic component which appears above 260 K is the signature of a transition to an orientationally disordered phase. This result agrees with the existence of a disordered phase before the recovering of the polymer phase at higher temperature as predicted by different authors (3-5).

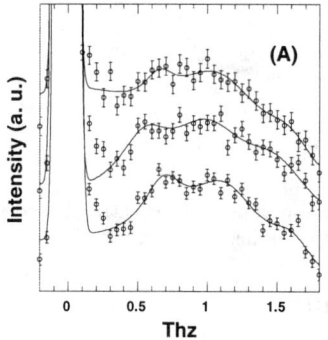 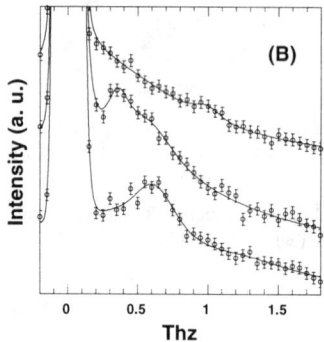

Figure 2. Inelastic neutron spectra . (A) CsC$_{60}$ polymer phase, (B) quenched CsC$_{60}$. bottom T=150 K, middle T= 210 K, top T= 260 K

CONCLUSION

Raman and inelastic neutron scattering investigations of the sequence of phase transitions of slowly- and rapidly-cooled CsC_{60} have been performed. The librational signatures of several phases have been identified, especially the Raman-active chain libration mode in the polymer phase of the slowly cooled CsC_{60}, the $(C_{60})^-$ libration in the ordered low-temperature phase and the librational double-peak structure in the $(C_{60})_2^{2-}$ dimer phase of the quenched CsC_{60}.

REFERENCES

1. Chauvet O., Oszlanyi G., Bortel G., Faigel G., Tegze M., Pekker S. and Forro L., *Phys. Rev. Lett.* **72**, 2721 (1994)
2. Oszlanyi G., Bortel G., Faigel G., Tegze M., Granasy L., Pekker S., Stephens P.W., Bendele G., Dinnebier R., Mihaly G., Janossy L. ,Chauvet O. and Forro L., *Phys. Rev. B* **51**, 12228 (1995).
3. Kosaka M., Taginaki K., Tanaka T., Atake T., Lappas A. and Prassides K., *Phys. Rev. B* **51**, 12018 (1995).
4. Oszlanyi G., Bortel G., Faigel G., Granasy L., Bendele G.M., Stephens P.W. and Forro L., *Phys. Rev. B* **54**, 11849 (1996).
5. Zhu Q., Cox D.E and Fischer J.E., *Phys. Rev. B* **51**, 3966 (1995).
6. Bormann D., Sauvajol J.L., Goze C., Rachdi F., Moreac A., Girard A., Forro L. and Chauvet O., *Phys. Rev. B* **54**, 14139 (1996).
7. Schober H, Tolle A., Renker B., Heid R and Gompf F., *Phys. Rev. B* **56**, 5937 (1997).
8. Sauvajol J.L., Anglaret E., Girard A., Moreac A., Ameline J.C., Delugeard Y. and Forro L., *Phys. Rev. B* **56**, 13642 (1997).
9. Winter J. and Kuzmany H., *Phys. Rev. B* **52**, 7115 (1995).
10. Sauvajol J.L., Anglaret E., Aznar R., Bormann D.and Hennion B., *Solid State Comm.* **104**, 387 (1997).
11. Neumann D.A., Copley J.R.D., Kamitakahara W.A., Rush J.J, Cappelletti R.L., Coustel N., Fischer J.E., Mc Cauley J.P., Smith III A.B.,Creegan K.M. and Cox D.M., *J. Chem. Phys.* **96**, 8631 (1992).

Fingerprints of solid-state chemical reactions in the dynamics of fullerenes

B. Renker[1], H. Schober[2], R. Heid[3], B. Sundqvist[4]

[1]*Forschungszentrum Karlsruhe, INFP, D-76021 Karlsruhe Germany*
[2]*Institut Laue Langevin, B.P.156, F-38042 Grenoble Cedex, France*
[3]*MPI f. Physik komplexer Systeme, D-01187 Dresden, Germany*
[4]*Dep. of Experimental Physics, Umea Univ., S-90187 Umea, Sweden*

Abstract. Excitation spectra of polymerized fullerides as RbC_{60}, Na_4C_{60} and pressurized C_{60} are studied by inelastic neutron scattering and Raman spectroscopy in view of differences in the interfullerene bonding. Changes in the dynamics are followed by temperature dependent measurements. A detailed analysis is performed by model calculations.

The formation of interfullerene covalent bonds leads to an interesting new class of polymers. Single and double bonded systems (neighboring molecules are connected via a 4-membered ring of single bonds) as well as compounds with a linear chain structure (A_1C_{60} and orthorhombic (*o*) C_{60} polymer: o-pC_{60}) or materials polymerized within 2-dimensional (2D) sheets (Na_4C_{60}, rhombohedral (*rh*) and tetragonal 2D-(pC_{60})) can be synthesized (1-3). All peculiarities of these compounds are well reflected in their vibrational properties. Strong on-ball C-C bonds cause a rich spectrum of intramolecular modes within an energy region of 30-200 meV, whereas intermolecular bonds of intermediate strength give rise to lower lying librational and translational molecular modes. Finally the breaking up of polymer bonds enables a free rotation of fullerene molecules which shows up by the onset of quasielastic scattering. Results from inelastic neutron scattering (INS) which is sensitive to all excitations and which offers for good resolution at lower energies (≤ 30 meV) as well as results from Raman (R) measurements which allow for detailed studies of symmetry allowed optical modes will be presented in this contribution. INS data are shown in the form of a generalized phonon density of states (GDOS) or as $\omega^{-1}\chi''(\omega)$ where $\chi''(\omega)$ denotes the q-integrated imaginary part of the dynamical susceptibility. All INS measurements were performed in energy gain on the IN6 time-of-flight spectrometer at the HFR in Grenoble with 4.72 meV incoming neutron energy. Doped samples were produced from a mixture of A_6C_{60} and C_{60}, pressurized fullerene samples up to 1.2 GPa were prepared in Karlsruhe only the 2D polymer of C_{60} was obtained at 2 GPa at Umea (Sweden) both in a belt apparatus between pistons. Related work of the authors on these systems, experimental details and further references can be found in (4,5).

Fig.1 shows results obtained for the linear chain polymer o-RbC_{60} (INS measurements on a triple axis spectrometer). It can be seen that it is possible by the application of uniaxial pressure (1.2 Gpa) to obtain samples with aligned polymer chains. The polymer chains grow along a 110 direction of the high temperature (HT) fcc-cell which is closest to the direction of the applied pressure (the two left panels). The chains have metallic conductivity at temperatures T>50 K and the nature of the

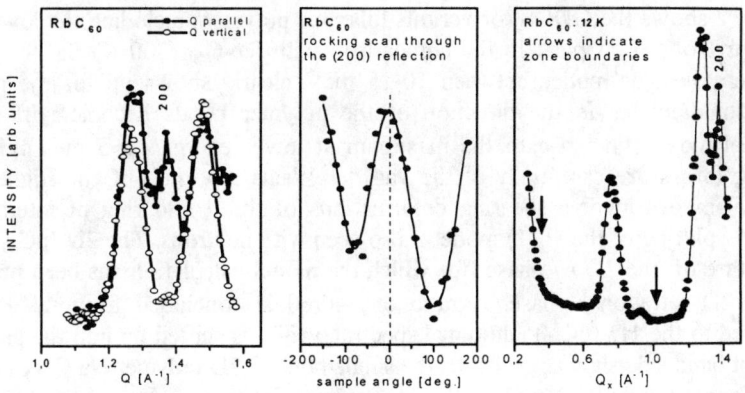

Figure 1. Alignment of polymer chains by application of uniaxial pressure for o-RbC$_{60}$, a: The 200 chain reflection disappears in a scan perpendicular to the chain direction. b: alignment of the chains, c: No AF-superstructure is visible in the insulating phase.

insulating groundstate is still a puzzling question. The right panel shows a longitudinal scan along the chain direction. As before for powder samples we could also for the present sample with aligned polymer chains not observe the expected antiferromagnetic (AF) ordering (occurrence of new magnetic bragg peaks at the zone boundaries of the HT-Brillouin zone below the ordering temperature).

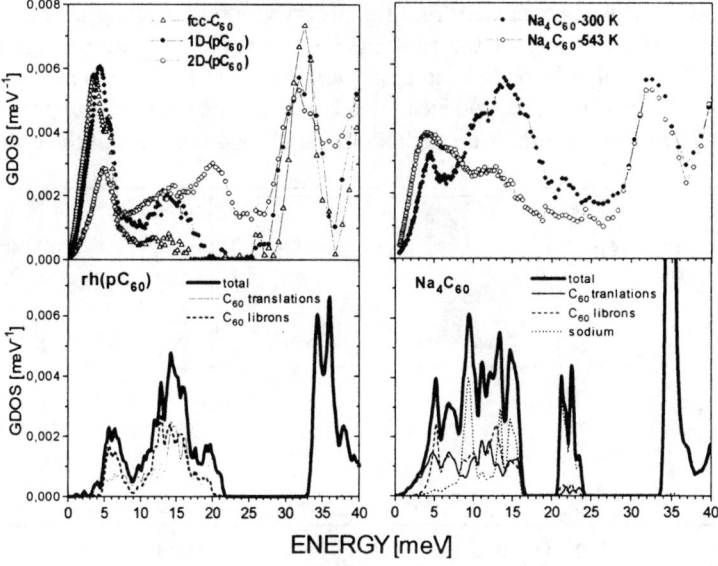

Figure 2: The generalized phonon density of states for different polymers, 1D-(pC$_{60}$): linear chains, o-structure, 2D-(pC$_{60}$): polymerized in 2D sheets, double bonded, mixture of rh- and tetragonal phases ($\approx$40/60), Na$_4$C$_{60}$: single bonded 2D-polymer. Model calculations show the contributions of particular modes.

Figure 2 shows the GDOS for various fullerene polymers including the lowest $H_g(1)$ molecular mode. Left part: In the comparison with fcc-C_{60} (300 K) the formation of new intermolecular modes between 10-25 meV clearly shows up for the polymers. Translational modes in the direction of the polymer bonds become stiff. In-phase librational modes give rise to the maximum at lower energies ($\approx$5 meV). Here, the restoring forces are essentially of the van der Waals type (1D-pC_{60}). Higher energy rotational modes involve shearing deformations of the 4-fold ring of intermolecular bonds. A splitting of the $H_g(1)$ mode is also seen with neutrons. The 2D-(pC_{60}) sample is a mixture of the 2D rh-phase (for which the model calculation has been performed) and the 2D tetragonal phase. A ratio of $\approx$40/60 is concluded from the R-spectra. Compared to the 1D-(pC_{60}) additional spectral weight is shifted up into the gap region. The right hand side shows results for the single bonded 2D polymer Na_4C_{60}. Compared to the double bonded 2D-(pC_{60}) the spectrum of intermolecular modes is softer. On heating (T>430 K) the polymer bonds start to break up and a large part of spectral density between $\approx$10-20 meV is shifted back to lower energies. Remaining modes are attributed to optical vibrations of sodium atoms. Although this material is single bonded a considerable distortion of the C_{60} cages is registered (see the R-spectrum of figure 3). Unfortunately there are no structural data with atomic resolution. A consideration of distortions in the calculation would result in smaller intermolecular distances (2.18 A now) and improve the agreement with the measuring results. For both compounds contributions of particular modes can be read off from the calculated spectra.

Figure 3 proves the existence of considerable molecular distortions in the polymer phase of Na_4C_{60}. This is concluded from the observed mode splitting in comparison to pristine C_{60}. A disappearing of the splitting for the $H_g(1)$ and the pentagonal pinch mode, as well as a huge increase in intensity for the $A_g(1)$ line is observed on heating (not shown). INS spectra give evidence for a larger charge transfer comparable to the A_3C_{60} phase (large spectral shifts around 40-50 meV). A down shift of the order of $\approx$40

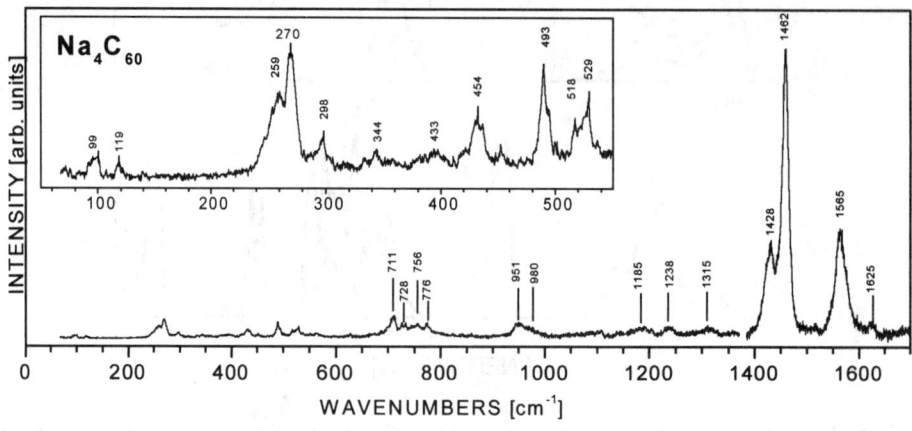

Figure 3. Raman spectrum of the single bonded 2D-polymer Na_4C_{60}. Distortions of the C_{60}-cages result in a symmetry reduction, hence mode splittings and a redistribution of intensities occur with respect to fcc-C_{60}.

cm^{-1} would be expected for the $A_g(2)$ pentagonal pinch mode. Although the Na_4C_{60} polymer is conducting (3) we do not observe indications for a strong electron-phonon coupling as for the A_3C_{60} phase (width of the $H_g(8)$ mode).

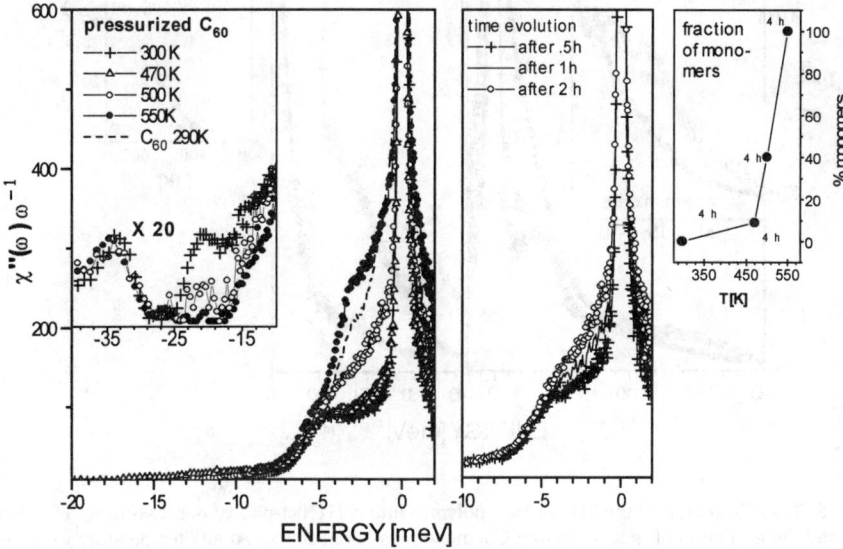

Figure 4. Thermal decomposition of the 2D-(pC$_{60}$). The breaking up of polymer bonds causes a disappearing of intermolecular modes (insert) and the rising up of a quasielastic scattering intensity which is due to a free rotation of C$_{60}$ monomers. The fraction of C$_{60}$ monomers is calculated from the measured intensities. The decomposition into fcc-C$_{60}$ is thermally activated (panel in the middle).

Finally the figures 4 and 5 show the decomposition of the polymers on heating. It can be seen for both samples that the modes connected with the formation of covalent intermolecular bonds disappear when the transition temperature is approached ($T_c \approx$ 500 K). In parallel a quasielastic component proving the free rotation of C$_{60}$ monomers arises. There are characteristic differences between both polymers. 2D-(pC$_{60}$) is metastable at lower temperatures and decomposes into fcc-C$_{60}$ in a thermally activated process. Estimations of an activation energy from the time and temperature dependent evolution of the quasielastic component give an order of $\approx$ 2 eV. The fraction of C$_{60}$ monomers has been deduced from the measured quasielastic intensities and is shown in the figures. For Na$_4$C$_{60}$ the transformation into the monomer phase (bct) is not thermally activated. The fraction of monomers at a particular temperature is stable and does not change with time (panel in the middle). Independent from the cooling rate from the HT bct-phase we find at 300 K identical spectra which are characteristic for the polymer (both INS and R experiments). Although the Na$_4$C$_{60}$ polymer is metallic, we do not observe an anomalous temperature behavior when the sample is cooled down to 80 K in difference to the A$_1$C$_{60}$ and A$_3$C$_{60}$ phases. Binding energies for both

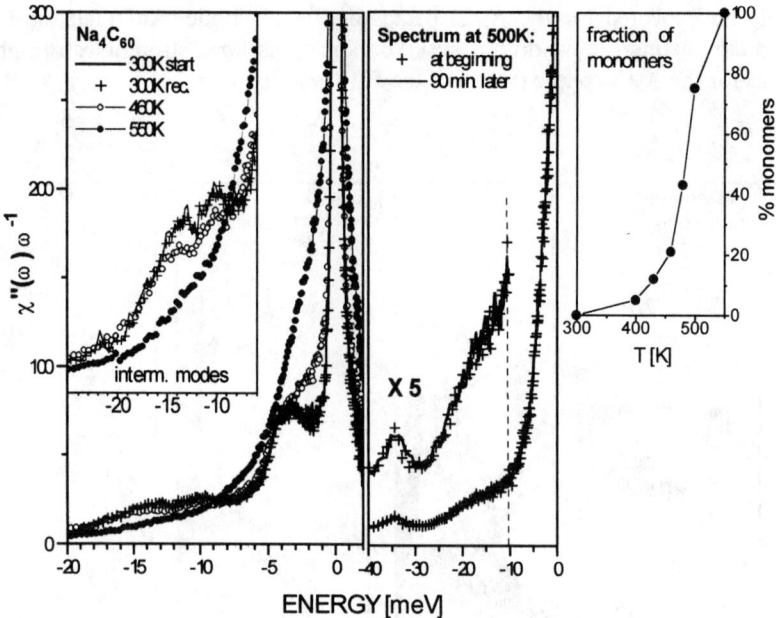

Figure 5. Transformation of the 2D-Na_4C_{60} polymer into a HT bct-phase. A breaking up of polymer bonds and the formation of freely rotating C_{60} monomers is registered. At any temperature we observe a stable ratio of molecules in the monomer and polymer phase (panel in the middle).

polymers have been deduced from DSC measurements: 23.8 kJ/mole for Na_4C_{60} and 25.2 kJ/mole for our 2D-(pC60) sample. Both values do not contain corrections for the rotational energy which is of the order of ≈ 5kJ/mole.

REFERENCES

1. Stephens, P.W., Bortel, G., Faigel, G., Tegze, M., Janossy, A., Pecker, S., Oszlanyi, G., Phys. Rev. Lett. **78**, 4438 (1997)
2. Nunez-Regueiro, M., Marques, L., Holdeau, J.L., Bethoux, O., and Perroux, M., Phys. Rev. Lett. **74**, 278 (1995)
3. Oszlanyi, G., Baumgartner, G., Faigel, G., Forro, L., Phys. Rev. Lett. **78**, 4438 (1997)
4. Schober, H., Tölle, A., Renker, B., Heid, R., and Gompf, F., Phys. Rev. **B56**, 5937 (1997)
5. Renker, B., Schober, H., Heid, R., v. Stein, P., Sol. State Comm. **104**, 527 (1997)

Fulleride Polymerisation at Ambient and Elevated Pressure

S. Margadonna[*], C. M. Brown[*†], A. Lappas[*], K. Kordatos[*],
K. Tanigaki[‡] and K. Prassides[*]

[*] *School of Chemistry, Physics and Environmental Studies, University of Sussex,
Brighton BN1 9QJ, U.K.*
[†] *Institute Laue Langevin, F-38042 Grenoble, France*
[‡] *Fundamental Research Laboratories, NEC Corporation, Tsukuba 305, Japan*

Abstract. The polymerisation of ternary and quaternary sodium fullerides, $Na_2Rb_{1-x}Cs_xC_{60}$ ($0 \leq x \leq 1$) is studied by neutron and synchrotron X-ray powder diffraction at both ambient and elevated pressure. All $Na_2Rb_{1-x}Cs_xC_{60}$ ($0 \leq x \leq 1$) salts, except Na_2CsC_{60}, polymerise at ambient pressure upon slow cooling. Na_2CsC_{60} polymerises only upon application of pressure, the cubic phase surviving to ~0.84 GPa. The monomer→polymer phase transition in Na_2RbC_{60} is followed in detail for both slow cooling and heating procedures, as well as a function of time at low temperature. High resolution powder neutron diffraction confirms the monoclinic ($P2_1/a$) structure of Na_2RbC_{60} at 2.5 K and allows the determination of the interball bridging C-C bond distance as 1.70(7) Å.

INTRODUCTION

At ambient pressure, ternary alkali fullerides $A_2A'C_{60}$ (A, A'= alkali metal) adopt various crystal structures. While K_3C_{60} and Rb_3C_{60} have a merohedrally disordered *fcc* (*Fm3m*) structure, Na_2AC_{60} (A= Rb^+, Cs^+) adopts an orientationally ordered primitive cubic (*pc, Pa3*) structure. The Na^+ ionic radius is smaller than the tetrahedral hole (~1.12 Å) and so the fulleride anions rotate in order to optimise the attractive and repulsive interactions. Polymerisation of C_{60} and its derivatives critically depends on the parent monomer structure, with the interball bonding depending on the charge state of the fullerene units [1,2]. For $(C_{60})_n$ and $(C_{60}^-)_n$, one-dimensional polymers are formed by [2+2] cycloaddition (*fcc* phase); for $(C_{60}^{3-})_n$, the formation of a single C-C covalent bond between the C_{60}^{3-} ions is favoured (*pc* phase). Synchrotron X-ray diffraction patterns of Na_2CsC_{60}, $Na_2Rb_{0.5}Cs_{0.5}C_{60}$, $Na_2Rb_{0.2}Cs_{0.8}C_{60}$, Na_2RbC_{60}, and Na_2KC_{60}, obtained at 200 K after slow cooling, show that polymer formation occurs in all systems, except Na_2CsC_{60} [3]. The latter remains strictly cubic on cooling, but the polymer is formed after applying a pressure ≥ 0.7 GPa. Here we present a summary of our study of the C_{60}^{3-} monomer to polymer transformation for Na_2RbC_{60} and Na_2CsC_{60}, as a function of temperature, pressure and time.

RESULTS

(i) Polymerisation as a function of temperature: Temperature-dependent synchrotron X-ray and neutron powder diffraction was used to study the monomer (pc) → polymer (monoclinic, $P2_1/a$) phase transition in Na_2RbC_{60} during both slow cooling and heating procedures. Neutron powder diffraction was performed on the D1b diffractometer, ILL, Grenoble [4]. The sample was cooled from 315 to 180 K at 5 K h^{-1} and the temperature was then fixed to 180 K for 12 hours before further cooling to ~2 K. Afterwards the sample was heated to room temperature at a rate of 16 K h^{-1}. Synchrotron X-ray powder diffraction patterns were collected on line A of the BM1 beamline at the ESRF, Grenoble with a 300 mm diameter Mar Research circular image plate. The sample was first cooled at 30 K h^{-1} from ambient temperature to 200 K, where it was kept for 8 hours. It was then heated to 310 K at the same rate. For both measurements, a $fcc \to pc$ transition is first seen in the temperature range 290-310 K. The percentage of polymeric phase starts to become significant at a temperature of 250 K (~10%) increasing up to ~53% on cooling to 200 K. On heating, hysteretic behaviour is observed with the monoclinic phase disappearing at ~270 K (Fig 1). The thermal expansivity (200- 270 K) along a, b, and c is found as $0.5(1.2) \times 10^{-6}$ K^{-1}, $1.79(9) \times 10^{-5}$ K^{-1} and $1.0(5) \times 10^{-6}$ K^{-1}, respectively, for the polymer, while it is $1.87(6) \times 10^{-5}$ K^{-1} for the primitive cubic phase.

A detailed structure determination of polymeric Na_2RbC_{60} was attempted from high resolution powder neutron diffraction data, collected at 2.5 K with the D2b diffractometer, ILL, Grenoble. Rietveld refinements were performed as a function of the rotation angle of the C_{60}^{3-} ions about the c axis. A minimum in R_{wp} (5.6%, R_{exp}= 5.2%) is found for a rotation angle of 82°; the fraction of polymeric phase is 70.4(7)% and the lattice parameters a= 13.703(3) Å, b= 14.464(3) Å, c= 9.372(2) Å, β= 133.63(1)°. The covalent bridge between the C_{60}^{3-} ions refines to a length of 1.70(7) Å and is tilted by ~7.7° to the c axis [4].

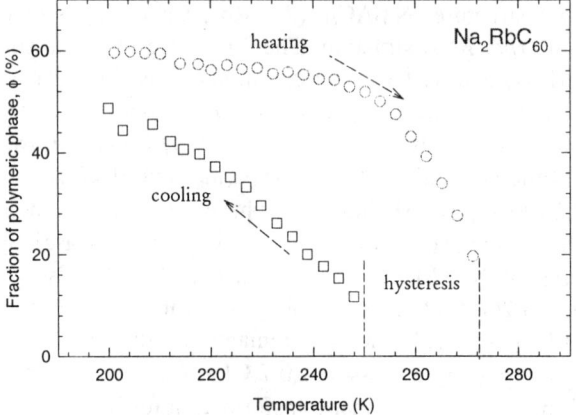

Fig 1: Temperature evolution of the fraction of polymeric phase on cooling and heating (synchrotron X-ray diffaction data)

Synchrotron X-ray powder diffraction patterns of Na_2CsC_{60} were collected at the ESRF under the same protocol used for Na_2RbC_{60}. The sample was cooled from 320 to 200 K at a rate of 60 K h^{-1}. As before, the *fcc→pc* phase is observed at ~290 K, but there is no evidence of the formation of the monoclinic phase down to 200 K. The thermal expansivity of the cubic phase is $2.20(6)\times10^{-5}$ K^{-1} between 200 and 265 K.

(ii) Polymerisation as a function of time: The fraction of polymeric phase, ϕ, reaches ~53% when the temperature is 200 K. After remaining for 7 hours at 200 K, ϕ has only increased by ~7% with its time evolution approximately described by a linear law: $\phi(t)$= 53.0(2) + 0.018(1)×t(/min) (Fig. 2). The extremely slow kinetics of the polymerisation reaction are clearly evident from these results and a complete transformation at this rate of the cubic to the polymer phase at 200 K would necessitate ~44 hours.

(iii) Polymerisation as a function of pressure: We have also studied the pressure dependence of the structure of solid Na_2CsC_{60} at ambient temperature up to 8.25 GPa by synchrotron X-ray powder diffraction experiments (ESRF, Grenoble). At ambient pressure, the structure is *pc* with a= 14.1329(3) Å (*Pa3*). When a pressure of 0.74 GPa is reached, the polymeric phase appears and coexists with the cubic one up to 1.05 GPa. The pressure evolution of the normalised monoclinic lattice constants a, b, and c, indicates that there is a substantial anisotropy in the compressibility along the three axes. The structure is least compressible along c which defines the polymeric chain and is most compressible along the interchain b direction. The pressure-volume curve for the monoclinic phase up to ~8.25 GPa has been fitted using the Murnaghan EOS (Fig 3). The average bulk modulus, K_o of the polymer is 28(1) GPa (compressibility, κ= $4.0(1)\times10^{-2}$ GPa^{-1}) with a pressure derivative of dK_o/dp = 11(1). The bulk modulus of Na_2CsC_{60} is slightly larger than that of pristine C_{60} due to the tighter crystal packing.

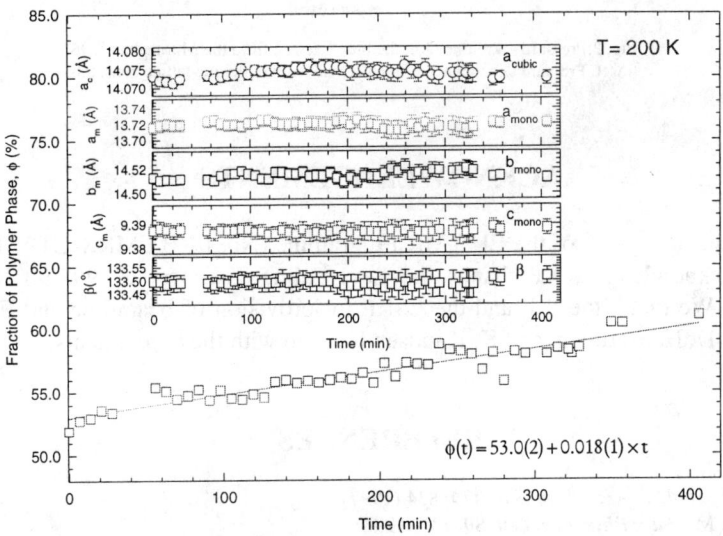

Fig. 2: Time evolution of the Na_2RbC_{60} polymeric phase fraction at 200 K.
Inset. Time evolution of the lattice constants.

CONCLUSIONS

Slow cooling of Na_2RbC_{60} results at ~250 K in the formation of a singly bonded polymer. More than half of the sample volume transforms to the polymer by 180 K. The polymer fraction increases by ~7% in 7 hours at 200 K. Hysteretic behaviour is observed on heating, the polymer disappearing at ~270 K. High resolution powder neutron diffraction at 2.5 K has been used to obtain information on the bridging geometry. The interball C-C bridge refines to 1.70(7) Å and makes an angle of ~7.7° with the c axis. Na_2CsC_{60} does not polymerise on cooling − it only shows the $fcc{\rightarrow}pc$ phase transition at ~ 290 K. However, it polymerises under pressure with the monoclinic phase appearing at ~0.74 GPa. From the equation of state (EOS), we find an average bulk modulus, K_o= 28(1) GPa with a pressure derivative of dK_o/dp= 11(1).

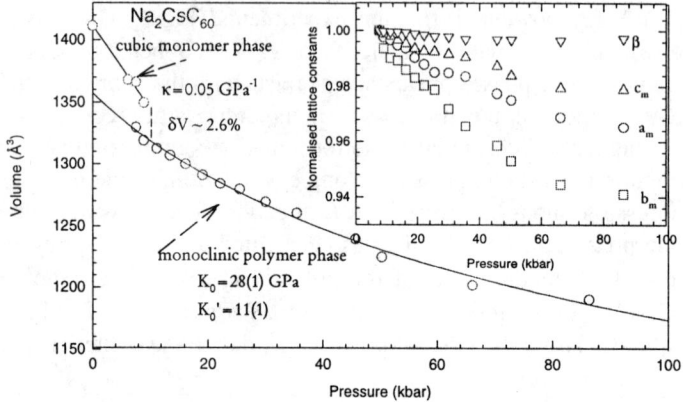

Fig. 3: Pressure - Volume plot: the blue line is a fit to the Murnaghan EOS.
Inset. Pressure dependence of the normalised monoclinic lattice constants.

ACKNOWLEDGEMENTS

Financial support of the EU TMR Programme (No. ERB FMRX-CT97-0155) is gratefully acknowledged. KP thanks the Leverhulme Trust for a 1997-98 Research Fellowship. We thank the ILL and the ESRF for provision of beamtime and E. Suard, A. N. Fitch, D. Häusermann and K. Knudsen for help with the experiments.

REFERENCES

1. Prassides, K. *et al.*, *J. Am. Chem. Soc.* **119**, 834 (1997).
2. Bendele, G. M. *et al.*, *Phys. Rev. Lett.* **80**, 736 (1998).
3. Prassides, K. *et al.*, *Physica C* **282**, 307 (1997).
4. Brown, C. M. *et al.*, submitted; Lappas, A. *et al.*, submitted.

First-Principles Study of Polymerized Alkali-Fullerene Compounds

T. Ogitsu*, K. Prassides[†], K. Tanigaki[‡], K. Kusakabe*, and S. Tsuneyuki*.

*Institute for Solid State Physics, University of Tokyo, Roppongi, Minato-ku, Tokyo 106, Japan
[†]School of Chemistry, Physics and Environmental Science, University of Sussex, Falmer, Brighton BN1 9QJ, United Kingdom
[‡]Fundamental Research Laboratories, NEC, 34-Miyukigaoka, Tsukuba 305 Japan

Abstract

We have studied the stable geometry of monoclinic Na_2RbC_{60} polymer by means of local density functional (LDF) calculations. Our optimised geometry shows an Rwp value comparable with the Rietveld models. However, the positions of the Na atoms are found to be off the tetragonal sites. The calculated bonding configuration around the bridging atom is much closer to an ideal sp^3 than that in the experimental geometry. The charge distribution is compared with that of $(C_{59}N)_2$, and the charge concentration the site comparable with nitrogen is found to be smaller.

1 Introduction

Alkali-doped fullerene polymers show variety in the bridging geometry[1][2] [3]. One-dimensional[1] as well as two-dimensional connections[2] between fullerenes are found as global as well as local stable structures of the alkali-fullerene compounds. A striking fact of this polymer family is that the charge state of the fullerene is apparently correlated with its bridging geometry. The A_1C_{60} (A=K, Rb), and the Na_2RbC_{60} polymers have one-dimensional interconnections, while the Na_4C_{60} polymer has two-dimensional interconnections. Even in the case of the one-dimensional polymers, we can find a remarkable difference between the A_1C_{60} (A=K, Rb) and the Na_2RbC_{60} polymers; the former are known to have two covalent bonds between the fullerenes[1], while the latter is deduced from its larger inter-molecular separation[2] to have one covalent bond. These dramatic differences in bonding which are controlled by the charge states give rise to the question as how the excess electrons on the fullerenes affect the polymer geometry. Recently, Ogitsu et al.[5] studied the band structure as well as the stable atomic configuration of the ortho-KC_{60} polymer by the local density functional (LDF) calculation method. They showed that the LUMO (t_{1u}) character of I_h C_{60} survives the polymerization, and the LUMO states show clear inter-chain bonding/anti-bonding character. In this study, we present a first-principles study of the monoclinic-Na_2RbC_{60} polymer.

2 Method

The calculation method is based on the LDF[4]. A plane-wave basis expansion and optimised pseudopotentials in separable form were adopted in this simulation. To obtain as precise a geometry as possible, we have examined the convergency of the geometry as a function of the cutoff energy for plane waves ranging from 30Ry to 80Ry with uniform sampling of either one (Γ) or eight k-points. The geometry was relaxed until all the forces acting on the atoms became less than 5×10^{-4} [Hartree/a_o].

3 Results and Discussion

Structural optimisation was performed on the Na_2RbC_{60} polymer, which has a monoclinic unit cell containing two C_{60} molecules, four Na atoms, and two Rb atoms. The cell parameters were fixed at the experimental values[1]. The Rietveld geometry derived from X-ray diffraction experiments was used as the initial atomic configuration[1]. The structural optimisation was then performed without symmetry constraint.

The optimised structure breaks the initial P1 21/a1 symmetry only slightly. The maximum deviation from the atomic configuration given by symmetry operations is negligibly small, 0.008Å. Except for the Na positions, our optimised geometry is consistent with the results from the Rietveld model; the Rwp value for the present geometry is 4.16%. In the original Rietveld geometry, the closest Na-C distance is 2.509Å, while our optimised geometry has a closest Na-C distance of 2.635Å. The Na atom is located off the tetragonal site by (-0.01587, 0.01570, -0.02413) in coordinates relative to the unit cell. In the optimised geometry, we find other small changes from the Rietveld geometry which seem to be physically reasonable. If we look at the bonding configuration around the bridging atom, we will see that the sp^3 character is strengthened in comparison with the experimental geometry. In the experimental geometry, we see considerable differences in bond lengths as well as bonding angles from an ideal sp^3 configuration such as occurs in diamond, 1.545Å

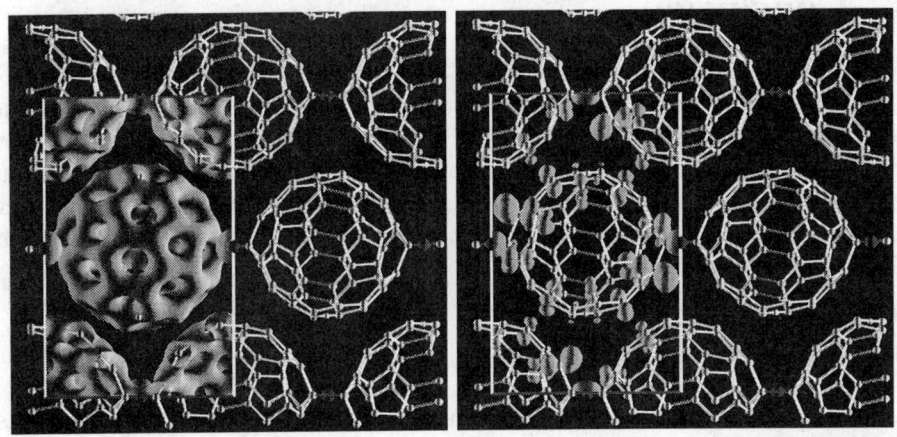

Figure 1: The total charge density (left side) and the density of the state corresponding to the HOMO of $(C_{59}N)_2$ (right side) is shown as isosurfaces, while the fullerenes are represented with ball and stick models. The viewpoint is normal to the chain. The iso-levels are set at 80/255 (left) and 20/255 (right) of the maximum values.

and 109 degrees. In our optimised structure, both the bond lengths and the bonding angles are slightly closer to that of the ideal sp^3. The bond lengths of C4-C3, C4-C4, C4-C12, and C4-C14, in the Rietveld model are, 1.5077Å, 1.5500Å, 1.5640Å, and 1.5687Å, respectively, while those in our optimised geometry are, 1.5297Å, 1.5473Å, 1.5489Å, and 1.5506Å, respectively. Here, C3, C4, and C12/C14 correspond to the hexagon-hexagon corner, bridging, hexagon-pentagon corner atoms, respectively. The bonding angles around bridging atom, C4-C4-C12, C4-C4-C3, C4-C4-C14, C3-C4-C12, C3-C4-C14, and C12-C4-C14, in Rietveld model are, 115.0, 115.1, 115.0, 106.4, and 106.4 degrees, respectively, while those in optimised geometry are, 111.9, 114.0, 112.9, 108.9, 108.7, 97.0, and 99.4 degrees, respectively.

Figure 1 depicts the total charge density of Na_2RbC_{60}. If we compare this charge density with that of $(C_{59}N)_2$[6], we find that the charge concentration on the nitrogen atoms of $(C_{59}N)_2$ is much higher than that of our polymer. However, this does not mean an absence of dangling bond states; if we look at the state corresponding to the HOMO of $(C_{59}N)_2$, clear dangling bond character is seen around the C3 atoms. The difference in the charge concentration arises from the difference in the number of valence electrons belonging to nitrogen and carbon.

4 Conclusion

The stable geometry of monoclinic Na_2RbC_{60} is studied by LDF calculations. The resultant structure shows an Rwp value comparable with that from a recent experiment[1]. We found considerable differences in the equilibrium positions of the Na atoms compared with the experimental geometry. The calculated bonding configuration around the bridging carbon atom is closer to an ideal sp^3 than that in the experimental geometry. We did not find the charge concentration seen in $(C_{59}N)_2$[6], but we did find an analogous dangling bond state.

5 Acknowledgements

The authors thank Professor Iwasa and Dr. Suzuki for stimulating discussions. T.O. thanks Dr. T. M. Briere for critical reading of this manuscript. T.O also thanks Prof. M. Fujita for leading him to this work. This work was performed on Fujitsu VPP500 supercomputers at the super computer center, Institute for Solid State Physics, Univ. of Tokyo and at the computer center, Kyushu University, and on the Hitach SR2201 at the computer centre, University of Tokyo. This work was conducted as part of the JSPS Research for the Future Program in the Area of Atomic-Scale Surface and Interface Dynamics.

References

[1] P. W. Stephens, G. Bortel, G. Faigel, M. Tegze, A. Jánossy, S. Pekker, G.Oszlányi and L. Forro, *Nature* **370**, 636 (1994)., O.Chauvet, G. Oszlányi, L. Forro, P. W. Stephens, M. Tegze, G. Faigel, and A. Jànossy, Phys. Rev. Lett. **72**, 2721(1994).

[2] K. Prassides, K. Vavekis, K. Kordatos, K. Tanigaki, G. M. Bendele, and P. W. Stephens, J. Am. Chem. Soc. **119**, 834(1997)., G. M. Bendele, P. W. Stephens, K. Prassides, K. Vavekis, K. Kordatos, and K. Tanigaki, Phys. Rev. Lett. **80**, 736(1998).

[3] G. Oszlányi, G. Baumgartner, G. Faigel, and L. Forró, Rhys. Rev. Lett. **78**, 4438(1997)

[4] M. C. Payne, M. P. Teter, D. C. Allen, T. A. Arias, and J. D. Joannopoulos, Rev. Mod. Phys. **64**, 1045(1992), and references therein.

[5] T. Ogitsu et al., submitted elsewhere.

[6] W. Andreoni, A. Curioni, K. Holczer, K. Prassides, M. Keshavarz-K., J.-C. Hummelen, and F. Wudl, J. Am. Chem. Soc. **118**, 11335(1996).

Two dimensional fulleride polymers

Mikael Christensen and Sven Stafström

Department of Physics and Measurement Technology, IFM, Linköping University, S-581 83, Linköping, Sweden

Abstract. A tetramer model system of the single bonded two-dimensional fulleride polymer phase was investigated theoretically. The optimized geometrical structure showed that the sp^3 hybridized carbon atoms forming the bonds between adjacent C_{60} molecule are pairwise identical as a consequence of the different polygon sequences between these carbon atoms. The bonding nature of the molecular orbitals of the model system shows that a charging of four electrons per C_{60} molecule stabilize the two-dimensional polymeric phase, other values of the charging destabilizes this phase. This charging also results in a split of the t_{1u} level of C_{60} and an electronic gap is created around the Fermi energy. The dispersive nature of the highest occupied band is, however, expected to reduce this gap for an increasing size of the system.

INTRODUCTION

Studies of polymeric phases of A_xC_{60} fullerides (A=Alkali metal) have shown that the type of covalent interfullerene bonding depend on the charge of the monomer ions. For instance, the [2+2] cycloadduct polymer[1,2] is the most stable phase for weakly charged monomers, while for an increasing charge, the formation of a single bonded polymer can be expected.[2]

The two-dimensional polymer Na_4C_{60} was the first fulleride polymer with singly bonded monomers to be identified.[3] By a study of the enthalpy of polymerization, this stoichiometry was also confirmed to be the most stable for the quadruple charged monomer.[2,4] The two-dimensional singly bonded polymer is formed in a body centered monoclinic structure.[3] The lattice parameters of the monoclinic unit cell are a=11.24 Å, b=11.72 Å, c=10.28 Å and $\beta = 96.2°$. The nearest interfullerene distance within the plane of the polymer is 9.28 Å, and the intra planar angle is 78.3°.

In this work, geometrical and electronic properties of this two-dimensional polymer are studied theoretically. The focus of our studies is on the bonding nature of the highest occupied molecular orbitals (MO's) and the possibility to form a metallic state.

METHODOLOGY

To model the polymer, a system of four C_{60} molecules was used (see Fig. 1). The system was terminated with tert-butyl groups in order to get a structure with four sp^3

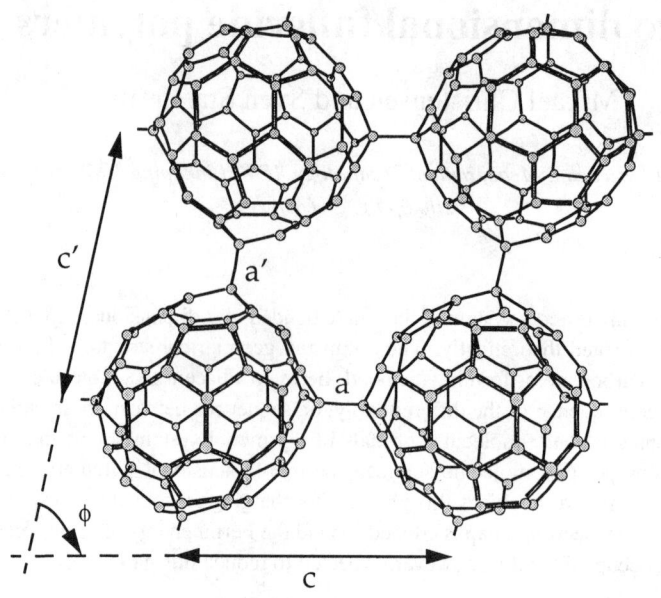

FIG. 1: Schematic structure of the C_{60} tetramer

atoms on each C_{60} molecule. The interfullerene bonding break the icosahedral symmetry of the C_{60} molecule.

Geometrical optimization of the structure was carried out using the AM1 Hamiltonian.[5] No symmetry restrictions or constrains on the atomic coordinates were used but the structure converged towards the expected S_2 symmetric state. The geometry optimizations of the tetramer were performed on a neutral system. To investigate the effect of charging on the system we performed dimer calculations (terminated with tert-butyl groups) on neutral and charged systems with and without counterions. The intermolecular $C-C$ bond lengths (a and a', see below) obtained for the neutral dimer are within 0.001 Å from those of the tetramer, therefore, reducing the system in this way is a minor source of error. The fact that most of the structural constants have converged already for the dimer also indicates that the tetramer is a very good model system for the infinite polymer, at least as far as the structure is concerned. By charging the dimer with 8 electrons the intermolecular bond lengths increase by 0.003 Å. The change introduced by the sodium counter ions (in tetrahedral lattice sites according to the crystallography data) is less than one thousand of an Ångström compared to the charged state. These results show that the geometry is quite insensitive to charging and that the geometry of the neutral system is a good approximation to the structural properties of Na_4C_{60}.

The electronic structure is obtained from Valence Effective Hamiltonian (VEH)[6] calculations. The method of performing VEH calculations using the AM1 optimized geometry is well established and has been proven to be very reliable.[7]

RESULTS AND DISCUSSION

The structure of the tetramer is shown in Fig. 1 and the corresponding optimized geometrical constants are given in Table 1. Clearly, both the interfullerene angle and the interfullerene separations are very close to the experimentally determined values. Note also that the interfullerene bond length a is ~ 0.008 Å shorter than in the direction given by the bond a'. This is an effect of the different polygon sequences between the sp^3 carbon atoms on the C_{60} molecule. Each sp^3 carbon is sited in the vertex of two hexagons and a pentagon. The hexagon-hexagon bonds adjacent to the a-bond point perpendicular to the plane of the polymer whereas for the a' bond they are oriented in this plane. Thus, the two-dimensional single bonded C_{60} polymer is slightly anisotropic with respect to the two axes shown in Fig. 1.

	a [Å]	a' [Å]	c [Å]	c' [Å]	α [°]
calc. values (AM1)	1.537	1.545	9.238	9.320	79.049
exp.values			9.28		78.3

TABLE 1: Optimized geometrical constants (Fig. 1).

MO	Energy (eV)	orbital character	
9	-4.925	Anti-bonding	Delocalized
8	-5.332	Bonding	Delocalized
7	-5.408	Bonding	Delocalized
6	-5.474	Non-bonding	Localized
5	-5.559	Non-bonding	Localized
4	-5.571	Non-bonding	Localized
3	-5.595	Non-bonding	Localized
2	-5.711	Anti-bonding	Delocalized
1	-5.728	Anti-bonding	Delocalized
0	-6.230	Bonding	Delocalized

TABLE 2: Character of molecular orbitals in the C_{60} tetramer system.

From VEH calculations we obtain the electronic structure of the tetramer shown in Fig.1. The corresponding MO's and their character are shown in Table 2. The Na_4C_{60} polymer has 4 excess electrons per C_{60} molecule. The tetramer model system of this polymer should therefore have 16 excess electrons occupying the 8 lowest unoccupied MO's of the neutral system (MO's 1 to 8 in Table 2, MO 0 corresponds to the highest occupied MO in the neutral system). The two lowest of these 8 MO's (MO 1 and 2) are observed to be anti-bonding in the interfullerene bonds. Thus, by adding 4 electrons (instead of 16) to the neutral system, i.e., one for each C_{60} molecule, the two dimensional polymer get destabilized. This is a clear indication that this type of structure will not be formed for the A_1C_{60} compound. Instead, it is well known that the [2+2] cycloadduct is the stable phase for this compound.

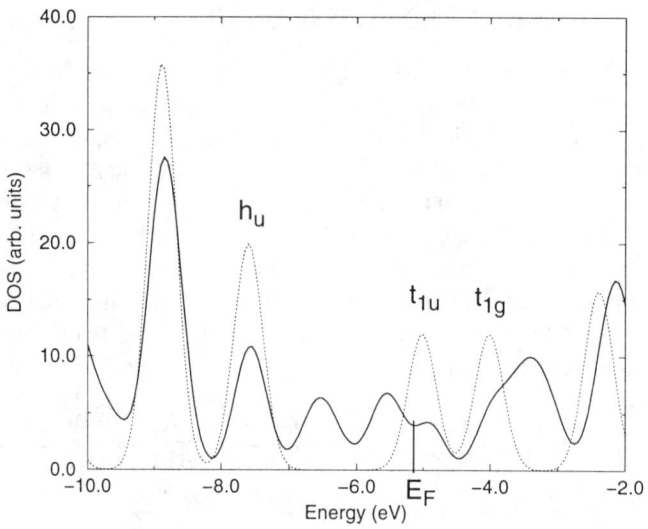

FIG. 2: Density of states (DOS) of the tert-butyl terminated tetramer (solid line) and of C_{60} (thin dotted line, scaled by a factor of four).

Occupying the succeeding four nearly degenerate non-bonding orbitals will not have any significant effect on the stability. Therefore, neither the A_2C_{60} nor the A_3C_{60} compounds stabilize in the two-dimensional polymer phase. However, filling of the next two MO's (MO 7 and 8) leads to a bonding contribution to the interfullerene bonds. This shows clearly that the only stoichiometry that get stabilized in the two-dimensional polymeric phase is A_4C_{60}, in complete agreement with experimental observations.[3] Furthermore, the anti-bonding MO's above the bonding orbitals imply that there is a decrease in the stability of the two-dimensional polymer for the A_5C_{60} compound.

The relation between the structure and the character of the MO's is a very direct method to find out the stability of the polymeric system. In fact this method can be used as a general tool to predict the stability of many different polymeric phases of C_{60}.[8]

The valence band density of states (DOS) is shown in Fig. 2. The DOS is obtained from the VEH eigen energies broadened with Gaussian functions (FWHM=0.2 eV). For comparison, the DOS of a single C_{60} molecule is also included in the figure. Away from the Fermi energy, the two curves are in rather close agreement. The h_u, t_{1u}, and t_{1g} levels, however, are strongly modified due to the distortions caused by the intermolecular bonds. By analyzing the various levels we find that there are 8 states that split off from the h_u level and appear as a new peak around -6.5 eV. These states are best characterized as surface states since they are associated with mixing between the π-system of the fulleride and the terminating tert-butyl groups. In the infinite polymer such surface states will not exist and consequently, the large destabilizing effect on the h_u level that we see in this calculation will be considerably smaller in that case.

The splitting of the t_{1u} level is an effect of the interfullerene bonds. In the manifold of the 12 t_{1u}-like states of the tetramer, 8 are occupied (see Table 2 above) in the A_4C_{60} stoichiometry and shift towards lower energies (peak at -5.6 eV in Fig. 2). The gap to the first unoccupied state is 0.41 eV, which is considerably larger than the average level spacing of 0.05 eV. However, we note that the highest occupied states are delocalized and the should give rise to a dispersive band. This will lead to a decrease in the band gap as the size of the system is increased. It will also prevent Mott localization since the band width is certainly larger in this case than in non-polymeric fullerides. The appearance of a dispersive valence band in the two-dimensional polymer is in contrast to the [2+2] cycloadduct bonded polymer, for which the orbitals around the Fermi energy are non-bonding, i.e., they localize away from the interfullerene bonds, and give rise to a very narrow valence band.[9]

Whether or not the dispersive nature of the highest occupied MO of the two-dimensional polymer is strong enough to lead to a metallic state is not clear from this type of study but we emphasize that the details of the electronic band structure has to be considered in studies focussed on the metal/insulating behavior of fulleride polymers.

ACKNOWLEDGMENTS

Financial support from the Swedish Research Council for Engineering Science (TFR) and the Swedish Natural Science Research Council (NFR) is gratefully acknowledged.

REFERENCES

[1] O. Chauvet, G. Oszlányi and L. Forró, Phys. Rev. Lett. **72**, 2721 (1994).
[2] S. Pekker, G. Oszlányi and G. Faigel, *Structure and stability of covalently bonded polyfulleride ions in A_xC_{60} salts*, unpublished.
[3] G. Oszlányi, G. Baumgartner, G. Faigel and L. Forró, Phys. Rev. Lett. **78**, 4438-4441 (1997).
[4] M. Christensen, *Structural and electronic properties of two-dimensional fullerene polymers*, Master's Thesis, Linköping (1998).
[5] M. J. S. Dewar, E.G. Zoebisch, E.F. Healy and J.J.P. Stewart, J. Am. Chem. Soc. **107**, 3902 (1985).
[6] J.-M. André, L. A. Burke, J. Delhalle, G. Nicolas and P. Durand, Int. J. Quantum Chem. Symp. **13**, 283 (1979).
[7] J.-M. André, J. Delhalle and J.-L. Brédas, *Quantum Chemistry Aided Design of Organic Polymers* (World Scientific, Singapore, 1991).
[8] S. Stafström, unpublished.
[9] J. Fagerström and S. Stafström, Phys. Rev. B **53**, 19 (1996).

Ferromagnetic Resonance and High Field ESR in a TDAE-C_{60} Single Crystal

D. Arčon, P. Cevc, A. Omerzu, and R. Blinc
J. Stefan Institute, Jamova 39, 1000 Ljubljana, SLOVENIA

M. Mehring, S. Knorr, and A. Grupp
2. Physikalisches Institut, Universität Stuttgart, Pfaffenwaldring 57, 70550 Stuttgart, GERMANY

A.-L. Barra and G. Chouteau
Laboratoire des Champes, CNRS, 25 Av. des Martys, BP 166, Grenoble, Cedex 9, FRANCE

Frequency variable ESR measurements have been performed on well annealed TDAE-C_{60} single crystals between 40 MHz and 245 GHz. A non-linear variation of the electron resonance frequency with the magnetic field has been observed below T_C=16 K in the radio-frequency region. The observed ferromagnetic resonance data are characteristic for a three-dimensional Heisenberg ferromagnet with a small positive uniaxial anisotropy field. The easy axis coincides with the crystal c-direction which is the direction of closest approach of the C_{60}^- ions.

Different models such as itinerant ferromagnetism, superparamagnetism, spin glass behavior, spin canted weak ferromagnetism as well as simple Heisenberg ferromagnetism have been proposed[1,2] to account for the magnetic transition in TDAE-C_{60} at T_C=16 K. In order to discriminate between the above models, we decided to perform frequency variable electron spin resonance (ESR) measurements on well annealed TDAE-C_{60} single crystals in the frequency range between 40 MHz and 245 GHz at different orientations and temperatures.

The ESR spectra recorder at $\vec{a} \parallel \vec{H}$ and T=5 K are shown in Fig. 1 for 40 MHz, 50 MHz, 65 MHz, 80 MHz, 95 MHz, 105 MHz, 140 MHz, 200 MHz, 1.1994 GHz, 9.6 GHz, 94.08 GHz, and 245 GHz. A zero field gap and a non-linear variation of the resonance frequency versus magnetic field (Fig. 2b) is seen below T_C for $\vec{a} \parallel \vec{H}$ in the radiofrequency region.[3] Above T_C, i.e. at T=20 K, this non-linear behavior disappears[3]. The resonance frequency versus magnetic field relation here becomes a straight line passing through the origin (Fig. 2c). For $\vec{c} \parallel \vec{H}$ the resonance frequency versus magnetic field relation is nearly linear. There is however for T<T_C again a zero field gap in the frequency region which disappears above T_C (Fig. 2d).

The above data are characteristic of an easy axis Heissenberg ferromagnet with an exceptionally small anisotropy field. The dip in the resonance frequency versus field relation occurs when $H=2H_K$. The easy axis coincides with the crystal c-direction which is also the direction of closest approach of the C_{60}^- ions. The present data also explain the high field ESR spectrum[4], where the external field is much larger than the anisotropy field so that the resonance frequency linearly increases with increasing field

(Fig. 2a) approaching an g=2 line. The angular dependence of the high field ESR spectra[2] can be as well explained by this model.

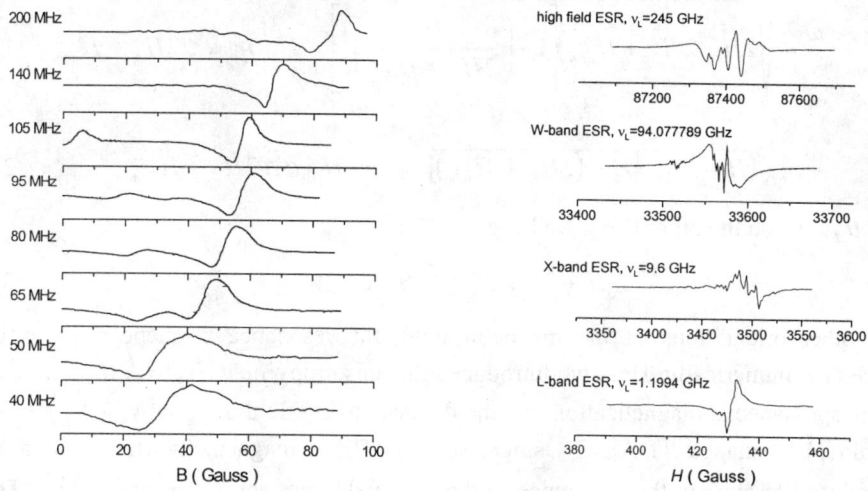

Figure 1: Electron magnetic resonance spectra of a well annealed TDAE-C_{60} single crystal at $\vec{a} \parallel \vec{H}$ and T= 5 K<T_C for different frequencies and $\vec{H}_1 \perp \vec{H}$. At 245 GHz the magnetic field was swept between 0 and 10 T but only the paramagnetic like g=2 line was detected around 8.7 T. Note the additional fine structure which is absent in the low field ESR spectra and which may be indicative of standing spin wave resonances.

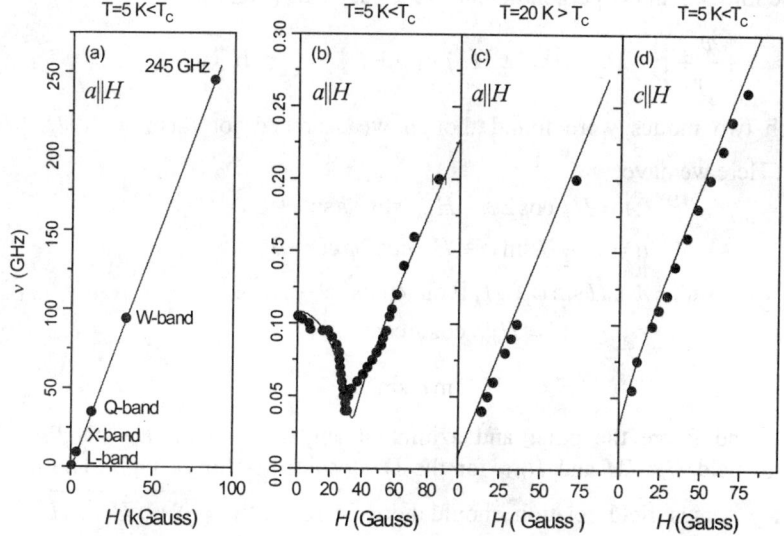

Figure 2: Experimental frequency versus field relation for TDAE-C_{60} in the radiofrequency and the high field regions for $\vec{H}_1 \perp \vec{H}$.

341

For $\vec{a} \parallel \vec{H}$, i.e. for a magnetic field applied perpendicularly to the easy axis ($\theta_H = \pi/2$) the ferromagnetic frequency is for $K>0$:

$$\frac{\omega}{\gamma} = \sqrt{2H_K(2H_K + H_{dem})\left[1 - \left(\frac{H}{2H_K + H_{dem}}\right)^2\right]} \qquad H \parallel a \leq 2H_K + H_{dem} \quad (1)$$

$$\frac{\omega}{\gamma} = \sqrt{(H - H_{dem})(H - (2H_K + H_{dem}))} \qquad H \parallel a \geq 2H_K + H_{dem} \quad (2)$$

For $\theta_H = 0$, on the other hand, we have

$$\frac{\omega}{\gamma} = H + 2H_K \qquad H \parallel c \quad (3)$$

For other orientations of the magnetic field, the resonance frequency has to be calculated numerically. Here we introduced the anisotropy field $H_K = KM$, where M is the spontaneous magnetization and the demagnetizing field $H_{dem} = (N_\parallel - N_\perp)M$ for a thin plate shaped TDAE-C_{60} single crystal. The demagnetizing field has to be introduced to explain the difference in the zero field gaps for $\vec{a} \parallel \vec{H}$ and $\vec{c} \parallel \vec{H}$. The data of Fig. 2 can be fitted with the above expressions (1) and (2) and with H_K=29 G and H_{dem}=-39 Gauss.

It should be stressed that the above data specifically exclude the previously proposed model of spin canted weak ferromagnetism[2]. Two resonance modes ω_1 and ω_2, one of which should show a dip around the spin flop field, should be observed in this case for $K>0$ and H perpendicular to the easy axis (Fig. 3):

$$\left(\frac{\omega_{1,2}}{\gamma}\right)^2 = (B \mp A)(C \pm H_E) - (G \mp F)^2 \qquad H \parallel a \quad (4)$$

No such two modes were found though we searched for them with $\vec{H}_1 \parallel \vec{H}$ and $\vec{H}_1 \perp \vec{H}$. Here we have

$$A = H_E \cos 2\varphi + H_{DM} \sin 2\varphi \sin \vartheta \quad (5a)$$
$$B = A + H \sin \varphi + H_K \cos 2\varphi \cos^2 \vartheta \quad (5b)$$
$$C = A + H \sin \varphi + H_K (\cos^2 \varphi \cos^2 \vartheta - \sin^2 \vartheta) \quad (5c)$$
$$G = H_{DM} \cos \varphi \cos \vartheta \quad (5d)$$
$$F = \frac{1}{2} H_K \sin \varphi \sin 2\vartheta \quad (5e)$$

where θ and φ are the polar and azimuthal angles and with H_E standing for the exchange field H_E=JM and H_{DM} for the Dzaloshinskii-Moriya field. The dip in the frequency versus field relation should here occur at $H = \sqrt{2H_K H_E - H_{DM}^2}$. This mode should be observable for $\vec{H}_1 \parallel \vec{H}$ whereas we observed a dip for $\vec{H}_1 \perp \vec{H}$. The observed anisotropy in the high frequency ESR is as well much too small to be consistent with the spin canted model.

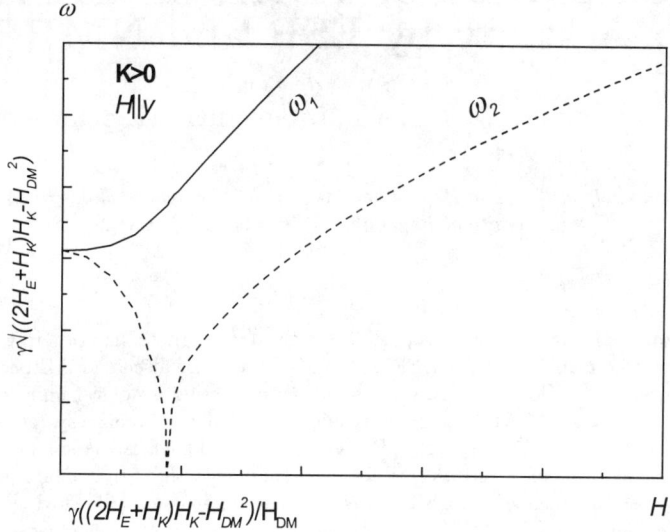

Figure 3: Theoretical frequency versus field relations for a weak spin canted ferromagnet with K>0 for $\theta_H = \pi/2$. The full line represents the mode which can be excited for $\vec{H}_1 \perp \vec{H}$ whereas the dotted line represents the mode excitable with $\vec{H}_1 \parallel \vec{H}$.

It should be noticed that our data also exclude an antiferromagnetic ground state. In this case two resonance modes should exist for all orientations of the crystal in the magnetic field as long as $H < \sqrt{2H_E H_K}$, i.e. as long as it is smaller than the spin flop field.

The above data do not however exclude the existence of a mixed ferromagnetic state[4,5] with non-zero random interactions and non-zero random fields as long as they are smaller than the mean exchange field.

References:

1. P. M. Allemand, K. C. Khemani, A. Koch, F. Wudl, K. Holczer, S. Donovan, G. Gruner, and J. D. Thompson, *Science* **253**, 301 (1991).
2. R. Blinc, K. Pokhodnia, P. Cevc, D. Arčon, A. Omerzu, D. Mihailović, P. Venturini, L. Golič, Z. Trontelj, J. Lužnik, Z Jagličič, and J. Pirnat, *Phys. Rev. Lett.* **76**, 523 (1996).
3. D Arčon, P. Cevc, A. Omerzu and R. Blinc, *Phys. Rev. Lett.* **80**, 1529 (1998).
4. R. Blinc, D. Arčon, P. Cevc, A. Omerzu, M. Mehring, S. Knorr, A. Grupp, A.-L. Barra, and G. Chouteau, to be published.
5. A. Lappas, K. Prassides, K. Vavekis, D. Arčon, R. Blinc, P. Cevc, A. Amato, R. Feyerherm, F. N. Gygax, and A. Schenck, *Science* **267**, 1799 (1995).

Perdeuteration of TDAE in [TDAE]C_{60}: A study by ESR and NMR

A. Schilder, W. Bietsch, J. Gmeiner, M. Schwoerer

Experimentalphysik II and Bayreuther Institut für Makromolekülforschung(BIMF)
University of Bayreuth, 95440 Bayreuth, Germany

Abstract. The mechanism responsible for the feromagnetic-like ordering at 16 K of the organic compound [TDAE]C_{60} is still a matter of discussion. Especially, the consequence of the donor TDAE for the magnetic ordering is not yet understood. All attempts to replace TDAE by similar organic donors failed in giving a system with such an outstanding magnetic property. Perdeuteration of TDAE is the smallest possible change of the system. AC-susceptibility measurements show that deuteration does not suppress the magnetism. ESR and NMR measurements on [TDAE-d_{24}]C_{60} give a further evidence that TDAE$^+$ is involved in electron spin exchange processes.

I INTRODUCTION AND SYNTHESIS

Perdeuteration of TDAE should be a very weak distortion of the [TDAE]C_{60} without changing the unusual magnetic properties. On the other hand it is known that perdeuteration could initiate or suppress phase transitions [1] because of slight changes in lattice constants which motivates to investigate [TDAE-d_{24}]C_{60}.
Perdeuterated TDAE was synthesized in a way described by Wiberg and Buchler [2]. The reaction time of dimethylamine-d_6 with chlorotrifluoroethene was about 8 hours at a temperature of 50 °C up to 80 °C. After a microdistillation to separate unwanted reaction products only three drops of TDAE-d_{24} were gained. [TDAE-d_{24}]C_{60} was precipitated as microcrystalline powder out of solution.

II AC-SUSCEPTIBILITY, ESR AND NMR

The in-phase component of the ac-susceptibility is plotted in figure 1 for both [TDAE-d_{24}]C_{60} and [TDAE]C_{60} . Obviously the magnetism is not suppressed by perdeuteration and the transition temperature shifts by less than 0.5 K as is best seen in the derivative (figure 1, inset). However, a small decrease of the peak susceptibility is observed. Due to the very small batch size the purification of TDAE-d_{24} was not perfect and could be the reason for the slightly smaller signal.
Results of temperature dependent electron spin resonance at 34 GHz are shown in

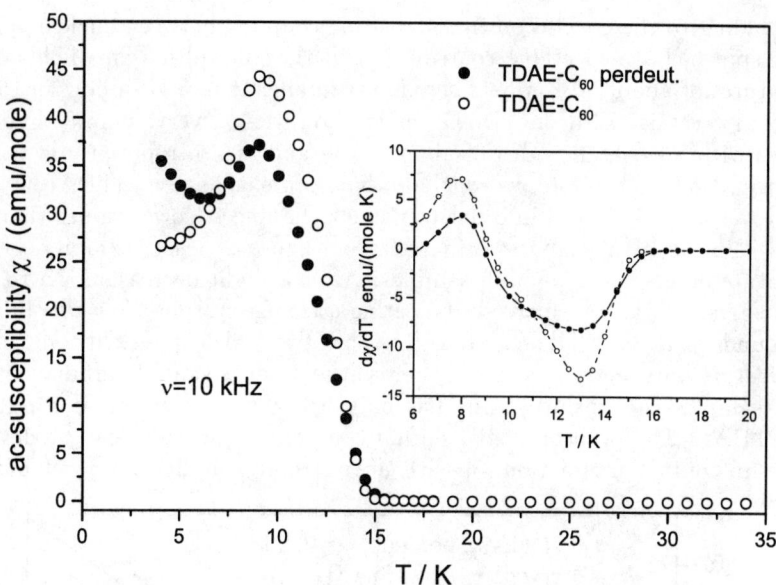

FIGURE 1. AC-susceptibility χ of [TDAE]C_{60} and [TDAE-d_{24}]C_{60} at 10 kHz; the inset shows the derivative $d\chi/dT$.

figure 2 for both samples. One would expect that the ESR-linewidth is smaller in the perdeuterated sample as the magnetic moment of deuterons is less by a factor of 6.5 compared with protons. However, no significant narrowing of the ESR-line

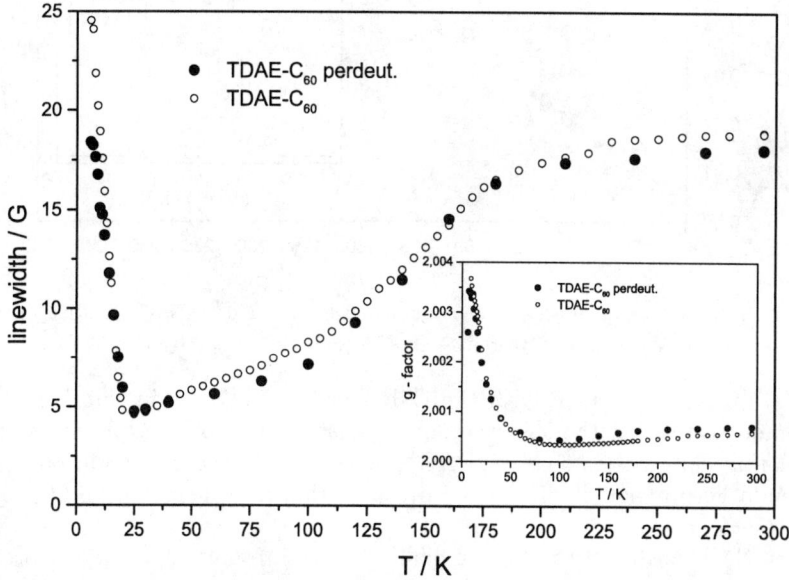

FIGURE 2. ESR-linewidth of [TDAE]C_{60} and [TDAE-d_{24}]C_{60} in Q-band. *inset:* g-factor

is seen leading to the conclusion that hyperfine coupling between nuclear spin and electron spin has a neglectable contribution to the linewidth. This holds also at temperatures of about 30 K where the line is smaller than 5 G in powder samples and 3 G in crystals. ESR on TDAE$^+$ in solution show a very complex hyperfine-structure with an overall width of about 15 G and an isotropic proton coupling constant of 3.2 G. Therefore we conclude that spin-exchange in [TDAE]C$_{60}$ must also involve the TDAE$^+$ [3] in order to narrow the line to lower values than measured in solution. An analysis of the g-factor (figure 2, inset) exhibits the same temperature dependence for both samples: After a slight decrease down to about 90 K the g-value starts to increase indicating a change in spin-orbit coupling.

It was found in ^{1}H-NMR [4] on [TDAE]C$_{60}$ that two NMR-lines exist. One line has a temperature independent position whereas the other is shifted paramagnetically. On our samples the unshifted line has only 6% of intensity and is ascribed to neutral TDAE. The position of the shifted component (figure 3) can be described by a Fermi-contact interaction [5] with an isotropic coupling constant of 3.0 G

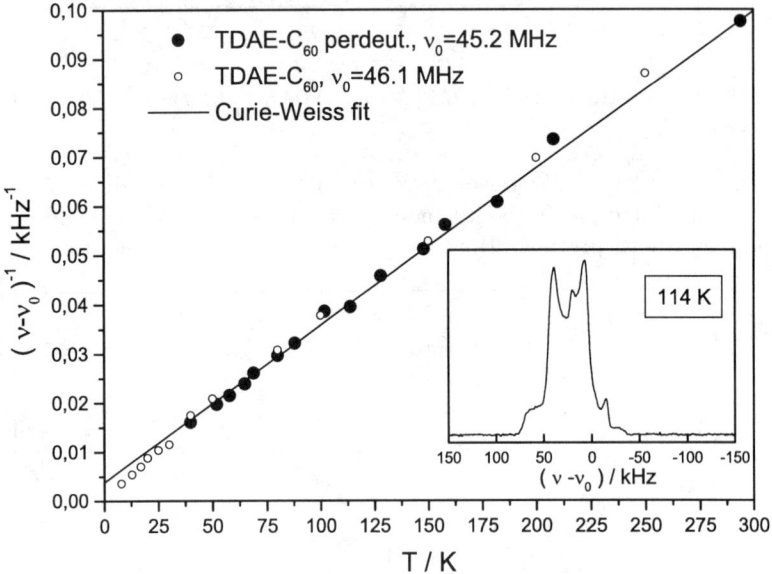

FIGURE 3. Inverse paramagnetic shift as measured by ^{1}H- and ^{2}H-NMR, relative to the unshifted line; a Curie-Weiss curve-fitting gives $\Theta \approx -15$ K. *inset:* ^{2}H-spectrum at 114 K.

which is in good agreement with the ESR result on TDAE$^+$ in solution. It further demonstrates that the observed paramagnetic shift can be explained solely by Fermi-contact interaction on the TDAE$^+$ molecule. There are no additional effects like pseudo Fermi-contact shift or spin-polarization from neighbouring C$_{60}^-$ necessary.

Deuteron NMR has the advantage of higher spectral resolution as dipolar broadening of deuterons in methylgroups is smaller due to the smaller gyromagnetic ratio:

Whereas the homonuclear interaction leads to a linewidth of 30 kHz in ^{1}H-NMR at 295 K, it is reduced by a factor of three in ^{2}H-NMR. ^{2}H-NMR on [TDAE-d_{24}]C_{60} gives evidence for the existence of deuterons in paramagnetic and diamagnetic molecules, too. Below 30 K the line gets too broad to be excited completely with our setup. Fitting a Curie-Weiss behaviour to the ^{2}H-shifts (figure 3) gives an isotropic hyperfine coupling constant of a_D=0.44 G. Therefore the ratio a_H/a_D=6.8 is in excellent agreement with the ratio of gyromagnetic factors $\gamma_H/\gamma_D = 6.5$. The lineshape of the shifted component is not a symmetric Pake-pattern and the complicated analysis will be published elsewhere.

III CONCLUSION

Our experiments show that perdeuteration of TDAE in [TDAE]C_{60} does not suppress the magnetic properties of [TDAE]C_{60}. ESR measurements on [TDAE-d_{24}]C_{60} give further evidence [3] that TDAE$^+$ takes part in electron-spin exchange processes and that hyperfine-interaction has no measurable contribution to the linewidth.

The higher spectral resolution of ^{2}H-NMR allows the measurement of paramagnetic shifts with higher resolution than ^{1}H-NMR. From coupling constants of ESR and NMR we conclude that Fermi-contact interaction of unpaired spin density, transferred by hyperconjugation from nitrogen-atoms, is the dominant mechanism for the paramagnetic line shift. Models based on spin-polarization from C_{60}^- to the protons/deuterons in TDAE$^+$ are not necessary to explain the observed shift.

ACKNOWLEDGEMENT

We want to thank Prof. E. Rössler for fruitful discussions and Sonderforschungsbereich 279 for financial support.

REFERENCES

1. S. Hünig, K. Sinzger, M. Jopp, D. Bauer, W. Bietsch, J.U. von Schütz, H.C. Wolf, *Angew. Chemie* **104**, 7 (1992)
2. N. Wiberg, J.W. Buchler, *Z. Naturforschung* **19b**, 5 (1964)
3. B. Gotschy, *Phys. Rev. B* **52**, 10 (1995)
4. R. Blinc, J. Dolinšek, D. Arčon, D. Mihailovič, P. Venturini, *Solid State Comm.* **89**, 487 (1994)
5. H.M. McConnell, R.E. Robertson, *J. Chem. Phys.* **29**, 6, 1361 (1958)

Searching For the Reactions of Fullerenes With Fe Compounds

P.Byszewski[1,2], E.Kowalska[1], J.Radomska[1], Z.Kucharski[3],
R.Diduszko[1], A.Huczko[4], H.Lange[4],
R.Kochkanjan[5], A.Zaritowskij[5], A.Bondarenko[5], V.Chabanenko[6].

1. *Institute of Vacuum Technology, ul.D³uga 44/50, 00-241 Warsaw, Poland,*
2. *Institute of Physics PAS, al.Lotników 32/46, 02-668 Warsaw, Poland,*
3. *Institute of Atomic Energy, 05-400 Œwierk, Poland,*
4. *Department of Chemistry, Warsaw University, ul. L.Pasteura 1, 02-093 Warsaw, Poland,*
5. *Institute of Physico-Organic and Coal Chemistry NAS, ul. R.Luxembourg 68, 340-114 Donetsk, Ukraine,*
6. *Physico-Technical Institute NAS, ul R.Luxembourg 72, 340-114, Donetsk, Ukraine.*

Abstract. Three types of chemical reactions aimed to bind atomic iron to fullerenes are outlined. The heat of formation of expected reaction products and structure of these molecules are discussed in terms of semiempirical quantum chemistry model ZINDO1. Experimental results of thermal analyses of the products are presented.

1. INTRODUCTION

Since the discovery of the production method of fullerenes in quantities sufficient for laboratory experiments only a few publications devoted to the properties of fulleride:iron compounds were published. The existence of endohedral $Fe@C_{60}$ and exohedral complexes $C_{60}Fe$ in a gas phase was detected by mass spectroscopy [1,2], thus proving that binding energy of iron to C_{60} was high. It was also shown that C_{60} cocrystallizes with ferrocene (Fn) at the composition of 1:2 [3], however ^{57}Fe Mössbauer spectroscopy (MS) proved that cyclopentadiens effectively shield iron atoms from interacting with fullerenes [4-6]. It seemed worth preparing materials where Fe would be in direct contact with fullerenes with free electrons originating from Fe 4s electrons and localized Fe 3d adopting various spin states.

Previously we doped fullerides with iron by thermal decomposition of Fn [4,7], here we present reactions standard in the organometallic chemistry [8] applied to fullerenes and Fn. To predict the reaction products and to estimate their stability we used semiempirical quantum chemistry model ZINDO1 [9], which well reproduces geometry of C_{60} giving the bonds length $d_{66}=0.1398$ nm and $d_{56}=0.1451$ nm.

2. CHEMICAL REACTIONS.

a: Nitration of Fullerenes.

Nitration of fullerenes was carried out in toluene solution to which mixture of nitric acid and acetate acid (the acid is miscible with toluene) was added. During annealing to 95° C the mixture gets yellow without any characteristic absorption bands in the UV/VIS region. The dried powder contains C_{60}, $C_{60}(NO_2)_{1\div3}$ and $C_{60}OH$ detectable by LSIMS and remains soluble in toluene: alcohol mixture. It readily reacts with Fn added to this solution, forming precipitate containing fullerenes and Fn derivatives. The DSC measurements exhibit an exothermic peak at 150÷250° C and TG measurements mass loss extending up to 380° C. the amorphous powder remains insoluble in standard organic solvents, its structure improves by annealing in vacuum to this temperature. There appears in the X-ray diffraction, strong reflection corresponding to diffusive scattering by objects of the size of 1nm, with no traces of iron grains. There is no feature in the Mössbauer spectrum, which can be attributed to iron or iron carbide clusters.

b: Reaction of Fullerenes with Ferricenium Ion.

Instead of activating fullerenes using nitric acid, it can be used to ionize Fn to +1 state in a complex Fn^{+1} NO_3^{-1}, there Fe binding is weaker. In this reaction iron is delivered to fullerenes still in the solution. This reaction was performed in toluene solution heated to 70° C for 2h, at the proportion of active components $HNO_3/Fn/C_{60}$ equal to 4/2/1. The proportion of substrates was chosen to minimize polymerization of fullerenes. The products still exhibit weak exothermic peaks (Fig.1) which we ascribe to polymerization induced by NO_3 groups of C_5H_5 and solvent molecules. To remove adducts and polymerized organic species the samples have to be annealed above 400° C (Fig.1). Mass spectroscopy, with electron impact ionization, in fresh samples revealed a sequence of peaks in the low mass range which might be ascribed

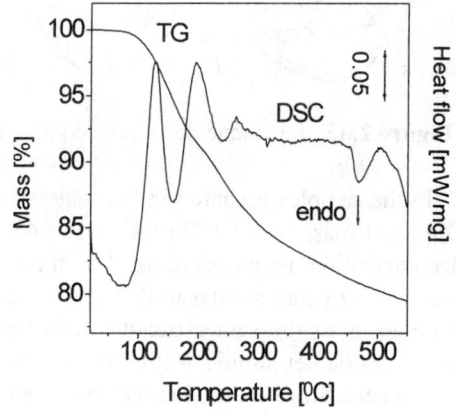

Figure 1. TG and DSC plots measured on C_{60}:Fn sample obtained by method b.

to fragmentation of $C_{11}H_{10}$ thus suggesting, that polymerization of cyclopentadiens occurred already in the toluene solution. In the X-ray diffraction measured on the powder, several reflections could be distinguished showing beginning of crystallization of the product, though the pattern was insufficient to identify the structure. We optimized several of the possible complexes and calculated their energy what showed that one ought to expect exothermic effects in this environment and that Fe bonding in the complexes is weak. Furthermore localization of HOMO at Fe site indicates that this fragment is chemically the most active one and iron may easily join other complex.

c: Ferrocene Ligands Exchange and Ferrocene Adducts to Fullerenes.

Boiling of Fn in benzene or toluene in the presence of catalyst: $AlCl_3$ [8] leads to substitution of one of cp by solvent molecule. This complex is stable as cation coordinated e.g. with $AlCl_4^-$. In case of fullerenes present in the solvent, C_{60}^{-1} can stabilize Fn derivative, it can substitute cp or Fn can form complex with C_{60}. Four of the complexes optimized by ZINDO1 method are depicted in Fig.2.

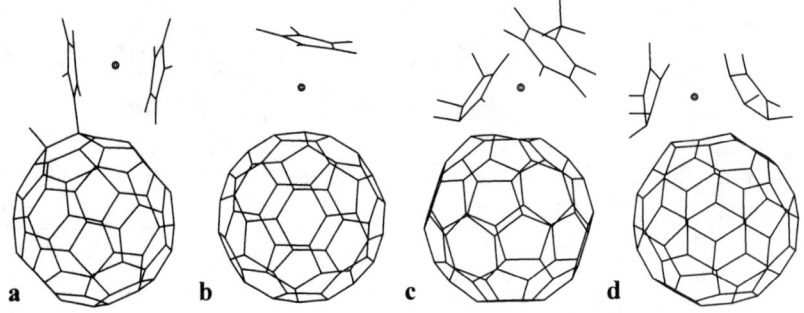

Figure 2. C_{60} ferrocene derivative complexes arranged by decreasing stability.

In the samples prepared in benzene solution the endothermic peak was detected by DSC and mass loss by TG method at 450°C (Fig.3). The effect occurs close to the decomposition temperature of Fn; therefore one can suppose that C_{60} derivative accepted structure similar to that of Fig.2a or 2b where cp binding of to the rest of the complex is of the same strength as in ferrocene. In the MS in samples annealed to 400°C a doublet similar to this in Fn was observed, suggesting structure resembling ferrocene. LSIMS method revealed in such samples ions corresponding to the mass of $C_{60}C_5H_5Fe$ in quantities much exceeding contamination by C_{70}. Samples annealed above 450°C still contain iron at a concentration of $1Fe/3\div5C_{60}$, as estimated by X-ray fluorescence, however form of the iron compound has not yet been identified.

It ought to be mentioned that progress of the reaction in toluene is easily controlled by measuring absorption of the solution, boiling the solution for 2 hours causes an

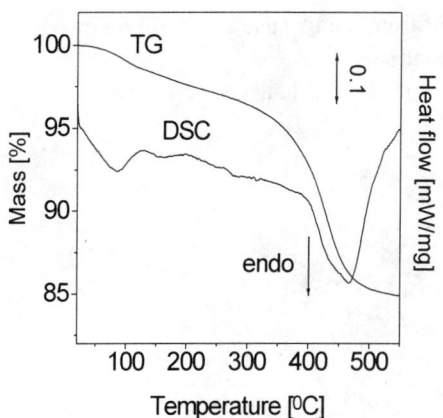

Figure 3. TG and DSC plots measured on the sample prepared by the catalytic method.

increase of the absorption in the range of 416 ÷ 543 nm of orbitally forbidden electronic transitions in pure C_{60}, moreover new absorption band emerges at ~431 nm. Both effects can originate from C_{60} symmetry lowering by adducts.

3. CONCLUSIONS.

We have presented three types of reactions aimed to bind iron atom to fullerenes at mild conditions excluding formation of clusters of iron, iron oxides or iron carbides. Though products of the presented reactions need more experimenting to identify bonding character, sites occupied by metal atom, Fe concentration, homogeneity etc. we think that reaction outlined in paragraphs **b** and **c** may give the required products.

ACKNOWLEDGEMENTS

This work was supported by State Committee for Scientific Research Grant No 7 T08A 016 10 and 3 T09A 087 13.

REFERENCES:

1. Roth, L.M., Huang, Y., Schwedler, J.T., Cassady, C.J., Ben-Amotz D., Khar, B., Freiser, B.S., J.Am.Chem.Soc. **113**, 8186, (1991).
2. Pradeep, T., Kulkarni, G.U., Vasanthacharya, N.Y., Guru Row, T.N., Rao, C.N.R., Indian J. of Chem. **31A&B**, F17 (1992).
3. Crane, J.D., Hitchcock, P.B., Kroto, H.W., Taylor, R., D. Walton, R.M., J.Chem.Soc., Chem. Commun. 1765 (1992).
4. Byszewski, P., Kucharski, Z., Suwalski, J., in *Progress in Fullerene Research*, edt. H.Kuzmany, J.Fink, M.Mehring, S.Roth by World Scientific, 1994, p.82.
5. Birkett, P.R., Kordatos, K., Crane J.D., Herber, R.H., J.Phys.Chem. 1997.
6. Vertes, A., Klencsar, Z., Gal, M., Kuzmann, E., Fullerene Science and Techn. **5**, 97 (1997).
7. Byszewski, P., Kucharski, Z., Fullerene Science and Techn. **5**, 1261 (1997).
8. F.Pruchnik, Organometallic Chemistry, transition elements, Warsaw:. PWN, 1991,

(in Polish); A.N.Nesmiejanov, Chemistry of iron, manganese and rhenium σ- and π-complexes, Moscow: Nauka, 1980 (in Russian).
9. We used commercially available package Hyperchem 5.1, license No 500-10002209.

On the Structure of Iron Fullerene Complexes

E. Kowalska[1], Z. Kucharski[2], P. Byszewski[1,3]

1. Institute of Vacuum Technology, ul. Dluga 44/50, 00-241 Warsaw, Poland,
2. Institute of Atomic Energy, 05-400 Swierk, Poland,
3. Institute of Physics PAS, al. Lotników 32/46, 02-668 Warsaw, Poland,

Abstract. ^{57}Fe Mössbauer spectra in C_{60}:Fe compound obtained by reaction of fullerenes with ferrocene were measured. To interpret these results, standard quantum chemistry calculations were used to determine the geometry and charge distribution in FeC_{60} complexes, which might be formed in the reaction. Besides our previous experimental results, where existence of $C_{60}FeC_{60}$ coordination compound was proved, here we find that iron can also be bound to the fullerene hexagon either inside or outside the carbon cage.

INTRODUCTION

Transition metals are much less commonly used to modify fullerides than other metals, and only few papers are devoted to this problem [1-8]. Synthesis of the first host-guest C_{60}(ferrocene)$_2$ compound [3] has opened iron - fullerene material for the experimental investigations. In our previous papers [4,7,10,11] we proved existence of $C_{60}Fe_2$ system where iron was facing two pentagons of C_{60} neighbors. The main goal of this research is to discuss other possible iron positions in fullerene lattice and evaluate chemical reactions that can be used to bind iron to fullerenes.

EXPERIMENTAL RESULTS

Iron to our previous C_{60}:Fe samples was introduced by decomposing ferrocene dissolved in fullerides. The method exploits that ferrocene molecule is unstable against polymerization of cyclopentadiens when heat in the range of 20 to 108 kcal/mol depending on the formed polymer. Iron may be released in the polymerization process. Preparation of those samples required high temperature what might have lead to contamination of the samples by iron clusters. Fullerenes had been first nitrated then reacted with ferrocene (at C_{60}:Fe=1:2 molar proportion), then the mixture was dried and annealed in vacuum to various temperatures. The method yields amorphous material insoluble in organic solvents. Both experiments and calculations show that solvents (toluene or benzene) are active in the reactions and can not be treated as an inert medium. The diffusive x-ray scattering evidenced, that fullerenes retained their shape and begin ordering, when residual hydrocarbon and nitric groups were gradually removed by annealing. Further details of the technology are described in [12].

The Mössbauer spectra were recorded in a standard transmission geometry using constant acceleration spectrometer coupled to a 50 mCi ^{57}Co/Rh source. The Mössbauer spectrum in the temperature range of 78÷300 K was measured at various stages of annealing to observe evolution of iron state in the compound. The typical spectrum measured at T = 78 K is presented in Fig. 1a. The spectra were approximated by Lorentzian lines using a least-squares routine to determine Mössbauer parameters: Quadrupole Splitting (QS) and Isomer Shift (IS). The quadrupole doublet with QS = 1.22 mm/s and IS= 0.44 mm/s is usually ascribed to iron Fe^{3+} and is very close to that found in ferrocenium ions. Beside the main component, a second doublet with QS = 2.37 mm/s and IS= 0.52 mm/s emerges after heating. The quadrupole splitting of the main component exhibits strong fluctuation around 260 K as previously we ascribe the fluctuation to the influence of fullerenes on charge distribution around Fe ion.

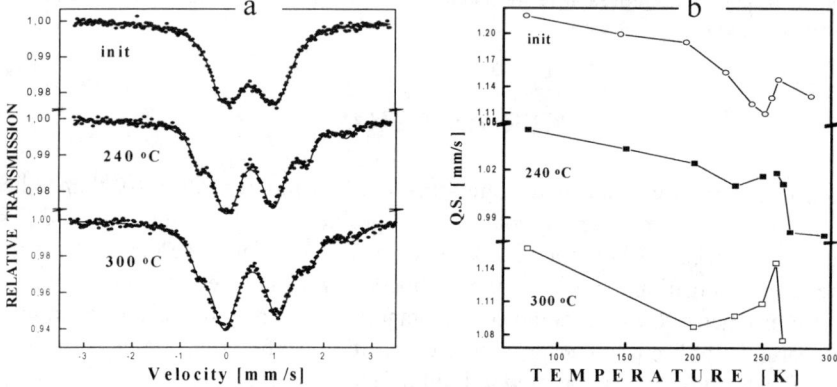

Figure 1. Mössbauer transmission spectra measured at 78 K (a) and temperature dependence of quadrupole splitting QS of the main component (b) of the sample in initial form (top), heated to 240 °C (central), heated to 300 °C (bottom).

DISCUSSION

To verify interpretation of experimental results we have performed semiempirical quantum calculation of C_{60}:Fe complex using ZINDO1 model implemented in commercially available package (Hyperchem 5.1). Although calculations were started with various configurations only three could be optimized, e.g., exohedral Fe opposite one of C_{60} hexagons, opposite pentagon and an endohedral complex opposite hexagon. The structural parameters and energy of the optimized complexes relatively to separated Fe and C_{60} are listed in Table 1.

The heat of formation of C_{60}Fe calculated for the isolated complex depends on the configuration of the whole molecule and its spin state. As it is seen from the Table 1,

Table 1. Structural parameters and heat of formation (relatively to isolated Fe and C_{60}) of FeC_{60} complexes in various configuration and spin states, bond lengths d in fragments closest to Fe (in nm), charge q in e, energy E (in kcal/mol).

		d_{Fe-C}	d_{5-6}	d_{6-6}	q	E
C_{60}			0.145	0.140		
FeC_{60} h	S=2	0.228	0.145	0.142	0.62	-191
	S=0	0.234	0.146	0.142	0.78	-219
$FeC_{60}p$	S=0	0.230	0.147	---	0.74	-188
$Fe@C_{60}$	S=2	0.250	0.146	0.141	0.51	-218
	S=0	0.246	0.146	0.141	0.54	-225

deformation of the fullerene cage under the influence of foreign atom and the charge distribution differ in the five considered cases; the most stable configuration corresponds to iron placed inside the fullerene cage. It is worth noticing that various spin states with almost the same energy, induces different de-formation of the fullerene

Table 2. Population of Fe atomic orbitals in ferrocene and FeC_{60} complex.

AO	s	p_x	p_y	p_z	d_z^2	$d_{x^2-y^2}$	d_{xy}	d_{xz}	d_{yz}
$FeC_{60}h$	1.72	0.51	0.24	0.28	1.78	1.1	1.48	0.52	1.16
Fn	0.34	0.55	0.54	0.27	1.98	1.99	1.99	0.19	0.19

cage, more-over such a small $Fe-C_{60}$ sphere distance allows to fit easily iron into the C_{60} based fcc lattice.

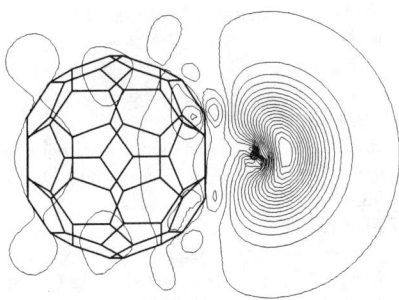

Figure 2. HOMO of the FeC_{60} complex in singlet state, of C_{3v} symmetry. Iron is bound to one of the hexagons.

The charge distribution around Fe determines the QS observed in the Mössbauer spectroscopy. In all samples containing iron, annealed above 450°÷500°C, there were observed very small QS's. Trying to explain the phenomena we used to assume, that Fe was in the +3 ionization state, by comparison with QS observed in (ferrocene)$^{+1}$. The low QS value might, however, be caused by charge distribution around iron atom other than in ferrocene. In fact population of Fe AO's differs in both cases, as it is shown in Table 2.

Such population produces charge distribution of $C_{\infty v}$ symmetry in the vicinity of Fe in ferrocene, while in FeC_{60} it is rather an octahedral symmetry. It is interesting to notice, that the anisotropy of this complex is compensated to some extent by the shape of HOMO (Fig. 2), which results from poor hybridization of Fe 4s electrons with C_{60} MO's, not participating in the bonding.

CONCLUSIONS

The calculations presented here can help to explain confusing experimental results. Quantum chemical calculations proved that iron can bind fullerene in few different configurations. These results are consistent with our preliminary Mössbauer experiment. Most of the configurations are metastable and thermal treatment can partially change final iron position. In all studied samples we always observed in ME spectra two doublets which relative intensity depends on the thermal treatment. However till now we have no experimental evidence of the iron location in the lattice.

ACKNOWLEDGMENTS

This work was supported by State Committee for Scientific Research Grant No 7 T08A 016 10 and 3 T09A 087 13.

REFERENCES:

1. Penicaud, A., Hsu, J., Reed, C.A., J.Am.Soc. **1991**, 113, 6698, (1991)
2. Stinchcombe, J., Penicaud, A., Bhyrappa, P., Boyd, P.D.W., Reed, C.A., J.Am.Chem.Soc. **1993**, 115, 5212, (1993)
3. Crane, J.D., Hitchcock, P.B., Kroto, H.W., Taylor, R., Walton, D.R.M., J.Chem.Soc., Chem.Commun. **1992**, 1765, (1992)
4. Kucharski, Z., Byszewski, P., Suwalski, J., Materials Science Forum 191 (1995) p. 31-36.
5. Roth, L.M., Huang, Y., Schwedler, J.T., Cassady, C.J., Ben-Amotz, D., Khar, B., Freiser, B.S., J.Am.Chem.Soc. **1991**, 113, 8186, (1991)
6. Pradeep, T., Kulkarni, G.U., Vasanthacharya, N.Y., Guru Row, T.N., Rao, C.N.R., Indian J. of Chem. **31A&B**, F17, (1992)
7. Byszewski, P., Kucharski, Z., Suwalski, J., in Progress in Fullerene Research, edt. H.Kuzmany, J.Fink, M.Mehring, S.Roth by World Scientific, p.82, 1994
8. Birkett, P.R., Kordatos, K., Crane, J.D., Herber, R.H., J.Phys.Chem. 1997
9. Ksari, Y., Chouteau, G., R.Yazami, Cherigui, A., J.M.M.M. **140-144**, 2061, (1995)
10. Byszewski, P., Diduszko, R., Kowalska, E., Recent Advances in Chemistry of Fullerenes and Related Mat. edt. K.Kadish, R.Ruoff by The Electrochemical Society v.24, p.1392, 1994
11. Byszewski, P., Kucharski, Z., Fullerene Science and Techn. **5**, 1261, (1997)
12. Byszewski, P., Kowalska, E., Radomska, J., Kucharski, Z., Diduszko, R., Huczko, A., Lange, H., Kochkanjan, R., Zaritowskij, A., Bondarenko, A., Chabanenko, V., submitted to the same issue.

Spectroscopy of C_{60} and C_{70} Complexes.

Dmitry V. Konarev, Natalia V. Drichko[1], Viktor N. Semkin[1],
Yury M. Shul`ga, Andrzej Graja[2], Rimma N. Lyubovskaya*.

Institute of Chemical Physics RAS, Chernogolovka, 142432, Russia.
[1] *Ioffe Physical-Technical Institute RAS, St.Peterburg, 194021, Russia.*
[2] *Institute of Molecular Physics PAN, Poznan, 60-179, Poland.*

Abstract. Complexes of C_{60} and C_{70} fullerenes with organic donors were studied by IR-, UV-VIS-NIR- and X-ray photoelectron spectroscopies. The IR spectra of single crystals of some complexes show the $F_{1u}(4)$ C_{60} mode splitting into two components which is attributed to the freezing of the rotation of C_{60} molecule in the coordination with donors. The degree of charge transfer (CT) on C_{60} molecule estimated from the shift of the frequency of the $F_{1u}(4)$ unsplit band shows only weak CT in the complexes. The UV-VIS-NIR- spectra of some complexes show the presence of weak CT bands. A linear dependence of CT absorption energy on vertical ionization potentials of the donors was obtained for C_{60} complexes with tetrathiafulvalenes in solid state. The XP-spectra of the complexes show the changes in the energies of S2p, N1s, Te3d peaks of donors by 0.1-1.6 eV with respect to individual ones.

Introduction. Fullerenes as electron acceptors form compounds varying from molecular complexes [1-3] to ion-radical salts [4] analogously to planar π-acceptors such as tetracyanoethylene (TCNE) and tetracyanoquinodimethane (TCNQ). Due to a spherical shape, large size, high symmetry and polarizability [5] of fullerene molecules donor-acceptor complexes of fullerene have some peculiarities: charge transfer from initially planar donors to spherical C_{60} molecules [5], the I_h symmetry breaking [6,7] and the freezing of the rotation of C_{60} molecules in their coordination with donors [7]. In this paper we present some results of IR-, UV-VIS-NIR- and X-ray photoelectron spectroscopic studies of these peculiarities of the fullerene complexes.

Experimental. Complexes of C_{60} and C_{70} fullerenes were obtained by evaporation of fullerenes and donors solutions in carbon disulfide, benzene or pyridine (Py) under argon. The composition of the complexes was determined by elemental, thermogravimetric and X-ray analyses [2,3,8,9].
The IR transmission spectra of single crystals were measured at room temperature by using a FT-IR Perkin-Elmer 1725X spectrometer equipped with a microscope (2 cm^{-1} resolution). KBr pellets were prepared with concentration of complexes 1:2000. UV-VIS-NIR absorption spectra were measured with a Lambda 19 Perkin Elmer UV-VIS-NIR spectrometer at room temperature in KBr pellets with concentration of complexes 1:4000 within 220 -2000 nm range. The CT bands were obtained in 600- 1300 nm range by the subtraction of a normalized spectrum of individual fullerene from that of the complexes. This procedure was possible due to the absence of donors absorption bands in this spectral range. XP -spectra were recorded on a VIEE-15 instrument and were calibrated against the C1s peak (285.0 eV). The spectral data for the complexes are presented in Table.

Table. The frequencies of the $F_{1u}(4)$ C_{60} mode in the IR spectra of the complexes (single crystals and KBr pellets*), the position of the CT bands in the UV- VIS-NIR-spectra and the shift the binding energy of donor heteroatoms in XP-spectra of the complexes relatively to the individual donors.

N	Complex	Frequency of $F_{1u}(4)$, cm^{-1}		CT band, nm	Energy shift, eV	
	C_{60} 250K[6]	1427.5	1430.5	-	-	
	293K[6]	1429.4				
1	BTX·C_{60}·CS_2	1428	1432	620	(Te)	0.4
2	BTX·C_{60}	1427		620	-	
3	DBTTF·C_{60}·C_6H_6	1430*[1]		735	(S)	0.3
4	DBTTF·C_{60}·Py	1430*[1]		750	(S)	0.1
5	(BEDT-TTF)$_2C_{60}$(Py)$_2$	1428		820	-	
6	DPhTTF·C_{60}·C_6H_6	1428*		895	(S)	0.0
7	(TMDTDM-TTF)$_2C_{60}$(CS_2)$_3$	1429*[1]		900	(S)	0.1
8	(EDT-TTF)$_2C_{60}$·CS_2	1427		900	(S)	0.1
9	BEDO-TTF·C_{60}·C_6H_6	1429		900	-	
10	EDT-TTF·C_{60}·C_6H_6	1428*		920	(S)	0.1
11	(DMDPhTTF)$_2C_{60}$·C_6H_6	1429*		940	(S)	0.1
12	EDY-BEDT-DT·C_{60}·C_6H_6	1428*		935	(S)	0.4
13	OMTTF·C_{60}·Py	1428*		980	-	
14	OMTTF·C_{60}·C_6H_6	1428		1040	-	
15	TPDP(C_{60})$_2$(CS_2)$_4$	1428*		1240	-	
16	(BEDO-TTF)$_2C_{60}$	1428.5	1432	-	(S)	0.7
17	BNDY·C_{60}	1429		-	(S)	0.4
18	(S_4N_4)$_{0.8}C_{60}(C_6H_6)_{1.2}$	1430		-	(S) 0.3 (N) 1.6	
19	TPC·C_{60}	1425	1431	-	-	
20	DAN·$C_{60}(C_6H_6)_3$	1426	1430	-	-	
21	BEDT-TTF·C_{70} CS_2	1428(C_{70})		860	-	
22	(DMDPhTTF)$_2C_{70}$·C_6H_6	1430(C_{70})*		1000	-	

[1]-A superposition of the $F_{1u}(4)$ C_{60} mode and the donor absorption band.
Abbreviations for donors: BTX - 9,9'-trans-bis(telluraxantenyle); DBTTF - dibenzotetrathiafulvalene (TTF); BEDT-TTF - bis(ethylenedithio)-TTF; DPhTTF - trans-4,4'-diphenyl -TTF; TMDTDM-TTF - tetramethylenedithio-4,5-dimethyl-TTF; EDT-TTF - ethylenedithio-TTF; BEDO-TTF - bis(ethylenedioxo)-TTF; DMDPh-TTF-trans-4,4'-dimethyl-5,5'-diphenyl-TTF; EDY-BEDT-DT- 2,2'-ethanediilidene-bis(4,5-ethylene-1,3-dithiol); OMTTF-octamethylene-TTF; BNDY-binaphto[1,6-d,e]-1,3-ditiin-2-ylidene; TPDP-3,3',5,5'-tetraphenyldipyranylidene; S_4N_4- tetrasulfur tetranitride; TPC - triptycene; DAN – dianthracene.

Results and discussion. IR spectroscopy was used in the studies of the changes in symmetry and electron densities of donor and acceptor molecules in a complex formation. C_{60} molecule has the four IR active threefold degenerated F_{1u} modes [6,11], the $F_{1u}(4)$ mode at 1429cm^{-1} being the most sensitive to the changes in charge [10] and symmetry [6,11] of C_{60} molecule. Thus only the changes in the $F_{1u}(4)$ mode were considered. It is known [6] that in C_{60} crystals molecules nearly free rotate at room temperature and occupy the sites with T_h symmetry. At T<260K the orientational-ordering phase transition is realized in crystals and only "ratchet" rotation of C_{60} becomes possible, the molecular symmetry of C_{60} being lowered to S_6 one. This results in the double [6] or triple [11] splitting of the $F_{1u}(4)$ mode.

The absorption bands corresponding to the $F_{1u}(4)$ C_{60} mode in the complexes are presented in Fig.1. Donors have no substantial absorption in this range. For the first group of complexes (1a-4a) the $F_{1u}(4)$ band is split into two components. These split bands were fitted by a sum of two Lorentzians with nearly the same bandwidths (~6 cm^{-1}) whose position are given in

Table. The observation of only two components for the threefold degenerated mode can be explained by large bandwidths (~6 cm^{-1}) and a comparatively low resolution of the experiment. The similar splitting of the $F_{1u}(4)$ mode into 3 components at room temperature was also observed in C_{60} complexes with amines [7]. We attributed this effect to the freezing of the rotation of C_{60} molecules at their coordination with donors analogously to the phase transition in C_{60} crystals at T<260K. The absence of free rotation of C_{60} molecules in DAN·C_{60}·$(C_6H_6)_3$ and BTX·C_{60}·CS_2 complexes at room temperature was confirmed by the data of X-ray analysis [2,12]. The freezing of the rotation lowers the C_{60} symmetry in the complexes relatively to that in C_{60} crystals at T>260K and results in the splitting of the $F_{1u}(4)$ mode. For the second group of complexes (1b-4b) the $F_{1u}(4)$ mode remains unsplit indicating free rotation of C_{60} molecules in these complexes at room temperature.

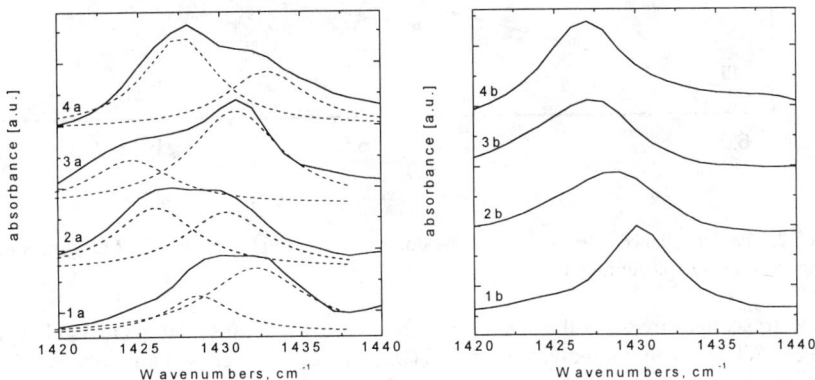

FIGURE 1. IR specrta of single crystals of the complexes in 1420-1440 cm^{-1} range at room temperature:
1a- (BEDO)$_2$C$_{60}$; 2a- DAN·C$_{60}$·(C$_6$H$_6$)$_3$; 3a- TPC·C$_{60}$; 4a- BTX·C$_{60}$·CS$_2$
1b- (S$_4$N$_4$)$_{0.8}$ C$_{60}$ (C$_6$H$_6$)$_{1.2}$; 2b- BNDY·C$_{60}$; 3b-(EDT-TTF)$_2$C$_{60}$·CS$_2$; 4b- BTX·C$_{60}$

The dependence of the position of the unsplit $F_{1u}(4)$ mode on the degree of C_{60} reduction is almost linear [10]. The following relationship for the estimation of the CT degree (δ) on C_{60} molecule is derived from this dependence [5] : $\delta \cong 0.03 \Delta\nu$, where $\Delta\nu$ is the shift of the $F_{1u}(4)$ mode position in the complexes (Table) relatively to that in individual C_{60}. The maximal shifts do not exceed 2 cm^{-1} and are within the experimental accuracy. The values of δ estimated from this relationship are close to zero for all complexes (δ<0.05). Thus the compounds obtained are neutral complexes in a ground state.

UV-VIS-NIR spectroscopy shows the presence of CT bands of weak intensity in 600-1300 nm range in some complexes. The CT absorption energy ($h\nu_{CT}$) in neutral complexes is defined by $h\nu_{CT} = I_P^v - E_A^v - E_c$, where E_A^v is electron affinity of the acceptor, I_P^v is vertical ionization potentials of the donor and E_c is the energy of electrostatic interaction in excited ionic state. Therefore for one acceptor with a series of donors the $h\nu_{CT}$ values depend linearly on donor I_P^v. It is seen in Fig.2 (circles) that this dependence for C_{60} complexes with OMTTF, BEDO-TTF, BEDT-TTF and DBTTF donors (I_P^v - 6.30, 6.46, 6.70, 6.81 eV, respectively [13]) is really quite linear and is approximated by $h\nu_{CT} = 0.82 \, I_P^v - 3.93$. This dependence enables the estimation of I_P^v values for other donors from the $h\nu_{CT}$ values in C_{60} complexes (Fig.2, crosses). The $h\nu_{CT}$ values for C_{70} complexes (Fig.2, stars) are ~0.08 eV lower than those for C_{60} ones with identical donors.

For the donors with equal I_p^v the $h\nu_{CT}$ values are ~0.6 eV higher in C_{60} complexes in solid state than those in TCNE ones [14]. This fact can be explained by two reasons. C_{60} is a weaker acceptor than TCNE. The E_c values in complexes depend on the distances between ions in excited state therefore the delocalization of the radical anion charge over a large C_{60} sphere results in the decrease of E_c values in C_{60} complexes relatively to TCNE ones.

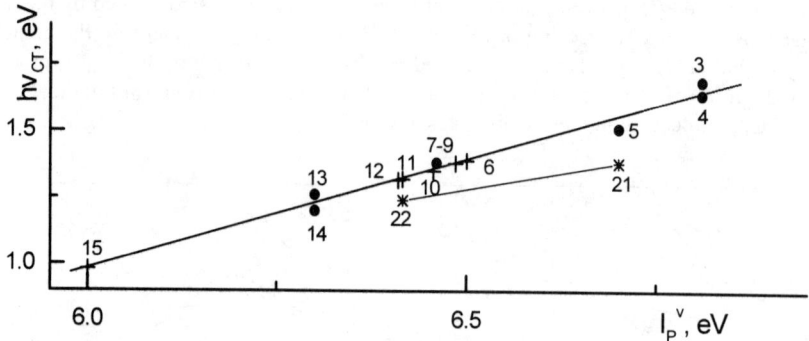

FIGURE 2. The dependence of $h\nu_{CT}$ vs. I_p^v of the donors for C_{60} and C_{70} complexes (the numeration of the complexes is given according to Table).

XP-spectroscopy shows that the energy of S2p, N1s, Te3d peaks of the donors changes by 0.1-1.6 eV relatively to the individual ones. These shifts can be caused not only by CT from the donor to C_{60} but the calibration against the C1s peak since the position of this peak can be different for the donor and the corresponding complex.

Acknowledgements

This work is supported by the Russian Program " Fullerenes and Atomic Clusters" and partially by the Polish grant N 7 T08A 003 12.

References

1. Saito G., Teramoto T., Otsuka A., Sugita Y., Ban T., et.al , *Synth.Met.*, **64**, 359-368 (1994).
2. Konarev D.V., Valeev E.F., Slovokhotov Yu.L., et.al., *J.Chem. Res.*, **12**, 442-443 (1997).
3. Konarev D.V., Lyubovskaya R.N., Roschupkina O.S., et.al, *Russ. Chem. Bull.*, **46**, 32-35(1997).
4. Penicaud A., Perez-Benitez A., Gleason R., et.al., *J.Am.Chem. Soc.*, **115**, 10392-10393 (1993).
5. Konarev D.V., Semkin V.N., Graja A., Lyubovskaya R.N., *J.Molecular Structure.*, accepted.
6. Winkler R., Pichler T., Kuzmany H., *Z.Phys.B*, **96**, 39-45 (1994).
7. Bagenov A.V., Maksimuk M.Yu., et.al., *Izv. Akad. Nauk, Ser. Khim.*, **6**, 1459-1463 (1996).
8. Konarev D.V., Zubavichus Y.V., Slovokhotov Yu.L., et.al., *Synth.Met.*, **92**, 1-6 (1998).
9. Konarev D.V., Valeev E.F., Slovokhotov Yu.L., Shul`ga Yu.M.,et al., *Synth. Met.*,**88**, 85-87 (1997).
10. Pichler T., Winkler R., Kuzmany H., *Phys.Rev. B*, **49**, 15879-15888 (1994).
11. Narasimhan L.R., Stoneback D.N., Hebard A.F., et al., *Phys.Rev. B*, **46**, 2591-2594 (1992).
12. Kveder V.V., Steinman E.A., et. al., *Chem., Phys.*, **216**, 407-415 (1997).
13. Lichtenberger D.L., Johnston R.L., Hinkelmann K., et al., *J.Am.Chem. Soc.*, **112**, 3302-3307 (1993).
14. Kobayashi M., Kinoshita H., Takemoto S., *J.Chem. Phys.*, **36**, 457-462 (1962).

ENDOHEDRALS

Study of N@C$_{60}$ and P@C$_{60}$

A. Weidinger[*], B. Pietzak[*], M. Waiblinger[*], K. Lips[**], B. Nuber[†], A. Hirsch[†]

[*] Hahn-Meitner-Institut Berlin, Glienickerstr. 100, D-14 109 Berlin, Germany
[**] Hahn-Meitner-Institut Berlin, Rudower Chausee 5, D-12 489 Berlin, Germany
[†] Universität Erlangen-Nürnberg, Institut für Organische Chemie, Henkestr. 42, D-91054 Erlangen, Germany

Abstract. This paper summarises the recent work of our group. The main achievement in the last year was the production and characterisation of endohedral phosphorus in C$_{60}$. Phosphorus in C$_{60}$ as nitrogen in C$_{60}$ keeps its atomic ground state structure and is suspended in the centre of the fullerene. The properties of P@C$_{60}$ are compared with those of the free atom and those of nitrogen in C$_{60}$. We also summarise the results obtained for the various adducts of N@C$_{60}$.

INTRODUCTION

N@C$_{60}$ (atomic nitrogen in C$_{60}$) is the first system in which a highly reactive atom is encapsulated in C$_{60}$ without a charge transfer to the C$_{60}$ shell (1-3). Nitrogen in C$_{60}$ is located in the centre of this highly symmetric fullerene and interacts extremely little with the surrounding carbon atoms. All other endohedral fullerenes contain either noble gases or the enclosed atoms leave the central position and form a bond with the fullerene shell. The ideal situation of two separate entities, i.e. the encapsulated atom and the C$_{60}$ shell, is realised only for noble gases and for nitrogen in C$_{60}$.

For a systematic study of this new category of endohedrals and for possible applications, it is important that further such systems exist in order that one can vary their properties and adjust them to the specific use in mind. Here, we report on the encapsulation of atomic phosphorus in C$_{60}$ (P@C$_{60}$) and compare it with N@C$_{60}$.

EXPERIMENTAL

The samples were prepared by ion implantation. In our set up, C$_{60}$ is continuously evaporated onto a substrate and bombarded at the same time by an intense ion beam from a remote ion source. For the production of phosphorus ions we used PH$_3$ gas in the source and care was taken to avoid contamination with this poisonous material. The deposition rate and the beam intensity were adjusted in such a way that approximately one ion per one C_{60} molecule arrived at the target. The ion energy has to be large enough for the ion to enter the cage but should not be too high in order to minimise the destruction of the C_{60} ball. We have chosen 40 eV as a compromise in these respects. After several hours of bombardment the irradiated sample was removed from the substrate, dissolved in toluene and filtered. Only the soluble part, which contained filled and empty fullerenes, was treated further.

The electron paramagnetic resonance (EPR) experiments were performed with a commercial cw spectrometer at 9.3 GHz (X-band).

EXPERIMENTAL RESULTS AND DISCUSSION

P@C_{60}

Figure 1 shows EPR spectra of P@C_{60} in solid form (powder) and in solution. The doublet splitting with a separation of 49.2 G is due to the hyperfine interaction between

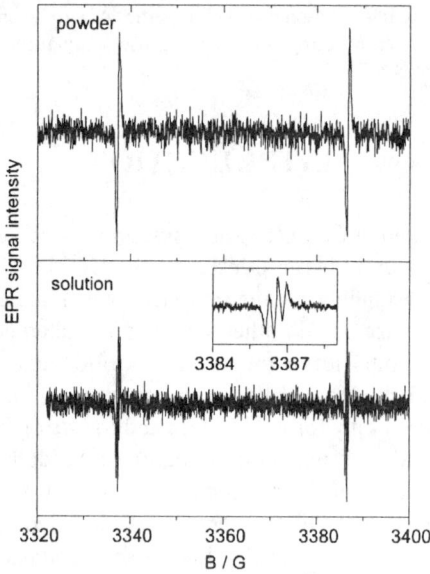

FIGURE 1. *EPR spectra of phosphorus in C_{60} for a solid sample (top) and in solution (bottom). The doublet splitting is due to the hyperfine interaction with the nuclear spin $I=1/2$ of ^{31}P. The insert in the bottom spectrum shows the triplet splitting due to the $S=3/2$ spin of the electron system.*

the spin of the electron system and the nuclear spin of ^{31}P which is $I = 1/2$. The further splitting of each line in three sub components as seen in the solution spectrum (insert in Fig.1) is a clear indication that the spin of the electron system is $S = 3/2$. In the high field limit these lines are degenerate but at the magnetic field used in the present experiment the splitting amounts to 0.36 G and due to the small line width is clearly visible. For N@C$_{60}$, because of the smaller hyperfine coupling, this splitting is two orders of magnitude smaller and could not be resolved in the cw EPR measurement. However, in the FT spin-echo experiment the resolution was sufficient to observe this splitting also in case of N@C$_{60}$ (2).

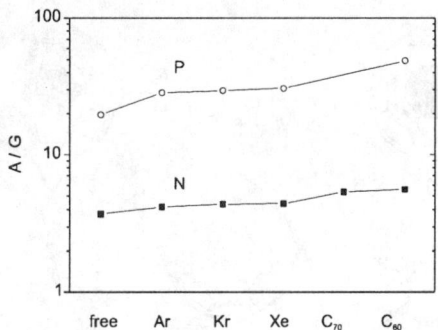

FIGURE 2. *Hyperfine interaction constant for nitrogen and phosphorus in the free state, matrix isolated in different noble gas solids (4) and encapsulated in fullerenes.*

Figure 2 shows the hyperfine interaction constant A of atomic nitrogen and atomic phosphorus for free atoms and for atoms matrix isolated in various inert gas solids (4) and encapsulated in fullerenes. For both atomic species, an increase of A from the free state via the various matrix isolation cases towards the fullerene encapsulation is observed. The reason is that the hyperfine interaction of N and P is much larger in the exited states than in the ground state. Therefore any interaction with the surrounding which leads to a mixing of atomic states causes an increase of A. The fact that the hyperfine coupling is largest for C$_{60}$ and that the increase is more pronounced for P than for N indicates that the wave function overlap between the enclosed atom and the surrounding C$_{60}$ shell increases in these directions. The increase of A from C$_{70}$ to C$_{60}$ is a consequence of the smaller available free space in C$_{60}$ compared to C$_{70}$. Similarly, the larger radius of P compared to N is responsible for the relatively larger increase of the hyperfine interaction for P@C$_{60}$ than for N@C$_{60}$.

Figure 2 shows that there is an appreciable interaction between the enclosed atom and the cage. However, the hyperfine interaction is very sensitive to such interactions and the drastic changes of the configuration observed for other endohedrals like charge transfer and bonding to the cage atoms are absent in the present case. In P@C$_{60}$ and to an even greater extend in N@C$_{60}$ the individuality of the enclosed atom remains preserved, the most convincing sign for that being that they remain completely spherical.

Chemistry with N@C$_{60}$

In collaboration with the Erlangen group of Prof. Hirsch, a series of chemical additions on the outside of the N@C$_{60}$ cage was studied. The different adducts together with the bare N@C$_{60}$ are schematically depicted in Fig. 4. We find that N@C$_{60}$ survives these drastic manipulations on the shell and all adducts are stable. The EPR

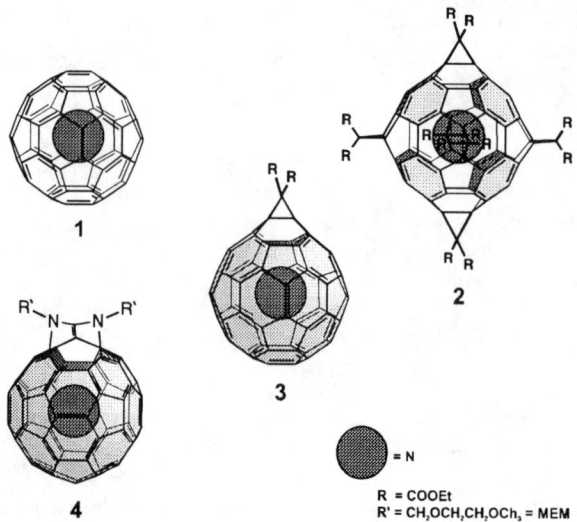

FIGURE 3. N@C$_{60}$ and different adducts derived from it: *1* bare N@C$_{60}$; *2* hexaadduct; *3* monoadduct and *4* bisadduct. These systems were studied by EPR.

spectra of the adducts in solution are almost identical with those of pure N@C$_{60}$. As an example we show the result for the bisadduct. A closer inspection shows that the isotropic hyperfine coupling constant increases in the 10^{-3} range with the addends. This increase is explainable by the slight increase of the available free space in the adducts.

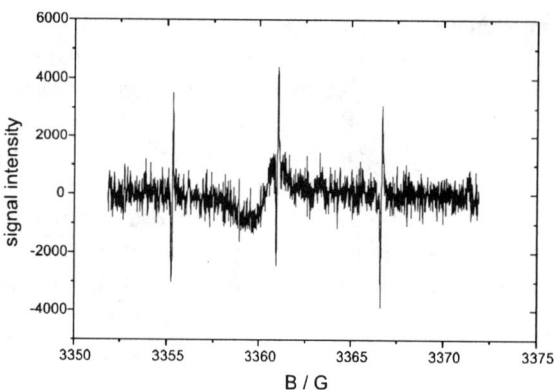

FIGURE 4. EPR spectrum of N@C$_{60}$(NMEM)$_2$. The brought line is due to an impurity.

In the solid, drastic changes in the EPR spectra were observed. In case of the monoadduct, the symmetry lowering leads to a fine structure interaction and as a consequence to additional EPR lines (5). In the hexaadduct, due to the symmetrical arrangement of the addends, the fine structure is absent again (6).

CONCLUSION

With P@C_{60} the second member after N@C_{60} in a new class of endohedrals was produced. The main characteristic of these new systems is that the encapsulated atoms keep their individuality in the symmetrical C_{60} cage which protects them from the outside. For P the hyperfine interaction is a priori larger than for N but also the change of the interaction due to encapsulation relative to the free atom is larger. This has to do with the larger radius of P and the weaker binding of the outer electrons relative to N.

ACKNOWLEDGEMENTS

This work was supported by the Bundesminister für Bildung und Wissenschaft, Forschung und Technologie and the Fonds der Chemischen Industrie.

REFERENCES

1. T. Almeida Murphy, T. Pawlik, A. Weidinger, M. Höhne, R. Alcala, J.M. Spaeth, *Phys. Rev. Lett.* **77**, 1075 (1996).
2. C. Knapp, K.-P. Dinse, B. Pietzak, M. Waiblinger, A. Weidinger, *Chem. Phys. Lett.* **272**, 433 (1997).
3. A. Weidinger, M. Waiblinger, B. Pietzak, T. Almeida Murphy, *Applied Physics A* **66**, 287 (1998).
4. G.S. Jackel, W.H. Nelson, W. Gordy, *Phys. Rev.* **172**, 176 (1968).
5. B. Pietzak, M. Waiblinger, T. Almeida Murphy, A. Weidinger, M. Höhne, E. Dietel, A. Hirsch, *Chem. Phys. Lett.* **279**, 259 (1997).
6. E. Dietel, A. Hirsch, B. Pietzak, M. Waiblinger, K. Lips, A. Weidinger, A. Gruß, K.-P. Dinse, *to be published*

Production, HPLC Separation and UV-vis spectroscopy of Li@C_{70}

N. Krawez, A. Gromov*, R. Tellgmann[§] and E.E.B. Campbell[#]

Max-Born-Institut für Nichtlineare Optik und Kurzzeitspektroskopie,
Rudower Chaussee 6, D-12489 Berlin, Germany
* present address: IFW Dresden, Helmholtzstr.20, 1069 Dresden
§: Debis Systemhaus, Magirusstr. 43, D-89077 Ulm
#: School of Physics and Engineering Physics, Gothenburg University and Chalmers University of Technology, S-41296 Gothenburg, Sweden

Endohedral Li$_n$@C_{70} (with n = 1, 2, 3) can be produced by low energy ion bombardment of C_{70} thin films. Considerably more multiple capture is found for C_{70} than was the case for C_{60}. The monomer endohedral species Li@C_{70} has been purified by HPLC. Unlike C_{60}, only one endohedral fraction is found. The UV-vis spectra of Li@C_{70} in CS_2 and toluene are reported and compared with C_{70}.

INTRODUCTION

Recently, we have developed a method for producing bulk amounts of endohedral fullerenes which involves low energy ion bombardment of thin fullerene films[1,2]. This method has the advantage over the conventional endohedral fullerene production procedures (such as arc-discharge or laser vaporisation of impregnated graphite) of limiting the obtained endohedral species to one definite fullerene cage (in our case C_{60} or C_{70}). Films produced by irradiating C_{60} layers with alkali metal ions (Li$^+$, Na$^+$, K$^+$, Rb$^+$) have been shown to contain on the order of a few percent endohedral molecules as determined by laser desorption mass spectrometry[2]. Our work up until now has concentrated on characterising Li@C_{60} by means of IR[3], Raman[4] and UV-vis[5] spectroscopy. The material can be dissolved in carbon disulphide and purified by HPLC[5]. Chromatographic analysis combined with laser desorption mass spectrometry showed the presence of two fractions, both of which give a mass peak at 727 u[5]. The fastest of these two fractions has been interpreted as being due to a dimer molecule, (Li@C_{60})$_2$. The rather weak single bond between the two fullerene cages would not be able to survive the extreme conditions existing during pulsed laser desorption thus giving rise to only the monomer mass peak.

In this contribution we shall report new results concerning the production, purification and spectroscopy of Li@C_{70}. There are some significant differences compared to the situation with Li@C_{60}.

PRODUCTION OF $Li_n@C_{70}$

Low energy Li-ion bombardment of C_{70} will form $Li@C_{70}$ under the same conditions used and optimised for $Li@C_{60}$ production. There is a clear threshold for the ion energy at about 6 eV below which no endohedral production is observed[1]. The maximum efficiency of endohedral fullerene production is found at about 30 eV for Li^+. The ratio of filled to empty fullerenes in the films increases with increasing ion:fullerene ratio[2]. However as the amount of lithium in the films increases it becomes increasingly difficult to dissolve them. The conditions under which the maximum absolute amount of $Li@C_{60/70}$ can be isolated from the films either by solution followed by HPLC[5] or by sublimation[6] correspond to an ion:fullerene ratio during deposition of 1:1. There is, within experimental error, the same probability for detecting an endohedral $Li@C_{70}$ as for $Li@C_{60}$ under identical deposition and irradiation conditions. This can be seen in Figure 1 where the laser desorption mass spectrum of a film containing a mixture of C_{60} and C_{70} and irradiated with 30 eV Li^+ ions is shown. In order to better see the endohedral peaks, this film was made with an ion:fullerene ratio of 6:1.

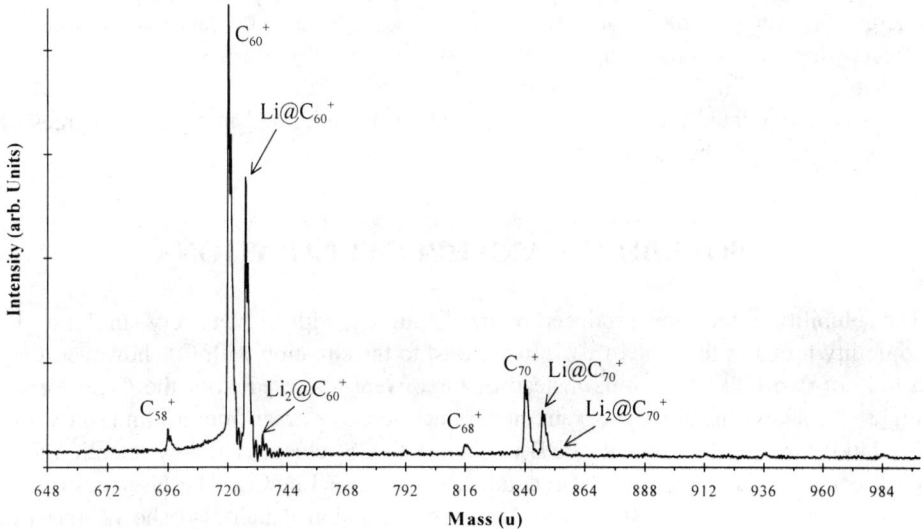

FIGURE 1. Laser desorption mass spectrum of a thin film containing endohedral Li fullerenes. Produced with an ion energy of 30 eV and an ion:fullerene ratio of 6:1.

Mass peaks corresponding to the capture of two and even three lithium ions can also be seen in this figure. Although the probability of capturing one lithium is very similar for both fullerenes, the probability for capture of two and three lithiums atoms is approximately twice and three times larger respectively in C_{70} compared to C_{60}. The ratios obtained from the mass spectrum of Figure 1 are plotted in Figure 2. A similar effect is observed for the single ion capture of the larger alkalis with almost six times more $K@C_{70}$ being detected than $K@C_{60}$. It is not yet clear whether these very dramatic effects are due solely to the larger volume inside C_{70} and thus reduced repulsion energy or whether there are also effects due to different detection probabilities in the mass spectrometer (related to differences in ionisation potential and number of degrees of freedom). This will be the subject of future investigations.

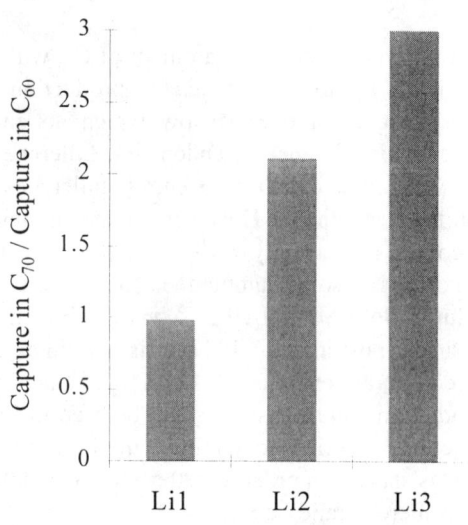

FIGURE 2. Ratio of capture of 1, 2 or 3 lithium ions in C_{70} to capture in C_{60} for the conditions in Figure 1.

SOLUBILITY AND HPLC SEPARATION

The solubility of the films produced by irradiating C_{70} with Li^+ was very similar to the solubility found in the case of C_{60}[5]. In contrast to the situation with C_{60}, however, only a total of two HPLC fractions instead of three were observed from the C_{70} material. Figure 3 a shows the chromatogram on passing the CS_2 extract from a film (30 eV Li^+, 1:1 ratio) through a Cosmosil 5PBB column with CS_2 eluent. The low retention time peak corresponds to C_{70} and the second, later peak, to $Li@C_{70}$. The laser desorption mass spectrum in Figure 3b is from the second fraction which also shows traces of oxidized fullerenes. This can be further purified by additional HPLC cycles. The retention time of the endohedral fraction corresponds to that of the second, later endohedral-containing fraction from films made with C_{60}. This is the $Li@C_{60}$ fraction which we have previously assigned to the monomer[5]. $Li@C_{70}$ is temperature sensitive and partially decays during solvent removal after the HPLC separation. $Li@C_{70}$ is significantly less soluble than C_{70} in toluene (about two orders of magnitude) and seems to be practically insoluble in n-hexane.

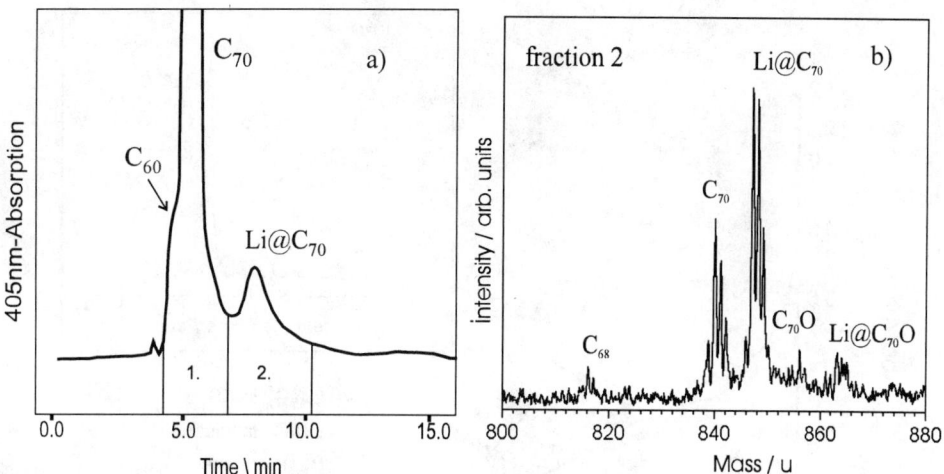

FIGURE 3. a) HPLC chromatogram of dissolved film material
b) Laser desorption mass spectrum of the fraction 2 (Li@C_{70})

UV-VIS SPECTRUM

The solution of Li@C_{70} in 1,2-dichlorbenzene, CS_2 and toluene is a red-brown colour and is not very different from C_{70}. Li@C_{60}, on the other hand, is a dark yellow colour and very different from C_{60}. This is perhaps not too surprising since the presence of an endohedral atom has a much less drastic effect on the symmetry of the system, and thus the lifting of level degeneracy, in the lower symmetry C_{70} (D_{5h}) than in the icosahedral C_{60}.

Contrary to Li@C_{60} fractions we could not obtain an electronic absorption spectrum of Li@C_{70} in n-hexane. Therefore, UV-vis absorption measurements were carried out between 300 nm and 1100 nm for the saturated endohedral fraction in toluene and also between 400 nm and 1100 nm in CS_2 because of better solubility. Figure 4 displays the spectra of isolated Li@C_{70} (the HPLC trace of the material after three-step separation is shown in the inset of the Figure 4) along with the absorption of the C_{70} fraction in toluene for comparison. The absorption peaks of Li@C_{70} at wavelengths above 400 nm resemble the features of the C_{70} absorption in this region, demonstrating electronic similarity. The peaks for wavelengths below 400 nm are significantly broadened and reduced in peak intensity, possibly due to a splitting into poorly resolved peaks. The absorption onset is at about 760 nm and is shifted by about 90 nm to higher wavelengths with respect to C_{70}.

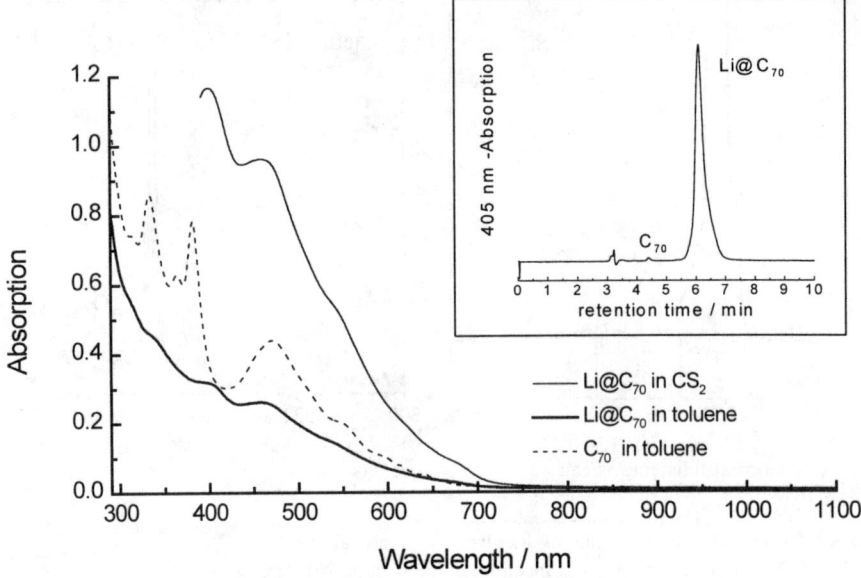

FIGURE 4. UV-vis absorption spectra of isolated Li@C_{70} fraction in toluene (saturated solution, 10mm), CS_2 (10mm) and the C_{70} fraction in toluene (diluted solution, 1mm). The inset shows the HPLC analysis of the Li@C_{70} fraction after three-step separation.

ACKNOWLEDGEMENTS

Financial support from the BMBF (13N6652/7), the EU (HCM Network "Formation, stability and photophysics of fulerenes") and the DFG (Ca 127/2) is gratefully acknowledged.

REFERENCES

1. R. Tellgmann, N. Krawez, S.-H. Lin, E.E.B. Campbell, I.V. Hertel, *Nature,* **382** (1996) 407.
2. E.E.B. Campbell, R. Tellgmann, N. Krawez, I.V. Hertel, *J. of Phys. & Chem. of Solids,* **58** (1997) 1763
3. N. Krawez, R. Tellgmann, E.E.B. Campbell, A. Gromov, W. Krätschmer, in "Molecular Nanostructures", ed. Kuzmany, Fink, Mehring, Roth (World Scientific, 1997) p.184
4. H. Jantoljak, N. Krawez, I. Loa, R. Tellgmann, E.E.B. Campbell, A.P. Litvinchuk, C. Thomsen, *Z. Phys. Chemie,* **200** (1997) 827
5. A. Gromov, W. Krätschmer, N. Krawez, R. Tellgmann, E.E.B. Campbell, *Chem. Comm.*, (1997) 2003
6. Ch. Kusch, N. Krawez, R. Tellgmann, B. Winter, I.V. Hertel, *Appl. Phys. A.*, **66** (1998) 293-298

Heterogeneous Electron Transfer at Endohedral Fullerenes

L. Dunsch, P. Kuran and M. Krause

*IFW Dresden, Abt. Elektrochemie und leitfähige Polymere
Helmholtzstr. 20, D-01069 Dresden, Germany*

Abstract. Endohedral fullerenes exist in a variety of redox states in dependence on the type and the number of metals incorporated. This redox state can be changed by heterogeneous electron transfer reactions. We studied the electron transfer at electrodes by cyclic voltammetry and spectroelectrochemistry. It is demonstrated in this paper for La@C_{82} and Eu@C_{74} that the two or three valency of the metal ion in endohedral fullerenes is caused by the type of the element and not by the shape and redox properties of the carbon cage. For La@C_{82} it is shown that the charge at the carbon cage is varied by heterogeneous electron transfer and in the higher cathodic region a new redox state of La inside of <u>La@C82</u> was found. At the Eu@C_{74} which was also studied by Raman spectroscopy and which has the metal in the 2+ state it is shown by ESR spectroscopy that the electron transfer does not result in a paramagnetic anion.

1. Introduction

Endohedral fullerenes have been shown to exist in a large variety of structures differing in the type of the carbon cage and the element included[1-3]. The first endohedral fullerenes like lanthanum-fullerene La@C_{82}[4], scandium fullerene Sc@C_{82}[5] and yttrium fullerene Y@C_{82}[2,6] had at least two different structures differing significantly in the stability of the isomers as measured by ESR spectroscopy in solution[7].

Our recent work[8-11] is focused on the relation between the structure and the physical properties of endohedral fullerenes. These fullerenes which were characterised by mass spectrometry, cyclovoltammetry, *ex situ* and *in situ* ESR as well as UV-VIS spectroscopy. The aim of the present paper is to describe the electron transfer at some of these fullerenes with respect to their redox state and type of incorporated metal.

2. Experimental

The preparation of the endohedral fullerenes by the arc burning method of Krätschmer and Huffman, the mass spectroscopic characterisation by a sector field mass spectrometer MAT 95 (FINNIGAN, Bremen) and by analytical HPLC after the second separation step have been described elsewhere[13].

Electron spin resonance spectra were recorded by an EMX X-band spectrometer (Bruker) with 100 kHz modulation at 9.5 GHz microwave frequency and

340 mT medium magnetic field strength. The *in situ* electrochemical technique used in this study was described elsewhere[14].

The electrochemical measurements were done at room temperature with a potentiostat PAR 273 (EG&G, USA) including the software PARC M270, version 3.0. A conventional voltammetric cell which was installed in a glovebox under inert conditions was used. As solvent o-dichlorobenzene and as supporting electrolyte tetrabutylammoniumtetrafluoroborate were applied. The working electrode was a platinum wire, the counter electrode a large platinum sheet and the reference electrode a silver-/silverchloride electrode in the same supporting electrolyte.

For the Raman measurements the fullerene solutions were dropped on a KBr disk, dried under ambient conditions and mounted on a probe rotation device rotating with a frequency of 3000 min^{-1}. The Raman spectrum was excited with 514.53 nm radiation of an argon ion laser Innova 305 (Coherent, USA) and analysed in a back scattering geometry with a triple spectrometer T 64 000 (Instruments S.A., France) equipped with a multichannel charge coupled device detector. The spectral slit width was set to 2.3 cm^{-1} and an accumulation time of one hour was used to record the whole spectrum.

3. Results

In o-dichlorobenzene solution La@C_{82} has a pronounced redox behaviour with both anodic and cathodic reactions (Fig. 1). The redox reactions of La@C_{82} are dominated by redox state of both the metal with Me^{3+} and the carbon cage with C_{82}^{3-}. It is to be shown what redox state results at any wave in the voltammogram. This is done by *in situ* ESR spectroscopic measurements. At the open cell potential which is at positive potential of the redox step at +200 mV the ESR signal has its maximum i.e. all the endohedral fullerene molecules are in the $La^{3+}@C_{82}^{3-}$ state. At the first reduction step at +180 mV the La@C_{82} is found to be ESR silent. The reduced state of the endohedral fullerene is therefore attributed to the $La^{3+}@C_{82}^{4-}$ state.

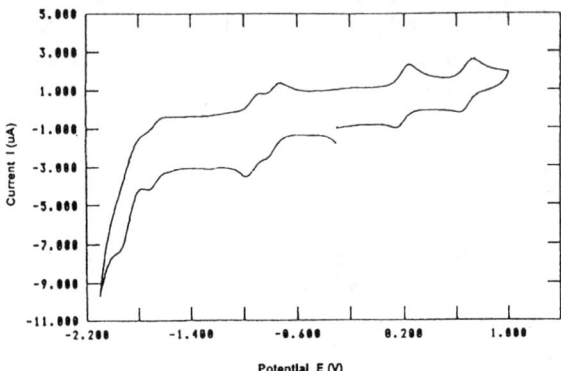

FIGURE 1: Cyclic voltammogram of La@C_{82} in a 10^{-4} mol l^{-1} o-dichlorobenzene solution at a platinum electrode. Scan rate: 100 mV/s.

La@C_{82} undergoes at more cathodic potentials two separated electron transfers. The *in situ* ESR study at potentials at the beginning of this wave (-650 mV) gives a new ESR-signal with the same splitting constant for La of 1.17 G and the g-factor of 2.0011 as in the initial state. This ESR signal is caused by the structure La^{3+}@C_{82}^{5-} and is difficult to obtain it in an undisturbed manner because a second electron transfer in a distance of 150 mV more cathodic gives rise of a new ESR signal of very different parameters. The mixed spectrum is given in Fig. 2 after an electrolysis at -800 mV. The g-value of the new species is changed to 2.0035 and the ESR signal appears to be a single line. The splitting of the La ion is diminished to a value which can not be detected under this experimental condition. The reason for this behaviour is expected to be a change in the position of the ion inside the carbon cage. As the redox state of La@C_{82} should change during the one electron transfer to La^{3+}@C_{82}^{6-} the formal new redox state might have no ESR signal. Therefore it should be taken into account that a different distribution of the transferred electron in the endohedral fullerene might be due to the structure La^{2+}@C_{82}^{5-}. This is only to be explained by an intramolecular electron transfer which was not found up till now at endohedral fullerenes. Otherwise the change in the redox state of the lanthanum ion to 2+ might be a further reason for the decrease of the interaction of the nuclear spin of the lanthanum with the electron spin of the unpaired electron on the carbon cage. Further studies on the state of La@C_{82} after multiple electron transfer in solution are needed.

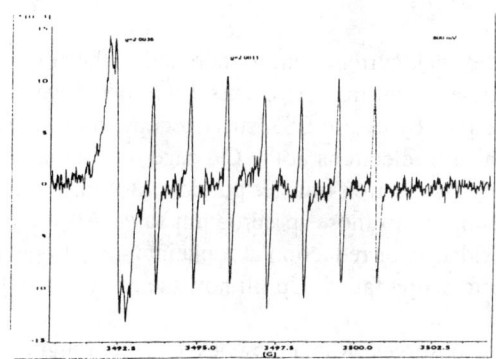

FIGURE 2: ESR spectrum of La@C_{82} in a 10^{-4} mol l^{-1} o-dichlorobenzene solution in potentiostatic electrolysis at a potenial of -800 mV (platinum electrode)

The cyclic voltammogram of Eu@C_{74} (Fig. 2) shows six redox peaks, four of them in the cathodic and two in the anodic potential range. Based on the even number of cathodic electron transfers previously found already for the twovalent Tm@C_{82} in contrast to an odd number for the threefold charged Y@C_{82} and La@C_{82} metallofullerenes, a twovalent electronic state of Eu@C_{74} is supported. Furthermore the cyclic voltammogram can be interpreted in terms of a molecular orbital study for

C_{74} leading to an A_2'-LUMO and an E''-LUMO+1 state[15]. The four cathodic redox peaks of Eu@C_{74} are nearly equidistantly separated and due to the subsequent occupation of the double degenerated LUMO +1 state[15] of C_{74}. This empty LUMO +1 level in the ground state of Eu@C_{74} further supports the two-valent Eu ion.

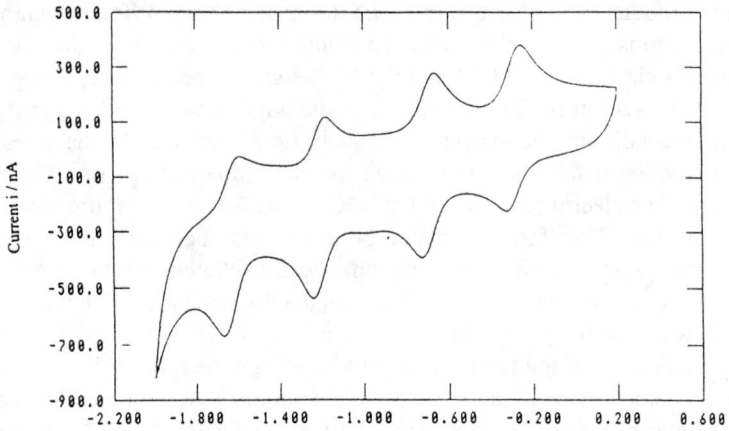

FIGURE 3: Cyclic voltammogram of Eu@C_{74} in o-dichlorobenzene solution at a platinum electrode in the cathodic range. Scan rate: 20 mV/s.

The analysis of the peak currents furthermore led to the conclusion that in both voltammetric peaks a one-electron transfer occurs. Therefore it was obvious to follow the electron transfer reactions by *in situ* ESR spectroscopy in cyclic voltammetry. The existence of an odd number of electrons at the C_{74} cage would make the measurement of paramagnetic structures by electron transfer possible assuming splitting constants of europium in fullerenes similiar to those in europium salts. All the tests to produce a paramagnetic state by oxidation or reduction at a platinum electrode and to detect this state by *in situ* ESR spectroscopy failed. Up till now the reasons for this behaviour are not fully understood.

For Eu@C_{74} the observed Raman modes can be roughly distinguished in internal cage modes between 200 and 1600 cm^{-1} and a special (europium metal - C_{74} cage) mode at 123 cm^{-1} (Fig. 1). The internal modes are extensively discussed elsewhere[16]. For the determination of the redox state of endohedral metallofullerenes the (europium metal - C_{74} cage) mode is of particular importance. Its wave number of 123 cm^{-1} can be compared with our previous results[17] for Tm@C_{82}, Y@C_{82}, La@C_{82} and Gd@C_{82}. A nearly linear increasing frequency was found for the three M^{3+}@C_{82}^{3-} (M = Gd, La, Y) metallofullerenes, the decreasing metal masses could account for. For the three Tm@C_{82} isomers A, B and C significantly lower frequencies of 118, 118 and 116 cm^{-1} were observed. The lower wave numbers were attributed to a lower bonding strength between the included thulium ion and the fullerene cage as the electronic state of these compounds is M^{2+}@C_{82}^{2-}. The wave number of 123 cm^{-1} for

Eu@C_{74} is very close to that of Tm@C_{82} but significantly lower than the value of 152 cm^{-1} found for the neighbour element in the periodic table gadolinium.

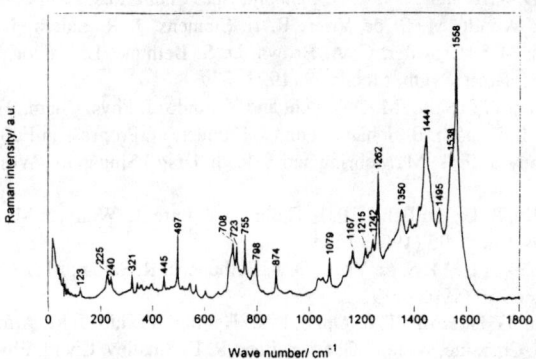

FIGURE 4: Raman spectrum of Eu@C_{74}

As a different cage geometry has little effect on the (metal ion - fullerene cage) Raman frequency and a site effect between different fullerenes is not obvious, the most probably interpretation for the (europium metal - C_{74} cage) frequency of 123 cm^{-1} is a similar bonding strength as in Tm@C_{82}. Therefore an electronic state of Eu^{2+}@C_{74}^{2-} has to be concluded. The sligthly higher frequency of Eu@C_{74} with respect to Tm@C_{82} can be explained by the mass differences between thulium and europium.

4. Conclusions

In the present study it has been shown that the endohedral fullerenes show a significant influence of the fullerene structure on their electrochemical behaviour. For the Eu@C_{74} endohedral structure the 2+ redox state of the metal inside the endohedral fullerene was expected from the absence of an ESR signal. The Raman spectra and the electrochemical properties of Eu@C_{74} are in good agreement with a two-valent electronic ground state of this endohedral Eu fullerene.

By *in situ* ESR spectroelectrochemistry it is shown for La@C_{82} that the charge at the carbon cage is varied by heterogeneous electron transfer and in the higher cathodic region a new state of the endohedral La@C_{82} is found.

5. Acknowledgement

The support of this work by the *Deutsche Forschungsgemeinschaft* and the German *Bundesministerium für Bildung und Forschung* is gratefully acknowledged. Furthermore we thank Dr. Anton Bartl for discussions on ESR-measurements and the technical assistence of Ulrike Feist, Brunhild Schandert and Frank Ziegs.

6. References

1. R. D. Johnson, H. S. de Vries, J. Salem, D. S. Bethune and C. S. Yannoni, Nature 355 (1992) 239; R. D. Johnson, D. S. Bethune and C. S. Yannoni, Acc. Chem. Res. 25 (1992) 169.
   C. S.Yannoni, H. R. Wendt, M. S. de Vries, R. L. Siemens, J. R. Salem, J. Lyerla, R. D. Johnson, M. Hoinkis, M S. Crowder, C. A. Brown, D. S. Bethune, L. Taylor, D. Nguyen, P. Jedrzejewski and H. C. Dorn, Synth. Metals 59 (1993) 279.
2. S. Bandow, H. Shinohara, Y. Saito, M. Ohkohchi and Y. Ando, J. Phys. Chem. 97 (1993) 6101
3. A. Bartl, U. Kirbach, L. Dunsch, B. Schandert and J. Fröhner, in: Progress in Fullerene Research; (H. Kuzmany, J. Fink, M. Mehring and S. Roth, Hrsg.) Singapure, World Scientific 1994, p. 112.
4. Y. Chai, T. Guo, C. Jin, R. E. Haufler, L. P. F. Chibante, J. Fure, L. Wang, J. M. Alford and R. E. Smalley, J. Phys. Chem. 95 (1991) 7564
5. C. S.Yannoni, M. Hoinkis, H. M. S. de Vries, D. S. Bethune, J. R. Salem, M S. Crowder and R. D. Johnson,.Science 256 (1992) 1191.
6. J. H. Weaver, Y. Chai, G. H: Kroll, T. R. Ohno, R. E. Haufler, T. Guo, J. M. Alford, J. Conceicao, L. P. F. Chibante, A. Jain, G. Palmer and R. E. Smalley, Chem. Phys. Lett. 190 (1992) 460
7. A. Bartl, L. Dunsch, J. Fröhner and U. Kirbach, Chem. Phys. Letters 229 (1994) 115; Y. Saito, S. Yokoyama, M. Inakuma and H. Shinohara, Chem. Phys. Letters 250 (1996) 80
8. U. Kirbach and L. Dunsch, Angew. Chem. Int. Ed. Engl. 35 (1996) 2380
9. A. Bartl, L. Dunsch and U. Kirbach, Appl. Magn. Res. 11 (1996) 301
10. A. Bartl, L. Dunsch, U. Kirbach, Solid State Comm. 94 (1995) 827
11. P. Kuran, M. Krause, A. Bartl, L. Dunsch, Chem. Phys. Lett. (1998) in prep.
12. K. Yamamoto, H. Funasaka, T. Takahashi, T. Akasaka, T. Suzuki and Y. Maruyama, J. Phys. Chem. 98 (1994) 12831
13. L. Dunsch, U. Kirbach, P. Kuran, in: "Recent Advances in the Physics and Chemistry of Fullerenes and related materials", Vol. 5 (K. M. Kadish and R. S. Ruoff, Eds.) Pennington, Electrochem. Soc. 1997, in press
14. L. Dunsch and A. Petr, Ber. Bunsenges. Phys. Chem. 97(1993) p. 436
    L. Dunsch, in: Progress in Fullerene Research; (H. Kuzmany, J. Fink, M. Mehring and S. Roth, Eds..) Singapure, World Scientific 1994, p. 482
15. S. Nagase, K. Kobayashi, Chem. Phys. Lett. 276 (1997) 55
16. P. Kuran, M. Krause, A. Bartl, L. Dunsch, Chem. Phys. Lett. (1998) in prep.
17. M. Krause, P. Kuran, U. Kirbach, L. Dunsch, Carbon (1998) in press

Far- and mid- infrared transmission for two isomers of the endohedral metallofullerene $Sc_2@C_{84}$

M. Hulman,[a] M. Inakuma,[b] H. Shinohara[b] and H. Kuzmany[a]

*a)Institut für Materialphysik, Universität Wien,
Strudlhofgasse 4, A-1090 Wien, Austria
b) Department of Chemistry, Nagoya University, Japan*

Abstract. We present IR transmission measurements for two isomers of the endohedral compound $Sc_2@C_{84}$ with D_{2d} (No.23) and D_2 (No.10) symmetry. The measurements were performed in the far- and mid-infrared region between 80 cm^{-1} and 4000 cm^{-1} and at temperatures from 80 K to 300 K. We identified several modes and investigated their dependence on the symmetry of the cage and on the temperature. The latter turned out to be rather strong and can be used to distinguish between modes of the metals in the cage and cage modes.

I INTRODUCTION

Although first samples of endohedral fullerenes were prepared in the early 90's, we have little knowledges about the physical properties of such materials. The difficulty in the preparation is the main reason for this situation. The endohedrals are available only in fractions of miligrams. Moreover they are often mixed-up of two or more isomers of the same cage. To obtain isomer-clean samples is even more difficult. In this contribution we report IR transmission measurements for two different isomers of the C_{84} cage filled with two scandium atoms.

There is some knowledges about the vibrational properties of the empty C_{84} cages as mid-IR measurements were performed previously [1]. As expected due to their lower symmetry many lines were observed distributed in an interval between 500 and 1800 cm^{-1}. Unfortunately no measurements have been reperted for the far-infrared which is the most interesting frequency range for us. The calculations for C_{84} tell us that there are no cage vibrations below $\approx$ 200 cm^{-1} [2]. One expects a gap between intramolecular and intermolecular modes. Due to the higher mass of the C_{84} molecule the modes must have an energy even lower than in C_{60} where the higest intermolecular mode is at $\approx$ 50 cm^{-1}. In the endohedrals the gap can be filled with modes originating from mutual metal - cage vibrations. This was observed very recently for metallofullerenes based on C_{82} [3] and confirmed by calculations [4]. Metall-cage modes were found at energies between 150 cm^{-1} and 190 cm^{-1} for different encapsulated atoms. The frequency scaled with the inverse square root of the reduced mass.

In the $Sc_2@C_{84}$ isomer with D_{2d} symmetry two scandium atoms are located on a twofold axes, 2.36 Å away from the carbon cage and 4.03 Å apart from each other. This geometry leaves the D_{2d} symmetry unchanged [5]. The scandium atoms can oscillate about their local energy minima but as a consequence of their strong separation they do not move as a dimer. Unfortunately no calculations of infrared spectra of $Sc_2@C_{84}$ were reported sofar and only one paper on FIR spectra was published up to now [6].

The temperature dependence of IR modes gives a deeper insight into the dynamics of the endofullerenes. It reflects anharmonic effects and can be used to distinguish between different mechanisms of interactions. One may observe order-disorder phase transitions as in C_{60}. In the $Sc_2@C_{84}$ both scandium atoms are in +2 state and an electrostatic interaction is the most important. The different nature of the bonding between the carbon atoms on the carbon cage and the metal atoms to the carbon leads to a different temperature dependence which we observed in our measurements.

II RESULTS AND DISCUSSION

We measured two isomers of $Sc_2@C_{84}$ with D_{2d} (No.23) and D_2 (No.10) symmetry. Both spectra are shown in Fig.1.in the far-infrared region.

The spectra are very similar. They consist of three well resolved parts. We see several peaks above 300 cm^{-1}, a large gap down to 100 cm^{-1} and a broad feature between 90 and 11 cm^{-1}. (The small feature at 225 cm^{-1} in the spectrum No.23 is not relyable reproduced from several runs.) Positions of the peaks are equal within a few wavenumbers. Thus, symmetry affects a vibration energy only very weakly. The most surprising fact is the large gap between 300 and 100 cm^{-1}. According to calculations the lowest cage mode is at $\approx$ 200 cm^{-1} in pure C_{84}. A presence of two atoms and a charge transfer to the cage can renormalize the energy as well as the number of the vibrations.

For us the most interesting part of the FIR spectra are the broad bands between 90 and 110 cm^{-1}. The exact frequency depends slightly on the symmetry. For a simple model where two scandium atoms are oscillating against the rigid cage two solutions with an energy difference of $\approx$5% were obtained in a harmonic approximation. The difference increases to 10% for a model in which each scandium atom has its own force constant (reflecting a different environment of the Sc atoms). In fact we observed for the isomer No.10 two components of the band with wavenumbers 98 and 107 cm^{-1} (difference 9%) and for the isomer No.23 two components with wavenumbers 90 and 102 cm^{-1} (difference 13%).

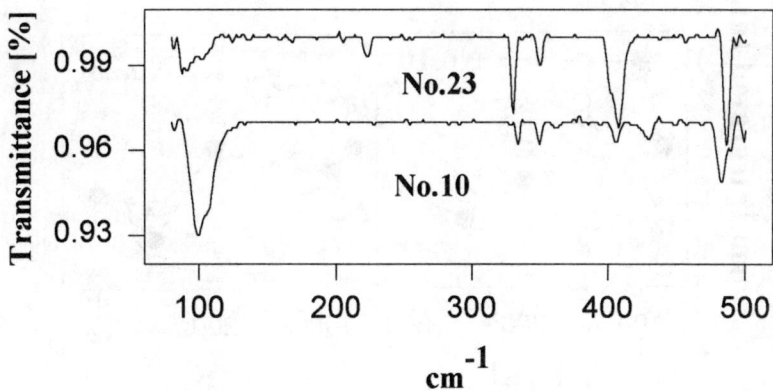

FIGURE 1. FIR transmission spectra of the isomers No.10 and 23 of $Sc_2@C_{84}$

Measurements of the temperature dependence of the spectra confirmed the conclusion that the peaks below 110 cm^{-1} originate from scandium-cage vibrations. Results are shown in Fig.2 where integrated intensities as a function of temperature are plotted for several modes in the far- and mid- infrared region.

The integrated intensities were analyzed rather then linewidths and peak positions of the modes since peaks exhibited a strong splitting into several components. The price one pays for this is a more difficult physical interpretration of the results.

For the isomer No.23 the integrated intensities of all cage modes investigated behaved similarly and exhibited a maximum at 100 K. As seen from Fig.2 the temperature dependence of the metal-cage mode is different. Similar results were observed for the isomer No.10 with the only difference that the maximum of the integrated intensity is shifted to 150 K. As mentioned above the physical interpretation is very difficult. For better understanding we performed a single line analysis for one cage mode.

The linewidth narrows exponentially below and above 150 K but with a markedly different rate (not shown here). The mechanism for the narrowing is the same in the whole temperature region and is assumed to be a coupling to thermal activation of diffuse motion. The endohedral molecules remain disordered at all temperatures and no order-disorder phase transition is observed. This agrees with temperature dependent X-ray diffraction measurements of C_{84} [7]. Authors found a fcc structure from room temperature down to 5 K and neither order-disorder nor structural phase transitions were observed. They also measured a lattice constant as a function of the temperature an found a linear dependence and a rate changing at 140 K.

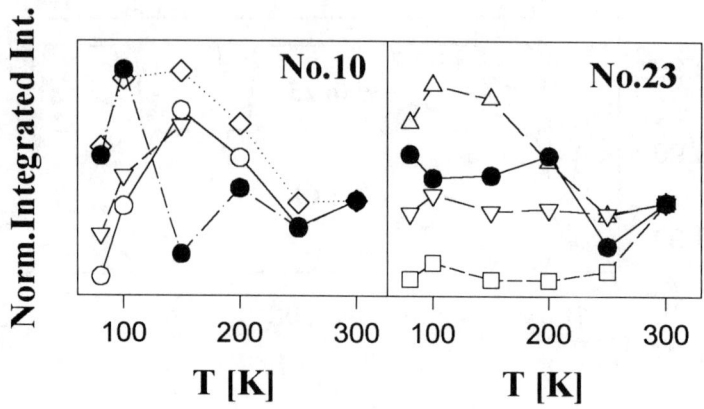

FIGURE 2. The normalized integrated intensity in the isomers No.10 (● 100 cm^{-1}, ◇ 406 cm^{-1}, ▽ 482 cm^{-1}, ○ 648 cm^{-1}) and 23 (● 100 cm^{-1}, ▽ 680 cm^{-1}, △ 764 cm^{-1}, □ 1130 cm^{-1})

Summarizing we investigated two isomers of the endohedral metallofullerene $Sc_2@C_{84}$. We found the metal-cage vibrations at low frquencies and explained their two component shape. The metal-cage character of the vibrations was further supported with the temperature dependent measurements.

ACKNOWLEDGMENT

Work supported by the FFWF in Austria, Project No. P 11943.

REFERENCES

1. A.G. Avent et al., J.Chem.Soc., Perkin Trans.2, 1907, (1997)
2. B. L. Zhang et al., Phys. Rev. B 47, 1643 (1993)
3. S. Lebedkin et al., Appl. Phys. A 66, 273 (1998)
4. W. Andreoni et.al., Appl. Phys. A 66, 299 (1998)
5. K. Kobayashi et al., Chem. Phys. Lett. 261, 502, (1996)
6. S. M. Grannan et al., Chem. Phys. Lett. 264, 359 (1997)
7. S. Margadonna, private communication

Electron Paramagnetic Resonance Investigation of Phosphorus and Nitrogen in [60]Fullerene

Claus Knapp*, Adnan Adla*, Norbert Weiden*, Hanno Käß*, Klaus-Peter Dinse*, Björn Pietzak**, Markus Waiblinger**, and Alois Weidinger**

*Phys. Chem. III, TU Darmstadt, D-64287 Darmstadt, Petersenstr. 20, Germany
**Hahn-Meitner Institut, D-14109 Berlin, Glienickerstr. 100, Germany

Abstract. Atomic Nitrogen and Phosphorus can be encapsulated in fullerenes acting as highly stable chemical traps. In solution, the neutral atoms in their quartet spin ground state are weakly coupled to the surrounding by zero-field-splitting interaction, resulting either from collision-induced deformations of the fullerene cage or by intrinsic deviations from spherical symmetry. From an analysis of the spin relaxation times, values for the correlation time and the mean square value of the interaction can be deduced.

INTRODUCTION

Recently it was shown that ion implantation is a convenient technique to generate endohedral fullerenes (1). As compared to the well established technique of using an arc discharge in a Helium atmosphere under addition of metal oxides or carbides, this new method has the tremendous advantage that the fullerene cage is well defined by choice of the substrate. Subsequent purification techniques are in principle only necessary to separate empty fullerenes from endohedral species, both of identical carbon mass. Ion implantation has successfully been used to incorporate noble gases (2), alkali ions (3) and group V atoms (1) into C_{60}, by which for the first time the most symmetric fullerene is available as „cage" in macroscopic quantities. Whereas EF, in which alkali or group IIIb atoms are incorporated, are always characterised by significant charge transfer from the centre atom to the carbon cage, in case of nitrogen or phosphorus no indication of charge or spin transfer is detected. From this it can be concluded that these highly reactive atoms are shielded from the environment, and that fullerenes can be envisaged as nearly perfect chemical traps. Magnetic resonance techniques can be used to detect the remaining weak interactions because the encased atoms in their spin quartet ground state are sensitive probes for the presence of rank 2 tensor interactions, analogue to probing for electric field gradients with nuclear quadrupole moments. Permanent deformations of the cage would lead to characteristic powder spectra when incorporating the EF in a solid matrix, whereas fluctuating deformations induced, e.g., by collisions with solvent molecules, would be detectable by measuring electronic spin relaxation rates in solution. For this study we investigated $N@C_{60}$ and $P@C_{60}$ in solution to obtain an order-of-magnitude information about the variance of the time-fluctuating zero-field-splitting (ZFS) in low-viscous solutions. Additionally we studied $N@C_{60}$ of different cage symmetry in solid matrices and in solution for a determination of the (permanent) ZFS tensor as well as of the correlation time of this interaction, results of which are given in an additional contribution of this volume (4).

EXPERIMENTAL

P@C_{60} was prepared by ion implantation using the same method as described previously (1). N@C_{60} and N@C_{70} were generated using a gas discharge set-up (5). EPR spectra were taken either from poly-crystalline powders or by preparing carefully degassed solutions. Pulsed and c.w. EPR spectra were recorded with a BRUKER ELEXSYS 680 X spectrometer.

RESULTS AND DISCUSSION

Fig. 1 shows c.w. EPR spectra of N@C_{60} and N@C_{70} in solution. The spectra are practically identical except for a small difference in their ^{14}N hyperfine coupling constant (hfc). The peak-to-peak line width in both spectra is mainly determined by a static magnetic field inhomogeneity of the order of 10 mG. The homogeneous width at room temperature as determined from a 2 pulse echo decay function was measured as 6.5 kHz for the C_{60} cage (in toluene) and 8(1) kHz (in CS_2) for the C_{70} cage, respectively.

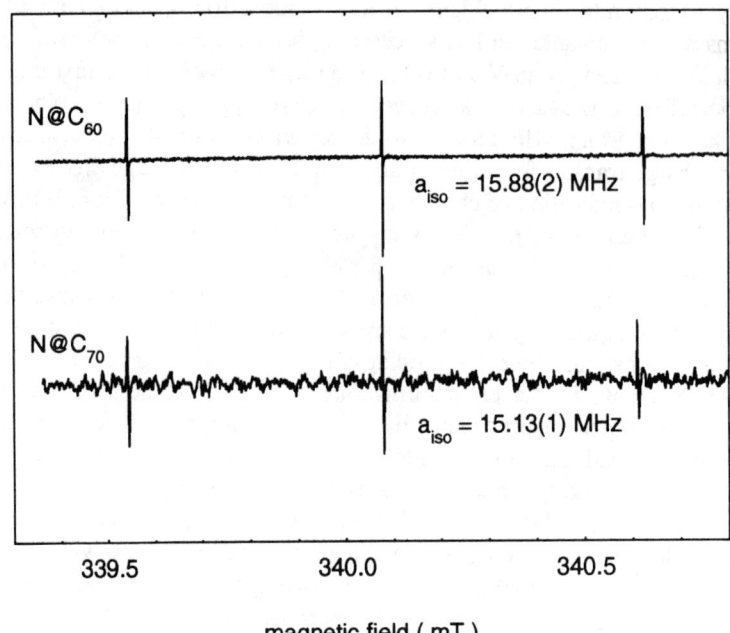

Figure 1. EPR spectra of N@C_{60} and N@C_{70} in CS_2 taken at room temperature. For better comparison, the spectra have been centered.

In Fig. 2, EPR spectra of N@C$_{60}$ and P@C$_{60}$ are compared. The interaction of the quartet electronic spin with the nuclear spin I = 1/2 of the ^{31}P nucleus leads to a line doublet which has additional fine structure, because transitions between (3/2, 1/2), (1/2, -1/2), and (-1/2, -3/2) electron spin levels, respectively, are no longer degenerate because of second-order hfi. To our knowledge these transitions have never been observed before separately, because homogeneous line broadening by strong ZFS interactions usually masked the splittings. (As was noted before (6), the same mechanism leads to incipient line broadening in N@C$_{60}$, observable via the reduced intensity of M_I = 1 components.)

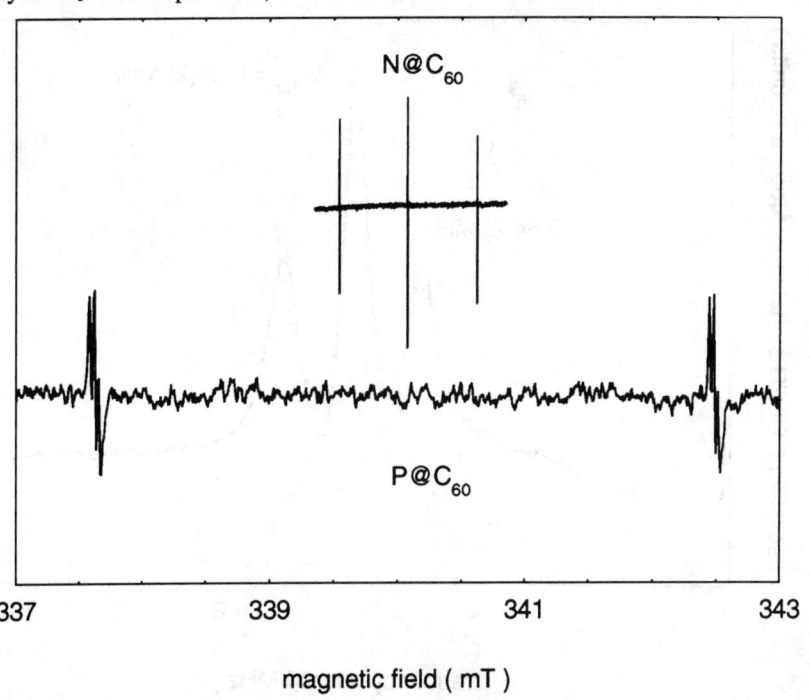

Figure 2. EPR spectra of N@C$_{60}$ and P@C$_{60}$ in CS$_2$ at room temperature

In contrast to N@C$_{60}$, in which significant inhomogeneous line broadening masks the homogeneous line width, necessitating enables spin echo experiments for a determination of T$_2$, in P@C$_{60}$ the lines can be described by Lorentzians, thus allowing for an easy extraction of T$_2$ of all allowed electron spin transitions. As is described in detail in a forthcoming publication (7), the relaxation rates of these transitions are given by Eq. 1, assuming relaxation by dominant ZFS interaction. Here, D denotes the principal value of the traceless ZFS tensor, E describes the asymmetry of the traceless interaction, ω is the Zeeman frequency, and τ gives the correlation time of the orientation of the ZFS tensor in its molecular frame with respect to the external field.

In the short correlation limit, the ratio of the subcomponent linewidth is predicted as 3/2, in good agreement with the experimental data recorded at room temperature. As is shown in Fig. 3, the absorption mode spectra (which were measured with

$$T_2^{-1}(1/2,-1/2) = \frac{4}{5}(D^2+3E^2)\left[\frac{\tau}{1+\omega^2\tau^2}+\frac{\tau}{1+4\omega^2\tau^2}\right]$$

$$T_2^{-1}(3/2,1/2) = T_2^{-1}(-1/2,-3/2) = \frac{4}{5}(D^2+3E^2)\left[\tau+\frac{\tau}{1+\omega^2\tau^2}+\frac{\tau}{1+4\omega^2\tau^2}\right] \quad (1)$$

FT-EPR) can be fitted with Lorentzians of width 370(20) and 540(20) kHz, respectively, At lower temperatures larger ratios are observed, allowing for a parameter-free determination of τ and subsequently of $D_{eff}^2 = (D^2+3E^2)$.

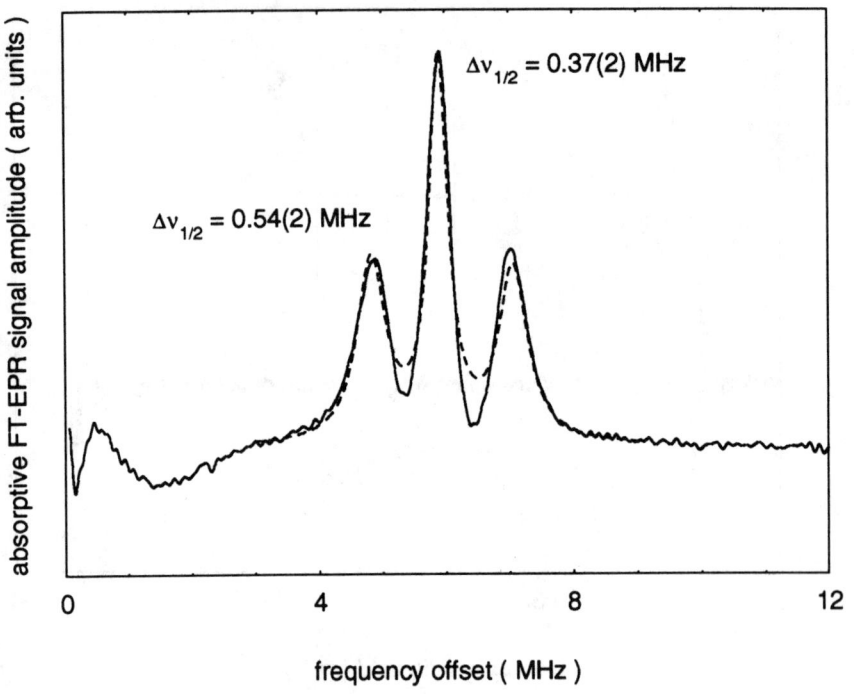

Figure 3. FT-EPR derived EPR absorption lines of the low field multiplet of P@C$_{60}$. The dotted line is a fit of the data assuming Lorentzian line shapes.

For the temperature range 193 to 293 K in toluene, τ ranges from 32 to 5 ps. Considering the large error which has to be attributed to the "high temperature" values which are calculated from data close to the limiting value of $T_2^{-1}(1/2, 3/2)/T_2^{-1}(1/2, -1/2) = 3/2$, the observed variation of τ by one order of magnitude compares well with a predicted variation of the translational correlation time $\tau_{trans} \sim \eta/T$ of toluene, which varies from 0.00155 to 0.032 cP/K in the same temperature interval. The finally deduced values of $D_{eff} = 60(5)$ MHz is a measure for the fluctuating deformation of the C$_{60}$ cage sensed by the 3p valence electrons.

In case of N@C$_{60}$ because of the lack of resolved second-order hyperfine splitting, τ and D_{eff} were determined by measuring T_1 and T_2, although quantum beats arising from the 255 kHz subcomponent splitting which modulates the spin echo decay lead to a significant uncertainty in the determination of T_2. From the measured reatio T_1/T_2 = 2.4(5), the correlation time of N@C_{60} in toluene at 300 K is estimated as τ = 10(2) ps. The difference of values determined for N@C_{60} and P@C_{60} in the same solvent is attributed to the error made in measuring T_2 from the partially overlapping lines. The value of D_{eff}(N@C_{60}) corresponding to τ = 10 ps is calculated as 6(1) MHz. This value is quite insensitive to the particular choice of τ because of its square-root dependence.

CONCLUSION

From an analysis of spin relaxation data it can be concluded that Nitrogen as well as Phosphorus in C_{60} can be described as atoms being confined in a highly stable and efficient "chemical trap". Interaction of the quartet state atoms with the surrounding molecules is promoted via a fluctuating ZFS, originating from solvent-induced random deformations of the fullerene cage. The dominance of this interaction was confirmed by performing a two-dimensional EPR experiment. The absence of cross peaks in an EXSCY type spectrum can be taken as evidence that nuclear spin-dependent terms, originating, e. g., from nuclear quadrupole interaction or electron-dipolar interaction, are practically absent.

ACKNOWLEDGEMENT

We gratefully acknowledge financial support from the "Deutsche Forschungsgemeinchaft" (Di182/19-1).

REFERENCES

1. Almeida Murphy, T., Pawlik, T., Weidinger, A., Höhne, M., Alcala, R., and Spaeth, J.-M., *Phys. Rev. Lett.* **77**, 1075 (1996).
2. Saunders, M., Cross, R. J., Jimenez-Vazquez, A., Shimshi, R., and Khong, A., *Science* **271**, 1693 (1996).
3. Tellgmann, R., Kravez, N., Lin, S.-H., Hertel, I. V., and Campbell, E. E. B., *Nature* **382**, 407 (1996).
4. Gruß, A., Knapp, C., Weiden, N., Dinse, K.-P., Dietel, E., Hirsch, A., Waiblinger, M., Pietzak, B., and Weidinger, A., *Proc. of the XIth Intern. Winterschool*, Kirchberg 1998.
5. Pietzak, B., Waiblinger, M., Almeida-Murphy, T., Weidinger, A., Höhne, M., Dietel, E., and Hirsch, A., *Chem. Phys. Lett.* **279**, 259 (1997).
6. Knapp, C., Dinse, K.-P., Pietzak, B., Waiblinger, M., and Weidinger, A., *Chem. Phys. Lett.* **272**, 433 (1997).
7. Knapp, C., Weiden, N., Käß, H., Dinse, K.-P., Pietzak, B., Waiblinger, M., and Weidinger, A., *Mol. Phys.* - submitted -

Thermal stability of N@C_{60}

M. Waiblinger*, B. Pietzak*, K. Lips**, A. Weidinger*

* Hahn-Meitner-Institut Berlin, Glienicker Straße 100, D-14109 Berlin, Germany
** Hahn-Meitner-Institut Berlin, Rudower Chausse 5, D-12489 Berlin, Germany

Abstract. N@C_{60} (atomic nitrogen inside C_{60}) is stable at ambient conditions for long periods. No loss of EPR signal intensity is observed after storage for several months in air. However, at approximately 480 K the EPR signal intensity starts to decrease. We present here an annealing study using EPR to measure the amount of N@C_{60} remaining in the sample at the respective temperature. We find an activation energy for the escape process of approximately 1.7 eV. The present data show that the disintegration occurs in different steps. Possible reasons will be discussed.

INTRODUCTION

N@C_{60} is produced by ion implantation as described earlier (1,2). The main characteristic of this system is that the encapsulated nitrogen keeps its atomic ground state configuration and that it is inert towards binding to the cage. Such features have been observed so far only for noble gas atoms which are inert by themselves (3). The remarkable stability of N@C_{60} manifests itself also in the fact that N@C_{60} survives chemical reactions (4). In this work the thermal stability of N@C_{60} is examined by temperature dependent EPR investigations.

EXPERIMENTAL

The encapsulated nitrogen is paramagnetic, therefore the EPR signal intensity (multiplied by the absolute temperature in order to account for the temperature dependent sensitivity of EPR) is a convenient measure of the amount of N@C_{60}. Both solid and liquid samples of N@C_{60} were investigated. The solid sample was prepared by filling N@C_{60} powder into an EPR-tube and evacuating the tube to 10^{-6} mbar in order to avoid oxidation effects. The liquid sample contains N@C_{60} in 1-Chloronaphthalene, which is a solvent with a high boiling point (532 K). The samples were inserted in a cw-X-band spectrometer and heated in a gas flow in a range from room temperature up to 600 K for the solid samples and 520 K for the liquid ones. EPR spectra were taken after every 2 K step in temperature.

RESULTS AND DISCUSSION

Figure 1 shows the quantity of $N@C_{60}$ derived from the EPR signal intensity. In this isochronal annealing experiment the temperature was increased with a rate of 1 K in 41 seconds. In the lower part of the figure the differentiated curve, which corresponds to the percentage loss of $N@C_{60}$ per Kelvin, is displayed. It can be seen that the major disintegration of $N@C_{60}$ starts around 490 K and is completed at 580 K. Surprisingly the data indicate that the disintegration occurs in at least two different steps. This will be discussed in more detail below.

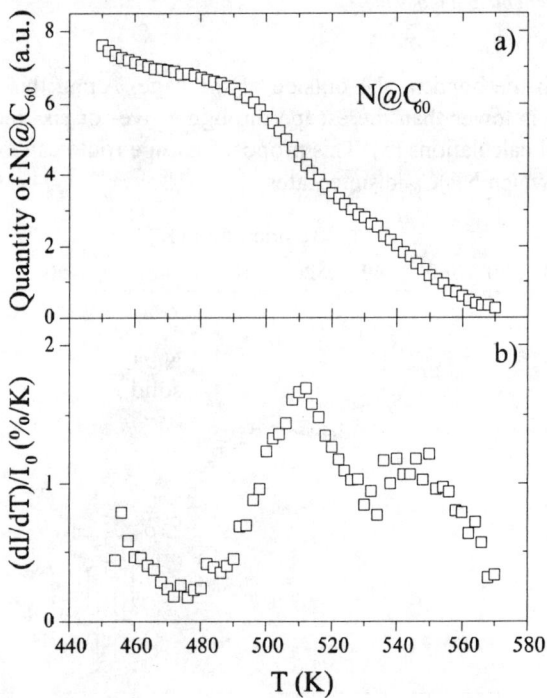

FIGURE 1. *Isochronal annealing of $N@C_{60}$. The heating rate was 1 K in 41 seconds. a) Quantity of $N@C_{60}$ measured by EPR at different temperatures. b) Differentiated curve which corresponds to the fractional loss of $N@C_{60}$ per K.*

The disintegration of $N@C_{60}$ occurs probably via the process depicted in Fig. 2: The nitrogen atom moves away from the centre, forms a bond with two carbon atoms and

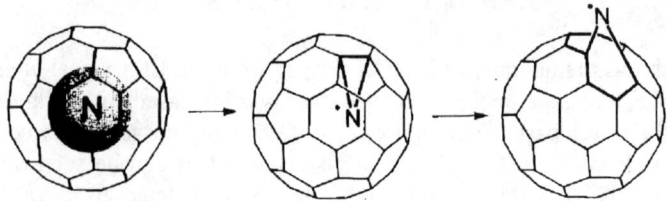

FIGURE 2. *Model for the escape of nitrogen out of C_{60}. In a first step the nitrogen forms an azabridge with two carbon atoms from the inside of the cage, and in a second step it leaves the cage by swinging through the bond to the outside.*

then swings through the bond to the outside of the cage. Along this way the barrier towards the outside is lower than the escape through a five- or six- membered ring as shown by theoretical calculations (5). This proposed escape route can explain the rather low temperature at which N@C_{60} disintegrates.

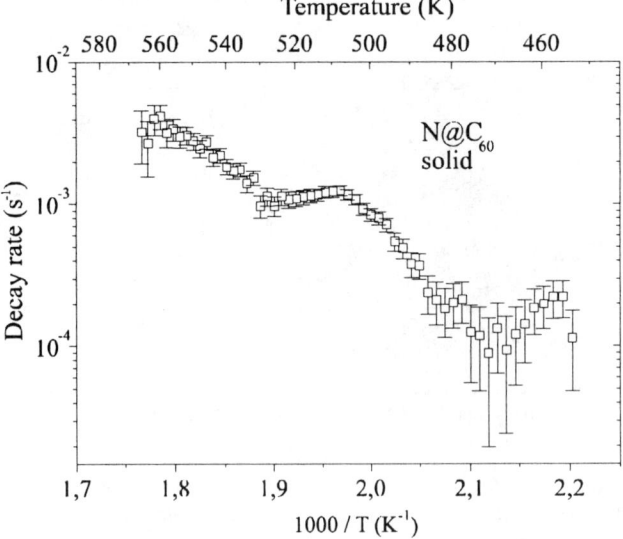

Figure 3: *Decay rate λ as function of the reciprocal temperature.*

Figure 3 shows the decay rate λ as a function of the reciprocal temperature. λ was calculated according to

$$\lambda = \frac{I(i) - I(i+1)}{I(i) \cdot \Delta t} \qquad (1)$$

where $I(i)$ is the EPR intensity for measurement number i and Δt the time between measurement number i and $i+1$. The different annealing steps mentioned above are again seen on this plot. These data can not be explained with two different escape mechanisms but require the assumption of different kinds of N@C_{60}. The increase of λ between 480 K and 500 K is attributed to species one of N@C_{60}, that between 530 K and 560 K to species two. The flat part between 500 K and 520 K arises from the fact that species one with a high decay rate dies out and species two with a lower decay rate progressively takes over and determines the effective decay rate in this temperature range.

The investigation of N@C_{60} in solution shows decomposition in the same temperature range as the solid sample beginning at around 480 K. The solution data indicate also that the decay occurs in different steps. However, these data are less reliable since, due to the thermal agitation in the solution, particles were ejected from the active area and deposit at other places in the EPR tube mimicking a decay of N@C_{60}. In general, however, the solution data confirm the results described above for the solid sample.

CONCLUSION

N@C_{60} disintegrates at a rather low temperature (above 500 K) compared to e.g. He@C_{60} which is stable up to 800 K. The reason is that the escape process is quite different for these two endohedral systems. Whereas He is a passive partner which takes the chance to escape after the opening of the cage by thermal excitation, nitrogen takes an active part in the opening of the shell by forming a bond with two carbon atoms and thereby lowers the barrier for the escape. Theoretical calculations show that this process results in the lowest barrier for the way out.

The present data show distinct annealing steps, both for the powder sample and for the sample in solution. These data indicate that two or more different species of N@C_{60} exist in the sample. The reason for these different species is not clear. At present we cannot decide whether these effects are intrinsic to the N@C_{60} molecule or induced by environmental effects. An interesting speculation is that the effect is due to the different isotopes of carbon. With a natural abundance of 1.1% ^{13}C, 47% of the C_{60} molecules contain no ^{13}C, 36% contain one ^{13}C, and 13% two ^{13}C. The influence of these ^{13}C atoms might show up in changes in the vibration modes leading to a change in the activation energy required for the escape reaction.

REFERENCES

1. T. Almeida Murphy, T. Pawlik, A. Weidinger, M. Höhne, R. Alcala, J.M. Spaeth, *Phys. Rev. Lett.*, 77, 1075 (1996)
2. A. Weidinger, M. Waiblinger, B. Pietzak, T. Almeida Murphy, *Applied Physics A* 66, 287 (1998)
3. M. Saunders, R.J. Cross, A. Jimenez-Vazquez, R. Shimshi, A. Khong, *Science* 271, 1693 (1996)
4. B. Pietzak, M. Waiblinger, T. Almeida Murphy, A. Weidinger, M. Höhne, E. Dietel, A. Hirsch, *Chem. Phys. Lett.* 279, 259 (1997)
5. H. Mauser, N.J.R. van Eikema Hommes, T. Clark, A. Hirsch, B. Pietzak, A. Weidinger, L. Dunsch, *Angew. Chem. Int. Ed. Engl* 36, No. 24, 2835 (1997).

Stabilisation of Atomic Elements Inside C_{60}

Harald Mauser, Timothy Clark and Andreas Hirsch

*Institut für Organische Chemie and Computer Centrum Chemie
Henkestr. 42, D-91054 Erlangen, Germany*

Abstract: In this study we present a topological explanation for the reactivity of elements encapsulated in C_{60}. Performing computations with semiempirical and DFT methods we investigated the influence of the charge distribution of the endohedral fullerene complexes on their chemical behaviour. In contrast to electropositive metals an N-atom is most stable uncharged in the centre of the fullerene. Due to the formation of an intermediate with an endohedral aza-bridge the barrier of extrusion of the N-atom through a CC-bond is comparatively low. The rigid concave shape of the interior of C_{60} and the absence of charge transfer to the fullerene cage are responsible for the astonishing inert behaviour towards N-atoms.

INTRODUCTION

Recently, the endohedral complex of nitrogen in C_{60} was described [1]. As conduced from EPR- and ENDOR-spectroscopic analyses the nitrogen atom does not undergo any reaction with the fullerene cage in contrast to electropositve metals encapsulated by C_{60} [2]. According to resent calculations [3] and EPR measurements on endohedral complexes of higher fullerenes [4] there is a charge transfer from the enclosed metal atom to the fullerene cage. In this computational study we compare the charge distribution in the endohedral complexes of Li, N, and F with C_{60} and discuss possible follow-up reactions. All semiempirical calculations were performed with the VAMP 7.0 [5] program using the PM3 [6] Hamiltonian within the unrestricted Hartree-Fock formalism.

CHARGE DISTRIBUTION AND REACTIVITY

There are three types of endohedral fullerene complexes represented by $Li^+C_{60}^-$, NC_{60} and $F^-C_{60}^+$. The polarisation of the fullerene cage is visualised in figure 1 [7]. The plotted surface is obtained by subtraction of the electrostatic potentials of the fullerene cage with and without the encapsulated guest atom. The geometries represent the optimised structures of the endohedral complexes. Whereas nitrogen and fluorine are most stable in the centre of the cage lithium is off-centred. This position is accompanied by a partial charge transfer of about 0.6 while the charge of the in-centre atoms N and F is nearly zero according to PM3-UHF. The density functional

UB3LYP/DZ95* basis set, however, provides a Mullikan charge of -0.53 for F and almost zero for ^{4}N [8].

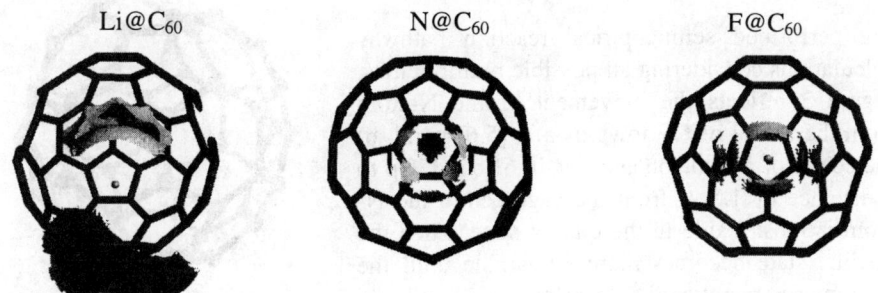

Figure 1

The chemical behaviour of these endohedral compounds depends on the charge transfer interaction between the guest-atom and the fullerene cage. In case of nitrogen (S=3/2) the unpaired electrons are symmetrically distributed in three degenerated orbitals. There is no charge transfer according to our calculations and the EPR measurements [1]. The chemical behaviour is indistinguishable from C_{60} [9]. Calculations done on Li@C_{60} [3, 10] show that the lithium ion inside C_{60} is off-centred due to the attractive interaction with a double bond. The transferred charge is basically located on two carbon atoms of a [6,6]-bond. Thus, the chemistry should be similar to C_{60}^- or the isoelectronic $C_{59}N$ which is known to form a dimer or to add H leading to $C_{59}NH$. In our calculation the formation of a dimer (scheme 1) is an exothermic reaction of -19.4 kcal/mol (PM3-UHF). Recent experiments of the Campbell group (these proceedings) corroborate these findings.

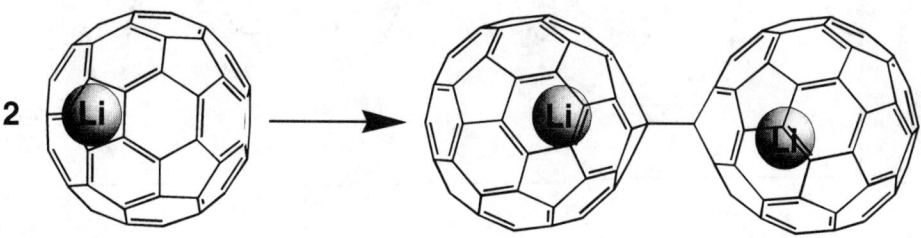

Scheme 1

For F@C_{60} we expect a different behaviour because, if any, there is an electron charge transfer from the cage to the encapsulated F-atom. Since C_{60}^+ is very reactive, a polarisation of the endohedral complex may induce an exohedral addition or electron impact to give FC_{60} or F@C_{60}X. In both cases a remarkable stabilisation of the endohedral fullerene complex is calculated (-20.7 kcal/mol for F@C_{60}H). Whereas in F@C_{60} the F-anion is located in the centre the formation of an endohedral C-F bond moves the F-atom away from the centre to the C_{60} shell (figure 2).

MECHANISM FOR THE EXTRUSION OF THE N-ATOM OUT OF THE CAGE

We performed semiempirical reaction pathway calculations considering all possible multiplicities. Figure 3 reflects the movement of the N-atom from the centre of C_{60} towards a) a [5,6]-bond, b) the centre of a hexagon and c) a [6,6]-bond. Up to a distance of 1.5 Å from the cage centre the N-atom is most stable in the quartet state. Then the doublet state becomes more favourable until the N-atom has penetrated a [5,6] or [6,6] bond and reaches its exohedral global minimum. After dissociation of the C-N bonds the quartet spin state is most stable again. Close to the transition states when bond opening occurs both spin states are very close in energy. A comparison of the energies of the stationary points with the DFT results is given in lit. [8b].

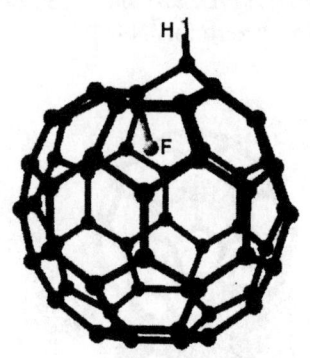

Figure 2

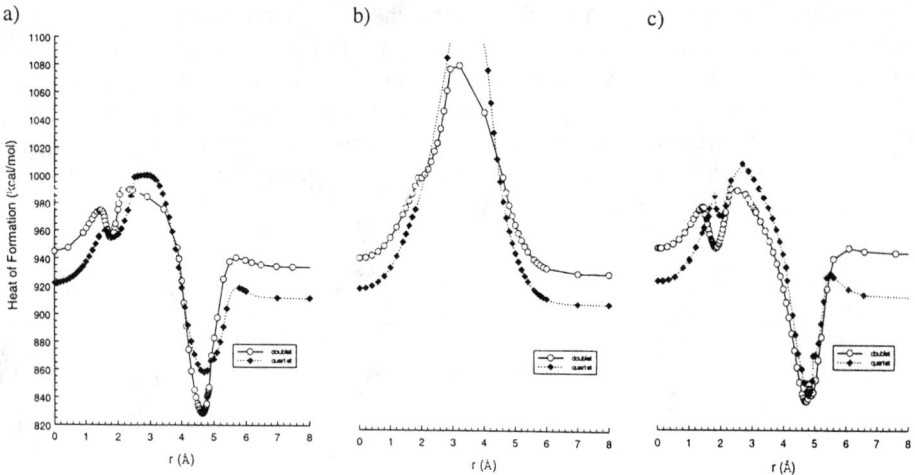

Figure 3

The barrier for the extrusion of an N-atom through a CC-bond is about 100 kcal/mol lower than that through the centre of a hexagon. Thus, the calculated activation energy of about 60 kcal agrees well with temperature dependent EPR experiments [8b] (scheme 2).

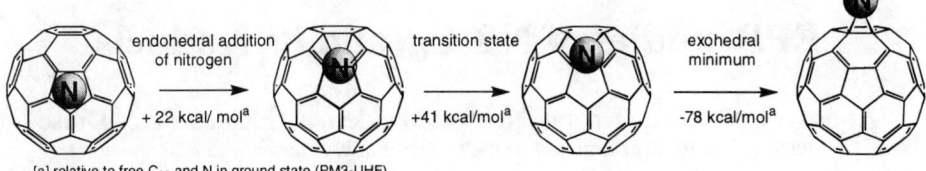

[a] relative to free C_{60} and N in ground state (PM3-UHF)

Schema 2

The barrier (40-50 kcal/mol PM3-UHF) for the extrusion of atomic oxygen through a [6,6]-bond is even lower. For the endohedral $O@C_{60}$ there are two nearly isoenergetic minima. The oxygen-atom is either located in centre in its triplet ground state or is bound to a double bond as a singlet epoxide.

CONCLUSION

We demonstrate that the reactivity of endohedral fullerene complexes depends both on the shape of the rigid carbon network and on the electron affinity of the guest atom. A very special situation is represented by $N@C_{60}$ were the atomic nitrogen does not undergo any reaction with the fullerene cage. The extrusion of enclosed N or O-atoms out of C_{60} can occur via low energy pathways due to the formation of an intermediate endohedral addition of the guest atom with a CC bond of the cage.

REFERENCES AND NOTES

1. Almeida Murphy, T., Pawlik, T., Weidinger, A., Höhne, M., Alcala, R., Spaeth, J. M., *Phys. Rev. Lett.*, **77**, 1075 (1996).
2. a) Bethune, D. S., Johnson, R. D., Salem, J. R., de Vries, M. S., Yannoni, C. S., *Nature*, **366**, 123 (1993); b) Tellgmann, R., Krawez, N., Lin, S.-H., Hertel, I. V., Campbell, E. E. B., *Nature*, **382**, 407 (1996).
3. Andreoni, W., Curioni, A., *Appl. Phys. A.*, **66**, 299-306 (1998).
4. a) Knapp, C., Weiden, N., Dinse, K.-P., *Appl. Phys. A.*, **66**, 249-256 (1998); b) Knorr, S., Krupp, A., Mehring, M., Kirbach, U., Bartl, A., Dunsch, L., *Appl. Phys. A.*, **66**, 257-264 (1998).
5. Clark, T., Alex, A., Beck, B., Chandrasekhar, J., Gedeck, P., Horn, A., Hutter, M., Rauhut, G., Sauer, W., and Steinke, T., *VAMP 7.0*, Oxford Molecular Ltd., Madawar Centre, Oxford Science Park, Standford-on-Thames, Oxford, OX4 4GA, England, 1997.
6. Stewart, J.J., *J. Comput. Chem.*, **10**, 209; 221. (1989).
7. To achieve a clear presentation only the potential surfaces with the highest or lowest charge are shown. The dark regions represent negative, the pale grey positive charged van der Waals surface.
8. a) Mauser, H., Hirsch, A., Hommes, N. v. E., Clark, T., *J. Mol. Model.*, **3**, 415-422 (1997); b) Mauser, H., Hommes, N. v. E., Clark, T., Hirsch, A., Pietzak, B., Weidinger, A., Dunsch, L., *Angew. Chem. Int. Ed. Engl.*, **36**, 2835-2838 (1997).
9. Pietzak, B., Waiblinger, M., Almeida Murphy, T., Weidinger, A., Höhne, M., Dietel, E., Hirsch, A., *Chem. Phys. Lett.*, **279**, 259-263 (1997).
10. Tománek, D.; Li, Y. S.; *Chem. Phys. Lett.*, **243**, 42-44 (1995).

EPR studies of N@C_{60} and its Adducts

Andrea Gruß, Claus Knapp, Norbert Weiden, and Klaus-Peter Dinse

Phys. Chem. III, TU Darmstadt, D-64287 Darmstadt, Petersenstr. 20, Germany

Elke Dietel and Andreas Hirsch

Org. Chem. II, Universität Erlangen-Nürnberg, D-91054 Erlangen, Henkestr. 42, Germany

Björn Pietzak, Markus Waiblinger, and Alois Weidinger

Hahn-Meitner Institut, D-14109 Berlin, Glienickerstr. 100, Germany

Abstract Atomic Nitrogen in its quartet ground state can be encapsulated in C_{60} and its derivatives of reduced symmetry. Three endohedral fullerenes were studied in solid matrices for a determination of the permanent zero-field-splitting tensors. In solution the lowering of the site symmetry by collision-induced deformations of the fullerene cage apparently leads to a time dependent zero-field-splitting of the quartet spin state. The variance of the zero-field-splitting and the correlation time of this interaction are estimated from the absolute values of T_1 and T_2.

INTRODUCTION

The endohedral complex N@C_{60}, nitrogen in its atomic ground state ($^4S_{3/2}$) encapsulated in C_{60}, can be derivatized by chemical reactions without a loss of the endohedral nitrogen. Recently, it was shown that a lowering of the I_h cage symmetry to C_{2v} by derivatization to the mono-adduct N@C_{61}(COOEt)$_2$ causes additional lines in the EPR spectra resulting from zero-field-splitting (ZFS) (1). Assuming that the nitrogen atom is still located at the centre of the cage, the ZFS should vanish for the quartet spin in the tetrahedral environment of the hexa-adduct C_{66}(COOEt)$_{12}$. In this paper, we studied nitrogen encased by fullerenes of different cage symmetries in solid matrices to determine the permanent ZFS tensors. Additionally, we investigated the fluctuating cage deformations by measuring electronic spin relaxation rates in solution.

EXPERIMENTAL

N@C_{60} was produced by ion bombardment (2). A mixture of the endohedral compound and empty C_{60} (ratio 10^{-5}:1) was used to synthesise these adducts by cyclopropanation with diethyl bromomalonate (3). Especially, the hexa-adduct T_h-N@C_{66}(COOC$_2$H$_5$)$_{12}$ and the perdeuterated one, T_h-N@C_{66}(COOC$_2$D$_5$)$_{12}$, were prepared by the template activation method developed previously (4). Different products were separated by HPLC. For EPR investigations, samples of poly-crystalline powders and carefully degassed solutions were sealed off on a high vacuum line. Pulsed and c.w. EPR spectra were recorded with a BRUKER ELEXSYS 680 X spectrometer.

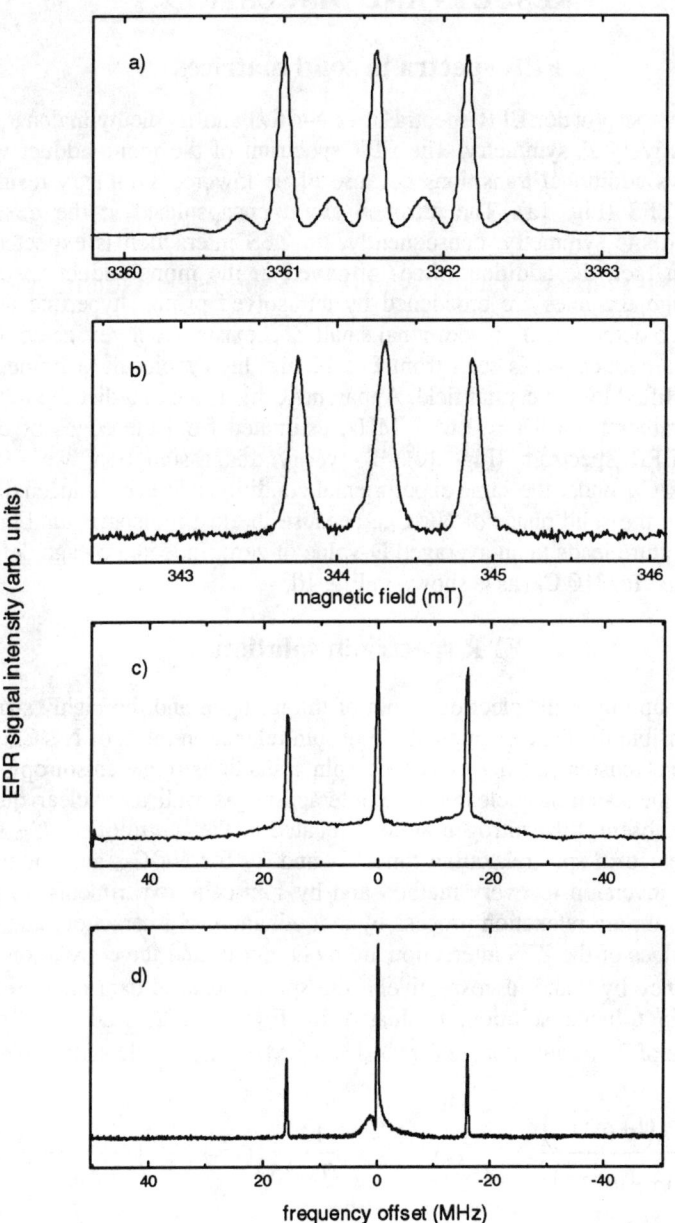

Figure 1. a) Integrated c.w. EPR powder spectrum of the mono-adduct C_{2v}-N@C_{61}(COOEt)$_2$ (95 GHz) Parameters derived by simulation: $|a_{iso}(^{14}N)| = 15.88$ MHz, $|D| = 8.4$ MHz, $|E| = 0.3$ MHz
b) Field swept EPR powder spectrum of T_h-N@C_{66}(COOC$_2$H$_5$)$_{12}$
c) Echo transformed powder spectrum of perdeuterated T_h-N@C_{66}(COOC$_2$D$_5$)$_{12}$
d) Echo transformed powder spectrum of I_h-N@C_{60}

RESULTS AND DISCUSSION

EPR spectra in solid matrices

Fig. 1 shows the powder EPR spectra of I_h-N@C_{60} and its diethylmalonate adducts with C_{2v} respectively T_h symmetry. The EPR spectrum of the mono-adduct with axial symmetry shows additional transitions because of its lowered symmetry resulting in a non-vanishing ZFS (Fig. 1a). The nitrogen atom, encapsulated in the hexa-adduct, retains its cubic site symmetry; consequently, no ZFS interaction is expected for the quartet spin. In fact, the additional lines observed in the mono-adduct spectrum are missing, although the lines are broadened by unresolved proton hyperfine interaction (hfi) (Fig 1b). To determine if an additional small ZFS exists, we investigated the totally deuterated hexa-adduct. As is seen from Fig 1c, the high molecular symmetry at the nitrogen site is lifted by the crystal field. Apparently, this leads to a distribution of zero-field-splitting values from |D| = 0 to 7 MHz, estimated from the edges of the broad "feet" in the EPR spectrum (Fig. 1c). To verify this assumption, we recorded a spectrum of N@C_{60} under the same experimental conditions. Matrix-induced distortions should vanish in the solid phase of N@C_{60}, because the fast molecular tumbling of C_{60} at room temperature leads to an averaged D-value of zero. Indeed, no cage deformation could be detected for N@C_{60} as is shown in Fig. 1d.

EPR spectra in solution

The weak coupling of the electronic spin of the nitrogen and the highly symmetrical cage are responsible for the exceptionally long spin relaxation times of N@C_{60}, because most of the mechanisms causing electronic spin transitions, e.g., anisotropy of the g matrix and of the electron-nuclear dipole interaction, as well as nuclear quadrupole interaction, are absent if the nitrogen atom is located at the centre of the C_{60} shell. We measured the electron spin relaxation times T_1 and T_2 for N@C_{60} and for two of its adducts by the inversion recovery method and by spin echo experiments, respectively (see Tab.1). Assuming relaxation process by a dominant ZFS interaction, and denoting the principal values of the ZFS interaction are by D and E, and the correlation time and Zeeman frequency by τ and ω, respectively, the spin relaxation data measured for the mono-adduct in toluene solution, evaluated by Eq.1 and 2, lead to a rotational correlation time of 25 ps and a value for |D_{eff}| = 8.5 MHz ($D_{eff} = \sqrt{D^2+3E^2}$). Nearly the

$$\frac{T_1}{T_2} = \frac{1}{2} \frac{\tau + \tau/(1+\omega^2\tau^2)}{\tau/(1+4\omega^2\tau^2)} \quad (1) \qquad \frac{1}{T_1} \approx \frac{8}{5}(D^2+3E^2)\frac{\tau}{1+4\omega^2\tau^2} \quad (2)$$

same value was determined by a simulation of the powder spectra shown above (Fig. 1). This indicates that the modulation of the permanent ZFS by rotational tumbling is the major source of spin relaxation in solution. In absence of a permanent cage distortion, as is the case of unmodified N@C_{60} or of the hexa-adduct, a ZFS tensor rigidly attached to the molecular frame is lacking and therefore rotational tumbling cannot induce spin relaxation. Instead, we have to consider collision-induced deformations of the carbon shell leading to a temporary ZFS tensor fluctuating in direction and magnitude (5). Table 1 summarises the experimental data and the correlation times of the interaction as well as the resulting effective ZFS parameter.

	$N@C_{61}(COOC_2H_5)_2$ [a]	$N@C_{66}(COOC_2H_5)_{12}$ [b]	$N@C_{60}$ [a]
T_2 [µs]	13(2)	24(2)	50(1)
T_1 [µs]	78(5)	116(3)	120(2)
τ [ps]	25	20	11
$|D_{eff}|$ [MHz]	8.5	6.8	5.8

Table 1. Spin relaxation times measured at 300 K, correlation times of the ZFS interactions, and permanent respectively temporary effective zero-field tensors for nitrogen encapsulated in cages of different symmetry. a) toluene solution, b) dichloromethane solution.

CONCLUSION

Evaluation of the electron spin relaxation data in solution and the analysis of powder spectra reveals the existence of a permanent ZFS tensor with axes rigidly attached to the molecular frame for the mono-adduct, which indicates an *intrinsic* distortion of the cage. In contrast, the distribution of ZFS interactions monitored by nitrogen in the hexa-adduct are of *extrinsic* nature and can be explained by matrix-induced distortions of the cage. In solution, in the absence of a permanent ZFS interaction, fluctuations of the ZFS tensor are the major source of spin relaxation. The root-mean-square value of this collision-induced fluctuating zero-field-splitting interaction as estimated from relaxation data for $N@C_{60}$ and the hexa-adduct is in the range of the ZFS interaction measured for the mono-adduct.

ACKNOWLEDGEMENT

We gratefully acknowledge financial support from the DFG (Di182/19-1).

REFERENCES

1. Pietzak, B., Waiblinger, M., Almeida-Murphy, T., Weidinger, A., Höhne, M., Dietel, E., Hirsch, A., *Chem. Phys. Lett.* **279**, 259 (1997).
2. Almeida-Murphy, T., Pawlik, T., Weidinger, A., Alcala, R., Spaeth, J.-M., *Phys. Rev. Lett.* **77**, 1075 (1996).
3. Bingel, C., *Chem. Ber.* **126**, 1957 (1993).
4. Lamparth, I., Maichle-Moessmer, C., *Angew. Chem.* **109**, 2858 (1997).
5. Rubinstein, M., Baram, A., Luz, Z., *Mol. Phys.* **20**, 67 (1971).

CLUSTERS

Giant gold-cluster compounds - Gaps in optical and charging spectra, and an electronic origin of abundance anomalies

Robert L. Whetten, M. M. Alvarez, T. Bigioni, J. T. Khoury,, B. E. Salisbury, T. G. Schaaff, Marat N. Shafigullin, I. Vezmar, Schools of Physics & Chemistry, Georgia Institute of Technology,, Atlanta, GA 30332-0430 USA

Over the past three years, an extraordinary series of as many as 9 abundance anomalies have been identified in the thiogold clusters ($Au_N SR_M$) with metallic (Au^0) cores. [1], [2], [3], [4] These anomalies, discovered at ever smaller sizes in a system lacking a small-molecule antecedent, now span a range from nearly 250 Au to fewer than 20 Au atoms. Several have been confirmed in other labs.

For each abundance anomaly identified, a new material phase is obtained, in high yield, through the optimized decomposition reaction of a polymer precursor (p-RSAu), followed by isolation through recrystallization, chromatography, or gel-electrophoresis methods. The result is a series of new (macro)molecular materials displaying remarkable homogeneity-uniformity and high robustness with respect to thermal-chemical-photochemical-electrochemical stress, fully justifying the analogy to their extended-surface asymptotes (s-Au:SR self-assembled monolayers, SAMs). The functionality achieved through the selection or exchange of R-groups has made it possible to incorporate the core gold-sulfur clusters into an ever growing variety of complex, hierarchical nanostructures.

Earlier hypotheses for the origin of these anomalies have been structural in nature, i.e. centered around the identification of unusually stable atom-packing structures and crystallite morphologies. A series of novel and very stable structures have been found, of truncated pentagonal-decahedral morphology, that explain the x-ray scattering functions [5] better than any others proposed (and also account for an earlier 1.4-nm Au:PR3:halide compound [6]). However successful, this explanation can not account for the full range of anomalies observed, particularly at small sizes.

An alternative source of special stability lies in the electronic structure, and in particular the *filling* of the Au-core conduction-electron levels. The statistical (Kubo) prediction [7] for the HOMO-LUMO gap Δ in Au clusters is ~ 8 N^{-1} eV, where N is the number of Au atoms in the neutral cluster. Predicted gaps are ~ 1 eV at N = 8, 0.4 eV (20), 0.2 eV (40), 0.11 eV (70). However, experiments on the clusters obtained reveal gaps that are typically three times greater. [8], [9]

The observed gaps are predicted much better by a very different picture, namely the (spherical) *electron-shell model*, which associates each stability-abundance anomaly with a distinct angular-momentum state, L, of high occupancy $2(2L+1)$.

[10] [See accompanying Table.] The first two such closings, at 2 and 8 (total) conduction electrons, are already known to account for the stable smaller Au cluster compounds, e.g. the well established $Au_{8+q}(X)_q$ clusters (X = halide) with $\Delta \sim$ 2.5 eV. [11] Six subsequent major shell closings are predicted to occur in the $18 < N < 198$ range, and have been thoroughly established, e.g. by atomic-cluster beam experiments. [12] Each higher shell-closing can account for the magnitude of the gaps found experimentally at the corresponding cluster size (mass), suggesting further that they may account for the preferential formation of certain $Au_N(SR)_M$ clusters of definite composition.

The gaps Δ are measured by optical transmission spectroscopy of dilute samples of rigorously purified clusters and with excellent dynamic range, so that even very weak transitions can be detected. [8] The spectra are highly structured, giving evidence of a rich set of electronic transitions. The clusters prepared with a natural thiol show rich circular-dichroism spectra as well, indicating a further decomposition into distinct transitions. [4] In each case, a dispersion (Kramers-Kronig) analysis is used to identify the energy gaps from the imaginary part (ϵ_2) of the dielectric response.

The optical measurements may nonetheless miss a highly dipole-forbidden transition (i.e. not an L - L+1 type expected from the above model). The gaps derived optically are therefore confirmed by measuring the sequence of charging transitions, using either double tunnel-junction or electrochemical methods. [9] In this case, the large-cluster response (a regular Coulomb staircase-like charging curve, indicating a well-defined, sub-aF capacitance) gives way for smaller clusters to a pattern with an increasingly large central gap. The magnitudes of these charging gaps are in reasonable agreement with those observed optically, once a charging contribution has been subtracted.

The finding of gaps in the obtained materials that are uniformly much larger than expected statistically suggests that a universal explanation for the enhanced abundance of certain metal cluster compounds lies in the ability to open a large electronic gap, *without* deforming to a non-globular shape (morphology). A large electronic gap can act to increase a cluster's relative inertness against growth or etching, leading to its accumulation. These principles can be taken to apply to the commercially available 0.8-nm and 1.4-nm Au cluster compounds (undecagold, nanogold), as 1st- and 4th-generation shell-closings. [13]

If this explanation holds, it could mark the beginning of a grand unification of the principles of metal-atom clusters (beam experiments and theory), metal-cluster compounds, small-particles materials research, and capacitive nano-electronics. However, there remains an pressing need to clarify several remaining uncertainties in this series of compounds, concerning: (i) the nature of the surface-chemical bond in Au:SR systems; (ii) unambiguous determination of the precise number of Au atoms in the clusters, from fragmentation-free mass-spectrometry; and (iii) total structure determination, e.g. by single-crystal x-ray diffraction. Work on these fronts is underway. [14]

REFERENCES

1. R. L. Whetten, J. T. Khoury, M. M. Alvarez, S. Murthy, I. Vezmar, Z. L. Wang, P. W. Stephens, C. L. Cleveland, W. D. Luedtke, U. Landman, Adv. Mater. 8, 428-433 (1996).
2. M. M. Alvarez, J. T. Khoury, T. G. Schaaff, M. N. Shafigullin, I. Vezmar, R. L. Whetten, Chem. Phys. Lett. 266, 91-8 (1997).
3. T. G. Schaaff, M. N. Shafigullin, J. T. Khoury, I. Vezmar, R. L. Whetten, W. G. Cullen, P. N. First, C. Gutièrrez-Wing, J. Ascensio, M. J. Jose-Yacaman, J. Phys. Chem. 101, 7885 (1997).
4. T. G. Schaaff, M. N. Shafigullin, R. L. Whetten, submitted.
5. C. L. Cleveland, U. Landman, T. G. Schaaff, M. N. Shafigullin, P. W. Stephens, R. L. Whetten, Phys. Rev. Lett. 79, 1873-6 (1997).
6. W. Vogel, B. Rosner, B. Tesche, J. Phys. Chem. 97, 11611 (1993).
7. R. Kubo, A. Kawabata, S. Kobayashi, Ann. Rev. Mater. Sci. 14, 49 (1984).
8. M. M. Alvarez, J. T. Khoury, T. G. Schaaff, R. L. Whetten, submitted.
9. R. S. Ingram, M. J. Hostetler, R. W. Murray, T. G. Schaaff, J. T. Khoury, R. L. Whetten, T. P. Bigioni, D. K. Guthrie, P. N. First, J. Am. Chem. Soc. 119, 9279-80 (1997). S.-W. Chen, R. Ingram, M. Hostetler, R. W. Murray, T. G. Schaaff, J. T. Khoury, M. M. Alvarez, R. L. Whetten, submitted.
10. W. A. deHeer, Rev. Mod. Phys. 65, 611-676 (1993).
11. For review, see D. M. P. Mingos, J. Chem. Soc. (Dalton), (1996).
12. For electronic shell-structure in monovalent metal-atomic clusters: [Ag] G. Alameddin, J. Hunter, D. Cameron, M. M. Kappes, Chem. Phys. Lett. 192, 122 (1992); [Cu] M. B. Knickelbein, ibid. 192, 129 (1992); [Na] E. C. Honea, M. L. Homer, J. L. Persson, R. L. Whetten, ibid. 171, 147 (1990); [Cs] H. Gšhlich, T. Lange, T. Bergmann, T. P. Martin, Phys. Rev. Lett. 65, 748 (1990); [Au] K. J. Taylor, C. L. Pettiette-Hall, O. Cheshnovsky, R. E. Smalley, J. Chem. Phys. 96, 3319 (1992).
13. Published documents of the Nanoprobes, Inc. corporation.
14. The authors acknowledge Prof. W. A. deHeer for stimulating discussions and suggestions, and the National Science Foundation for support of this research.

Electron shell-fillings vs. observed properties of gold-cluster compounds

Generation	Configuration	e-Count	Mass (kDa)	D_{eq} (nm)	Δ_{optic} (eV)	Δ_{charg} (V)
0th	S^2	2	----	----	> 3	---
1st (o)	S^2P^6	8	~ 2	~ 0.7	~ 2.5	---
2nd (e)	[ ]$D^{10}S^2$	20	~ 4	~ 0.9	~ 1.4	> 1
3rd (o)	[ ]$F^{14}P^6$	40	~ 8	~ 1.1	~ 0.9	~ 1.2
4th	[ ]$G^{18}D^{10}S^2$	70	~14	~ 1.35	~ 0.6	~ 0.8
5th	[ ]$H^{22}F^{14}P^6$	112	22	~ 1.5	-------	-------
6th	[ ]I^{26}	138	28	~ 1.65	< 0.6	~0.4
6'	[ ]$G^{18}D^{10}S^2$	168	34	~ 1.8	-------	~ 0.35
7th	[ ]J^{30}	198	39	~ 1.9	-------	"
7'	[ ]$H^{22}F^{14}P^6$	240	47	~ 2.1	-------	-------

Proton NMR of a Manganese-Ion Cluster Nanomagnet

F. Milia, R. Blinc*, R.M. Ashey[†], and N.S. Dalal[†]

NCSR Demokritos, 153 10 AG. Paraskevi Attiki, Athens, Greece
*J. Stefan Institute, Jamova 39, 1000 Ljubljana, Slovenia
[†]Dept. of Chemistry and National High Magnetic Field Laboratory,
Florida State University, USA

Abstract. The Manganese ion cluster compounds allow for a study of the transition from molecular paramagnetism to bulk ferromagnetism. They also show bistability on a molecular level and are thus potentially useful for information storage. In view of a large energy gap the system is ESR silent in the X and Q band but not in high field ESR. Here we report on the first proton NMR study of the Mn 12 acetate compound.

INTRODUCTION

The manganese ion cluster compounds show a bistability on a molecular level and thus are very interesting compounds for information storage. In addition they allow for a study of the transition from molecular paramagnetism to bulk ferromagnetism.
The chemical formula of the compound studied in this work is
$[Mn_{12}O_{12}(CH_3COO)_{16}(H_2O)]2CH_3COOH4H_2O$
The twelve Mn ions within a cluster, form a high total spin S=10 ground state which is well separated from excited levels.
$S(Mn^{3+})=2$
$S(Mn^{4+})=3/2$
so that the ferromagnetic ordering of Mn spins leads to:
$S=8*2-4*3/2=10$
The ground multiplet is split by the tetragonal symmetry so as to leave the M=±10 states laying lowest.
Studying the low temperature properties one sees the following characteristics:
(a) Slow relaxation of the magnetization similar to the blocking temperature of supermagnets
$\tau=\tau_0 Exp[A/kT]$
where $\tau_0=2.1 \times 10^{-7}$ sec. and A/k=61 K.

(b) At low temperatures, the Mn cluster behaves like a single molecule magnet showing hysterisis effects.

(c) For T<2K the relaxation time becomes temperature independent which means that we have a quantum tunneling of the magnetization between M=-10 and M=10 states.

Nevertheless powder neutron diffraction in an applied magnetic field shows that the above mentioned simple model of coupled Mn-Mn ions is inadequate.

The metal site spin-density population 2.2 for Mn^{4+} and 4.0 for Mn^{3+} are much larger than the values expected for octahedrally coordinated ions of 1.5 and 2.0 spins respectively. As a result there is a substantial, generally negative, spin density on the bridging ligands. They obtain:

O(2)=O(3)= -1.0
C(1)OO=0.5
C(3)OO=2.0
C(5)OO=C(7)OO= -1.4

The magnetization of spin differences within the molecule is traditionally described as resulting from "spin polarization" and is a manifestation of electron electron correlation effects.

The above results imply that the unpaired electron spin density is delocalized over the whole molecular cluster.

If this is indeed the case there should be also some unpaired spin density at the protonic positions. This would lead to isotropic contact Fermi shifts of the proton lines which would be clearly evident even in a powder sample.

In view of that we decided to look for the possible delocalization of the electron spin density implied by the neutron data by a proton NMR.

EXPERIMENTAL WORK AND CONCLUSIONS

Proton NMR Fourier Transform spectrum at T=290K and 270MHz shows two pronounced peaks shifted to lower fields in Fig. 1

The salient features of the spectrum are:

1. The strongly shifted A peak is most likely due to the four H_2O molecules which are bonded to the Mn-core. The Oxygen of these H_2O molecules is directly attached to the Mn ion so that the corresponding protons are only one bond length away.

2. The CH_3 peaks are less shifted because in this case the protons are 3 bonds away from the end oxygen that is attached to the Mn ion.

3. An additional contribution to the less shifted peak comes from the protons of the soliated molecules that are present in the crystal lattice but not attached to the Mn-core. These are the other four H_2O molecules and the two CH_3COOH molecules.

4. One thus expects an over-all intensity ratio of 1:8 of the less shifted peaks. This roughly agrees with the observed integrated intensities.

ACKNOWLEDGMENTS

The authors would like to thank Dr. D. Arcon from J. Stefan Institute for his help during the preparation of this manuscript

Note: After this study was finished we were made aware of a similar work on the same compound that has been published by Lascialfari A. and co-workers, on spin-lattice and spin-spin relaxation time measurements.

REFERENCES

1. Lascialfari A., Borsa F.,Shastri A., Jang Z.H., and Carretta P. *Phys. Rev. B.* **57**, 514-520 (1998)
2. Galteschi A., Caneschi L., Pardi L., and Sassoli R. *Science* **265** 1055 (1994)
3. Awschhalom D.D., and DiVncenzo D.P., *Phys. Today* 48, No. 4, **43**, (1995)
4. Sassoli R. Gatteschi D., Caneschi A. And Novak M.A., *Nature* (London) **365**, 141, (1993)
5. Friedman R. Sarachik M.P., Tejada J., and Ziolo R., *Phys. Rev. Lett.* **76**, 3830 (1996)
6. Politi P., Rettori A., Hartmann-Boutron F., and Villain J., *Phys. Rev. Lett.* **75**, 537 (1995)

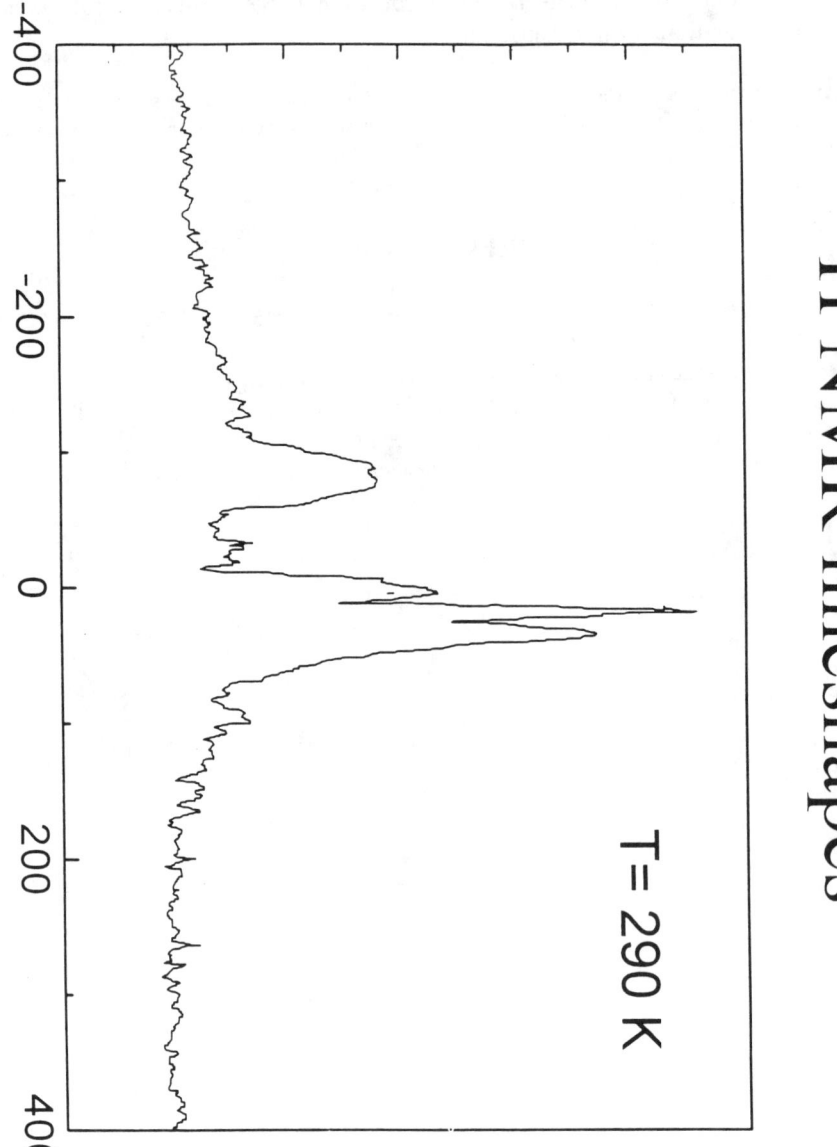

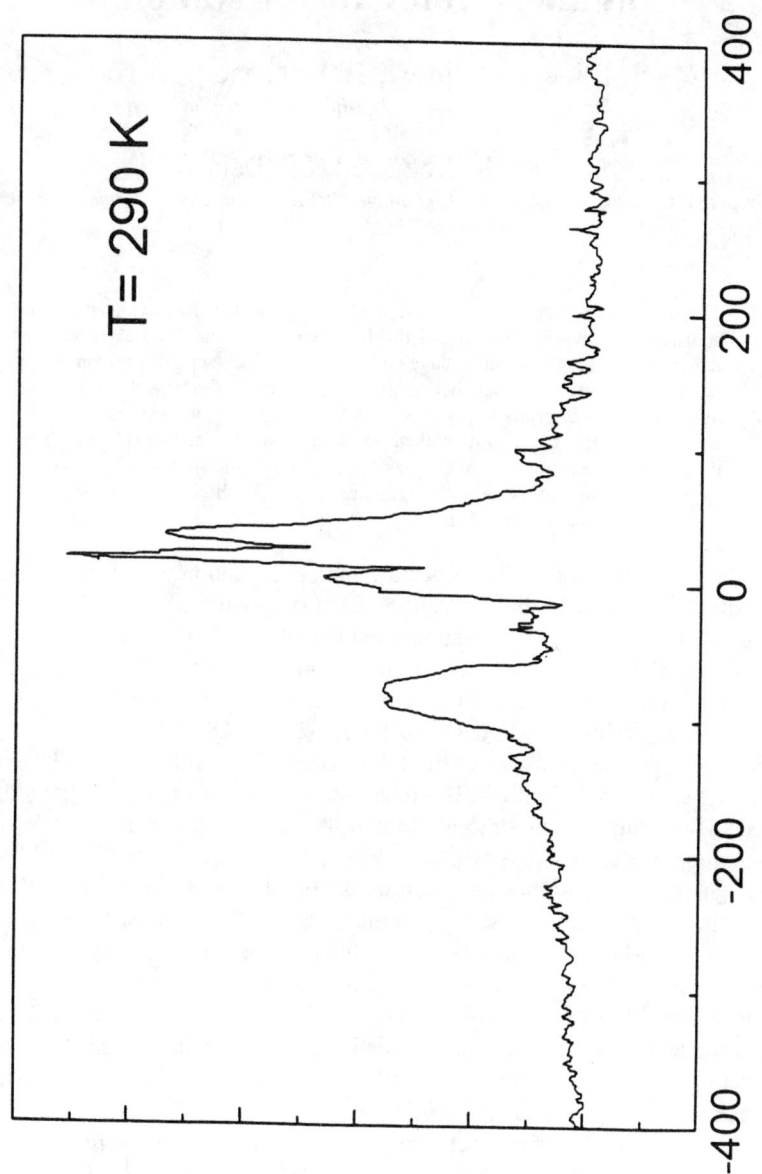

Deuteron Magnetic Resonance and Relaxation in a Manganese-Ion Cluster Nanomagnet

R. Blinc, J. Dolinšek, T. Apih, and D. Arčon
J. Stefan Institute, Jamova 39, 1000 Ljubljana, SLOVENIA

R.M. Achey and N.S. Dalal
Department of Chemistry and National High Magnetic Field Laboratory, Florida State University, Tallahassee, FL 32306-4390, USA

The temperature dependence of the deuteron NMR spectra and the deuteron spin-lattice relaxation time T_1 of the fully deuterated manganese ion cluster nanomagnet $Mn_{12}O_{12}$-Acetate have been measured. Two sets of deuteron peaks have been observed which both shift towards lower fields with decreasing temperature. The results imply that the unpaired spin density extends over the whole molecular cluster rather than being concentrated at the Mn ions. The deuteron magnetization recovery was found to be of the stretched exponential type and the spin-lattice relaxation time parameter T_1 showed two pronounced BPP type minima around 70 and 10 K.

Magnetic clusters formed by Mn_{12} acetate molecules (abbreviated as $Mn_{12}O_{12}$) offer the possibility to study the transition from molecular paramagnetism to bulk ferromagnetism[1]. They show superparamagnetic behavior and bistability on a molecular level so that they may be important for information storage. Another possible application is quantum computing. The chemical formula is

$$[Mn_{12}O_{12}(CH_3COO)_{16}(H_2O)_4]\cdot 2CH_3COOH\cdot 4H_2O$$

The molecular cluster is made of four tetrahedrally coordinated Mn(IV) ions with S=3/2 at the center and eight Mn(III) ions with S=2 at the outside of the cluster. It is assumed that in the ground state all Mn(III) spins are up and all Mn(IV) spins are down resulting in a total spin S=10.

To get some additional an information on the possible delocalization of the unpaired spin density and the spin dynamics we decided to study the deuteron NMR spectra and spin-lattice relaxation of a fully deuterated polycrystalline $Mn_{12}O_{12}$ powder sample at a Larmor frequency $\omega_L/2\pi=42.5$ MHz.

The observed deuteron NMR spectra are shown in Fig. 1. Below 80 K two sets of peaks, designated as A and B, emerge. The B peak is practically unshifted down to 14 K, whereas the position of the A peak follows a Curie law (Fig. 2). The total shift of the A peak amounts to ≈ 1000 Gauss≈ 650 kHz between 192 K and 14 K. The shifts are due to the isotropic Fermi contact hyperfine coupling. The strongly shifted A peak seems to be due to the four D_2O molecules bonded to the Mn core. The less shifted B peak seems to be due to the $-CD_3$ deuterons which are three bonds away from the Mn ions and the solvated D_2O and CD_3COOD molecules. This proves that the unpaired spin density is distributed over the whole molecular cluster including the organic framework and not localized at the Mn cluster skeleton.

The recovery of the deuteron spin magnetization of the less shifted B peak was found to be of the stretched exponential type. This is not surprising in view of the large number of deuterons contributing to this peak.

The temperature dependence of the deuteron spin-lattice relaxation time parameter T_1 is shown in Fig. 3. Two BPP T_1 minima, one around 70 K and another one around 10 K, are clearly discernible. The high T minimum is present in the proton T_1 data[2] too. At the position of the two minima $\omega_L \tau_1 = 1$ and $\omega_L \tau_2 = 1$ respectively. Here τ_1 and τ_2 are the correlation times for the fluctuations in the hyperfine coupling responsible for the occurrence of the two minima.

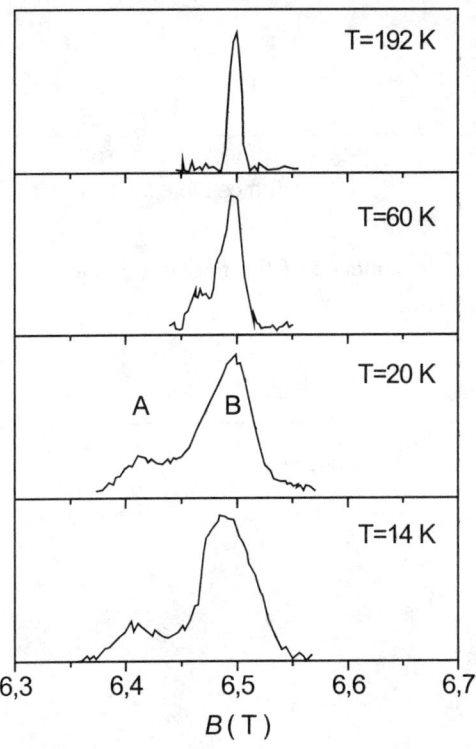

Figure 1: Deuteron NMR spectra of polycrystalline $Mn_{12}O_{12}$ at different temperatures.

It should be noted that the fluctuations in the hyperfine coupling can be due to fluctuations in the magnetization or sublattice magnetization of the whole $Mn_{12}O_{12}$ cluster, to single Mn spin excitations, or due to the reorientations of the $-D_2O$ or CD_3 groups which take the spin from a position with one value of the hyperfine field to another position with a different hyperfine field.

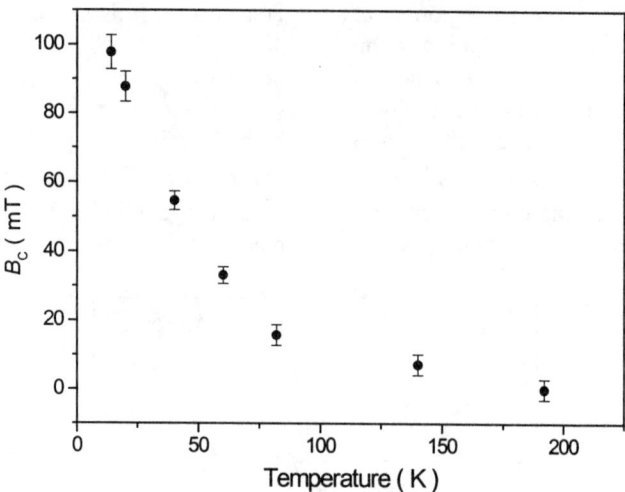

Figure 2: Temperature dependence of the position of the deuteron NMR A peak in polycrystalline $Mn_{12}O_{12}$.

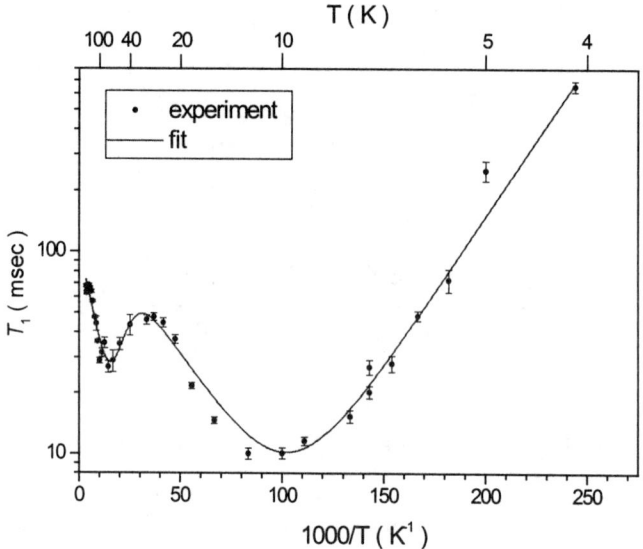

Figure 3: Temperature dependence of the deuteron spin-lattice relaxation time parameter T_1 in $Mn_{12}O_{12}$.

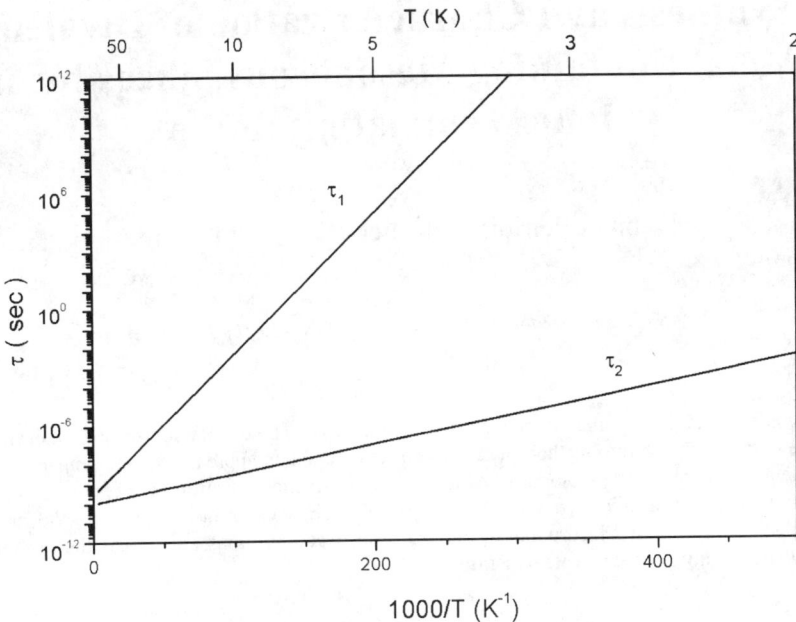

Figure 4: Temperature dependence of the two correlation times τ_1 and τ_2 responsible for the deuteron T_1 minima in $Mn_{12}O_{12}$.

The temperature dependence of the two correlation times τ_1 and τ_2 are shown in Fig. 4. Both are thermally activated and orders of magnitude faster than the correlation time τ_M for the reorientation of the macroscopic magnetization[3]. This mode however seems to contribute to the deuteron T_1 at room temperature.

REFERENCES:

1. D. Gattechi, A. Caneschi, L. Pardi, and R. Sessoli, *Science* **265**, 1055 (1994).
2. A. Lascialfari, D. Gatteschi, F. Borsa, A. Shastri, Z.H. Jang, and P. Carretta, *Phys. Rev.* **B 57**, 514 (1998).
3. R. Sessoli, D. Gatteschi, A. Caneschi, and M.A. Novak, *Nature* (London) **365**, 141 (1993).

Synthesis and Characterization of Divalent Metal Containing Mesoporous Silicas by an Ionic Templating Route

Dimitris Petridis and Michael A. Karakassides

N.C.S.R. "Demokritos", Ag.Paraskevi, Attikis,153 10, Athens, Greece

Abstract. The ability of organofunctional silicon alkoxides to bind metal ions enables their use in a new room temperature method based on an ionic template approach for the preparation of divalent metal containing mesoporous silicas. X-ray diffraction, electron spin resonance (ESR), transmission electron microscopy (TEM) and surface area measurements, provide evidence for the Cu^{2+}, Ni^{2+} and Co^{2+} incorporation into the framework positions and reveal the hexagonal mesoporous nature of the silicate structure.

INTRODUCTION

The recent advent of inorganic mesoporous solids possessing well-defined and designable mesopores with a relative narrow pore size distribution has triggered both academic and industrial interest owing to their use in high economical impact potential applications (1). Hexagonal MCM-41, is one of these materials (2), that recently has been prepared at room temperature by the co-condensation of tetraethoxysilane and organosiloxanes in the presense of surfactant templates (3,4). In this work, combining the sol-gel approach and the ability of organosiloxanes to bind metal ions, it was possible to prepare stable mesostructures under ambient conditions which containing catalytically active metals such as Cu, Ni and Co.

EXPERIMENTAL

In a typical preparation the N-[3-(Trimethoxysilyl)propyl]-ethylenediamine (1mmol) were dropwise added to an alcoholic solution of cupric chloride (0.5mmol in 5ml methanol). The resulting dark blue solution was then added under stirring to $Si(OCH_3)_4$ (23mmol) and the mixture was left for 3-4min. The obtained clear solution was slowly added to a stirred solution of cetyltrimethyl ammonium chloride (3 mmol) in 37ml distilled water with 0.325ml 50wt% NaOH (5.7mmol) and 8ml methanol. After few seconds a blue solid was precipitated. The product, designated as Cu-MCM-41, was allowed to age for 5hours, and then filtered, washed extensively with distilled water and with methanol, dried in air and finally calcined at 600°C for 12 hours to remove the alkylamino groups and the template surfactant. The same synthetic route was also used for the preparation of nickel and cobalt containing-MCM-41 materials.

RESULTS AND DISCUSSION

The X-ray diffraction pattern of the calcined metal containing MCM-41 materials are shown in Fig.1. All spectra show a characteristic strong diffraction peak at the low scattering angles 2θ, which correspond to a d(100) spacing from 31Å to 33Å. The repeat distance a_o, between the pore centers of the porous structure was calculated with the formula $a_o = 2d_{100}/\sqrt{3}$ (Table I). The spectra show additional 110 and 200 reflections in the 2θ range from 4.0 to 6.0° (fig.1) which indicate the existence of the hexagonal phases. Fig.2 presents the N_2 adsorption-desorption isotherm for the Co-MCM-41 sample and the corresponding pore size distribution curve (inset) calculated using a cylindrical pore model (5). The shape of the isotherms is typical of the previously reported mesoporous silica samples.(1) A well defined step in the adsorption curve was observed near the 0.2 value of P/Po, indicative of the filling of the framework- confined mesopores. The calcined Co-MCM-41 sample has a surface area of 1010m²/g and a pore size of 18Å. The uniform porous structure of Co-MCM-41 is evident in transmission electron microscopy (TEM) lattice image shown in fig.3.

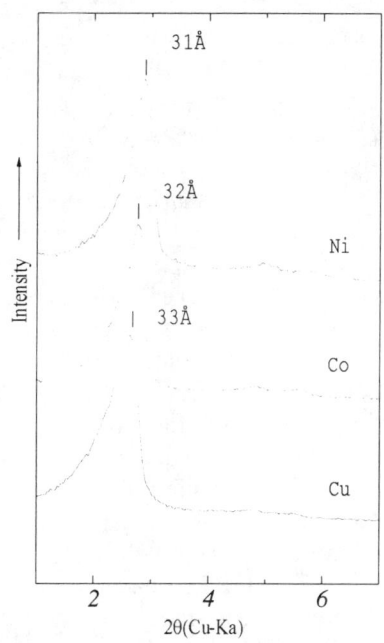

FIGURE 1. Powder X-ray diffraction pattern of MCM-41 samples calcined at 600°C for 12 hours

Table I Chemical Composition (atomic ratio) and structural properties of MCM-41 materials

Sample	metal %wt	XRD, d_{100} d-spacing (Å)	a_o (Å)	Surface area BET(m²/g)
Cu-MCM-41	2,4	33	38	980
Ni-MCM-41	1,96	31	37	695
Co-MCM-41	1,01	32	36,5	1010

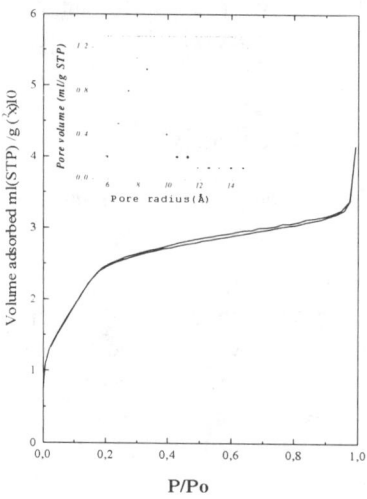

FIGURE 2. Nitrogen adsorption-desorption isotherms for calcined Co-MCM-41.

FIGURE 3. TEM micrograph for calcined Co-MCM-41

In an effort to elucidate the environment and the geometry of the catalytic metal centers in the silicate framework the ESR spectra of the as-synthesized (a) and the air-heated Cu-MCM-41 sample (b) were examined (fig.4). The spectrum of the as-synthesized Cu-MCM-41 can be described by an axial spin Hamiltonian with parameters $g_{//}=$ 2.27, $g_\perp=$ 2.035 and $A_{//}=$ 0.016. On the basis of these parameters and previous assignments (6,7) the axial signal may be attributed to a distorted octahedral Cu(II) symmetry, similar to the geometry reported for the cupric- tetrammine complexes introduced in silica gels by a cation exchange procedure (8). The number of nitrogen atoms coordinated to cupric ion may be four or less with framework oxygens competing for the coordination environment of copper. Turning to the spectrum of the calcined Cu-MCM-41 sample, fig.4b, we observe that the E.S.R. parameters are different from those of the as-synthesized sample. The spectrum of the calcined sample can also be described by an axial spin Hamiltonian with $g_{//}=$ 2.395, $g_\perp=$ 2.07, $A_{//}=0.012 cm^{-1}$.

These ESR parameters differ only slightly from the $g_{//}$ and $A_{//}$ for the $[Cu(H_2O)_6]^{2+}$ complex found in hydrated Cu^{2+} ion exchange MCM-41($g_{//}$ = 2.4 and $A_{//}$=0.0141)(6) or hydrated clay minerals ($g_{//}$ = 2.38 and $A_{//}$=0.0145) (9). In line with these studies we propose a similar octahedral geometry with tetragonal elongation for the coordination of Cu(II) in Cu-MCM-41.

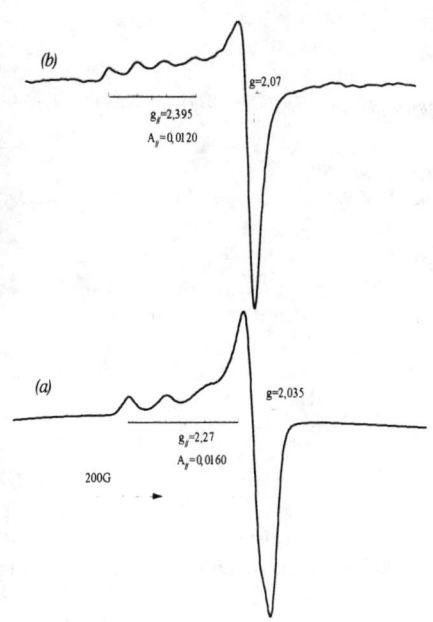

In conclusion we present a general route for the preparation of mesoporous divalent metal containing MCM-41 materials based on ionic template approach. The x-ray diffraction data in conjunction with the surface area measurements reveal that the metal containing mesoporous material exhibits similar structural characteristics and physical properties with the hydrothermal method obtained MCM-41 materials. The E.S.R. results provide evidence that the cupric ions centers in the silicate framework occupy octahedral sites.

FIGURE 4. E.S.R. spectra of as sythesized Cu-MCM-41 (a) and calcined (b)

REFERENCES

1. Tanev P. T. and Pinnavaia T. J., *Acces in Nanoporous Materials*, ed. Pinnavaia T. J. and Thorpe M. F., Plenum Press, New York, **1995**, p13.
2. Beck J. S., Vartuli J. C., Roth W. J., Leonowicz M. E., Kresge C. T., Schmitt K. D., Chu C. T-W., Olson D. H., Sheppard E. W., McCullen S. B., Higgins J. B.; Schlenker J. L., *J.Am.Chem.Soc.*,**1992**,114,10834.
3. Anderson M.T., Martin J.E., Jodinek, P.Newcomer, *Acces in Nanoporous Materials*, ed. Pinnavaia T. J. and Thorpe M. F., Plenum Press, New York, **1995**, p29.
4. Burkett S.L., Sims S.D.and Mann S., Chem.Commun., **1996**, 1367.
5. Rierce C., *J. Phys. Chem.*, **1953**,57,149. Poppl A., Newhouse M., and Kevan L., *J.Phys.Chem.*,**1995**, 99, 10019.
6. Poppl A., Newhouse M., and Kevan L., *J.Phys.Chem.*,**1995**, 99, 10019.
7. Luka V., Maclachan D.J., Bramley R. and Morgan K., *J.Phys.Chem.*,**1996**, 100, 1793.
8. Tominaga H., Ono Y., Keii T., *J. Catal.*, **1975**, 40, 197.
9. McBride M.B., Pinnavaia T.J. and Mortland M.M., *J.Phys. Chem.*, **1975**, 79, 2430.

CARBON CLUSTERS – SIZE EFFECTS OF PROPERTIES

G. Seifert, K. Vietze*
Institut für Theoretische Physik, Technische Universität Dresden, D-01069 Dresden, Germany
P.W. Fowler
University of Exeter, Department of Chemistry, Stocker Road, Exeter, EX4 4QD, UK

Abstract

The stabilities of carbon clusters C_n, $30 \leq n \leq 82$ with fullerene like cage structures are discussed on the basis of theoretical calculations. The outstanding stability of clusters with $n = 32, 36, 44, 50, 60, 70$ as "magic numbers" is emphasized. Over a wide size range between a few dozen and a few hundred atoms general trends in stability and electronic structure of clusters with different topologies (fullerenes, graphitic sheets, tubes) are discussed.

It has been known for a long time that carbon forms 1D structures (linear chains) for $n < 10$ and the size region between $10 \leq n \leq 20$ is dominated by monocyclic rings[1,2,3]. As it was shown by systematic DFT investigations[4] and also by ion chromatography[5], ion drift experiments[6] and photoelectron spectroscopy[7,8] the size region $20 \leq n \leq 30$ may be viewed as the transition region from 2D cyclic structures to 3D fullerene cage structures. Calculations[4] have shown also that graphitic like structures for $C_{20...30}$ are energetically rather stable but they may be suppressed under the high temperature conditions of the experiments by entropic factors[9]. Undoubtedly, the fullerene structure is the most stable structure of carbon clusters in the size range beyond $n = 30$[7,10,8]. The existence of fullerenes for $30 < n < 60$ was indicated in the mass spectra[11,13] and proofed by photoelectron spectra[7,8]. Besides the most prominent fullerene C_{60} a number of fullerenes larger than C_{60} has been isolated[14]. However, until now no fullerene material could be synthesized with $C_n, n < 60$. Very recent experimental results, as for example reported at this meeting[15], indicate the possibility for the synthesis of fullerene material with fullerenes $C_n, n < 60$[15,16].

These experimental findings may be supported by theoretical calculations. The energetics of a cluster (C_n) with respect to fragmentation ($C_n \rightarrow C_{n-k} + C_k$) and further growth ($C_n + C_k \rightarrow C_{n+k}$) can be described by its (relative) stability (δ):

$$\delta = E(n+k) + E(n-k) - 2E(n).$$

Several experimental studies[13,17,18] and also theoretical models[19] suggest consideration of a process with k=2 for fullerenes. The calculated stabilities of fullerene structures between C_{30} and C_{84} are drawn in fig. 1. The method we have applied is a density functional based tight binding scheme (DF-TB)[21]. The initial structures of the fullerenes are those of the so called "fullerene road"[19,20]. These initial structures, which are minimal pentagon-adjacency fullerenes at each nuclearity and are connected by a C_2 insertion process, were optimized using conjugate gradient algorithms. The calculated stabilities confirm the outstanding stabilities of the two most prominent fullerenes (C_{60}, C_{70}). The results agree also nicely with the experimentally determined relative binding energies of fullerene ions for $n \geq 46$ [18]. Furthermore, the calculations strongly support the high stabilities of several fullernes below $n < 60$: $n = 50, 44, 36, 32$ as "magic numbers".

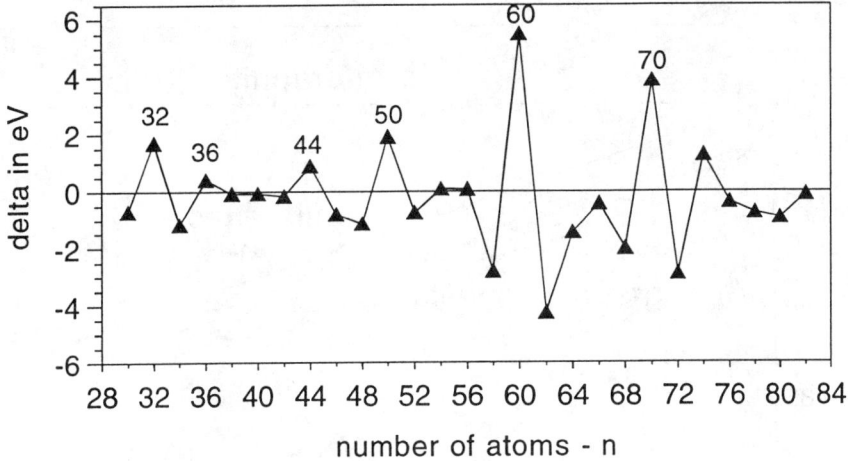

Fig. 1: Calculated stabilities ($\delta(n)$) for fullerene structures.

In fullerenes as C_{60} the binding energy (per atom) is still clearly smaller than in graphite owing to the curvature of cages. One can describe the size dependence of the binding energy of the fullerenes as [22]:

$$E(n)/n = \varepsilon_\infty + \gamma(n),$$

where ε_∞ is the binding energy per atom in a graphite monolayer and $\gamma(n)$ is the curvature energy. Assuming a spherical geometry of the fullerene cage, $\gamma(n)$ goes like $1/R^2$. The number of atoms in a hollow sphere increases proportional to R^2. Therefore one can write:

$$E(n)/n = \varepsilon_\infty + c/n.$$

Graphitic sheet structures of carbon clusters are also less stable than a infinitely large graphitic monolayer, and they are less stable than the corresponding fullerenic structures, at least for $n \geq 30$ - see fig. 2. The deviation of the binding energy of a graphitic sheet cluster from that in bulk graphite is mainly due to the "unsaturated" surface atoms. The ratio of surface atoms (n_S) to the total number of atoms (n) in the cluster may be used as a measure of γ in these clusters. A graphitic sheet with a radius R covers an area $A = 4\pi R^2$ and has a circumference $U = 2\pi R$. The area is covered by all atoms: $A = nA_i$ and the circumference is determined by the surface atoms: $U = n_S d_i$. Therefore the ratio n_S/n scales as $1/R$ and like $1/\sqrt{n}$. This means the size dependence of graphitic sheets may be described by:

$$E(n)/n = \varepsilon_\infty + c/\sqrt{n}.$$

As one can see from fig. 2 these size dependences are confirmed by calculations of a large series of fullerenes and graphitic sheets over a wide size range ($n \sim 20 \rightarrow n \sim 500$).

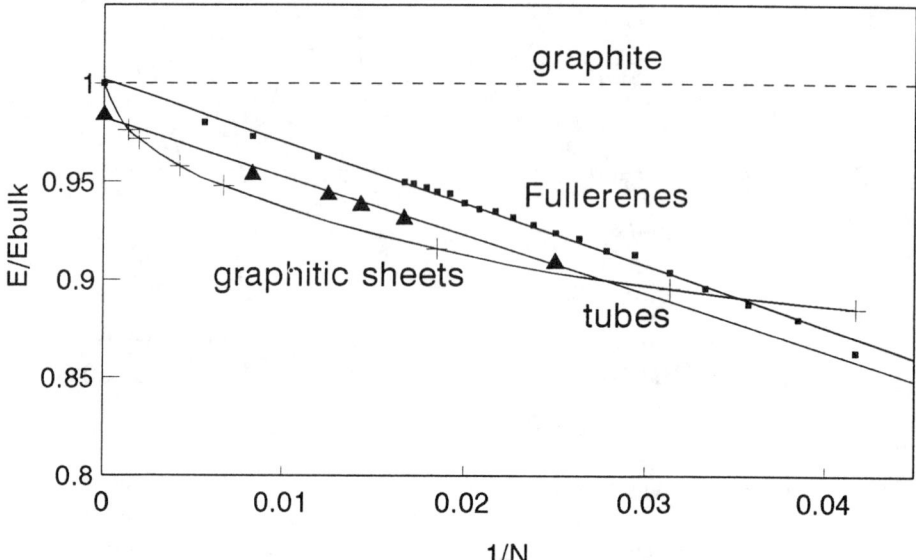

Fig. 2: Calculated binding energies per atom ($E(n)/n$) of fullerenes (dots), graphitic sheets (crosses) and tubes (triangles) divided by the binding energy of a graphite monolayer (ε_∞) versus $1/n$.

As one can see also from fig. 2 tubular structures (energies of (5,5)-tubes are drawn as an example) are generally less stable than the corresponding fullerenes, but they are more stable than graphitic sheets within the size region $50 < n < 500$ for the tubes considered here. The deviation for $n \rightarrow \infty$ from the graphite value describes the strain energy for a tube of a given radius (for the (5,5)-tubes $R \approx 3.5 \text{Å}$).

As shown in [22] the deviation of the ionization energy ($I_p(n)$) of a fullerene from the work function of a planar infinite graphite-like plane (Φ_∞) can be described roughly by:

$$I_p(n) = \Phi_\infty + 1/2R \Longrightarrow \Phi_\infty + a/\sqrt{n}.$$

The same holds for the electron affinities ($E_a(n)$). It is necessary only to change the sign in the correction ($-1/2R, -b/\sqrt{n}, a = b$). The calculated ionization energies and electron affinities for a large number of fullerenes are drawn versus $1/\sqrt{n}$ in fig. 3. One can clearly see that for $n < 100$ the abovementioned size dependence has superimposed "structural" effects, but the extrapolation for the large fullerenes gives for the affinities and for the ionization energies a value close to the work function of graphite. The difference between the affinity and the ionization energy is related to the HOMO-LUMO gap. The gap shows significant fluctuations again for $n <$ 100. These fluctuations will be discussed and compared with experimental data elsewhere[16]. The extrapolations from $n > 100$ indicate clearly that the gap is closing - see dashed lines in fig. 3. For $n > 100$ the graphitic sheets and the tubes have considerably smaller gaps (< 0.5 eV) than the fullerenes. Details will be given in [23].

In fig. 3 there is drawn also the average of electron affinity and ionization energy, which might be viewed as the electronegativity of the fullerenes ($\chi(n)$). Interestingly this quantity does not vary much, going from small clusters up to $n \to \infty$. Certainly, this rough qualitative approach deserves a more detailed discussion and refinement[23]. For example, the trend lines (dashed lines) in fig. 3 show a slightly different slope for the electron affinities and the ionization energies ($a \neq b$) and $\chi(n)$ tends to decrease slightly with increasing n.

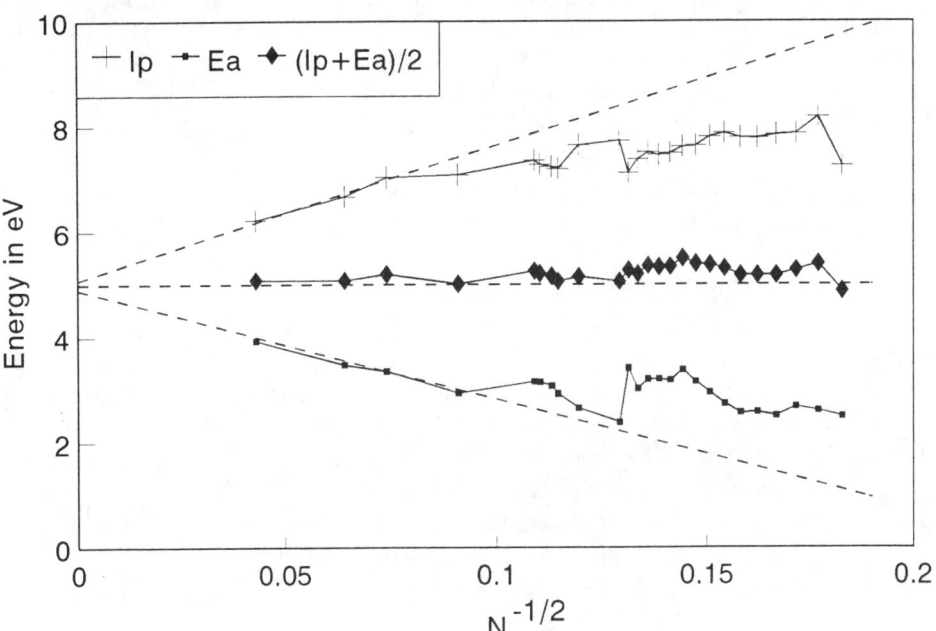

Fig. 3: Calculated electron affinities (crosses), ionization energies (dots) and $\chi(n) \equiv 1/2[E_a(n) + I_p(n)]$ for fullerenes versus $1/\sqrt{n}$.

Acknowledgement: The work was supported by the "British-German Academic research collaboration programme" (project-No. 868).
* Present address: Max–Planck–Institut für Physik komplexer Systeme, Nöthnitzer Straße 38, D-01187 Dresden, Germany

References

1. K.S. Pitzer and E. Clementi, *J. Am. Chem. Soc.* **81**, 4477 (1959).
2. G. Seifert, S. Becker and H.J. Dietze, *Int. J. Mass Spect. and Ion Proc.* **84**, 121 (1988).
3. J. Hutter, H.P. Lüthi and F. Diederich, *J. Am. Chem. Soc.* **116**, 750 (1994).
4. R.O. Jones and G. Seifert, *Phys. Rev. Lett.* **79**, 443 (1997).
5. G. von Helden, M.T. Hsu, N.G. Gotts, P.R. Kemper and M.T. Bowers, *Chem. Phys. Lett.* **204**, 15 (1993).
6. J.N. Hunter, J.L. Fey and M. Jarrold, *J. Chem. Phys.* **97**, 3460 (1993).
7. S. Yang, K.J. Taylor, M.J. Craycraft, J. Conceicao, C.L. Pettiette, O. Chesnovsky and R.E. Smalley, *Chem. Phys. Lett.* **144**, 431 (1988).
8. H. Handschuh, G. Ganteför, B. Kessler, P.S. Bechthold and W. Eberhardt, *Phys. Rev. Lett.* **74**, 1095 (1995).
9. C.J. Brabec, E.B. Anderson, B.N. Davidson, S.A. Kajihara, Q.M. Zhang, J. Bernholc and D. Tomanek, *Phys. Rev.* **B46**, 7326 (1992), G. Galli, F. Gygi and J.C. Golaz, *Phys. Rev.* **B57**, 1860 (1998).
10. M. Feyereisen, M. Gutowsky and J. Simons, *J. Chem. Phys.* **96**, 2927 (1992).
11. H.W. Kroto, J.R. Heath, S.C. O'Brien, R.F. Curl and R.E. Smalley, *Nature* **318**, 162 (1985).
12. see e.g. H.W. Kroto, *Science* **242**, 1139 (1988).
13. S.C. O'Brien, J.R. Heath, R.F. Curl and R.E. Smalley, *J. Chem. Phys.* **88**, 220 (1988).
14. see e.g. C. Thilgen, F. Diederich and R.L. Whetten in "Buckminsterfullerenes" W.E. Billups, M.A. Ciufolini (eds.), VCH Publishers, New York 1993.
15. A. Zettl et al., this volume
16. H. Kietzmann, R. Rochow, G. Ganteför, W. Eberhard, G. Seifert, K. Vietze and P.W. Fowler, *Phys. Rev. Lett.* , submitted for publication.
17. M. Foltin, O.Echt, P. Scheier, B. Dünser, R. Wörgötter, D. Muigg, S. Matt and T.D. Märk, *J. Chem. Phys.* **107**, 6246 (1997).
18. P.E. Barran, S. Firth, A.J. Stace, H.W. Kroto, K. Hansen and E.E.B. Campbell, *Int. J. Mass Spect. and Ion Proc.* **167/168**, 127 (1997).
19. P.W. Fowler and D.E. Manolopoulos, "An Atlas of Fullerenes", Clarendon Press, Oxford 1995.
20. D.E. Manolopoulos and P.W. Fowler, "The chemical physics of the fullerenes 10 (and 5) years later", Ed. W. Andreoni, Kluwer Academic Press, Dordrecht (1996) 51.
21. G. Seifert, D. Porezag and Th. Frauenheim, *Int. J. Quant. Chem.* **58** , 185 (1996), D. Porezag, Th. Frauenheim, Th. Köhler, G. Seifert, and R. Kaschner, *Phys. Rev.* **B51**, 12947 (1995).
22. G. Seifert, K. Vietze and R. Schmidt, *J. Phys.* **B29**, 5183 (1996).
23. G. Seifert, K. Vietze, to be published.

Laser Induced Emission Spectroscopy of Carbon Clusters in Solid Argon

I. Čermák, M. Förderer, S. Kalhofer, I. Čermáková, W. Krätschmer

Max-Planck-Institut für Kernphysik, Postfach 103980, D-69029 Heidelberg, Germany

Carbon clusters C_n ($n \leq 20$) play an essential role in the formation of larger carbon structures like fullerenes and nanotubes. To prepare these species, we applied the matrix isolation technique. Graphite rods were evaporated in vacuum and the carbon vapor molecules (C_1 - C_3) were trapped in argon matrices. By thermal annealing, the molecules can form clusters of up to about C_{20}. We used FT absorption spectroscopy and high-sensitivity emission spectroscopy to characterize the vibrational modes of the of these species. The time-resolved dispersed fluorescence and phosphorescence spectra were recorded by a gated photodiode array spectrometer. For selective excitation of the matrix-isolated molecules, a pulsed dye laser system was used. Such studies are technically extremely demanding since the emissions of larger carbon molecules are very weak.

1. INTRODUCTION

Carbon molecules are of predominant interest in fullerene science. They also play an essential role in combustion research and, last not least, as possible constituents of the interstellar medium. There is a plenty of evidences that buckminsterfullerene C_{60} and the other closed structures are formed in the gas phase from small precursor molecules,[1] such as the species C_1, C_2, C_3 which are abundant in carbon vapor.[2] The structures and the spectroscopic properties of these small molecules and that of the fullerenes starting from C_{60} are rather well known. The geometric structures of molecular ions in the intermediate size range have become apparent by ion chromatography.[3] However, corresponding data for neutral carbon species are very much limited, even though an impressive progress has been made in identifying the absorption spectra of the linear molecules ranging up to about C_{15}.[4,5] Theoretical predictions[6] and also experimental data[3,7] indicate that carbon clusters above a certain size should occur not only as linear chains but also as monocyclic rings. The electronic transitions of the carbon rings, however, could not be identified so far.

Linear carbon molecules exhibit dangling bonds at their periphery and thus are highly reactive. Moreover, the intermediate size clusters can be produced in the gas phase only in moderate amounts. Hence, the spectroscopical studies of these species are accompanied by many problems. In our previous report on matrix-isolated species, we presented the strategies by which the molecular identification and structure determination can be achieved.[8] In this paper, we give a short overview about the recent developments in this field. We report on our success in obtaining the fluorescence and phosphorescence emission spectra of the matrix isolated carbon species. This kind of spectroscopy allows to determine the energies of the symmetrical (i.e. Raman active and infrared inactive) vibrational modes in the ground states of linear molecules. In conjunction with the infrared absorption data, the vibrational modes of the ground states can be completely characterized. The emission spectra yield also important informations about the structures of the molecules. The corresponding excitation spectra allows us to selectively study the electronic transitions of the matrix isolated species. By such methods, we aim to clarify the structure of the intermediate-sized carbon clusters, search for cyclic isomers, and ultimately also for fullerene-like species.

2. EXPERIMENTAL

Carbon-vapor molecules (C_1 - C_3) produced by evaporation of graphite in vacuum were deposited together with an excess of argon (C/Ar-ratio of 0.1 - 1 mol%) onto a cold (10 - 20 K) rhodium-plated sapphire mirror.[8,9] The evaporation process was monitored by a mass spectrometer and the molecular beam was purified by a differential pumping stage before reaching the cold substrate. Larger clusters were synthesized in the matrices by chemical reactions induced by controlled matrix annealing. In case of large C/Ar-ratios, the molecular growth took place already during deposition, and no thermal annealing was required. To identify the produced molecules, high-resolution absorption spectra of the prepared matrices were recorded in the infrared and uv-visible spectral ranges by FT-spectrometers. The clusters embedded in the matrices were selectively excited by a pulsed dye laser system and time-resolved dispersed fluorescence and phosphorescence spectra were obtained with a gated photodiode array spectrometer.[10]

3. ABSORPTION SPECTROSCOPY

Figure 1 shows an infrared absorption spectrum of a freshly prepared matrix. The C/Ar-ratio during the deposition was kept relatively high (about 1 mol%) so that already after matrix preparation larger clusters were synthesized. The most intense features in the infrared spectrum are at 1956, 1998 and 2039 cm^{-1} and come from linear C_6, C_9 and C_3, respectively. The spectrum shows absorption features of all so far identified linear carbon molecules ranging up to C_{11}.[9] The actual cluster size distribution is, however, wider, and extents probably up to linear C_{20}. Beside the linear isomers, also a carbon ring could recently be identified. Based on isotopic substitution experiments and *ab-initio* calculations, Wang and co-workers assigned the absorption at 1695 cm^{-1} to cyclic C_6.[7] On the other hand, we never found absorption lines at frequencies below 700 cm^{-1}, that would be typical for the radial modes of fullerene-cage vibrations. Very likely, carbon cluster growth at cryogenic conditions does not lead to fullerenes. This result confirms the widely accepted conjecture that fullerenes only form efficiently under high-temperature conditions where the molecular structures have the chance to relax into the minimum-energy configurations, i.e. to fullerenes.

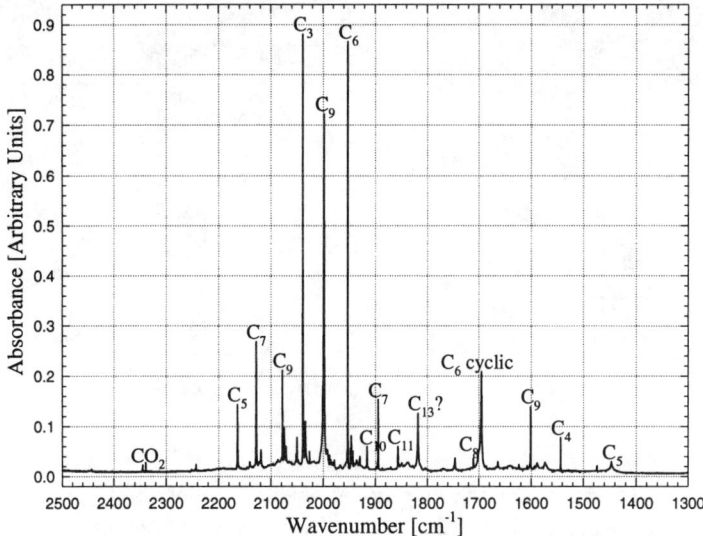

Figure 1. Infrared absorption spectrum of small carbon clusters trapped in an argon matrix. The molecular absorbers are indicated. Most of the absorption lines originate from linear molecules.

4. EMISSION SPECTROSCOPY

The high-sensitivity emission spectroscopy is a very powerful tool to analyze molecular vibrations and structures. We demonstrate the impressive potential of this method by our data obtained on the linear C_3 molecule.

4.1. Dispersed Emission Spectra

Figure 2 shows an example of an emission spectrum: the complete dispersed fluorescence ($\tilde{A}^1\Pi_u \to \tilde{X}^1\Sigma_g^+$) spectrum of C_3 embedded in solid argon. The insert in the figure shows the excitation wavelengths labeled by arrows over the well known C_3 absorption ($\tilde{A}^1\Pi_u \leftarrow \tilde{X}^1\Sigma_g^+$) spectrum. The emission spectrum is plotted in semilogarithmic scale. The very high sensitivity of our setup enabled us to observe emission bands up to vibrational energy of about 8000 cm^{-1}. The intensity of the observed emission peaks extends over about seven orders of magnitude.[10]

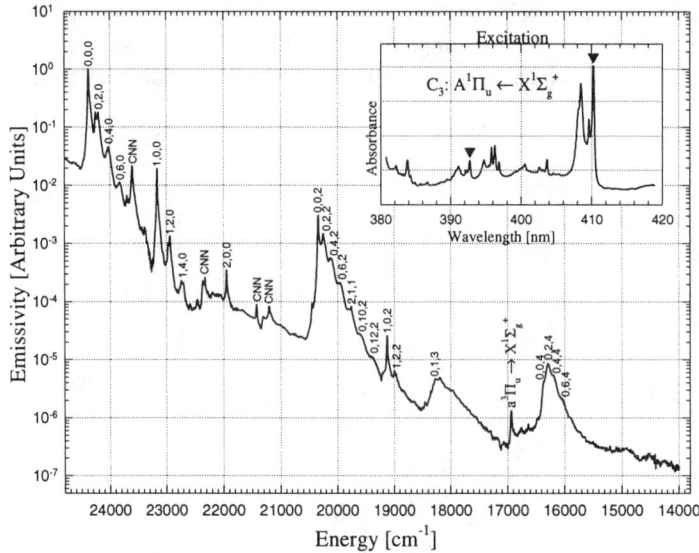

Figure 2. Dispersed fluorescence emission spectrum of C_3. The emission peaks of C_3 are labeled by corresponding vibrational quantum numbers (v_1'', v_2'', v_3''); several fluorescence emission peaks are due to the CNN molecule. See text for details.

The emission spectrum shows evidences for fast vibrational relaxation in the excited state $\tilde{A}^1\Pi_u$ caused by the matrix environment. We never observed emission bands occurring from the vibrationally excited electronical state $\tilde{A}^1\Pi_u$. The vibrational relaxation in the excited state $\tilde{A}^1\Pi_u$ is obviously much faster than the lifetime of the state itself. The emission spectrum shows also evidences for the linear structure of C_3 in the ground state $\tilde{X}^1\Sigma_g^+$. The observed emission peaks exactly obey the selection rules for linear molecules (only transitions to even multiples of $v_2''+v_3''$ can occur). Thus, our measurements clearly answer the question discussed several years ago, whether C_3 is linear in the ground state or not.

4.2. Excitation Spectra

Scanning the laser emission wavelength over the absorption structures and observing the corresponding emission intensities, excitation spectra of the matrix-isolated molecules can be obtained. Depending on the spectral resolution of the spectrometer and on the width of the laser emission line, some of the matrix-induced effects can be resolved, as e.g. matrix site effects and also to some degree the inhomogeneous line broadening due to the matrix. Generally, excitation spectra yield additional informations about the molecules: they reflect the vibrational structure of the excited states of the studied species. Because of the selectivity of the emission spectroscopy, the excitation spectra provide - in comparison with the absorption spectroscopy - a more reliable method to determine the absorption lines of a specific molecule, even in a sample containing a mixture of different clusters.

5. OUTLOOK

We have developed an apparatus by which extremely weak emission lines with quantum yields lower than 10^{-8} can be detected. Besides C_3, we recorded for the first time emission spectra of larger linear species. At present, we are analyzing the emission spectra of these molecules. Using the emission spectroscopy, we are searching for cyclic carbon isomers in the matrices.

ACKNOWLEDGMENTS

The authors thank the Deutsche Forschungsgemeinschaft and the HCM program of the EC for financial support.

REFERENCES

[1] e.g. in: J.M. Hawkins, *Acc. Chem. Res.* **25**, 150 (1992).
[2] P.D. Zavitsanos, *Carbon* **6**, 731 (1968).
[3] e.g. in: G. von Helden et al., *J. Phys. Chem.* **97**, 8182 (1993).
[4] J.P. Maier, *Chem. Soc. Rev.*, 21(1997).
[5] J.R. Heath and R.J. Saykally in *On Cluster and Clustering* (ed. P.J. Reynolds) Elsevier Science Publishers, 7 (1993).
[6] J.M.L. Martin and P.R. Taylor, *J. Chem. Phys.* **100**, 6047 (1996).
[7] S.L. Wang et al., *J. Chem. Phys.* **107**, 6032 (1997).
[8] I. Čermák et al., in *Fullerenes and Fullerene Nanostructures* (eds. H. Kuzmany, J. Fink, M. Mehring, S. Roth) World Scientific, ISBN 981-02-2853-8, 327 (1996).
[9] I. Čermák et al., in *Advances in Molecular Structure Res.* No 3, ISBN: 0-7623-0208-9, 117 (1997).
[10] I. Čermák et al., *J. Chem. Phys.* **24** (1998) *in press*.

Carbon Onions Produced by Ion-Implantation

T. Cabioc'h, M. Jaouen, M.F. Denanot, J. P. Rivière, J. Delafond
and J. C. Girard

*Laboratoire de Métallurgie Physique, Université de Poitiers, UMR 6630 CNRS,
SP2MI, Téléport2, Bd3, BP 179, 86960 Futuroscope cedex, France*

Abstract. Carbon films were produced by high dose carbon ion-implantation into polycrystalline copper and silver substrates held at high temperature (> 400°C). The carbon layers so obtained were characterized by Transmission Electron Microscopy and Atomic Force Microscopy. Numerous carbon-onions (50-200 nm in diameter), randomly distributed into a graphitic film were obtained after the implantation into copper whereas a continuous film of carbon onions (10-30 nm in diameter) was observed onto the silver substrates. We discuss the growth mechanisms of the carbon clusters and propose that the onion formation is simply based on a carbon precipitation in the substrate volume. We propose that the here presented process is a powerful way to synthesize a high density of carbon onions, especially in silver substrates.

INTRODUCTION

Carbon onions, usually described as concentric spherical carbon shells having a graphite-like structure, have been subject of interest when Ugarte (1) discovered a reproducible technique to synthesize such carbon clusters. This author irradiated a carbon soot (composed of fullerenes, nanotubes and graphite particles) with a high current electron beam in a Transmission Electron Microscope (TEM). Surprisingly, he observed the destruction of the various carbon clusters composing this soot and their rearrangement in an onion-like structure. Nevertheless, the inability of such a method to synthesize carbon onions in macroscopic quantities let explain that one has a poor knowledge of their physical properties.

More recently, other processes have been demonstrated to be efficient for the carbon onion synthesis. Annealing experiments performed on diamond nanoparticles (2) or on carbon black (3) have in this way been successfully used to obtain numerous carbon onions of about 3-10 nm in diameter. However, the so formed carbon onions are often polyhedral. Another kind of experiments that have been recently demonstrated to be powerful for onion synthesis consists in high dose carbon ion-implantation into copper or silver. Below, we will focus our attention onto the results obtained using such an approach.

These last experiments were first performed to find a new route for diamond thin film synthesis using an original approach. Carbon and copper are poorly miscible and Prins (4) proposed that during a high temperature carbon ion-implantation into copper, C atoms will diffuse towards the surface where they will precipitate. Copper and diamond have close structures and a very small lattice mismatch. Thus Prins proposed that carbon could precipitate onto the copper surface in the diamond phase. In fact, using such an approach we have never been able to characterize the presence of diamond onto the implanted copper substrates(5) and this whatever the implantation conditions were. It was only detected a uniform turbostratic graphite film seeded with numerous carbon onions randomly dispersed onto the surface. To obtain a better understanding about the growth mechanisms of these spherical clusters, we perform the same kind of experiment but into silver, in which carbon has also a very low solubility.

In this paper, we present the results obtained for the different carbon phases formed by carbon-ion implantation into copper and silver. Analysis and comparison of carbon onions formed during implantation in copper and silver are presented. The mechanisms of growth of the different carbon phases we obtain and the potentiality of the presented implantation technique are then discussed.

EXPERIMENTAL

Copper and silver substrates (purity > 99.99%) are shots (3 mm in diameter) mounted onto a furnace that can reach temperatures higher than 1000°C. The 120 keV C_{12}^+ ion implantations have been performed in a vacuum better than 10^{-5} Pa. The substrates' temperature has been varied for copper in the range 700-1000°C and 500-800°C for silver. We also varied the ion-fluxes (2 to 45 µA cm^{-2}s^{-1}) and implanted high doses (0.5 to 5x10^{17} cm^{-2}). To prepare the samples for TEM observations, the backside of the substrate was polished (the implanted face being protected with a plastic film) with a chemical jet. Near the hole so created, areas where only the carbon thin film free of silver (or copper) is present are obtained. The different synthesized carbon phases were studied by Transmission Electron Microscopy (TEM) and High Resolution TEM (HRTEM). The surface of the implanted samples was examined by Atomic Force Microscopy (AFM).

RESULTS AND DISCUSSION

Typical bright field micrographies of the carbon film so produced are presented in figure 1. Spherical micrograins (50-200 nm in diameter) embedded into a continuous carbon film are observed onto copper substrates. This result differs from the one obtained for the implantations into a silver matrix since only spherical carbon clusters are observed in this last case (see fig. 1b). The Selected Area Electron Diffraction Patterns (SAEDP) obtained are typical of polycrystalline graphite. We performed dark field observations of the spherical clusters by selecting a part of the most intense ring (corresponding to a distance of 0.34 nm between graphene layers) as shown in figure 2. The fact to observe right segments of disks indicates that the graphene layers of these clusters are parallel from each part of the center. Such an observation is easily explained if one considers that these micrograins have an onion-like structure.

High resolution Transmission Electron Microscopy experiments assess that carbon onions have been elaborated after implantation into both copper and silver. Nevertheless, it is important to note that carbon-ion implantation in silver is much more promising since only carbon onions are obtained. Indeed for copper substrates, if some areas present a high density of onions, a uniform film of turbostratic graphite free of carbon onions is observed on many other areas. Let's also note that the density of onions formed into silver is so high that the observation of individual onions is only possible on the vicinity of the hole created by the chemical polishing. Furthermore, we annealed some silver samples in vacuum (800°C, 10^{-5} Pa) to evaporate the silver and thenafter observed that uniform films of carbon onions were formed in these substrates.

The positions of the onions versus the implanted surface have been determined using AFM observations. On Figure 4, a high density of onions is clearly observed onto the implanted surface for both silver and copper substrates after high fluence implantations ($\geq 3 \times 10^{17}$ ions/cm^2). These experiments confirm the TEM observations since it attests

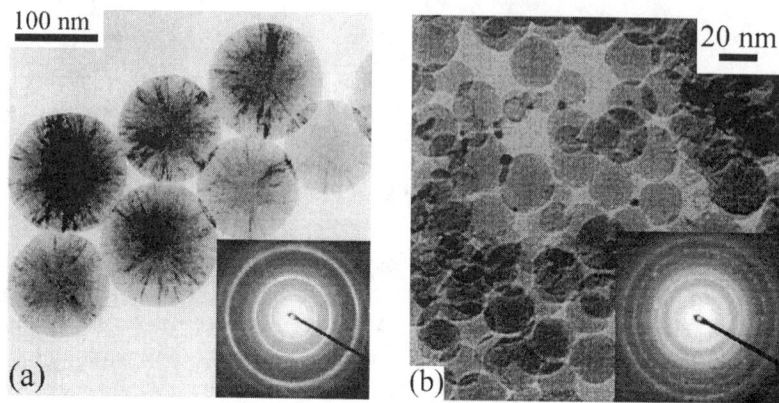

FIGURE 1. TEM bright field micrographies (and corresponding SAEDP) of carbon film formed by implantation : (a) into copper (dose : 5.10^{17} cm^{-2}, ion-flux :2 µA.cm^{-2}, temperature : 800°C) and (b) into silver (Bragg spots on the SAEDP are due to some silver residues) (dose : 3.10^{17} cm^{-2}, ion-flux :2 µA.cm^{-2}, temperature : 600°C)

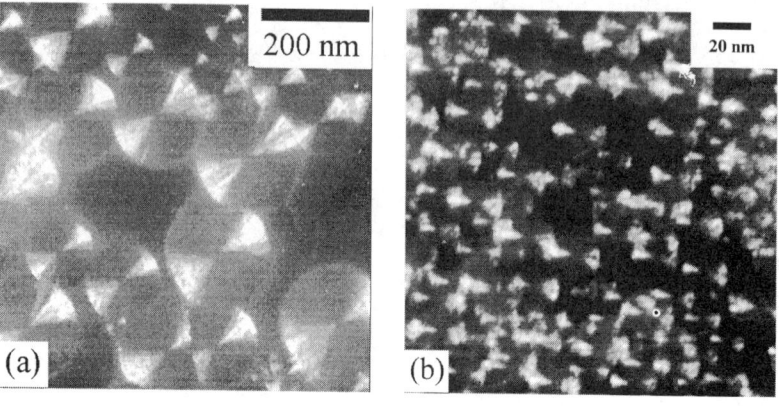

FIGURE 2. TEM dark field micrographies of the carbon film formed : (a) into copper and (b) into silver (same implantation parameters than for the figure 1)

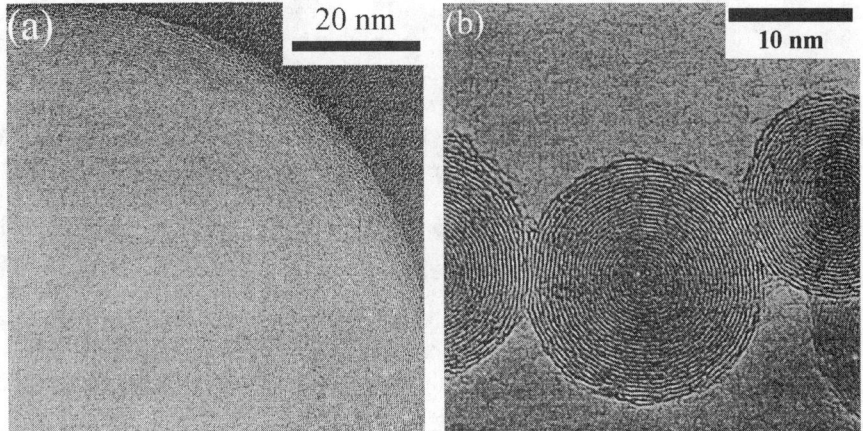

FIGURE 3. HRTEM micrographies of some carbon onions synthesized : (a) into copper (dose : 5.10^{17} cm^{-2}, ion-flux :2 µA.cm^{-2}, temperature : 800°C) and (b) into silver (dose : $2.5.10^{17}$ cm^{-2}, ion-flux :45 µA.cm^{-2}, temperature : 500°C).

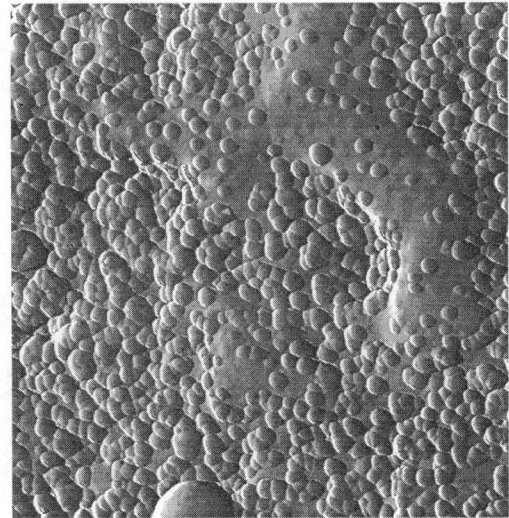

FIGURE 4. AFM observation (signal error image) of carbon onions synthesized into silver (dose : 3.10^{17} cm^{-2}, ion-flux :2 µA.cm^{-2}, temperature : 500°C).(image size : 1x1 µm^2)

of the formation of a high density of carbon onions for implanted silver substrate. Furthermore, on many areas, only the top of some onions can be observed; the main part of the carbon onion film being still embedded inside the silver volume. Onto copper substrates, the density of onions is heterogeneous : isolated onions as well as clusters of some tenth of onions are randomly distributed onto the surface (not shown here).

In view of AFM experiments, we propose that the carbon onions formed into the bulk of the silver substrates is due to a carbon precipitation. During this three dimensional precipitation, one can expect spherical precipitates to be formed. Furthermore for such dissolution-diffusion-precipitation mechanism, it is well documented that some metals (as Cu, Ag, Ni, Co,...) have a catalytic effect onto graphitization (6-9). More especially, during experiments related to the study of the catalytic effect of these metals, graphitization occurs at low temperature, the graphene layers being parallel to the C/metal interface. Thus one can expect the spherical carbon precipitates to have their graphene layers parallel to the metal interface and this would lead to the formation of an onion-like structure. Moreover, if such a structure is not directly formed in the bulk bulk, irradiation effects combined with the temperature effect can allow the transformation of carbon precipitate into onions (10). The onions being formed into the bulk of the metal substrates, we propose that copper (or silver) preferential sputtering will play a significant role by letting appear the top of the onions on the metallic surface.

From our results, carbon-ion implantation into copper and silver seems to be a powerful technique to synthesize numerous carbon onions. Especially, the formation of a continuous thin carbon onion film onto silver is of interest if one consider the poor knowledge that we have about onion properties. We can then think to perform many new experiments to characterize tribological, magnetic and electrical properties of carbon onion films. Furthermore, the medium range temperature process we have presented has the advantage to be a one step process and high current implantors can contribute to cover large areas with carbon onions in short times. For example, carbon onion thin film formation has been observed after ion implantation into silver with a high ion-flux (45 mA.cm^{-2}). With such high ion-fluxes we could then obtain carbon onions' films in only some tenth of minutes. Moreover, the onion formation by carbon ion-implantation seems to be a general feature for substrates in which carbon is not miscible. Thus, one can expect to obtain identical results by performing high fluence carbon-ion implantations into gold, lead, etc....

REFERENCES

(1) D. Ugarte, *Nature*, **359**, 707-09, (1992)
(2) V.L. Kuznetsov, A.L. Chuvilin, Y.V. Butenko, I.Y. Mal'kov and V.M. Titov, *Chem. Phys. Lett.*, **222**, 343-46, (1994)
(3) D. Ugarte, *Carbon* **32**, 1245-47 (1994)
(4) J.F. Prins and H.L. Gaigher, *Proceedings 2nd International Conference New Diamond Science and Technology, edited by R. Messier, J.T. Glass, J.E. Butler & R. Roy, Materials Research Society, Pittsburgh*, p.561-66, 1991.
(5) T. Cabioc'h, J.P. Rivière, J. Delafond, M. Jaouen and M.F. Denanot, *Thin Solid Films* **263**, 162-68 (1995)
(6) A. Oya and S. Otani, *Carbon* **17**, 131-35 (1979).
(7) R. Sinclair, Conference in Université de Poitiers (1995)
(8) R. Lamber, N. Jaeger and G. Schulz-ekloff, *Surf. Science* **197**, 402-07 (1988)
(9) T.J. Konno and R. Sinclair, *Acta Metall. Mater.* **43**, 471-84 (1995)
(10) P. Wesolowski, Y. Lyutovitch, F. Banhart, H. D. Carstanjen and H. Kronmüller, Appl. Phys. Lett. 71, 1948-50 (1997)

Molecular Dynamics Study of Carbon Structures

István László

Department of Theoretical Physics, Institute of Physics
Technical University of Budapest, H-1521 Budapest, Hungary

Abstract. Tight binding molecular dynamics calculations were performed for the study of carbon structures. Under helium atmosphere cage-like structures were obtained independent of the initial arrangement of the carbon atoms. The final cage C_{60} cluster contained 11 pentagons, 19 hexagons, 1 tetragon and 1 heptagon. When the helium atoms were removed, the final bulk structure was determined by the densities of the carbon atoms. In the case of amorphous and graphite densities of bulk simulation mostly threefold coordinated carbon atoms were obtained, and when diamond density was used the final structure contained mostly fourfold coordinated carbon atoms.

INTRODUCTION

Molecular dynamics [1] can provide detailed insight into the formation process of fullerenes although the simulation time of several psec is small compared to the real time of formation. In the laser evaporation [2] and arc deposition [3], for example, the estimated average time between collisions of He, C or C_{60} is the order of nanoseconds [4]. This drawback is usually avoided by using higher pressure and increased temperature in the simulation process. In all of these calculations, however, there are applied some artificial conditions in order to obtain the closed cage like structures. Ballone and Milani [5] kept the carbon atoms on the surface of a sphere for $T > 4000K$, Chelikowsky [6] removed and replaced randomly the energetically unfavorable atoms and Wang et al. [7] confined them into a sphere and obtained closed cage only for R = 3.832 Å.

In the proceedings of the 1997 Kirchberg Winterschool [8] we presented our preliminary molecular dynamics calculations on the formation of the C_{60} molecule, where tight-binding molecular dynamics calculations were performed for sixty carbon atoms in helium atmosphere. In the present work we increased the time of simulation from 11.2 psec to 30.8 psec and studied the influence of the different initial carbon structures on the final results. We studied also the formation of amorphous graphite and diamond structures when the helium atmosphere was replaced with the periodic boundary condition on the carbon unit cell.

I THE METHOD

In our simulation the carbon carbon interaction was calculated with the help of the

$$E = \sum_I \frac{P_I^2}{2m} + \sum_n^{occupied} <\psi_n|H_{TB}|\psi_n> + U_{rep} \qquad (1)$$

tight-binding total energy expression of Xu et al. [9].

The helium-helium and carbon-helium interaction was modeled with a pairwise 6-12 Lennard-Jones potential

$$U_{ij} = 4\varepsilon_{ij}\left(\frac{\sigma_{ij}^{12}}{r_{ij}^{12}} - \frac{\sigma_{ij}^6}{r_{ij}^6}\right), \qquad (2)$$

where $\varepsilon_{HeHe}/k_B = 10.8 K$, $\sigma_{HeHe} = 2.57 Å$ [10] and $\varepsilon_{CC}/k_B = 33.25 K$, $\sigma_{CC} = 3.47 Å$ [11,12]. The carbon-helium parameters were calculated with the relations $\varepsilon_{HeC} = (\varepsilon_{HeHe}\varepsilon_{CC})^{0.5}$ and $\sigma_{HeC} = (\sigma_{HeHe} + \sigma_{CC})/2$.

In the first part of our calculation we put 60 carbon atoms in a cube of side 25.0 Å with randomly dispersed 1372 helium atoms. Four initial carbon arrangements were used, as Kroto's four-deck sandwich model 6:24:24:6, and other random structures with amorphous, graphite and diamond densities. These carbon clusters were exploded with an initial temperatures of 13000 K. A Nosee-Hoover thermostat controlled the T = 4000 K temperature of the helium gas. Periodic boundary condition was applied for the cube of side 25.0 Å.

In the second part of our calculation the helium atmosphere was removed and the periodic boundary condition was used for the unit cells of side 8.424 Å, 8.00 Å and 6.4Å in the case of amorphous, graphite and diamond densities of 60 carbon atoms.

II RESULTS AND DISCUSSIONS

Figure 1 shows the evolution of the carbon structures in helium atmosphere.

In the first 0.35 psec the disintegration of the various carbon structures were found. Until 1.4 psec new poligons were created and for 5.6 psec cage like structures were developed for each case of the simulation. The calculation was continued for the four-deck model and after 30.8 psec a cage structure was obtained with 11 pentagons, 19 hexagons, 1 tetragon and 1 heptagon (Figure 2).

Any artificial condition was not used in the simulation.

In the case of bulk simulation the time of simulation was 5.6 psec. The number of onefold, twofold, threefold and fourfold coordinated sites are in order 0.0%, 10.0%, 86.7% and 3.3% for the amorphous; 3.3%, 10.0%, 86.7% and 0.0% for the graphite; and 0.0%, 0.0%, 10.0% and 90.0% for the diamond densities.

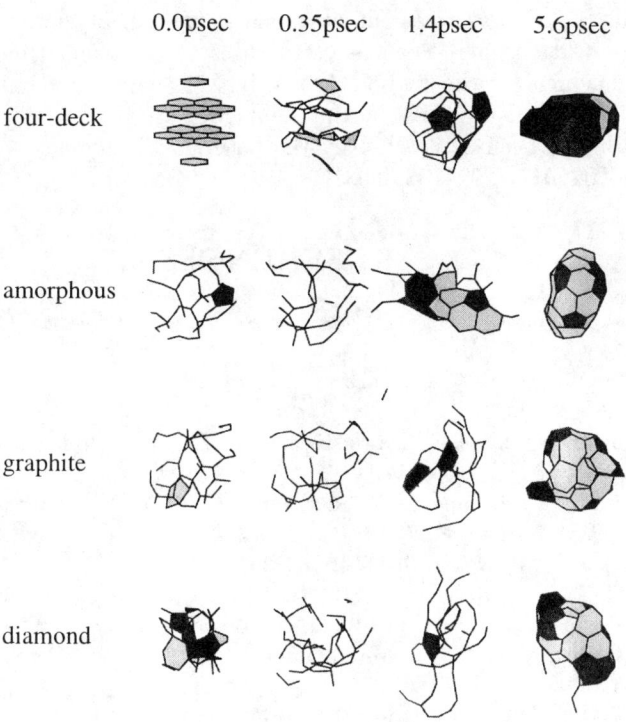

FIGURE 1. Snapshots at 0.0, 0.35, 1.4, and 5.6 psec of the four-deck, amorphous, graphite and diamond structures.

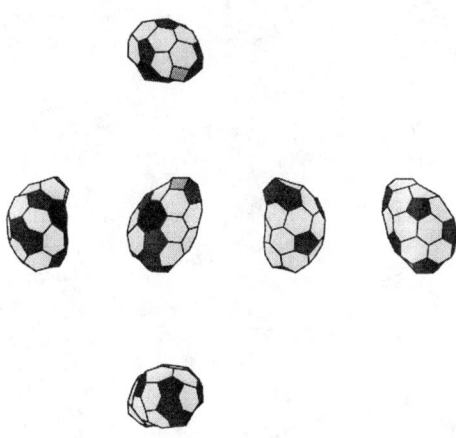

FIGURE 2. The final structure after the simulation time of 30.8 psec. The structure is seen from the six faces of the cubic unit cell.

The results of the present computational study can be summarized as follows: Under helium atmosphere cage like structures were obtained independent of the initial arrangement of the carbon atoms, but in the case of bulk simulation the final structure was determined by the atomic densities.

This work has been supported by the Országos Tudományos Kutatási Alap (Grant No. T025017, T024138, T021228).

REFERENCES

1. Allen M. P., Tildesley D. J., *Computer Simulation of Liquids*, Oxford: Clarendon Press, 1996.
2. Kroto H. W., Heath J. R., O'Brien S. C., Curl R. F., and Smalley R. E., *Nature* **318**, 162, (1985).
3. Krätschmer W., Lamb L. D., Fostiropoulos K., and Huffman D. R., *Nature* **347**, 354, (1990).
4. Robertson D. H., Brenner D. W., and White C. T., *J. Phys. Chem.* **96**, 6133, (1992).
5. Ballone P. and Milani P., *Phys. Rev. B* **42** 3201 (1990).
6. Chelikowsky J. R., *Phys. Rev. Let.* **67** 2970 (1991).
7. Wang C. Z., Xu C. H., Chan C. T., Ho K. M., *J. Phys. Chem.* **96** 3563 (1992).
8. László I., in *Molecular Nanostructures* , Eds. Kuzmany H., Fink J., Mehring M., Roth S. Singapore, New Jersey, London, Hong Kong: World Scientific, 1998, pp. 143-146.
9. Xu C. H., Wang C. Z., Chan C. T., Ho K. M., *J. Phys. C* **4**, 6047, (1992).
10. Reed T. M., Gubbins K. E., *Applied Statistical Mechanics*, New York: McGraw-Hill, 1973.
11. Girifalco L. A., *J. Phys. Chem* **96**, 858, (1992).
12. Girifalco L. A. and Lao R. A., *J. Chem. Phys* **25**, 693, (1956).

Plasmon Excitations in Carbon Onions: Model vs. Measurements

Thomas Stöckli, Zhong Lin Wang[†], Jean-Marc Bonard,
Pierre Stadelmann[‡], André Châtelain

*Institut de Physique Expérimentale, Département de Physique,
Ecole Polytechnique Fédérale de Lausanne, CH - 1015 Lausanne, Switzerland*

[†]*School of Materials Science and Engineering, Georgia Institute of Technology,
Atlanta, GA 30332-0245, USA*

[‡]*Centre Interdépartemental de Microscopie Electronique,
Ecole Polytechnique Fédérale de Lausanne, CH - 1015 Lausanne, Switzerland*

Abstract. Non-relativistic local dielectric response theory has proven successful in the interpretation of Electron Energy Loss data of nanometer-size isotropic particles of different geometries. In previous work, we have adapted this model to take into account anisotropy as encountered in the case of carbon onions. We have shown that this anisotropy needs to be taken into account since important deviations with respect to an isotropic model can be observed. In this contribution, we report on the first energy filtered images of carbon onions and compare intensity profiles across the spheres to our calculations.

INTRODUCTION

Electron Energy Loss Spectroscopy (EELS) in a High Resolution Transmission Electron Microscope (HRTEM) allows at the same time the characterization of the geometrical parameters and the investigation of the electronic properties of one single nanometer-size particle. This technique therefore does not rely on the necessity to dispose of high purity samples, which is interesting in the case of carbon nanostructures such as tubes or onions, since even though important progress has been made (1), the purity of the samples still is a problem. In fact, some measurements using this technique have already been reported, but a qualitative interpretation of the results has not yet been possible due to the lack of theoretical background.

Recent calculations (2) based on non-relativistic local dielectric response theory (3) have now evened a way for a detailed qualitative analysis of the experimental data. We report on the analysis of energy filtered transmission electron micrographs of carbon onions based on those calculations.

MODEL OF THE EXCITATION OF PLASMONS

The model of the plasmon excitations in carbon onions was developed in the frame of non-relativistic local dielectric response theory (3). This approach consists in deducing surface and volume plasmon excitation probabilities from the dielectric tensor of the bulk material, taking into account the geometry of the particle. In the case of carbon onions, it is necessary to make an assumption about the dielectric properties of the particle. In our calculations, this was done following the scheme proposed by Lucas et al. (5) which consists in projecting the dielectric tensor of planar graphite into spherical coordinates (figure 1a). Accordingly it is assumed that locally the dielectric properties of a carbon onion can be described by the dielectric tensor of planar graphite.

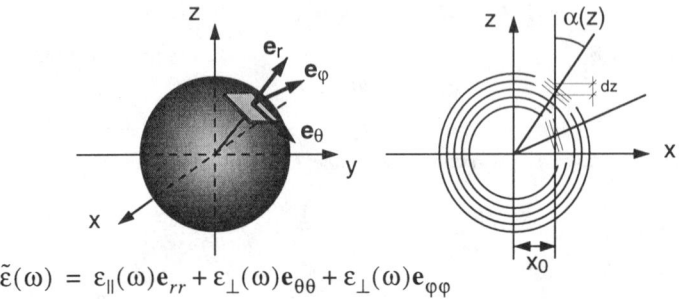

$$\tilde{\varepsilon}(\omega) = \varepsilon_\|(\omega)\mathbf{e}_{rr} + \varepsilon_\perp(\omega)\mathbf{e}_{\theta\theta} + \varepsilon_\perp(\omega)\mathbf{e}_{\varphi\varphi}$$

FIGURE 1. (a) Model used for the electronic properties of a carbon onion. Locally, the electronic properties are supposed to be identical to those of planar graphite. (b) As the electron moves along its trajectory, it passes infinitely small layers of planar graphite oriented at an angle α with respect to the optical axis of the microscope.

Non-relativistic local dielectric response theory allows to calculate the excitation of the surface and volume plasmons separately, the total plasmon excitation probability being just the sum of the two. In our approach, the volume contribution is deduced from the excitation probability per unit path length of a planar sheet of graphite as calculated by Wessjohann (4). In fact, as the electron is travelling through the carbon onion, it crosses infinitely thin layers of graphite which continuously change their orientation as the electron moves on (see figure 1b). The volume plasmon excitation probability is therefore simply given by the integral of the orientation-dependent excitation probability given by Wessjohann along the path in the onion (Eq. 1). q_p^2 and q_c^2 are the projection of the transferred momentum on the plane perpendicular and on the direction parallel to the c-axis of the graphitic layer of thickness dz, respectively.

$$\frac{dP^{volume}(\omega)}{d\omega} = \frac{e^2}{4\pi\varepsilon_0\hbar v^2}\int_{-z_0}^{z_0}dz\int_0^{\theta_c}\theta d\theta\int_0^{2\pi}d\varphi\ \mathrm{Im}\left[\frac{-q_0^2}{q_p^2\varepsilon_\perp(\omega) + q_c^2\varepsilon_\perp(\omega)}\right] \quad (1)$$

In order to calculate the surface plasmon excitation probability, the expression for surface plasmon excitation given in the review article by Wang has been used (3).

$$\frac{dP^{surface}(\omega)}{d\omega} = \frac{e}{\pi\hbar v^2}\int_{-\infty}^{\infty} dz' \int_{-\infty}^{\infty} dz\, Im\left\{e^{i\omega(z'-z)/v}V^{ind}(\mathbf{r}, \mathbf{r}_0)\Big|_{\mathbf{r} = (x_0, 0, z')}\right\} \quad (2)$$

$V^{ind}(\mathbf{r}, \mathbf{r}_0)$ is the induced electric potential at position $\mathbf{r}$ caused by a stationary electron located at position $\mathbf{r}_0$. It is the homogenous part of the solution of the equation $\nabla \bullet [\tilde{\varepsilon}(\omega)\nabla V(\mathbf{r}, \mathbf{r}_0)] = (e/\varepsilon_o)\delta(\mathbf{r}, \mathbf{r}_0)$. The details of how this induced potential is calculated can be found in ref. (2).

MEASUREMENTS

The EELS measurements have all been carried out on a Philips CM 300 field emission microscope equipped with a Gatan Imaging Filter. The onions were produced *in situ* by intense irradiation of polyhedral closed shell graphitic particles (6) frequently found in the deposit of a conventional arc discharge used for the production of carbon nanotubes (7). Figure 2 shows a series of energy filtered images of a carbon onion of 12.5 nm radius (insets (b) through (e)).

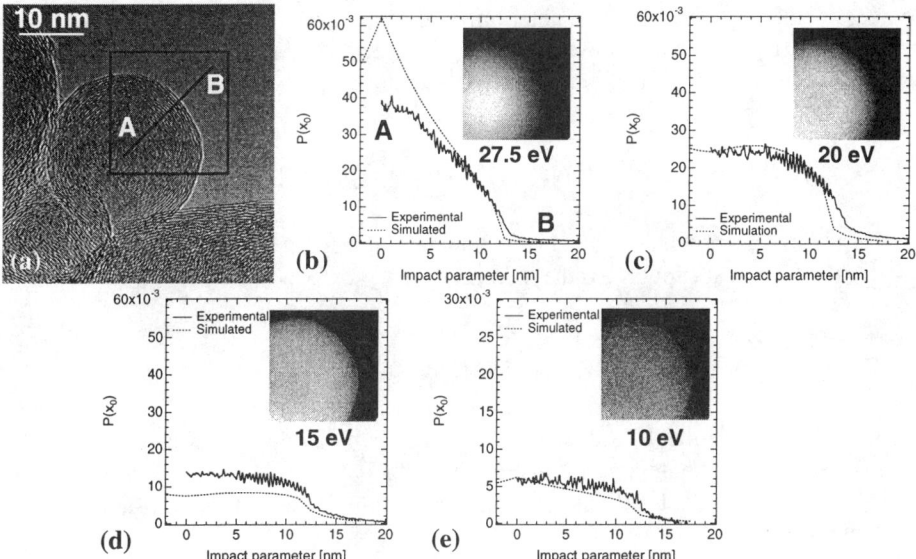

FIGURE 2. Non filtered HRTEM image (a) of a carbon onion of 12.5 nm radius on which the experiment was carried out. Figures (b) through (e) display intensity profiles across the energy filtered images shown in the inset (solid lines). The energy window was 4 eV wide and centred at the energies indicated in the corresponding figure. The profiles were taken from the centre of the sphere (A) to the point (B) indicated in the non-filtered image. Also shown are the simulated intensity profiles as obtained from the non-relativistic local dielectric response theory.

The experimental intensity profiles (solid lines) taken from points A to B are compared with the simulated profiles obtained from our model (dotted line).

It has to be noted that the observed intensities have not been scaled in any way in order to fit to our simulations, and that the simulations have been obtained without taking into account the characteristics of the electron beam (energy distribution and angular convergence). It can be seen that there is an excellent agreement between the experimental and theoretical curves at 10 and 20 eV, whereas at 15 eV the simulations slightly underestimate and at 27.5 eV they slightly overestimate the actual excitation probability. Further experiments and simulations are currently undertaken in order to clarify whether this difference between model and measurement arises from the experimental broadening of the plasmon excitation probability due to the finite energy resolution of the experimental setup or if it is due to either the effect of curvature of the layers or to the finite size of the particle.

CONCLUSIONS

The preliminary analysis of our experimental data shows that non-relativistic local dielectric response theory is adapted for the interpretation of experimental electron energy loss data. A detailed analysis of the EEL spectra and energy filtered TEM images should therefore allow to determine the differences between the intrinsic properties of a carbon onion and those of planar graphite and hopefully contribute to a better understanding of the physical properties of those novel, intriguing form of carbon.

ACKNOWLEDGEMENTS

The authors acknowledge the Centre Interdépartemental de Microscopie Electronique for the technical support for the measurements.

REFERENCES

1. Bonard, J.-M., Stora, T., Salvetat, J.-P., Maier, F., Stöckli, T., Duschl, C., Forró, L., de Heer, W., and Châtelain, A., *Adv. Mater.* **9**, 827 (1997).
2. Stöckli, T., Wang, Z. L., Bonard, J.-M., Stadelmann, P., and Châtelain, A., *Phys. Rev. B*, in press.
3. Wang, Z. L., *Micron* **27**, 265 (1996).
4. Wessjohann, H. G., *Z. Phys.* **269**, 269 (1974).
5. Lucas, A. A., Henrard, L., and Lambin, P., *Phys. Rev.* **B 49**, 2888 (1994).
6. Ugarte, D., *Nature* **359**, 707 (1992).
7. Ebbesen, T. W. and Ajayan, P. M., *Nature* **385**, 220 (1992).

HIGHLY SYMMETRIC BORANE CLUSTERS AS FULLERENE ANALOGS

Zsolt Szekeres and Péter R. Surján

*Eötvös University, Dept. Theoretical Chemistry,
H-1518 Budapest 112, POB 32, Hungary*

Abstract. Although borane closo-clusters have been widely studied by chemists and their analogy to fullerene clusters has been emphasized by Lipscomb in 1992, they have not yet been closely investigated by material scientists. We undertake a theoretical study of several highly symmetric borane clusters analyzing their ground state geometry and electronic structure. The calculations confirmed the fact that the stable clusters are doubly ionized. Energetics and geometry relaxation in several ionized states has been investigated paying particular attention to Jahn-Teller distortions occurring due to degeneracies. It has been found that, in comparison with the corresponding data for fullerenes, the distortion energies are significantly larger, falling in the range of a few eV. We also study the extent of electron transfer between metal dopants and borane clusters. Preliminary investigations predict a moderate charge flow from potassium atoms to small borane clusters.

ELECTRONIC STRUCTURE OF CLOSOBORANES

Analogy between highly symmetric fullerene cages and closed borane clusters has been advocated some time ago by Lipscomb [1], who applied geometrical duality (face to vertex interconversion) to derive closo-boranes from fullerenes. For example, the first known species among closoboranes, $B_{12}H_{12}$, has icosahedral symmetry and it derives from C_{20}. The simplest boron cluster, B_4H_4, is an analog of C_4. These latter two clusters have not been actually synthesized yet, but the next members of the series, from B_6H_6 to $B_{12}H_{12}$, do exist [2] in charged states (see below). Oppositely to fullerenes, preparation of higher borane clusters seems to be difficult while smaller ones can be synthesized by standard inorganic techniques [2].

Existence of small borane clusters is advantageous as they are available for more sophisticated quantum chemical calculations. In spite of this, most computational results for these systems were obtained so far by semiempirical calculations [3]. In this work, we report our recent studies with ab initio calculations in various basis sets at the Hartree-Fock (HF) level augmented by second order Møller-Plesset (MP2) perturbation calculations to account for electron correlation. Geometries of the clusters have been optimized at the HF level.

In course of these calculations we had to face to difficulties coming from orbital degeneracies resulting from the high point group symmetries (D_{4d}, T_d, O_h). An example is shown in Fig 1 for B_6H_6 (O_h geometry) exhibiting doubly degenerate HOMO and LUMO levels with a partially filled shell. Exact description of such degenerate systems would be possible by expensive multi-configuration calculations.

We used here a simple approximate scheme [4] which treats the problem by fractional occupancies of open-shell levels. We have extended here this philosophy to MP2 calculations as well and applied the formula

$$E^{(2)} = -\frac{1}{2}\sum_{pqrs} n_p n_q (1-n_r)(1-n_s)\frac{[pq|rs]([pq|rs]-[pq|sr])}{\varepsilon_r + \varepsilon_s - \varepsilon_p - \varepsilon_q} \quad (1)$$

in terms of spin-orbitals p, q, r and s. Here n-s stand for (integer or fractional, $0 \leq n \leq 1$) occupation numbers, ε-s for orbital energies while $[pq|rs]$ are two-electron integrals in the usual notation.

Performing the calculations for the neutral species, we found that, in agreement with previous semiempirical studies [3], B_4H_4 and B_9H_9 have closed-shell ground states, while B_6H_6, B_7H_7, B_8H_8, $B_{10}H_{10}$, $B_{12}H_{12}$ possess partially filled degenerate HOMO. These latter systems can either accept two extra electrons forming $B_nH_n^{2-}$ observed actually in inorganic chemistry since decades [5], or they will undergo a geometrical distortion to resolve degeneracies.

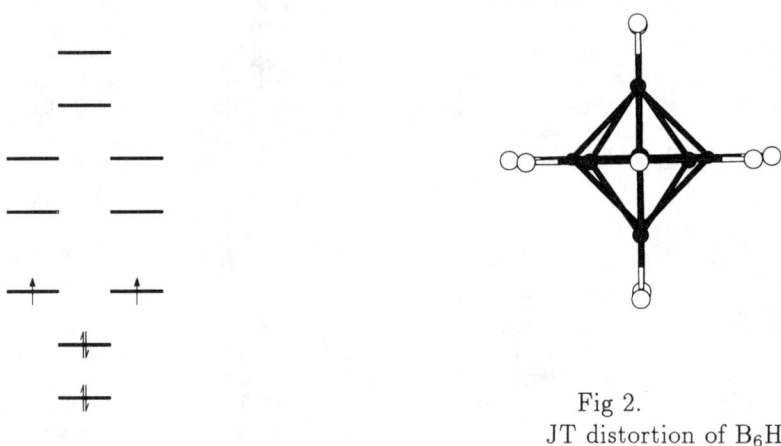

Fig 1.
Energy level diagram of $B_{10}H_{10}$

Fig 2.
JT distortion of B_6H_6

JAHN-TELLER DISTORTIONS

Jahn-Teller (JT) distortions [6] may occur for clusters with non-Abelian point group symmetry if they possess a partially filled (open-shell) level. Distortion is performed along the normal coordinates towards the state with lower symmetry and one-dimensional representation. Similar distortions are well known in fullerenes [7, 8, 9].

According to the results of the electronic structure calculations mentioned above, the singly, doubly and triply charged anions of B_4H_4 are JT active, while the neutral and fourtuply charged ions are described by nondegenerate wave functions.

The B_9H_9 clusters are stable in neutral and doubly ionized states while JT active in singly and triply ionized states. In contrast, the rest of the clusters investigated in this study are nondegenerate if they are doubly charged. Other ions as well as the neutral clusters will distort.

In order to monitor the JT distortions we have carried out SCF and MP2 calculations for the open-shell states of $B_nH_n^{x-}$ for various xs. Because, even if the degeneracies are slightly resolved by distortions, we can count with low-lying excited states, the damped MP2 formulae [10] were used which can handle quaside-generacies. The corresponding method will be referred to as QD-MP2.

Symmetry breaking geometries have been obtained by applying a manual distortion to the desired subgroup of the original (parent) point group, then re-optimizing the geometry by the aforementioned quantum chemical methods. The result of such distortions is schematically illustrated for B_6H_6 in Fig 2.

System	distortion	ΔE_{SCF}[eV]	E_{QD-MP2} [eV]	basis
$B_4H_4^-$	$T_d \to D_{2d}$	3.25	3.03	6-21G
$B_4H_4^{2-}$	$T_d \to D_{2d}$	5.19	4.39	6-21G
$B_4H_4^{3-}$	$T_d \to D_{2d}$	3.32	3.13	6-21G
$B_6H_6^0$	$O_h \to D_{4h}$	1.44		6-21G*
$B_6H_6^{2-}$	$O_h \to D_{4h}$	1.00		6-21G*
$B_6H_6^{3-}$	$O_h \to D_{4h}$	2.09		6-21G*
$B_7H_7^-$	$D_{5h} \to C_{2v}$	3.35		STO-3G
$B_7H_7^{3-}$	$D_{5h} \to C_{2v}$	0.71		STO-3G
$B_{10}H_{10}^{3-}$	$D_{4d} \to C_{2v}$	2.75		STO-3G

Table 1: Jahn-Teller distortion energies in some closoboranes

Table 1 presents the computed distortion energies at the SCF and QD-MP2 levels. Comparing to similar data for fullerene clusters, we found JT energies which are larger by at least an order of magnitude (e.g., for triplet C_{60} one gets $\Delta E_{JT}(I_h \to D_{5d})$=180 meV). The relatively small differences between HF and QD-MP2 results indicate that correlation effects, though not negligible, do not change the situation qualitatively. Distortion energies, however, are sensitive functions of the charge states for each cluster.

Similarly to the huge JT energies, we found extraordinarily large changes in atomic distances upon JT distortion. Typical bond length changes in neutral B_6H_6 are 0.15 Å (for comparison in C_{60} these are $\approx$ 0.01 Å).

DOPING OF CLOSOBORANES

Extra electrons which are necessary for stabilizing most clusters may come from alkali or alkali-earth atoms. Such a process is not a usual doping as neutral versions of the borane clusters do not exist. Nevertheless, we may keep this term for brevity.

To model doped structures, we performed ab initio STO-3G calculations on clusters containing one (two) potassium atoms at optimized distances. Atomic charges on K were computed by Mulliken's population analysis. We have found that the charge q on K was 0.88 (0.81) a.u. For comparison, a similar calculation on KC_{60}, when the K atom is placed at 7 Å from the center of C_{60}, resulted $q=0.73$ a.u. while for the KCl diatomic we got $q=0.64$ a.u. These numbers indicate that the electron accepting ability of borane clusters is similar to, possibly even larger than, that of fullerenes.

Acknowledgments

We appreciate fruitful discussions with B.Csákvári (Budapest). Financial support from the grants OTKA T023052 and AKP 96/2-462 are gratefully acknowledged.

References

[1] W.N.Lipscomb, Inorg.Chem. **31**, 2297 (1992).

[2] N. N. Greenwood and A. Earnshaw, *Chemistry of the elements* (Pergamon press, Oxford, 1989).

[3] J. Joseph, B. Gimarc, and M. Zhao, Polyhedron **12**, 2841 (1993).

[4] C.C.J.Roothaan and P.Bagus, Methods Comput.Phys. **2**, 47 (1964).

[5] M. F. Hathworne and A. R. Pittochelli, J. Am. Chem. Soc. **82**, 3228 (1960).

[6] J. A. Jahn and E. Teller, Proc. Roy. Soc. (London) Ser. A **A161**, 220 (1937).

[7] F. Negri, G. Orlandi, and F. Zerbetto, Chem. Phys. Letters **144**, 31 (1988).

[8] P. R. Surján, L. Udvardi, and K. Németh, J.Mol. Struct. (THEOCHEM) **311**, 55 (1994).

[9] N. Koga and K. Morokuma, Chem. Phys. Lett. **196**, 191 (1992).

[10] P.R.Surján and Á.Szabados, J.Chem.Phys. **104**, 3320 (1996).

Pulse ESR of Triplet States of Large Molecular π Systems

S. Knorr*, A. Grupp*, M. Mehring*,
M. Wehmeier[†], P. Herwig[†], V. S. Iyer[†], and K. Müllen[†]

*2. Physikalisches Institut, Universität Stuttgart
Pfaffenwaldring 57, 70550 Stuttgart, Germany
[†] Max-Planck-Institut für Polymerforschung
Ackermannweg 10, 55128 Mainz, Germany

Abstract. We report on the electronic properties of flat, disc-type molecules with extended π systems, which are designed as well-defined graphite subunit molecules by a new synthetic route. In our contribution, we have investigated the excited states of three benzenoid hydrocarbons — namely hexa-*peri*-hexabenzocoronene and two higher homologues — with time-resolved electron spin resonance (ESR) and optical techniques. After applying a laser flash to shock-frozen solutions of these molecules, we find excited states with lifetimes in the range of seconds. We have examined the fine-structure parameters of the triplet states in two molecules and furthermore observed the luminescence in the third one.

EXPERIMENTAL

Toluene solutions of the benzenoid hydrocarbon compounds $C_{42}H_{12}[C(CH_3)_3]_6$, $C_{60}H_{18}[C_{12}H_{25}]_4$, and $C_{72}H_{18}[C(CH_3)_3]_8$ [1] — abbreviated as "C_{42}", "C_{60}", and "C_{72}" — are investigated by pulse ESR and optical techniques. Since all three molecules show optical absorption in the range 300–400 nm, we used a frequency-tripled Nd:YAG laser ($\lambda = 355$ nm) for photo excitation. All experiments were performed in frozen solution at low temperatures.

PULSE ESR OF "C_{42}" AND "C_{72}"

Spin-polarized triplet states

After applying a laser pulse to solutions of "C_{42}" and "C_{72}", we find phosphorescence with decay times of 9.4 s ("C_{42}") and 4.1 s ("C_{72}") at $T = 6$ K. Due to these long lifetimes, we could generate not only a single spin echo per laser pulse but a series of echoes using a Gill-Meiboom pulse sequence. When integrating the intensity of the echoes as a function of the magnetic field, we yield the spin-polarized triplet spectra shown in Fig. 1.

Due to the molecular symmetry, the intersystem crossing causes a selective population of the triplet energy levels resulting in emissive transitions at the low-field part and absorptive transitions at the high-field part of the spectra.

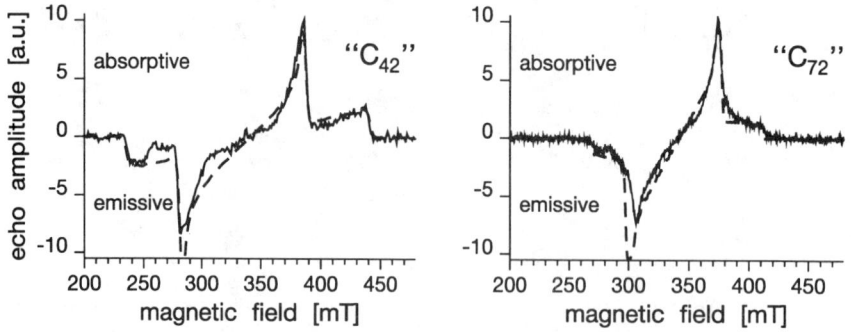

FIGURE 1. Comparison of the triplet lineshapes of "C_{42}" (left) and "C_{72}" (right) in frozen solution at $T = 6$ K. The solid lines are experimental data; the dashed lines correspond to simulations with parameters as summarized in Table 1.

Results and discussion

From simulations, we obtain the zero-field splitting parameters D and E as summarized in Table 1. We compare the obtained values with those of smaller aromatic molecules [2–4].

TABLE 1. Zero-field splitting parameters as derived from the simulations in Fig. 1 and comparison with smaller aromatic molecules.

	$|D|$ [mT]	$|E|$ [mT]		$|D|$ [mT]	$|E|$ [mT]
benzene [2]	169	6.9	naphthalene [3]	108	15
coronene [3]	103	0	anthracene [4]	78	9
"C_{42}"	103	0	"C_{72}"	72	2.5

Interestingly, the D and E values of "C_{42}" and "C_{72}" differ not much from those of coronene and anthracene, respectively. This is quite remarkable since the size of the conjugated π system in "C_{42}" and "C_{72}" is much larger.

We also calculated the spin-density distributions of the benzenoid hydrocarbons using the extended Hubbard Hamiltonian with geometry optimization by the program *XHuge* of P. Surján [5]. The results are displayed in Fig. 2.

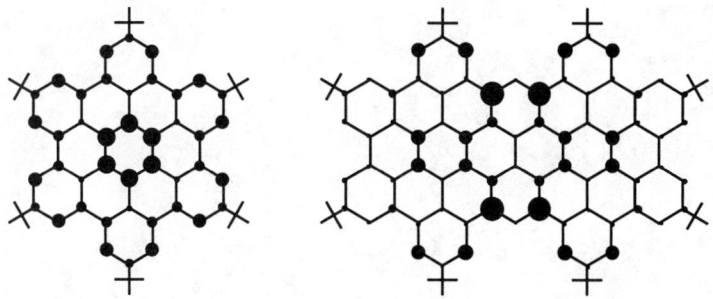

FIGURE 2. Spin-density distributions of the triplet states of "C_{42}" (left) and "C_{72}" (right). The spin densities are proportional to the radii of the circles.

The spin densities are concentrated in the central parts of the molecules. This corresponds with our experimental finding of rather large D values like those reported for small aromatics. The spin distribution in "C_{42}" has a six-fold symmetry axis (like the molecule itself) consistent with the experimentally observed E value. In "C_{72}", the symmetry of the wavefunction is reduced leading to a finite E value.

OPTICAL INVESTIGATIONS OF "C_{60}"

Luminescence

We find longliving luminescence phenomena in frozen solution of "C_{60}" at $T = 77$ K. There are two multi-line spectral regimes with maxima at 657 nm and 415 nm (see Fig. 3). The lifetimes of the emission maxima are 1.6 s (red) and 1.9 s (blue), respectively. Independent of the excitation intensity, the decay curves are monoexponential.

Results and discussion

We observe different luminescence intensities for the red and blue spectral regime with respect to the laser excitation intensity. At low excitation intensities the red emission shows a linear behaviour and seems to saturate for laser pulse energies > 1 mJ. In contrast, the blue emission exhibits a superlinear dependence starting with an almost quadratic intensity dependence at low intensities.

We assign the red emission to the T_1–S_0 phosphorescence. The blue emission is not yet understood. The quadratic dependence hints at a two-photon absorption process. The excitation into a higher triplet state is unlikely since fast internal conversion to the lowest triplet state T_1 would be expected. We are not quite sure that the blue luminescence is an intrinsic property of "C_{60}"; it might be originating from precursor compounds of the synthesis or photo fragments. This question is still under investigation.

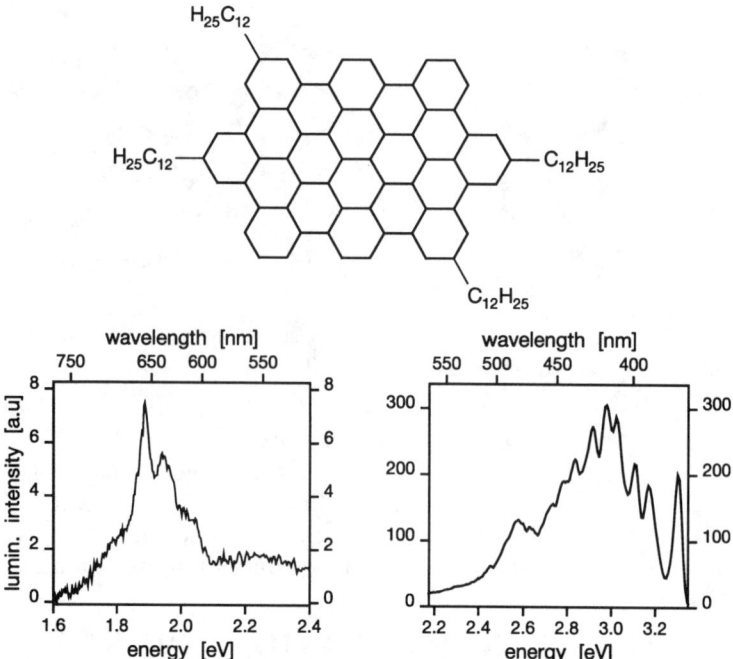

FIGURE 3. Structure of "C_{60}" (top); red (bottom, left) and blue (bottom, right) luminescence at $T = 77$ K after laser excitation at $\lambda = 355$ nm.

ACKNOWLEDGEMENT

We are grateful to A. Maier for help with the optical measurements. We would like to acknowledge G. Wäckerle for helpful discussions. We are indebted to P. R. Surján for allowing us to use his program *XHuge*. Furthermore, financial support by the Deutsche Forschungsgemeinschaft (SFB 329) is gratefully acknowledged.

REFERENCES

1. P. Herwig, C. W. Kayser, K. Müllen, and H. W. Spiess, *Adv. Materials* **8**, 510 (1996).
2. M. S. de Groot, I. A. M. Hesselmann, and J. H. van der Waals, *Mol. Phys.* **16**, 45 (1969).
3. S. P. McGlynn, T. Azumi, and M. Kinoshita, *Molecular Spectroscopy of the Triplet State*, Prentice-Hall, New Jersey (1969), p. 342.
4. P. B. Ayscough, *Electron Spin Resonance in Chemistry*, Methuen, London (1967), p. 411.
5. P. R. Surján, L. Udvardi, and K. Németh, *J. Mol. Struct. (THEOCHEM)* **311**, 55 (1994).

Metallic, Insulating and Superconducting States in κ-ET$_2$X Systems, where ET is the BEDT-TTF (bis(ethylenedithio)tetrathiafulvalene) molecule

Valery A. Ivanov, Elena A. Ugolkova and Mikhail Ye. Zhuravlev

N. S. Kurnakov Institute of General & Inorganic Chemistry of the Russian Academy of Sciences, 31 Leninskii prospect, Moscow, 117 907 Russia

Abstract. An electronic structure and normal and superconducting properties are reviewed for layered organic materials on the basis of bis(ethylenedithio)tetrathiafulvalene molecule (BEDT-TTF, hereafter ET) with essential intraET electron and cross-dimer κ-packing in ET-plane. The metall-insulator phase transition is derived for realistic model of κ-ET$_2$X salts. Based on the Fermi-surface topology and electron correlations the d-symmetry of superconducting order parameter is obtained with interplay between its nodes on the Fermi surface and superconducting phase characteristics. The results are in agreement with measured nonactivated temperature dependencies of NMR-relaxation rate of central carbon ^{13}C spins in ET and superconducting specific heat.

TIGHT-BINDING ELECTRONIC BAND STRUCTURE

Irrespectively of the similarity of electronic and crystal structure and the same carrier concentration of half a hole per ET molecule the κ-ET$_2$X family includes semiconductors, normal metals and superconductors with highest critical superconducting temperatures as much as $T_c \sim 13$ K. Its crystal motif is made by dimers, ET$_2^+$ arranged in a crossed-dimer manner in ET-layers, separated by alternating polymerized anion sheets, X$^-$, with a sheet periodicity of about 15 Å. The ET$_2$ dimers are fixed in sites of plane lattice closed to triangular one and its elementary cell is rectangular a-by-$\sqrt{3}a$ including two dimers ET$_2$ (hereafter the lattice constant a=1). The intermolecular distance within the ET$_2$ dimer is 3.2Å whereas the separation between the neighboring dimers ET$_2$ is about 8Å.

In the tight-binding approach [1] for triangular lattice electron energy dispersion relations are such as $E_p^\pm = \pm t_0 \pm \varepsilon_p$, where

$$\varepsilon_p = t_2 \cos p_y \pm \cos(p_y/2)\sqrt{t_1^2 + t_3^3 + 2t_1 t_3 \cos(\sqrt{3}p_x)} \qquad (1)$$

With taking into account a small interlayer carrier hopping τ along c axis perpendicular to ET_2 layers one can derive the general energy dispersion relations as follows ($t_{1,2,3} \equiv t$):

$$\omega_p^{1,2} = \varepsilon_p^+ \pm \frac{\tau}{t}\cos\frac{p_z c}{2}\sqrt{3 + 2\varepsilon_p^+}, \quad \omega_p^{3,4} = \varepsilon_p^- \pm \frac{\tau}{t}\cos\frac{p_z c}{2}\sqrt{3 + 2\varepsilon_p^-}. \qquad (2)$$

It is easily seen from Eq. (2) that effective carrier hopping $t_\perp = \tau\cos(p_z c/2)/\cos p_z c$ increases with increase of interlayer separation c in agreement with experimental visualization [2]. A metallic dimerized ET_2 layer in κ-ET_2X can be represented by a lattice of sites $ET_2 = ET_a ET_b$ with degenerated energy levels of orbitals, a and b, namely the doubly degenerated Hubbard model with a hole concentration, n, of around unity [3]. Then the energy dispersions (1)-(2) are narrowed due to the correlation factor $f = (4 - 3n)/4 = 1$. Energy dispersion relations of Eq. (2) provide three-dimensional corrugated Fermi surface in contrary to the cited conventional two-dimensional Fermi surface of κ-ET_2X metals.

INSULATOR-METAL PHASE TRANSITION

The κ-ET_2X insulators and metals can be described by the half-filled Hubbard model with two ET_2^+ sites per a unit cell of a triangular ET_2- lattice. We employ the X-operator machinery for generalized Hubbard - Okubo operators (e. g., Ref. [3]), $X_A^B = |B\rangle\langle A|$, projecting multielectron A state of a unit cell to B state. From tight-binding correlated energy bands it follows that intra ET_2^+ electron interactions U_{ET2} are responsible for the insulating gap:

$$\Delta = \left[\sqrt{9 + (2\varepsilon/t_2)^2} + \sqrt{(3/2)^2 + (2\varepsilon/t_2)^2} - 9/2\right] t_2/2. \qquad (3)$$

With the assumptions $\varepsilon = U_{ET2}/2 \sim t_0 \sim 0.2\text{eV}$ and $t \sim 0.1$ eV this magnitude is in agreement with the measured activation energy $E_g = \Delta/2 \sim 10^2$meV in κ-ET_2X semiconductors [4].

The Mott-Hubbard phase transition is governed by singularities of the two-particle vertex Γ_{ab} for small momentum transfer [5]. From solution of the Bethe-Salpeter system of Eqs. for vertices Γ_{ab} one can derive the critical point of the insulator-metal phase transition as follows [3, 6]

$$\left(\frac{t_2}{\varepsilon}\right)_{crit} = \left(\frac{t_2}{U_{ET_2}/2}\right)_{crit} = \frac{4\pi}{\sqrt[4]{3}\sqrt{15\pi^2 + 64}} = 0.66. \qquad (4)$$

Taking into account the antibonding bandwidth, $4.5t$ (c. f., Eq. (1)), and the effective intradimer correlations of carriers, U_{ET2}, the calculated phase critical point of Eq. (4) can be converted to the noticed empirical ratio such as $W/\Delta E = 1.1 - 1.2$ [7] in terms of the conducting bandwidth, W, and dimer splitting, $\Delta E = 2t_0$. From knowledge of evaluated magnitudes W and ΔE one can conclude that all known κ-ET$_2$X salts are located around the calculated insulator-metal phase boundary, Eq. (4), dividing regions of semiconducting and metalic κ-ET$_2$X compounds. Recently it was reported that κ-ET$_2$Cu[N(CN)]$_2$Cl insulator manifest the phase transition to metal and superconductor at a moderate hydrostatic pressure [8].

SUPERCONDUCTING PAIRING IN THE ET$_2$-LAYER MODEL

The band structure effects are important when the Fermi surface is near the Bz (this is the κ-ET$_2$X family case) where the influence of the crystal potential is strong. For realistic ET$_2$ triangular lattice of κ-ET$_2$X compounds the nonspecified attractive interaction depends on the momenta difference of pairing fermions, namely

$$V_{\mathbf{p}-\mathbf{p}'} = 2V[\cos(p_y - p_y') + \cos\frac{\sqrt{3}(p_x - p_x') + p_y - p_y'}{2} + \cos\frac{\sqrt{3}(p_x - p_x') - (p_y - p_y')}{2}],$$

and conserves a symmetry of carrier dispersions beyond the discussion about the nature of superconductivity. It can be expanded over six basis functions of irreducible representations of triangular lattice symmetry [9]. From the system of nonretarded BCS like equations we obtain the equations on T_c for each type of pairing. The symmetry of the representation with the largest T_c is the case according to developed approach to κ-ET$_2$X superconductors [3].

In view of the same V value and assuming that the cut-off energy parameter, ω_c, equals to the interdimer hopping integral $\omega_c = t \sim 0.1$ eV, in logarithmic approximation, we can estimate the superconducting transition temperature T_c at 10 K for d_{xy}-wave pairing with order parameter as $\Delta_p = \Delta_0 \sin(p_y/2)\sin(\sqrt{3}p_x/2)$. The T_c magnitude has a reasonable value for κ-ET$_2$X 10K class superconductors. Also the zeroth Knight shift measurements evidence in the favour of singlet pairing.

CHARACTERISTICS OF THE ANISOTROPIC SUPERCONDUCTING PHASE

For superconducting phase with order parameter of d_{xy}-wave symmetry the density of electronic states (DOS) is as following:

$$\rho_s^{\pm}(E) = \frac{\sqrt{3}}{4\pi^2} \int_{-\pi}^{\pi} dp_y \int_{-\pi/\sqrt{3}}^{\pi/\sqrt{3}} \delta\left(E - \sqrt{(\xi_p^{\pm})^2 + \Delta_p^2}\right) dp_x, \qquad (5)$$

where $\xi_p^{\pm}$ is the one-particle energies measured from the chemical potential μ. The magnitude Δ_p is small quantity in the neighbourhood of four nodes on the Fermi surface inside the first Bz near the $p_x = 0, p_y = 0$. Expanding the $\xi_p^{\pm}$ and Δ_p magnitudes over variations of thereof values at nodes of the order parameter, i. e. $\Delta_p=0$, on the Fermi surface μ (p_x, p_y) one can calculate the superconducting DOS around the node $p_x = 0, p_y = 2\arccos\left(\sqrt{3/4 + \mu/2ft} - 1/2\right)$ on the electronic portion, $\xi_p^+ = 0$, of the Fermi surface as

$$\rho_s^+(E) = \frac{E}{2\pi \Delta_0 f t \sin^2(p_y/2)(2\cos(p_y/2)+1)} = \beta_+ E \cdot \qquad (6)$$

Around the other node $p_y = 0, p_x = (2/\sqrt{3})\arccos(1/2 - \mu/2ft)$ on the hole portion of the Fermi surface the DOS is as follows:

$$\rho_s^-(E) = \frac{E}{2\pi \Delta_0 f t \sin^2(\sqrt{3}p_x/2)} = \beta_- E. \qquad (7)$$

In contrary of the conventional s-wave pairing picture from Eqs. (6) and (7) it is follows that superconducting DOS is linearly proportional to an energy nearby the nodes of the derived anisotropic order parameter with d_{xy}-wave symmetry. Noteworthy that the Fermi surface portions with different curvatures contribute different coefficients $\beta_-/\beta_+ \cong 3$ for the calculated magnitude $\mu/tf = -0.415$ (the correlation factor f = 1/4).

The obtained superconducting DOS allows to calculate a variety of superconducting condensate quantities. The linear behaviour of DOS on energy leads to quadratic temperature dependence of electronic specific heat, namely

$$C_s = 2\int_0^{\infty} (\beta_+ + \beta_-) E^2 \left(\frac{\partial N_F}{\partial T}\right) dE = 9(\beta_+ + \beta_-)\varsigma(3)T^2 = 10.8(\beta_+ + \beta_-)T^2 \qquad (8)$$

for a unit cell of ET$_2$ layer. Here $\zeta(3)$ is the Riemann ζ-function. Putting into Eq. (23) the reasonable parameters for T$_c$ = 10K and Δ_0 = (2.5÷3.5) T_c [10, 11] one can derive the superconducting specific heat per mole as $C_m = \alpha T^2$ with variation of the factor $\alpha = 10.8 N_A (\beta_+ + \beta_-) k_B^3/2$ (N_A is Avogadro's number and k_B is Boltzman's constant) in the range 1.59-2.23 mJ/K^{3}mole. The latter is in agreement with experimental data such as α = 2.2 mJ/K^{3}mole and less than 3.53 mJ/K^{3}mole for superconductors respectively κ-(ET)$_2$Cu[N(CN)$_2$]Br with T$_c$ = 11.6K and κ-(ET)$_2$Cu(NCS)$_2$ with T$_c$ = 10K [12].

In central fragment of ET molecule the ^{13}C nuclear magnetic momentum damps out through the conduction electrons in ET$_2$-plane. The corresponding NMR relaxation rate $R = 1/T_1$ is defined in superconducting phase by

$$\frac{R_s}{R_n} = \frac{2}{T}\int_0^w \frac{(\beta_+ + \beta_-)^2 E^2}{[\rho(\mu)]^2} \cdot \frac{\exp(E/T)}{(\exp(E/T)+1)\cdot(\exp((E+\nu)/T)+1)} dE, \quad (9)$$

where the radio-frequency $\nu \sim 10$MHz $\sim 10^{-4}$K for oscillating magnetic field, ρ denotes the normal DOS on the Fermi level and w is the cut-off energy for an energy proportional superconducting DOS. At low temperatures, $T \ll w < T_c$, the result is

$$\frac{R_s}{R_n} = \frac{2T^2 \beta^2 \zeta(2)}{\rho^2(\mu)} \cdot \left(1 - \frac{\nu}{T} \cdot \frac{\ln 2}{\zeta(2)}\right). \quad (10)$$

For normal phase $R_n \sim T$ according to the Korringa law, and the superconducting spin-lattice relaxation rate has cubic temperature dependence such as like as $R_s \sim T^3$. The following regulation can be derived from (8) and (10) for zero magnetic field:

$$\frac{R_s \cdot C_n^2}{R_n \cdot C_s^2} = \frac{4 \cdot \zeta^3(2)}{81 \cdot \zeta^2(3)} = 0.29 \quad (11)$$

at assumption that normal specific heat is $C_n = 2 \cdot \zeta(2) \cdot \rho(\mu) \cdot T$.

The derived d_{xy}-wave superconducting order parameter effects on the temperature dependence of the nuclear relaxation rate differently than the conventional BCS s-wave pairing also just below T_c. Averaging over the Fermi surface in Eq. (9) leads to the softening of the singularity in superconducting DOS and to disappearance of the Hebel-Slichter coherence peak at T_c for R_s according to Maleyev scenario in Ref. [13].

We acknowledge the support of the Russian Foundation for Basic Research and the Ministry of Science and Technologies of Russian Federation.

REFERENCES

1. Ivanov, V., Yakushi, K., and Ugolkova, E., *Physica* C **275**, 26-36 (1997).
2. Wosnitza, J., Goll, G., Beckmann, D. et al., *J. Phys. I (France)* **6**, 1597-1602 (1996).
3. Ivanov, V., *Phil. Magazine* B **76**, 697-713 (1997).
4. Williams, J. M., Kini, A. M., Wang, H. H. et al., *Inorg. Chem.* **29**, 3274-3278 (1990).
5. Zaitsev, R. O., and Ivanov, V. A., *Sov. Phys. Solid State* **27**, 2147-2158 (1985).
6. Ivanov, V., and Murayama, Y., "Electronic structure, insulator-metal transition and superconductivity in κ-(BEDT-TTF)$_2$CuX salts", in *Advances in Superconductivity* IX (eds. S. Nakajima, M. Murakami), Tokyo: Springer-Verlag, 1997,pp. 315-320.
7. Saito, G., Otsuka, A., and Zahidov, A. A., *Molec. Cryst. Liq. Cryst.* **3**, 284-289 (1996).
8. Ito, H., Ishiguro, T., Kubota, M., and Saito G., *J. Phys. Soc. Jpn.* **65**, 2987-2993 (1996).
9. Ivanov, V., Ugolkova, E., and Zhuravlev, M., *Zhurn. Eksperim. Teor. Fiz.* (in Russian) **113**, 715-733 (1998).
10. Dressel, M., Klein, O., Gruner, G. et al., *Phys. Rev.* B **50**, 13603-13615 (1994).
11. Lang, M., Toyota, N., Sasaki, T., and Sato, H., *Phys. Rev. Lett.* **69**, 1443-1446 (1992).
12. Nakazawa, Y., and Kanoda, K., *Phys. Rev.* B **55**, 8670 - 8673 (1997); *Physica* C **282-287**,1897-1900 (1997).
13. Maleyev, S.V., Yashenkin, A.G., and Aristov, D.N., *Phys. Rev.* B **50**, 13825-13828 (1994).

STM Studies of Synthetic Peptide Monolayers

David J. Bergeron[*], Wilfried Clauss[*†], Denis L. Pilloud[‡],
P. Leslie Dutton[‡], and Alan T. Johnson[*]

[*]*Department of Physics and Astronomy*
[‡]*Johnson Research Foundation, Department of Biochemistry and Biophysics, School of Medicine*
University of Pennsylvania, Philadelphia PA 19104
[†]*On leave from Institute of Applied Physics, University of Tübingen, 72074 Tübingen, Germany*

Abstract.
We have used scanning probe microscopy to investigate self-assembled monolayers of chemically synthesized peptides. We find that the peptides form a dense uniform monolayer, above which is found a sparse additional layer. Using scanning tunneling microscopy, submolecular resolution can be obtained, revealing the alpha helices which constitute the peptide. The nature of the images is not significantly affected by the incorporation of redox cofactors (hemes) in the peptides.

INTRODUCTION

Synthetic peptides are an exciting new form of engineerable electronic material. Chemical synthesis allows peptides to be synthesized with arbitrary amino acid sequences. It is therefore possible to design molecules with minimal complexity, yet which fold to a desired structure and can incorporate redox cofactors.

We have studied a redox peptide based on the prototype of Robertson *et al.* [1] with minor modifications [2]. The peptide consists of a pair of 2-helix dimers, each joined by disulfide bonds (see Figure 1). The sulfurs allow the peptide to form a self-assembled monolayer (SAM) [3] when dispersed onto a gold substrate. Alternatively, a peptide SAM can be deposited onto a SAM of dimercaptoalkane linker. The interior of the four-helix bundle contains four ligation sites for binding metalloporphyrin hemes. Heme redox centers found in natural proteins play an important role in electron transport [4,5]. In viewing synthetic peptides as a novel electronic material, we consider hemes as the basis for engineering the material's electronic properties. Using a scanning probe microscope, we have investigated the morphology of peptide SAMs, both on linker and on bare gold, with various numbers of hemes incorporated into the peptides.

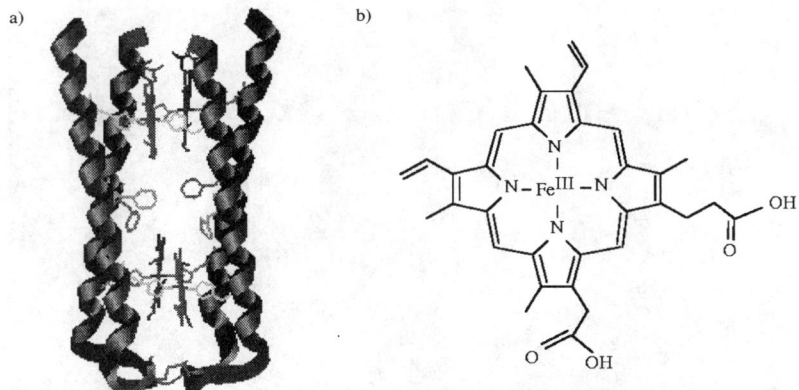

FIGURE 1. (a) Schematic of the four-helix peptide, shown with 4 hemes ligated. (b) The iron(III) protoporphyrin heme molecule.

EXPERIMENTAL METHOD

Substrates for self-assembly are made by evaporating gold onto cleaved mica squares which have been heated for 3 hr to 300 °C. Annealing these in a gas flame produces atomically flat Au(111) terraces with typical sizes around 100 nm.

When a linker layer is used, the substrates are immersed in a solution of dimercaptoalkanes, which form a self-assembled monolayer on the gold surface. The samples are rinsed in isopropanol, and then immersed overnight in a 200 μM solution of peptide. Alternatively, the peptide self assembly step is performed directly on the bare annealed gold. are Our primary tool for studying the peptide SAMs is an Omicron Beetle Scanning Tunneling Microscope (STM). In this work, all images were taken under ambient conditions, using tunneling impedances ranging from 1 to 15 GΩ. The Omicron Beetle can also be operated as an Atomic Force Microscope (AFM). AFM is not sensitive to the conductivity of the sample and forces exerted by the tip can be substantially lower than in STM.

RESULTS

As seen in Figures 2 and 3, our STM data show a uniform monolayer of peptides. The separation between nearest neighbors is typically 4-5 nm, while the observed vertical corrugation of the layer is only a few Ångstroms. Occasional depressions in the layer suggest voids in the monolayer, but are more likely etch pits in the gold substrates. Such etching is characteristic of thiol self-assembly processes [6].

Our images (see Figure 2) show that there is a sparse second layer of peptides which are probably not covalently bound into the monolayer, and which are mobile under the influence of the STM imaging. In the STM images, this layer appears as horizontal streaks up to 5 nm in length which persist for one or more scan lines of

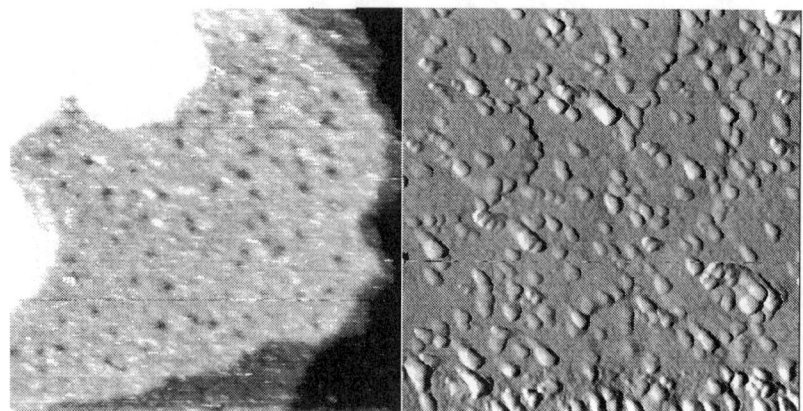

FIGURE 2. Left image: 100x100nm STM image of a peptide SAM. The Au(111) terrace structure dominates the image. The grains on the largest visible terrace are peptides in the SAM. The monolayer is punctuated by dark spots and bright horizontal streaks. The spots are attributed to etch pits in the gold layer, while the streaks are caused by a mobile second layer of peptides. V=1.1 V, I=220 pA. Right image: 1 μm x 1 μm AFM image of another peptide SAM showing sparse second layer of peptides. The image is created from the feedback signal (phase), which yields an image with a 'shaded' appearance. The peptides appear enlarged due to finite tip size. Some of the underlying gold grain structure is also visible.

the image. We infer that a peptide molecule is present in a location long enough to be imaged for a few scan lines, and then is removed by the tip to another location. By imaging with AFM, we can reduce the forces disturbing molecules, and we find that the second layer remains immobile on the surface. Figure 2 shows the manifestations of this layer in both STM and AFM images.

We have studied peptide SAMs using low 'conventional' tunneling impedances of 1-10 GΩ, as well as relatively high values of 13-15 GΩ. By imaging at higher impedances, we raise the tip further above the sample surface. The effect of this on our images is to reveal the 4-helix substructure of the peptides. The α helices appear as round features separated by 2-3 nm from their nearest neighbors (Figure 3). Images taken with various current setpoints and bias voltages suggest that the enhanced resolution is due to raising the impedance (and therefore the tip-sample separation), and not due to the increase in voltage alone. We conclude that at low impedances, the bottom of the STM tip is actually passing within the peptide layer, whereas at larger impedances, it is raised to near the surface of the monolayer.

When a linker layer is present between the peptide monolayer and the gold substrate, we find that imaging at low tunneling impedances leads to etching of the gold substrate. This is probably a result of linker molecules which bind to the Pt/Ir STM tip, and then pull gold atoms out of the the substrate. At higher tunneling impedances (above 13 GΩ), the tip is further above the linker layer, and this

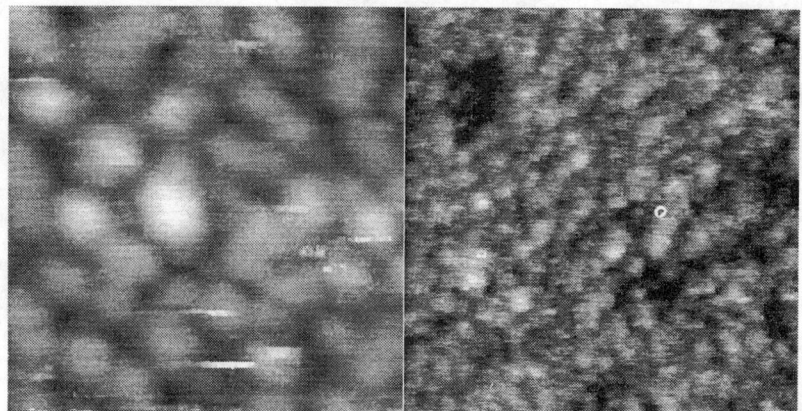

FIGURE 3. Images of peptide SAMs at low (left) and high (right) tunneling impedances, showing increased resolution at higher tip-sample distance. Features in the left images are full peptides. In the right image, individual α helices are observed. Most streaks due to the mobile layer of peptides have been removed in both images. Left: V=1.5 V, I=200pA. Right: V=2.3V, I=120pA. Both images are 43nm x 43nm with a vertical range of 0.6nm.

process is not observed. In this case, the images are essentially identical to those obtained without a linker present.

Finally, we have investigated peptide SAMs with the number of hemes per peptide varying from 0 to 4. Furthermore, a monolayer was made with a mixture of both peptides containing hemes and those without. Until now, we have not observed any effect in the images or tunneling spectra which could be attributed to the presence of hemes. Observation of such an effect will be the subject of future research.

This work is supported by Penn's Laboratory for Research on the Structure of Matter, and NSF MRSEC IRG #DMR-96-32598. W.C. thanks the Deutsche Forschungsgemeinschaft. P.L.D. receives funding from NIH grant #GM-41048. A.T.J. is supported by a David and Lucile Packard Foundation Fellowship.

REFERENCES

1. Robertson, D. E. et al., *Nature* **368**, 425 (1994).
2. Gibney, B. R. et al., *J. Am. Chem. Soc.* **119**, 2323 (1997).
3. Ulman, A., *Introduction to ultrathin organic films: from Langmuir-Blodgett to self-assembly*, Boston: Academic Press, 1991.
4. Moser, C. C. et al., *Nature* **355**, 796 (1992).
5. Moser, C. C. and Dutton, P. L., "Outline of theory of protein electron transfer," in *Protein Electron Transfer*, 1996, pp. 1-21.
6. Poirier, G. E., *Langmuir* **13**, 2019 (1997).

Self-assembly of Ropes of Cyanine Dye Molecules

S. Blumentritt, M. Burghard, S. Roth

Max-Planck-Institut für Festkörperforschung, Heisenbergstr. 1, 70569 Stuttgart, Germany

Cyanine molecules self-assemble into molecular ropes. Through the aid of a polyelectrolyte, which is oppositely charged compared to the dye, this ability is even enhanced and leads to formation of individual ropes with lengths up to a few hundred nanometres, widths on the order of ten nanometers, and a few nanometers in height. These ropes were attached to chemically modified surfaces of atomically flat gold and substrates bearing a gold/palladium electrode pattern.

1 Introduction

In the late 30ies, Scheibe [1] and Jelly [2] discovered that cyanine molecules are able to self-assemble into rope-like structures. The formation of such molecular ropes does not only occur in solution but also at solid-liquid interfaces. For a review on these so called Scheibe or J-aggregates see Ref. [3]. J-aggregates are characterized by a narrow absorption band which is shifted to longer wavelengths with respect to the monomer absorption. Due to their strong light absorption, these aggregates are of great interest for, e.g., the spectral sensitation in photographic processes. These properties suggest their use as photo or dark conductors over small dimensions.

In this paper, we describe the self-assembly of 1.1'-diethyl-2.2'-cyanine on different chemically modified gold surfaces.

2 Experimental details

For all the experiments described here, purified water (Milli-Q) was used. Prior to the assembly of the molecular fibers, electrode structures were fabricated on GaAs and SiO_2 substrates. On the GaAs substrates the oxide layer was removed [4] prior to electron beam lithography: The samples were first dipped into H_2O_2:HCl:H_2O 1:1:40 for 10 sec and then immersed in a solution of 10 ml CH_3CSNH_2 and 250 µl HCl for 15 min, at 90 °C. The samples were rinsed with water for 5 min and baked at 240 °C for 30 min. The SiO_2 substrates were cleaned ultrasonically using aceton and 2-propanol. With the electron beam, patterns of parallel wires (100 nm in width, 100 nm apart from each other) were defined in a double layer PMMA resist followed by evaporation of approximately 12 nm AuPd (3:2) and lift off.

Prior to assembling the molecules, functional groups were introduced to the gold surface: To achieve this, the GaAs sample was immersed in an aqueous 1 mM 3-mercaptopropionic acid (3-MPA) solution for 3 hours. Afterwards, the sample was rinsed with water and put into a freshly prepared cyanine solution for 1 hour (0.11 mg of cyanine solved in 2 ml water/ethanol 1:1 and one-molar equivalent of Na-polyacrylate added). The Na-polyacrylate solution, prepared from 1 mg polyacrylic acid Na salt and 1 ml water, was used as well for all self-assembly experiments.

For Scanning Tunneling Microscopy (STM) investigations, we prepared a 2nd set of samples by assembling the molecules on atomically flat gold surfaces as described for the GaAs substrate.

The surface of the gold electrodes on the SiO_2 substrate was modified with a solution of 1 mg 2-aminoethanethiol hydrochloride solved and 1 ml water for 6 - 7 hours. Afterwards, the samples were rinsed with purified water. Then, one sample was immersed in a solution containing cyanine molecules for 130 sec (1 mg cyanine dissolved in 4 ml water and 2 ml 2-propanol with 210 μl Na-polyacrylate solution added). A different sample was immersed in a more concentrated cyanine solution for 120 sec (1 mg cyanine dissolved in 1 ml water and 0.5 ml 2-propanol with 210 μl Na-polyacrylate solution added).

The samples were investigated using either a Hitachi S-800 Scanning Electron Microscope (SEM), or a Nanoscope IIIa Atomic Force Microscope (AFM) in tapping mode with commercially available Si_3N_4 tips. An OMICRON STM and mechanically cut Pt/Ir tips were used for STM investigations.

3 Results & Discussion

We successfully assembled ropes of cyanine molecules onto different chemically modified gold surfaces. For the electrode structures on GaAs and atomically flat gold substrates, the gold surface was treated with 3-MPA. This treatment attaches negatively charged carboxylate groups to the surface to which positive cyanine molecules are expected to bind. For electrodes on SiO_2 surfaces, binding is achieved through the negatively charged polacrylate binding to the positive amino-groups attached to the gold surface. In the following sections, we describe the results of these self-assembly experiments.

3.1 Assembly of cyanine ropes on carboxylic acid modified gold surfaces

Fig. 1a shows an SEM micrograph of a GaAs sample with molecular ropes bridging over the electrode pattern. Bending of the cyanine fibers over the electrode is observed. Carbon contamination on the substrate originating from the SEM investigations makes this method not suitable for characterization prior to transport measurements.

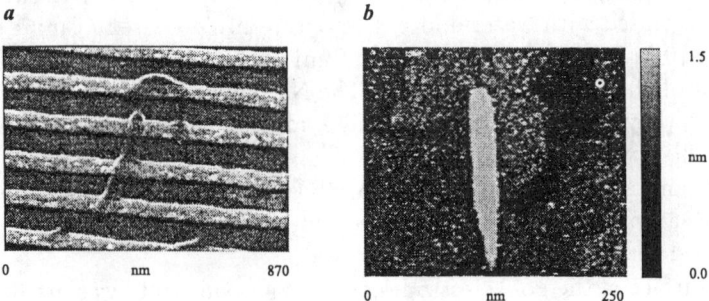

Figure 1: Molecular ropes attached to conducting substrates. SEM graph of molecular ropes attached to GaAs (a) and STM graph of a molecular rope (170 nm long, 20 nm wide and 3.2 nm high) attached to atomically flat gold surface (b). The concentration of the cyanine solution was not altered.

Fig. 1b displays an STM image of a molecular rope assembled on an atomically flat gold surface. We observed molecular ropes with lengths ranging from 65 nm to 195 nm. The height of the ropes were 1.3 to 3.2 nm, which suggests a vertical stacking of 2 to 4 molecules per rope. Similar widths of the ropes observed with the STM (about 20 nm) are attributed to tip size effects. Scanning Tunneling Spectroscopy performed at different positions on the cyanine rope and away from it was described elsewhere [5].

3.2 Assembly of cyanine ropes on amino-functionalized gold surfaces

Fig. 2a and 2b shows the assembly obtained with two different concentrations of the cyanine solution. Fig. 2b displays a sample where the dye solution was 4 times as concentrated as the one used for Fig. 2a. As the concentration of the dye solution increases, fibers of cyanine molecules interconnect to a network. This allows us to control the assembly of single fibers by adjusting the concentration of the dye solution.

4 Conclusion

Our results demonstrate that cyanine ropes not only self-assemble on atomically flat gold surfaces but can also be bridged over electrode structures on GaAs and SiO_2 substrates. This offers the possibility of transport measurements in a device configuration.

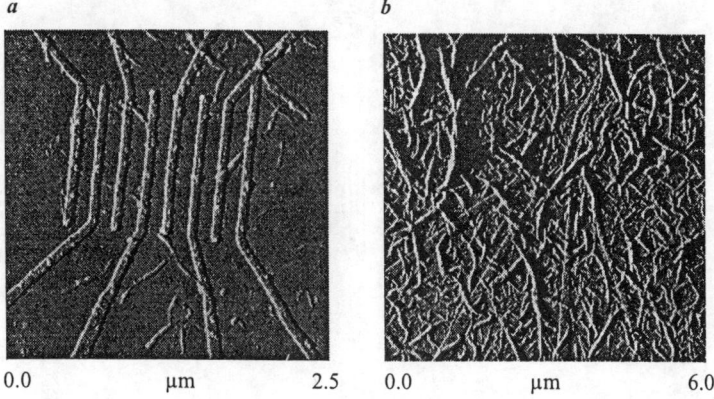

Figure 2: Assembly of molecular ropes on insulating substrates performed with two different concentrations of the dye solution: The concentration is 4 times higher for the image b.

Acknowledgements

The authors would like to acknowledge U. Waizmann and J. Muster for the help with SEM resp. AFM.

References

1. G. Scheibe, Angew. Chem. **49**, 563 (1936).
2. E.E. Jelly, Nature **138**, 1009 (1936).
3. D. Moebius, Adv. Mat. **7**, 437 (1995).
4. E.D. Lu et al., Appl. Phys. Lett. **69**, 2282 (1996).
5. S. Blumentritt et al., Surf. Science Lett. **397**, 289 (1998).
6. S.M. Lindsay in *Scanning Tunneling Microscopy and Spectroscopy: Theory, Techniques, and Applications*, (VCH Publishers, Inc., New York, Weinheim, 1993).

APPLICATIONS

Physical properties of carbon nanotubes

Jean-Paul Salvetat, Jean-Marc Bonard, Revathi Bacsa, Thomas Stöckli, László Forró

Physics departement, Swiss Federal Institute of Technology, CH-1015 Lausanne.

Abstract. Carbon nanotube science is a new exciting subject for all the carbon community. We now have in hand 1D graphite prototypes opening a new field for basic research and increasing the technological potential of traditional carbon fibers. In addition to many open fundamental questions, one of the main difficulties resides on the technological side, since large scale synthesis, high purity samples, and manipulation at the nanoscale are not yet fully developed. In this paper, we present recent development on different nanotube aspects: preparation and purification, electronic transport properties, electron spin resonance, mechanical behaviour of individual nanotubes, and field-emission.

MATERIALS

Preparation of nanotubes

The arc-discharge MWNTs were produced in a carbon arc apparatus using the method described by Ebbesen and Ajayan [1], i.e., by arc discharge between two graphite electrodes ($U = 16$ V, $I = 80$ A) in a 350 mbar He atmosphere. The SWNTs were produced by arc discharge under 500 mbar He static pressure using pure graphite electrodes 20 mm (cathode) and 5 mm (anode) in diameter. A 3 mm hole was drilled in the anode and filled with a graphite-Ni-Y mixture with weight proportion 2:1:1 [2]. The voltage and current used were approximately 25 V and 100 A. The nanotubes were predominantly found in the webs, as opposed to the cathodic deposit. The disordered nanotubes were produced by catalytic decomposition of acetylene on reduced cobalt oxide that was deposited on a silica substrate [3]. The nanotube powder was collected after dissolution of the silica substrate and catalyst in acid. In this case, the nanotubes show a high density of structural defects.

Purification of multiwall and singlewall nanotubes

The arc-discharge yields, along with carbon nanotubes, nanoparticles consisting of nested closed graphitic layers of polyhedral shape, as well as larger graphitic flakes which make up the largest part in weight of the deposit. The situation is even worse with SWNT where the catalyst remains in the deposit in the form of metallic nanoparticles embedded in amorphous or graphitized carbon. On the other hand, it is important for characterization to have good quality samples with as little "foreign" material as possible.

For MWNTs, we have developed a soft purification method which uses the properties of colloidal suspensions [4]. We started the purification with a suspension prepared from 500 ml distilled water, 2.5 g SDS (sodium dodecyl sulfate, a common surfactant) and 50 mg of MWNT arc powder sonicated during 15 minutes. Sedimentation and centrifugation (at 5000 rpm for 10 minutes) removed all graphitic particles larger than 500 nm from the solution, as confirmed by low magnification SEM observations. We then add surfactant to the solution to reach 12 CMC. At such surfactant concentrations, micelles form and induce flocculation, i.e., the formation of aggregates. These aggregates contain mainly large objects, while smaller objects remain dispersed, and sediment after a certain time, typically a few hours. After decanting the suspension after about one week, we repeat the procedure once or twice. Fig. 1 shows scanning electron microscopy images of a MWNT deposit after the separation procedure. The as-deposited material contains a large proportion of nanoparticles (typically 70 % in number and 40 % in weight.). The material remaining in suspension consisted nearly exclusively of nanoparticles, while the sediment contained nanotubes with a content of over 90 % in weight. This procedure allows us thus to eliminate nearly all nanoparticles that can be separated from the tubes by sonication.

For SWNT, the purification of the raw soot has been carried out by oxidative dissolution of the carbon encapsulated metal particles with concentrated acid which ensures maximum efficacity of the process of metal elimination. A weighed amount of the raw soot was sonicated in an ultrasonic bath at 25 °C with concentrated nitric acid for a few minutes and subsequently refluxed for 4-6 h. Thick brown fumes of oxides of nitrogen were seen, indicating the rapid oxidation of carbon to carbon dioxide. After cooling, water was added so as to leave the samples in 6M HNO_3 for the next 8-12h after which it was centrifuged several times and the supernatant rejected until the pH of the solution was around 6.5. High resolution transmission electron micrographs of this solution show the presence of long ropes of bundled nanotubes accompanied by small amounts of carbon-coated metal particles. Parallel examination of the unpurified soot showed that more than 80% of the metal had been dissolved.

The SWNT in suspension were stabilized by using a surfactant such as SDS and left undisturbed for 3-5 days until the slow aggregation of the nanotubes allowed their separation from the nanoparticles in solution. The nanotube suspension was filtered through a polycarbonate membrane (1 μm pore size) in order to eliminate most of the particles. On drying, the sediment on the filter paper peeled away to form a self supporting sheet of carbon nanotubes. Scanning electron micrograph of such a sediment, as on Fig. 1(b) shows a network of pure nanotubes.

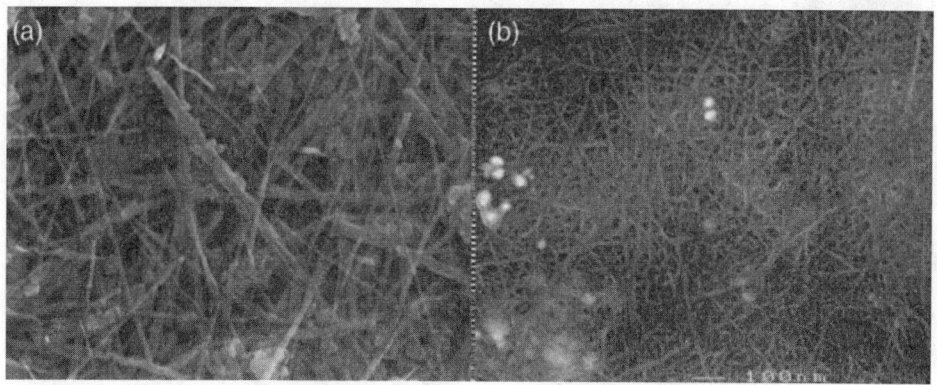

FIGURE 1. Scanning electron (SEM) micrographs of (a) a MWNT deposit and (b) a mat of SWNTs after purification. A few nanoparticles and embedded catalyst particles, respectively, are still present.

ELECTRONIC PROPERTIES

Transport properties of SWNTs

The study of the transport properties of SWNT attracts lot of attention since effects due to chirality, ballistic transport, low-dimensionality, twistons, non-Fermi liquid behaviours etc. are expected. Although in several laboratories nanolithographical technology permits resistivity measurements on individual tubes and/or ropes, the intrinsic nature of the charge transport is far from being clear even for a priori metallic nanotubes. This is due to the strong influence of the substrate on the electronic states of the SWNT and possibly to the interaction of the tubes of different helicity within a rope. Furthermore, there is also the technical problem of purification of SWNTs, to remove not only the catalytic particles, but also the amorphous carbon often covering the surface of the ropes which precludes the proper contacting of the tubes. In some of the measurements on single ropes the resistivity has shown a metallic behaviour in a limited temperature range, and below a T* temperature a non-metallic behaviour has been observed. There are several theoretical models which have been proposed to account for this metal-insulator transition. Rather surprisingly, a bulk sample of SWNT ropes often shows a metallic-like behaviour, and it was suspected that a percolating path of metallic nanotubes not interacting with any substrate gives such a positive slope. The Berkeley-group has nicely demonstrated that the T* is just a side effect in the network of loosely connected nanotubes with long localisation length, and at low temperature the resistivity follows a 3D variable range hopping behaviour [5]. Our resistivity measurements performed on a submicron thick film of purified SWNTs show also that the so-called T* might be just an effect of oxygen absorption in the tube-tube contact region. This is exemplified in Fig. 2(a). When the sample was cooled down (Fig. 2(a), bottom trace) the resistivity showed a minimum at a "T*" of 270K. This minimum was shifted downwards on heating the sample back to room temperature (top trace). After the sample space was evacuated (not shown) we did not observe any minimum in the resistivity on cooling down the sample again. Furthermore, this measurement has shown that even the slope of the log R vs $T^{-1/3}$ had changed (see Fig. 2(b)), demonstrating the importance of the tube-tube contacts in the hopping process.

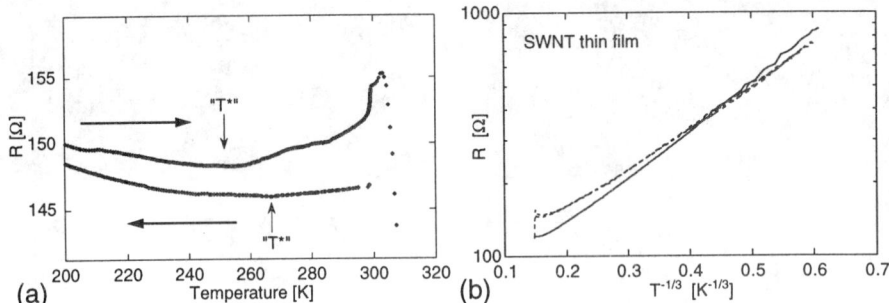

FIGURE 2. Resistivity as a function of temperature for a purified SWNT film before and after heating and evacuating the sample stage; (a) room temperature region (b) whole temperature range as log(R) vs $T^{-1/3}$ (2-D variable range hopping representation).

Electron Spin Resonance

Electron Spin Resonance spectroscopy (ESR) has proven to be a powerful tool for the investigation of graphitic materials. ESR allows one to determine three important quantities: the absorption intensity, which is proportional to the spin susceptibility; the g-factor, which is sensitive to the chemical environment and/or to the band structure; and the linewidth (ΔH) which depends on the spin dynamics. The absorption intensity is usually composed of two contributions arising from conduction carriers and localized spin centres due to imperfections, respectively. The total spin susceptibility is the sum of the Pauli component from the metallic density of states and of the Curie component from paramagnetic defects. Due to strong exchange interaction and rapid motion of conducting carriers, the two spin systems interact strongly, resulting in a single symmetric line. The two contributions can be separated by performing measurements on a large temperature range (4 - 300 K), which permits to estimate the density of states and the defect concentration. A comparison of the samples with standard polycrystalline graphite gives information about the degree of graphitization of the multiwall nanotube material.

An important result is the nearly flat spin susceptibility between 10-300K, similarly to graphite which behaves as a semi-metal (a zero gap semiconductor should give an activated susceptibility). The absolute value of the susceptibility per gram of material is also similar to graphite. Morover, no difference in susceptibility is observed between nanotubes and nested polyhedral nanoparticles [6]. This means that this parameter is insensitive to the structural arrangement between graphene segments, and that nano-graphite is semi-metallic no matter if the stacking is turbostratic or not. As for graphite the density of states is modified when boron atoms are incorporated in the graphitic planes [7].

However, some differences with single crystal graphite are observed on the g-factor and on the linewidth: the average g-factor is slightly lower (2.0010 instead of 2.0018) and ΔH shows a different temperature dependence. Due to the anisotropy of graphite, both g-value and ΔH are influenced by subtle motional averaging over the Fermi surface of different crystallites. In our case the differences are probably due to the turbostratic structure of multiwall nanostructures. Similar differences are observed on lampblack-base graphite [8] and graphitized carbon blacks [9].

ESR and purification

-case of multiwall nanotubes: ESR is also a probe of the sample purity. TEM observations of raw powder revealed that it is composed of a mixture of large graphite flakes and nested nanostructures (amorphous carbon particles, as found in soot, is rarely observed in the cathode deposit). Due to the slight yet significant difference in g-value for the two different graphitic structures, we can separate their ESR signals by dispersing the raw powder in paraffin. After purification only the nanotube line is measured.

-case of singlewall nanotubes: the raw powder signal is characterized by a broad line (3000 G) corresponding to the ferromagnetic resonance of catalyst nanoparticles. No additional line is detected near the free electron value due to internal field shift and relaxation by ferromagnetic spins. Purification treatment of the raw powder in acid drastically decreases the concentration of ferromagnetic nanoparticles. SEM/TEM observations demonstrate that only metallic particles embedded in graphitic shells survived the acid treatment. The ferromagnetic signal was effectively attenuated by orders of magnitude with respect to the pristine material. Quite surprisingly, no graphitic line superimposed to the low ferromagnetic background was detected near the free electron value, even when the powder was diluted in paraffine. To investigate the influence of the remaining catalyst particules, we incorporated well-characterized MWNT powder to the treated soot. In this case we observed the MWNT signal slightly shifted by the small internal magnetic field. We see two possible explanations: either SWNT have an intrinsic broad ESR signal completely different from graphite, or some catalyst remains trapped inside nanotube ropes or in the amorphous carbon that often covers the ropes, thus relaxing the graphitic spins. In fact, it is possible to remove completely the catalyst by annealing at high temperature (>1600°) under inert gas. No ferromagnetic background was then detected, and a narrow line was found near g=2.0023. These new results will be discussed in a forthcoming paper.

Anomalous contact effect in MWNT

An interesting effect of dilution on linewidth was also measured. When dispersed in paraffin or quartz powder, the MWNT nanotubes have a linewidth of ~10 G. However, for compacted thin films ΔH can be as large as 30 G. This effect cannot be attributed to motional averaging effect (as observed in carbon blacks by Arnold and Mrozovski [9]) since such an effect should narrow the line with increasing contact between particles, and it should also affect the g-shift which was not observed. In metals, the Elliott theory [10] predicts that the linewidth is governed by the g-shift (due to spin-orbit coupling) and the carrier relaxation time τ following the relation $\Delta H \propto \Delta g^2/\tau$. An increase of the linewidth while g remains constant reflects thus an increase of the resistivity. This theory proved to be well suited for isotropic metals. In graphite, no consensus has been yet found on the applicability of Elliot's analysis. In our case, if we assume that both in-tube and inter-tube conductivity play a role in the spin relaxation time, the increase in ΔH may be due to the increase in the intertube hopping component as described by Movaghar [11]. Note that the spin diffusion length in graphite is a few microns, i.e., larger than the average nanotube dimension. Intertube hopping offers then new relaxation paths for the spins. From our point of view, this effect can be qualified as mesoscopic since it disappears for larger graphitic particles (it is absent in 70 nm carbon blacks, yet still present in 40 nm arc-grown particles).

MECHANICAL PROPERTIES

Graphite is extensively used in low-tech and high-tech industry for its special mechanical and electrical properties. In fiber geometry, one takes advantage of the high in-plane modulus to realize strong but light composites, for example in aircraft industry. In cast-iron or steel, lamellar graphite induces efficient damping due to friction between planes. The low inter-plane shear modulus is also responsible for the lubrification properties of small crystallites diluted in oils.

The large anisotropy of graphite is however an inconvenience in microscopic fibers since it favors crack propagation initiated by inevitable imperfections. This is one explanation for the paradoxical effect that increasing the modulus decreases the tensile strength. The situation should be different in nanometer size fibers such as nanotubes where fewer imperfections are present and extended dislocations are absent.

Tensile strengh is an important technical parameter. Its measurement is however seldom achieved for nanometer-sized fibers, since a traction nanomachine remains to be invented. However, with the aid of an atomic force microscope, experiments can be done giving qualitative and quantitative information. Large reversible bending of nanotubes on an adhesive substrate provides an estimate of the strength in bending [12][13]. Small bending deformation of cantilevered or suspended nanobeams can yield quantitative values for the elastic modulus. Here we describe a novel method which permits a reliable measurement of the elastic modulus for different nanostructures. We compared ordered and disordered multiwall nanotubes, and analysed the elastic response of SWNT ropes.

One method to measure the elastic modulus of a material is to fabricate a beam that is subsequently clamped or left simply supported at each end and to measure its vertical deflection versus the force applied at a point midway along its length. Although the AFM naturally suggests itself as a means to study the mechanical properties of nanotubes, we are confronted with the problem of configuring a nanotube suspended over empty space. The solution is to deposit the nanotubes on a well polished alumina ultrafiltration membrane. On such a substrate nanotubes occasionally lie over the pores, either with the most of the tube in contact with the membrane surface (Fig. 3(a)), or with the tube suspended over a succession of pores. The attractive interaction between the nanotubes and the membrane acts to clamp the tubes to the substrate. Hence, we assume that tubes with long lengths in contact with the membrane are rigidly clamped.

The AFM is then used to apply a force and to measure the resulting deflection of the "beam". The deflection δ of a beam as a function of the applied load is known from small deformation theory to be $\delta = FL^3/\alpha EI$, where F is the applied force, L is the suspended length, E is the Young modulus, I is the beam's moment of inertia, and α equals 192 for a clamped beam [14]. For a hollow cylinder, $I=\pi(R_o^4-R_i^4)/4$; R_o and R_i being respectively the outer and inner radii.

We used an AFM in contact mode with Si3N4 tips for all measurements. To minimize the uncertainty of the applied force, we calibrated the spring constant of each AFM cantilever, usually ~0.1 N/m, by measuring its resonance frequency.

Once a suspended nanotube is located with the AFM, we determine its diameter, suspended length, and deflection midway along the suspended length from data like those of Fig. 3(b). The apparent tube width is a convolution of the tube diameter and the tip radius. Height remains a reliable measure of the tube diameter. The reversibility of the tube deflection and the linearity of the δ-F curve show that the nanotube response is linear and elastic. The slope of the curve, typically ~0.5m/N, directly gives the Young modulus of the MWNT once L and R_O are known.

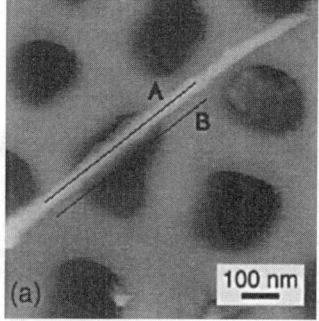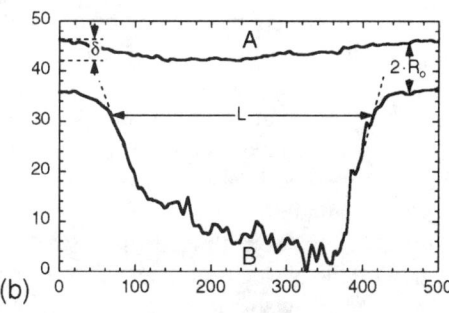

FIGURE 3. (a) AFM image of a MWNT adhered on the polished ultrafiltration alumina membrane, with a portion bridging a pore of the membrane; (b) cross-sectional profiles of the nanotube (A) and corresponding pore (B) depicted in image (a).

It is worth mentioning that the deflection is nonzero for zero nominal force, i.e., for zero deflection of the AFM cantilever. Two reasons can be evoked for this observation. First, the filter surface is rounded so that the tube can be slightly bent by the moment exerted by adhesion forces on the membrane. Second, the absolute force exerted on a surface by the tip is the sum of a net attractive force F_a and the nominal force due to cantilever bending F_n. Our measurement rely on the assumption that F_a does not depend on F_n. This is true as long as the surface is not deformed by the tip, i.e., if the contact surface is the same.

Let us examine the precision of the method. The precision on the tube diameter is about 0.5 nm; the relative error decreases for large tubes. The problem of convolution with the tip shape affects also the measurement of the suspended length. In fact we determine a minimum value of the suspended length. Then a minimum value for E is calculated. A further source of uncertainty arises from the inner diameter which cannot be measured by AFM. An extensive HRTEM study by Bacsa et al. [15] demonstrated that it can vary from 1 to 6 nm for a tube of 10 nm outer diameter, with the most probable value being approximately 2 nm. We assume $R_i = 1$ nm, with the consequence that we calculate the minimum value for E. If in fact the inner diameter were 6 nm, the error would be 13%, less than our probable error.

The Young's modulus for eleven individual arc-grown carbon nanotubes, nearly independent of tube preparation, diameter, suspended length, or clamping to the substrate, is 810±410 GPa, which should be taken as a minimum value. Our result compares favourably with simulations, and with the calculated value for the in-plane direction of graphite, 1.06 TPa [16].

For catalytically grown nanotubes a singular decrease of the elastic modulus is observed as compared to arc-grown nanotubes. Our values ranged from 10 to 50 GPa. This behaviour can be explained by the high anisotropy of graphite. The variation of the elastic modulus with the angle to hexagonal axis is strongly influenced by the shear effect between planes. A minimum value of about 10 GPa is obtained when the crystallites are disoriented by 45° from the hexagonal axis.

The characterization of SWNT ropes is more difficult [17]. The ropes are not always symmetric in cross-section, and may be covered with amorphous carbon. It was therefore not possible to obtain linear and reproducible results on all ropes. We checked that the measured diameter was the same on both sides of the hole to verify that the rope was homogeneous. The regularity of the profile measured with AFM ensures that the rope is not covered by amorphous carbon beads. The nearly cylindrical shape of the ropes we measured was verified by comparison with multiwall nanotubes. For small ropes it is probable that we underestimate the number of tubes because the actual shape should be a trapezoid. Then the Young modulus for small ropes may be overestimated.

Dealing with the elastic properties of ropes we feel that the problem will be different than for nested multiwall nanotubes or catalytic nanofibers. The low intertube cohesion will facilitate bending as compared to an isotropic beam obtained by cross linking neighbouring tubes. This is actually why an elongated crystalline array of SWNT is called a rope...

To modelize the deformation of the rope in bending, one may be tempted to consider the SWNT completely free in a rope. The clamped beam formula can then be used with the applied force divided by the number of tube in a rope. Doing so gives a Young modulus higher than 10 TPa for a single tube, which is far higher than the more optimistic predictions. We have clearly to improve the model. Fortunately the problem of deflection of anisotropic beams have been considered by mechanical engineers a long time ago. In fact, the shear deformation of the beam must be included in addition to the pure bending effect [17].

It appears that shear deformation can be neglected for long ropes of small diameter and we measured directly the Young modulus, obtaining about 1 TPa. This is in agreement with theoretical calculations [18] and with other experiments [19]. Using this value we can also extract the shear modulus for large ropes, which amounts to ~1 GPa.

To conclude this section it is worth stressing the good correlation we found between the elastic properties and the structure of the different beams. The good agreement between these experiments and theory show that nanofibers have an enormous technological potential which only asks to be exploited.

FIELD EMISSION

Although important insights have been obtained recently on field emission from carbon nanotubes, little is yet known about the emission mechanism, although experimental evidence strongly suggests that the electrons are not emitted from a metallic continuum as in usual metallic emitters [20][21]. We address here this still open question by studying the field emission from three different perspectives: measurements of the energy spectra of the emitted electrons, observations of light emission during field emission, and field emission microscopy.

Emitted electron energy distribution

According to the Fowler-Nordheim theory (F-N theory, i.e., elastic tunneling through a triangular barrier, with the electron distribution described by Fermi-Dirac statistics [22]), the energy distribution of the emitted electrons is $J_{FN}(E) \propto \exp((E - E_f)/b(F/\phi^{1/2}))f(E - E_f)$, where E is the electron energy, E_f the Fermi energy, ϕ the work function, F the electric field, and f(E) the Fermi-Dirac distribution. We show in Fig. 4(a) a field electron energy spectra obtained on a MWNT film, along with the best fit obtained with the F-N distribution. Obviously, the distribution doesn't match the measured spectra. The F-N theory predicts exponential slopes on both sides of the distribution which arise from the tail of the Fermi-Dirac distribution and from the increase of the barrier width, respectively.

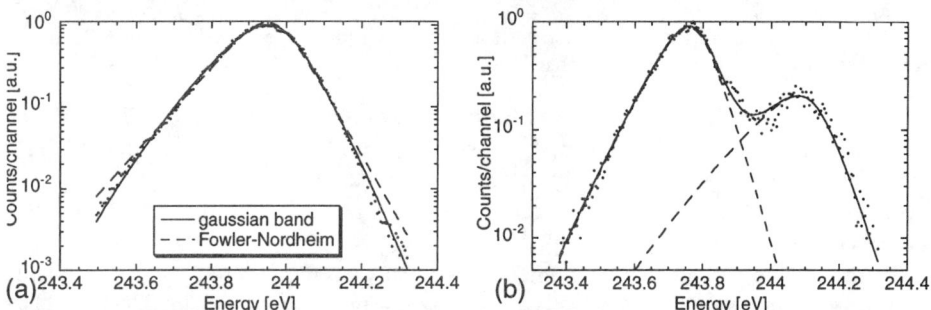

FIGURE 4. Field electron energy spectra obtained on a MWNT film (dots), showing (a) a single peak, along fits obtained with the F-N distribution and with the modified F-N distribution including a gaussian band of states; (b) two peaks along with the deconvolution in two gaussian distributions (dashed lines) and the resulting total distribution (solid lines).

To obtain reasonable agreements with the measured spectra, we considered a gaussian band of states at the tip of the tubes instead of the usual metallic density of states (DOS), resulting in a distribution $J(E) \propto J_{FN} \times \exp(-(E-E_c)^2/(\Delta E)^2)$, i.e., the Fowler-Nordheim distribution times a gaussian band of width ΔE centered at energy E_c. With this distribution, the tube body is taken as metallic, i.e., with a DOS described by the Fermi-Dirac statistics, and it supplies the tip states (gaussian bands) with electrons. This modified formula fits well the spectra, as can be seen in Fig. 4(a), and allows one to estimate the width of the gaussian (typically 0.2 - 0.4 eV) and the Fermi temperature of the electrons (300 - 400K). We found the presence of several peaks on some spectra as displayed in Fig. 4(b), where the two peaks have been deconvoluted with two gaussian distributions. Such a situation may arise from one tube if the local DOS at the tube tip has two bands not too far apart in energy, or from another tube. In the latter case, the difference in energy between the peaks wouldn't come from a potential difference, but because the bands are not located at the same energy from one tube to the next.

Field emission induced luminescence

We observed luminescence induced by electron field emission on single- and multiwall nanotubes films as well as on single MWNT emitters. The light emission occurred in the visible part of the spectrum, and could sometimes be seen with the naked eye. No light emission was detected without applied potential.

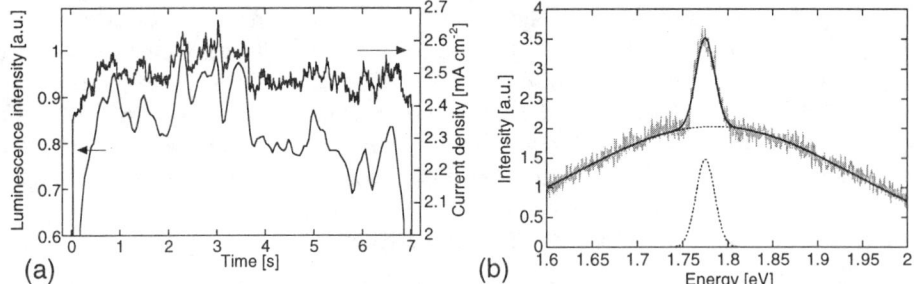

FIGURE 5. (a) Emitted light intensity and current density as a function of time for a MWNT film; (b) Spectra of field emission induced luminescence for one MWNT along with the two gaussians fit.

The emitted light intensity followed furthermore closely the variations in emitted current, as can be assessed in Fig. 5(a), where the emitted current and emitted light intensity have been simultaneously measured. Actually, the emitted intensity depends critically on the current, since the relative variations are 3-4 times higher for the luminescence as compared to the current.

To investigate further this phenomenon, we analysed the spectral repartition of the emitted light for single MWNTs. A typical spectrum is displayed in Fig. 5(b). The spectra could be described with very good accuracy as a sum of two gaussian functions, with peak energies, widths, and relative intensities that varied with experimental conditions. Fig. 5(b) shows an example of a fit, with the measured spectrum in grey, the two gaussians in dashed lines, and the resulting fitted spectra in solid line. The full width at half maxima (FWHM) in the case of Fig. 5(b) were ~0.34 eV and ~22 meV for the broad and the narrow gaussian, respectively, with an integrated intensity ratio of typically 20. No significant changes in the shape of the spectra was observed when the current was varied apart from a small shift (<25 meV) of the broad gaussian. The position and width of the narrow gaussian remains nearly constant from one tube to the next. As for the broad gaussian contribution, we observed peak intensities and widths varying between 1.73 and 1.83 eV and between 0.3 and 1 eV, respectively. Finally, light was emitted at higher energies from SWNTs as compared with MWNTs.

There has been one report of observed luminescence on opened nanotubes [23] but it was attributed to an incandescence of carbon chains at the tip of the tube provoked by resistive heating. Our results however strongly suggest that the light emission is directly coupled to the field emission. The narrowness of the luminescence lines and the very small shifts with varying emitted current (< 5 meV) show that we are not in presence of blackbody radiation or of current-induced heating effects, but that photons are emitted following transitions between well-defined energy levels. This trongly suggests the presence of localized states at the tip.

We estimate that one emitted photon corresponds to at least 10^6 field emitted electrons. With localized states at the tip, the greatest part of the emitted current will arise from occupied states with a large local density of states located near the Fermi level. Other, more deeply located electronic levels may also contribute to the field emission. In this case, the emitted electron will be replaced either by an electron from the semimetallic tube body with an energy comparable to the level energy, or by a tip electron from the main emitting state. Clearly, the second alternative may provoke the emission of a photon. Even if the tunneling probability for electrons from deep state is several orders of magnitude lower than for the main emitting state, it will be readily sufficient to cause the observed light intensities.

The luminescence intensity I_p as a function of the emitted current I_e follows a power law $I_p \propto I_e^\alpha$ with $\alpha = 1.4 \pm 0.2$. This dependence can be reproduced by a two-level model [24]. The distribution of localized states at the nanotube tip is simplified to a two level system, with the main emitting level at energy E_1 below or just above the Fermi energy, and a deep level at $E_2 < E_1$. When an electron is emitted from the deep level, it is replaced either by an electron from the tube body, or by an electron from the main level which can provoke the emission of a photon. From the Fowler-Nordheim model, the transition probability $D(E)$ can be evaluated for each level, and in the frame of our model, $I_e \propto D(E_1)$, $I_p \propto D(E_2)$. It appears that I_p varies as a power of I_e with an exponent that depends on the separation of the levels, and that amounts to 1.51-1.65 for the energies observed here (typically 1.8 eV), which corresponds well to the experimental observations.

Field emission microscopy

Field emission microscopy (FEM) is an extension of field emission characterization, which offers a unique possibility of visualizing the emitter by replacing the counterelectrode with a phosphor screen. We performed FEM on SWNT films with the emission area reduced to ~0.1 mm diameter. A simple 20 x magnificating electrostatic lens, held at about 1 mm distance, was used as counterelectrode, and a phosphor screen located 5 cm above the film-lens assembly allowed us to visualize the field emission patterns on a phosphor screen.

The FEM patterns observed on the screen reflect directly the emitted current distribution. Since the tunneling electrons have very small kinetic energy, they follow the lines of forces, which diverge in first approximation radially from the emitter surface. The emission pattern at the tip, reflecting the spatial distribution of the emitted density, is thus enlarged before hitting the screen. A pattern detected by FEM on carbon nanotubes can be induced either by adsorbates, by local changes in the emitted current due to preferential emission sites (surface steps that result in a locally enhanced field amplification, protusions of amorphous carbon present on the tip surface, or atomic wires), or spatial variations of the electronic density.

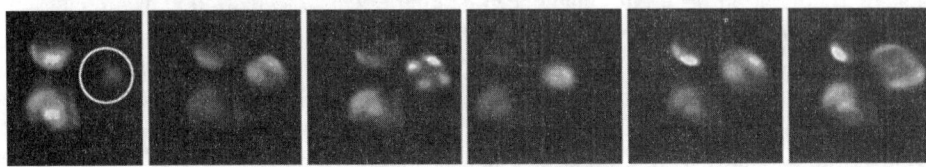

FIGURE 6. Sequence of field emission patterns obtained on a SWNT film.

Fig. 6 shows a sequence of field emission patterns recorded at ~1 s intervals. Usually, several spots were simultaneously visible on the screen. Beside single spots and elongated and/or circular features without any distinctive shape, some well-defined patterns were observed. Most of them acted definitely as a unit. Like the one outlined in the first frame of Fig. 6, they abruptly changed shape, rotated or librated, or disappeared suddenly. This behaviour makes it highly plausible that each of these patterns is due to a single tip, or to a single adatom/admolecule. Individual patterns of two-fold and four-fold symetry were frequently observed evolving from simple spot pattern, but no three- or five-fold symetry patterns were detected. More complicated figures consisting of a series of fringes (up to four) showing twofold symmetry, with some fringes divided in two leafs, were also observed.

Because of the observed symetries, it is very improbable that the different spots on the patterns were caused by preferential emission from protusions of amorphous carbon, single atomic wires, or surface steps. On metallic tips, absorbates also give raise to two- or four-fold leaf patterns, appearing bright and superimposed on the usual tip pattern. However, complicated patterns like the ones on Fig. 6 were only very rarely detected, whereas they systematically appear at high current for SWNT, and no additional tip pattern was observed in superposition. Furthermore, experiments carried out in ultra-high vacuum conditions yielded comparable results. It is thus most probable that the observed patterns are caused solely by spatial variations of the electronic density, i.e., that they reflect the electronic density of the emitting states at the tip. The fact that the electronic distribution from a single tube shows a non-homogenous structure points again to the fact that the electrons are emitted from electronic states localized at the tip, and are not delocalized conduction-band electrons as in metals.

Field emission mechanism

We deduce from the above results that the field emission behavior of carbon nanotubes cannot be understood in terms of emission from a metallic tip. The luminescence and the energy distribution of the emitted electrons indicate that the electrons are emitted from narrow band states of ~0.3 eV half-width. In fact, theoretical calculations show that the local density of states at the tip presents sharp localized states that are correlated to the presence of pentagonal defects [25]. The observed luminescence strongly suggests that although the greatest part of the emitted current comes from occupied states with a large density of states near the Fermi level, other, deeper levels also contribute to the field emission. The position of the tip states with respect to the Fermi level influences directly the field emission properties of the tube. Indeed, only tubes with a band state close under or just over the Fermi level are good candidates for field emission.

We conclude that the presence of such localized states instead influences greatly the emission behaviour. At and above room temperature, the body of carbon nanotubes behave essentially as graphitic cylinders. This means that the carrier density at the Fermi level is very low, i.e., on the order of $5 \cdot 10^{18}$ cm^{-3}, which is 3 orders of magnitude less than for a metal. Simulations show that the local density of states at the tip reaches values at least 30 times higher than in the cylindrical part of the tube. Since the field emission depends directly on this carrier density, the field emission current would be far lower without these localized states for a geometrically identical tip. The superiority of closed over open or disordered MWNTs is in this respect an additional indication [21], since the position and intensity of the localized states are strongly influenced by the crystalline structure. A disordered, or worse, a missing tip would consequently lead to inferior performances, as was observed for open and catalytic MWNTs [21].

ACKNOWLEDGMENTS

The authors wish to thank G. Beney, A. Briggs, N. Burnham and A. Kulik (IGA-EPFL) for their contribution to the nanomechanics investigation, and Karine Méténier, Sylvie Bonnamy and François Béguin (CNRS-CRMD Orléans) for the growth of catalytic nanotubes. The electron microscopy was performed at the Centre Interdépartemental de Microscopie Electronique of EPFL. Financial support was provided by the Swiss National Science Foundation

REFERENCES

1. T. W. Ebbesen and P. M. Ajayan, Nature 358 , 220 (1992).
2. W.K. Maser et al., Synthetic Metals 77, 243 (1996).

3. H. Alvergnat et al., Proc. Eur. Carbon Conf. (Carbon'96, Newcastle, UK), 715 (1996).
4. J.-M. Bonard et al., Adv. Mater. 9 , 827 (1997).
5. J. Hone et al., proceedings of this conference.
6. J.-P. Salvetat et al., in preparation.
7. A. Huzcko et al., unpublished results
8. L.S. Singer and G. Wagoner, J. Chem. Phys. 37, 1812 (1962).
9. G. Arnold and S. Mrozovski, Carbon 6, 243 (1968).
10. R.J. Elliott, Phys. Rev. 96, 266 (1954).
11. B. Movaghar et al., Phil. Mag. B37, 683 (1978).
12. M. R. Falvo et al, Nature 389, 582 (1997).
13. E. W. Wong, P. E. Sheehan, and C. M. Lieber, Science 277, 1971 (1997).
14. J. M. Gere and S. P. Timoshenko, in Mechanics of Materials (PWS-KENT Publishing, Boston, 1990)
15. W. S. Bacsa et al., in Fullerenes, Vol.3, edited by K.M. Kadish and R.S. Ruoff (Electrochemical Society, Pennington, 1996).
16. B. T. Kelly, in Physics of graphite (Applied Science Publishers, London, 1981).
17. J.-P. Salvetat et al., submitted to Science.
18. J. P. Lu, Phys. Rev. Lett. 79, 1297 (1997).
19. N. Shopra and A. Zettl, private communication.
20. J.-M. Bonard et al., Ultramicroscopy (in press).
21. J.-M. Bonard et al., Proceedings of the 193th ECS symposium (1998).
22. J. W. Gadzuk and E. W. Plummer, Reviews of Modern Physics 45, 487 (1973).
23. A.G. Rinzler et al., Science 269, 1550 (1995).
24. J.-M. Bonard et al., submitted to Phys. Rev. Lett.
25. D.T. Carroll et al., Phys. Rev. Lett. 78, 2811 (1997); J.-M. Bonard et al., submitted to Phys. Rev. Lett.; J.-C. Charlier et al., private communication.

Hydrogen Storage in Carbon Materials - Preliminary Results

Ludwig Jörissen[#], Holger Klos[*], Peter Lamp[†], Gudrun Reichenauer[‡] and Victor Trapp[§]

[#]ZSW Ulm, D-89081 Ulm, [*]Mannesmann Pilotentwicklungsgesellschaft mbH, D-81737 Munich, [†]ZAE Bayern, Div.Energy Conversion & Storage, D-85748 Garching, [‡]ZAE Bayern, Div. Thermal Insulation and Heat Transfer, D-97074 Würzburg, [§]SGL Technik GmbH, D-86405 Meitingen

Abstract. Recent developments aiming at the accelerated commercialization of fuel cells for automotive applications have triggered an intensive research on fuel storage concepts for fuel cell cars. The fuel cell technology currently lacks technically and economically viable hydrogen storage technologies. On-board reforming of gasoline or methanol into hydrogen can only be regarded as an intermediate solution due to the inherently poor energy efficiency of such processes. Hydrogen storage in carbon nanofibers may lead to an efficient solution to the above described problems.

THE GINA PROJECT (GAS STORAGE IN NANOFIBERS)

Large hydrogen storage capacities (HSC) of carbon nanofibers (CNF) were previously reported by a group of Boston researchers (1). Other research groups have been working on this topic and also obtained some interesting results (2). In reflection of such promising HSC in carbon-based nanostructures, a German group of two industrial partners and two R&D institutes have joined to follow up on this issue.

This project comprises the screening of various types of carbon materials, putting an emphasis on CNF, with regard to their HSC. Furthermore, a CNF synthesis apparatus was set up enabling us to a focussed production of certain CNF types with promising HSC. The hydrogen storage tests were carried out by applying two different methods, i.e. gravimetric measurements and volumetric tests. While in the initial phase it appeared to be difficult to obtain reliable test results, both methods prove to provide highly reproducable data. Although the tested materials were also characterized by a variety of analytical methods, mainly results of the HSC tests are reported here.

Within the scope of this project, a wide range of carbon materials were obtained from several providers from the scientific community and commercial sources. These initial HSC tests of different carbon materials were neccessary in order to get an understanding of the fundamentals of the relevant hydrogen storage mechanisms especially in CNF.

Hydrogen Storage Capacity Tests

The volumetric HSC test was carried out with a high-pressure apparatus (Fig. 1) by determining the gas pressure drop. The entire test device can be flooded with various gases at room temperature up to a pressure of 200 bars, and can be evacuated down to 10^{-7} mbars by means of a molecular pump.

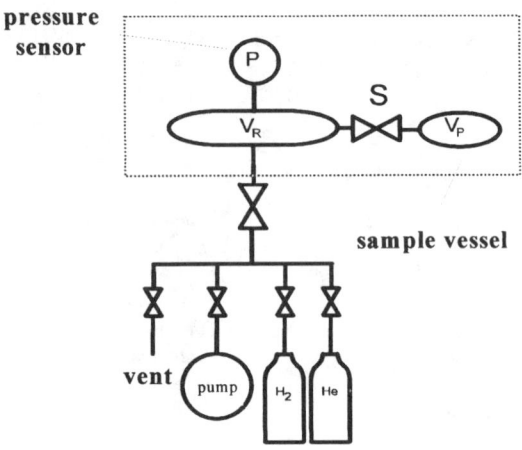

FIGURE 1. Schematic experimental set-up for volumetric HSC tests (ZAE division Garching).

The actual test volume (dotted square) is devided by the valve S into two volumes V_R (reference volume) and V_S (sample vessel). The pressure sensor P measures the pressure in V_R, and two temperature sensors monitor the ambient temperature and the gas temperature inside the sample vessel. After loading the sample vessel with a carbon material, the apparatus is evacuated and heated up to 450 °C for several hours. Subsequently, volume V_R is filled with hydrogen gas up to a final pressure of approx. 100 bars. Opening the valve S results in a pressure drop caused by expansion of the gas into V_S and uptake of hydrogen by the sample. The first contribution can be taken into account by applying the ideal gas law to the set of known values of p, V_S, V_R. The remaining pressure drop has then to be attributed to hydrogen storage in the sample. For further verification, a reference measurement is taken with helium which does not significantly adsorb on the sample surfaces.

Gravimetric tests of HSCs were based on a system (Fig. 2) employing a SATORIUS micro-balance S3D-P. This device allows to measure mass yields at the sample with a sensitivity of 0.1 µg at pressures up to 150 bars. Crucible S contained the sample and the reference crucible R was loaded with quartz powder. The HSC tests were carried out isothermally at 23 °C. First, the balance volume was evacuated to $2*10^{-6}$ bars until the pressure had stabilized for 30 min. Then the entire gas volume was flushed 3 times with hydrogen gas at ambient pressure. The HSC test was carried out stepwise up to 125 bars. Gas pressures were measured with a pressure sensor and were kept at a constant value after the balance had stabilized. The same procedure was applied during the subsequent desorption steps.

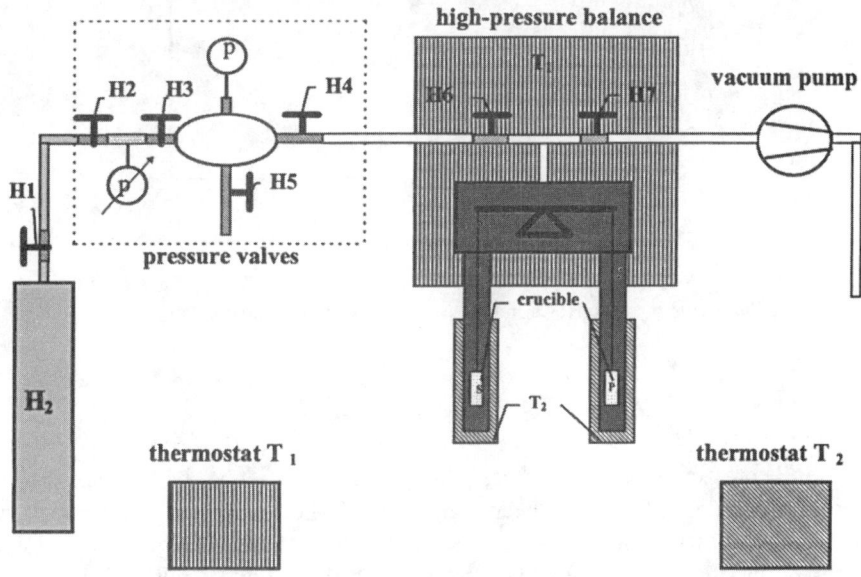

FIGURE 2. Set-up of gravimetric HSC tests (ZSW Ulm).

Preliminary Results

So far, none of the investigated carbon nanofibers and nanotubes had HSCs exceeding values of 2 w/w%. In many cases it was difficult to judge how much nanofibers the obtained sample actually contained because it consisted of an inhomogenous mixture of amorphous materials and various forms of nanofibers and nanotubes. In such cases, it could be applicable to determine the fraction of e.g. nanofibers in the specimen and to extrapolate the found HSC to 100% nanofiber content. However, such an approach and was not yet applied within the scope of our project.

Interestingly, a row of HSC tests on activated carbon blacks (ACB) with various micropore volumes gave some understanding on the important role of micropores for the HSCs. Figure 3 shows a diagram emphasizing the dependence of the HSC of various ACBs. Clearly, the HSC is increasing with greater micropore volume up to a maximum of about 0.6 w/w% at around 0.75 ml/g micropore volume. Since the investigated ACBs have not yet been fully characterized, we are so far unable to draw a more detailed conclusion of these test results.

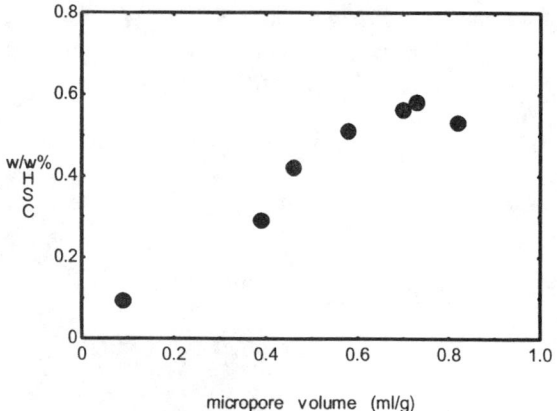

FIGURE 3. Relationship between HSCs and the ACB micropore volume (experimental data).

Outlook

In the initial project phase, a sound basis for various HSC test has been established and the first carbon nanofibers have been synthesized in-house. The following steps will comprise a focussed nanofiber synthesis, aiming at certain preferred morphological aspects. Furthermore, cleaning and activation procedures will be pursued. It will also be attempted to get a better understanding of the hydrogen storage mechanisms.

ACKNOWLEDGMENTS

The funding by the Bavarian Ministry of Economics (Wasserstoffinitiative Bayern - WIBA) is gratefully acknowledged. Our work would not have been possible without the samples provided by Prof. Beguin, Dr. Bonnamy, Prof. Cazorla-Amorós, Prof. Cheng, Prof. Figueirdo, Prof. Gadelle, Prof. Kushinov, Prof. Lineares-Solano, Prof. Nagy, Prof. Rößler, Prof. Schlögl, Dr. Wojtowicz, and other people who supported us.

REFERENCES

1. Hill, S., *New Scientist*, 21./28. December, 20 (1996).
2. Heben M.J. et al., *Nature* **386**, 377 - 379 (1997).

Pressure Dependent ^{1}H-NMR-Measurements of Activated Carbon and Carbon Nanofibers

Walter Schütz, Holger Klos

Mannesmann Pilotentwicklung,

Chiemgaustraße 116, 81549 München, Germany

Abstract: Pressure dependent ^{1}H-NMR-spectroscopy was used with the aim to study the microscopic mechanism of gas storage in different carbon materials. Investigations were performed on activated carbon, arc discharge produced soot, carbon-nanotubes and carbon-nanofibers in comparison to metal hydrids. We rebuilt a standard NMR-spectrometer, which allows hydrogen pressure dependent measurements up to 100 bar. Interesting parameters are the line width and the line position in comparison to pure hydrogen gas. Carbon-nanofibers exhibit large line shift and line width.

INTRODUCTION

Hydrogen is the most promising energy resource for the future. Fuel cells convert hydrogen and oxygen gas into electrical energy. While fuel cells are close to market entrance, there is still need for powerful hydrogen storage systems. Hydrogen can be stored in pressure vessels, but the storage capacity is comparatively low. Storage as liquified hydrogen requires a lot of energy and an expensive cryogenic tank system. Metal hydrids for hydrogen storage are not practicable, because they are too heavy and too expensive. Maybe carbon materials can help to solve these problems: Different press releases reported about enormous storage capacities in carbon nanofibers (Ref. 1, 2). This storage mechanism is not understood until now.

Using pressure dependent ^{1}H-NMR measurements it is possible to get information about the microscopic environment of the protons and to distinguish between physical adsorption and chemical absorption phenomena.

The line intensity is proportional to the amount of the stored hydrogen. The line width gives evidence about the interactions between the hydrogen molecules and the adsorption material. Line broadening is caused by dipolar interactions, which are independent of the outer magnetic field and by magnetic anisotropy or anisotropic chemical shift, which are proportional to the applied magnetic field. The line position depends on the local magnetic field. For example, in activated carbon it can be shown, that line shift is found, when the pores, where the hydrogen is adsorbed, are surrounded by anisotropic diamagnetic material and depends on the form of the pores and grains (Ref. 3).

EXPERIMENTAL SETUP

^{1}H-measurements were performed with a 20 MHz Bruker minispec spectrometer. The sample handling of this spectrometer typ is easy, because no special probe head is needed.

The sample is placed in a special glass tube (Fig. 1), which is connected to a gas capillar system. A hydrogen gas vessel is fixed at the other end of the capillar system.

The hydrogen pressure range is variable between 0 and 100 bar. This system is submitted as patent.

Figure 1. NMR glass tube.

RESULTS

NMR of metal hydrids generates spectra, which are typical for protons in solids (Fig. 2). In metal hydrids, hydrogen is stored in interstitial positions. Due to strong dipolar interactions, the line width is about 100 KHz. The sharp peak in the center of the signal is caused by gaseous hydrogen in the sample tube.

The spectra of the different carbon materials are shown in Figure 3. The line shift and the line position are summmarized in Figure 4. Both are independent of the applied pressure in contrast to the line intensity: The hydrogen pressure is directly proportional to the number of gas particles. In a wide range the ideal gas law is valid (up to 150 bar). The NMR-line intensity is proportional to the number of gas particles. (In Fig. 3 the line intensities of the different carbon materials can not be compared.)

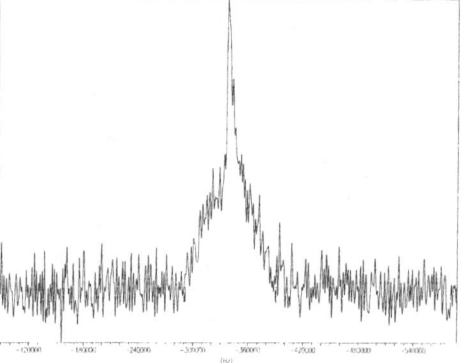

Figure 2. NMR-spectra of metal hydrid.

In activated carbon we found a line width, which is typical for this class of material (Ref. 3). The lines are broadened in comparison to free hydrogen gas (Fig. 4): This line broadening is caused by the distribution around the middle value of the local magnetic field. Dipolar interactions are middled out by fast reorientation of the hydrogen molecules. The line position is also typical for activated carbons and can be explained by a model, where the hydrogen molecules are adsorbed on a graphitic plane (Ref. 3).

In arc discharge produced soot, the signal is similar to pure hydrogen gas. No line shift can be detected, the line width remains nearly constant. This means, that almost no adsorption on the surface takes place.

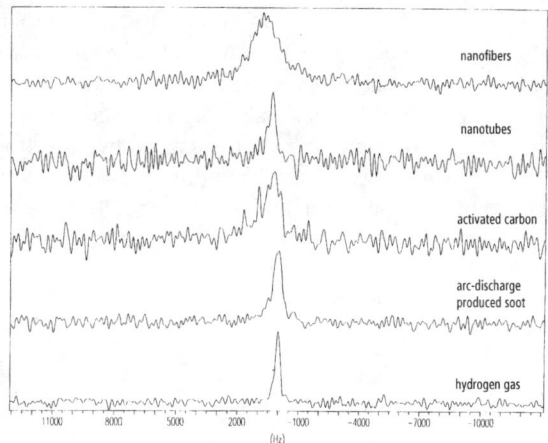

Figure 3. NMR spectra of different carbon materials (60 bar)

The NMR-signal of carbon nanotubes (MWNT) exhibits a shift, which is comparable to activated carbon. It can be explained with hydrogen adsoption on an aromatic surface (Ref. 3). The lines are narrower as in activated carbons and depend on the orientation of the aromatic plane.

In carbon nanofibers we detected unusual large shift and line width. At present an interpretation of these signals is not possible.

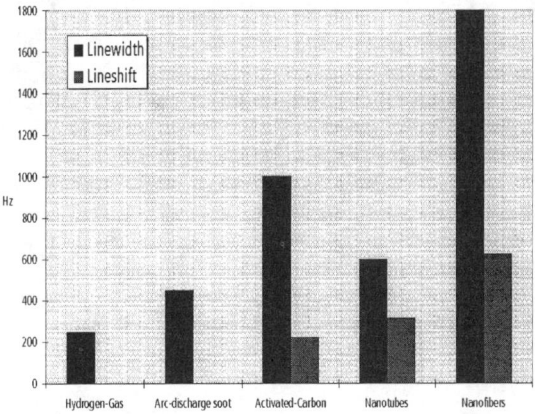

Figure 4. Line width and line shift of the different carbon materials.

DISCUSSION

NMR-spectra of metal hydrids exhibit that hydrogen is absorbed in interstitial positions. These spectra are typical for solids and have a line width of about 100 KHz due to dipolar interactions. We studied different carbon materials and found that hydrogen is just adsorbed on the surface. The NMR line widths are in the same range as those of pure hydrogen gas. The broadening by a factor of 5 to 10 is caused by the diamagnetic anisotropy of the carbon materials. The lines are shifted to higher fields in comparison to hydrogen gas. As well, this shift is caused by anisotropic diamagnetism. When hydrogen is absorbed between the graphitic platelets, which is expected in carbon nanofibers, the lines should get much broader, when the molecules become immobile. Until now, we could not confirm this hypothesis. However, we found an unusual line shift and line width in carbon nanofibers. Experiments with hydrogen pressure above 100 bar are in progress.

SUMMARY

We rebuilt a conventional ^{1}H-NMR spectrometer for pressure dependent measurements up to 100 bar and demonstrated that NMR is a powerful tool to study adsorption and absorption phenomena. All carbon materials we studied, showed adsorption on the surface in contrast to metal hydrids, where the hydrogen occupies interstitial positions.

ACKNOWLEDGEMENTS

The authors thank Prof. M. Schwoerer, Prof. E. Rößler, Prof. P. Gadelle, Prof. R. Schlögl and Dr. M. Wojtowicz for support and fruitful discussion.

REFERENCES

1. *Hydrogen & Fuel-Cell Letter*, Feb. 1997, p. 1
2. *New scientist*, 21/28 Dec. 1996, p. 20
3. S. Gradsztajn, J. Conrad, H. Benoit, *J. Phys. Chem. Solids*, Vol. 31, 1121-1135, 1997

Encapsulation of Ferromagnetic Metals into Carbon Nanoclusters

S. Seraphin, J. Jiao, C. Beeli*, P.A. Stadelmann*, J.-M. Bonard**, and A. Châtelain**

Dept. of Materials Science & Engineering, University of Arizona, Tucson, AZ 85721, U.S.A.
*Centre Interdepartemental de Microscopie Electronique, Ecole Polytechnique Fédérale de Lausanne, CH-1015 Lausanne, Switzerland; **Institut de Physique Experimentale, Ecole Polytechnique Fédérale de Lausanne, CH-1015 Lausanne, Switzerland

Abstract. Electron holographic measurements of the magnetization of carbon-coated ferromagnetic particles of different sizes in the nanometer range are reported. Values of a ratio of remanent to saturation magnetization of 0.3 are found for the entire assembly by Vibrating Sample Magnetometer (VSM). If determined for the individual particle by electron holography, however, values in between 0.75 and 0.30 are determined for diameter range in between 30 and 100 nm, with the highest value relating to the smallest diameter. Exposing the sample assembly to a 2 Tesla external magnetic field increases the magnetization by 10%. The small size of the particles as well as the interface of the ferromagnetic material with the carbon coating are cited in speculations on the interpretation of the results.

INTRODUCTION

The interest in magnetic properties of the carbon coated ferromagnetic particles was stimulated first by their technological promise. Recording technologies in particular are searching for very fine-grain magnetic materials that could be tailored in their magnetic properties and are resistant to oxidation. A second, more fundamental reason is given by the fact that the small size and subsequently limited number of atoms of a particle could support novel physical properties that are not observed in the macroscopic bulk. Several studies have measured the integral magnetic properties of these particles [1-4]. The only study on the magnetization measurements of individual carbon coated nanoparticles of Co, Ni, Dy using a Superconducting Quantum Interference Device (SQUID) focused on ellipsoidal particles of 15-30 nm diameter [5]. The results suggested single domain character. This study reports on the measurements of the magnetization of spherical carbon-coated Co and Ni particles as a function of their diameters by electron holography in the transmission electron microscope (TEM). The technique was used to determine the remanent magnetization of thin Co and Ni nanowires as a function of diameter and length-to-diameter ratio [6]. Recently, the magnetization reversal behavior of submicron-sized Co thin films was also studied *in situ* using electron holography [7]. The understanding of the formation of remanent states is of importance for the design of magnetic elements for device applications.

SAMPLE PREPARATION

The carbon-coated Ni and Co nanoparticles were prepared in a modified arc discharge chamber with two vertical electrodes and a perpendicular jet of helium gas [1]. The products consisted mainly of spherical particles without nanotubes or amorphous debris as shown by the scanning electron image of Ni in Fig.1(a). The size of the particles ranged from 20 to 80 nm. TEM revealed that these particles were coated with thin layers of graphitic carbon (Fig.1(b)). The coating layer became thicker and completely graphitic when the sample was annealed at 900 C for 1 hour (Fig.1(c)). The XRD analysis showed that these particles are elemental Ni and Co without changing to carbides. This is in contrast to most of the products from the conventional arc discharge which sustained a much higher carbon-to-metal ratio in the process. Our previous studies showed that the average size of the particles can be controlled by changing the size of the metal pool in the anode [8]. It was found that the larger the metal pool the bigger the average size of the particles.

MAGNETIC CHARACTERIZATION

The integral hysteresis loops of the samples taken by a Vibrating Sample Magnetometer (VSM) showed a ratio of remanent to saturation magnetization, M/M_s ~0.3 [1]. In this study, the electron holography was used to determine remanent magnetization as a function of individual particle size when they are attached to each other in chains as shown in Fig.2. The electron holography was performed at 100 kV with the objective lens current switched off in a HF-2000 FEG TEM equipped with an electrostatic biprism. Two sets of samples for each metal type were studied, one as-deposited and the other after applying a 2 Tesla magnetic field parallel to the grid plane. The profiles (Fig. 4) of the phase difference across the chain as indicated in the phase image (Fig.3) corresponded to the total magnetic flux leaking from the chain [6], thus determining the remanent magnetization. The direction of the phase change across the chain (after the reduction of a 2π phase wrapping) from positive to negative values or vice versa determines the direction of the magnetic moment in the chain.

Fig. 5 shows the ratio M/M_s as a function of the average Ni particle diameters. In the diameter range 30 to 100 nm the remanent magnetization decreases with increasing particle diameter. The highest relative magnetization obtained in this sample is around 0.75, and the lowest 0.30. After being exposed to the 2 Tesla external magnetic field, the magnetization of the particles increased by about 10% with the dependence on the particle size staying essentially unchanged. Compared to the integral value of 0.3 obtained from the VSM technique, the results from holography provide more detailed information on the magnetization. A magnetization much lower than in the bulk may be due first to size effects, and secondly caused by the presence of the carbon coating interfacing the metal core. The reduction in magnetization of larger particles may be due to the internal flux closure in the large particles. A TEM equipped with a Lorentz lens is required to provide the phase contours lines of constant magnetization inside the particles of these sizes. Fig. 6 shows the relative magnetization values of carbon coated Co nanoparticles of smaller values (0.55 to 0.16) but show the same trends as Ni. The results attach for the first time a magnetization value to a particle that can be identified individually in its graphite-coated existence. Further investigation is necessary to verify the cause of the size dependence of the magnetization and determine at what size range will the remanent magnetization increases again to reach the bulk values.

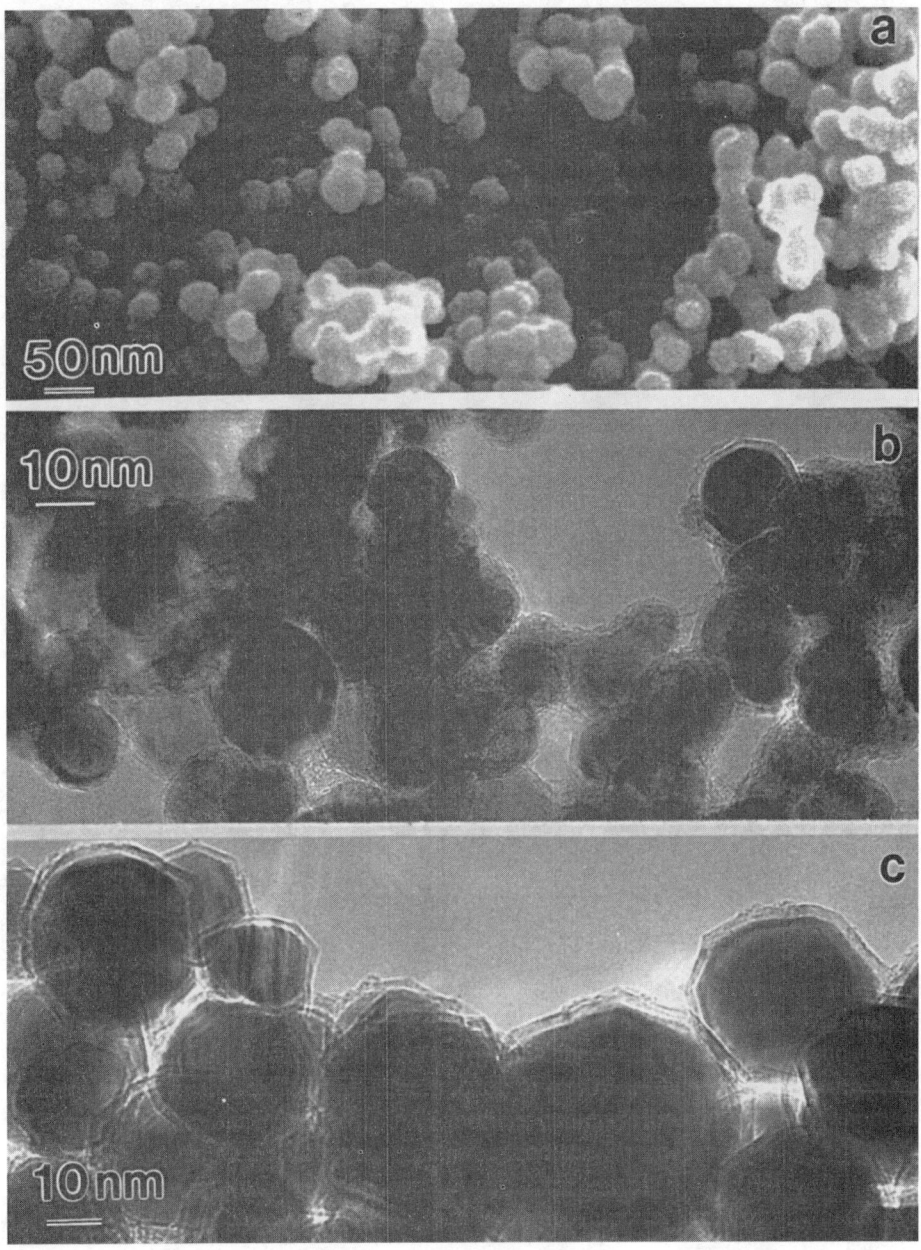

Fig. 1 (a) A scanning electron image of as-made Ni nanoparticles without any nanotubes.
(b) A TEM image of as-made Ni particles with a thin carbon coating layer.
(c) A TEM image of Ni particles after annealing at 900 C with thicker coating layers.

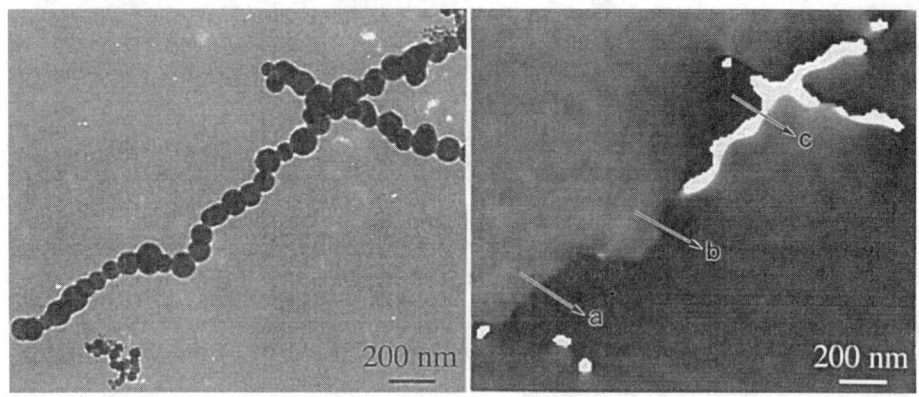

Fig.2 A chain of spherical nickel particles.　　Fig.3 Phase image of the Ni chain shown in Fig.2.
　　　　　　　　　　　　　　　　　　　　　　　　Arrows indicate the locations of line-scans.

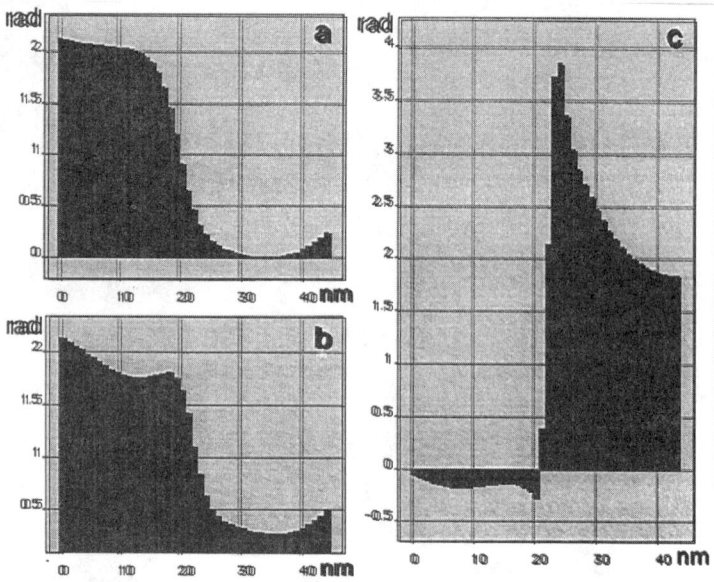

Fig.4　　Phase profiles of the three line-scans shown in Fig. 3. Note the
　　　　 sign inversion in c compared to a, b indicating a reversal of magnetization.

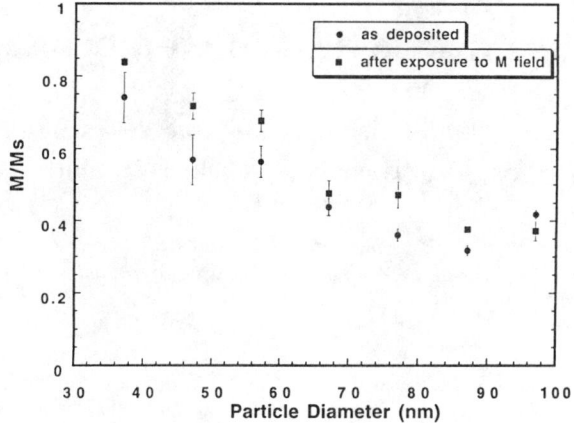

Fig.5 Relative remanent magnetization, M/M_S, of Ni particles as a function of average particle diameter. The external magnetic field (upper trace) increases the magnetization of the particles by 10% over that of the as-deposited particles.

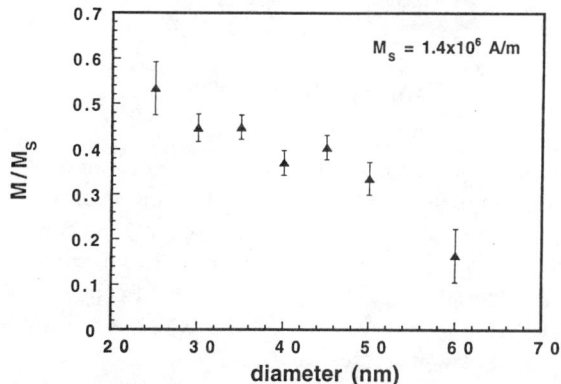

Fig.6 Relative remanent magnetization, M/M_S, of Co particles as a function of average particle diameter.

REFERENCES

1. J. Jiao and S. Seraphin, *J. Appl. Phys.* **80**, 103 (1996).
2. Z. Turgut, M. Huang, K. Gallagher, M.E. McHenry, and S.A. Majetich, *J. Appl. Phys.* **81**, 4039 (1997).
3. T. Hayashi, S. Hirono, M. Tomita, and S. Umemura, *Nature* **381**, 772 (1996).
4. J.J. Host, J.A. Block, K. Parvin, V.P. Dravid, J.L. Alpers, T. Sezen, and R. LaDuca, *J. Appl. Phys.* **83**, 793 (1998).
5. W. Wernsdorfer, E. B. Orozco, K. Hasselbach, A. Benoit, B. Barbara, N. Demoncy, A. Loiseau, H. Pascard, and D. Mailly, *Phys. Rev. Let.* **78**, 1791 (1997).
6. C. Beeli, B. Doudin, J-Ph. Ansermet, P.A. Stadelmann, *Ultramicroscopy* **67**, 143 (1997).
7. R.E. Dunin-Borkowski, M.R. McCartney, B. Kardynal, and D. J. Smith, *J. Appl. Phys.*, in press (1998).
8. J. Jiao and S. Seraphin, *J. Appl. Phys.* **83**, 2442 (1998).

Fullerene Incorporated Nanocomposite Resist Systems for Practical Nano-fabrication

Tetsuyoshi Ishii, Tomohiro Shibata,
Hiroshi Nozawa, and Toshiaki Tamamura

*NTT Opto-electronics Laboratories,
3-1, Morinosato Wakamiya, Atsugi, Kanagawa, Japan*

Abstract. We propose a nanocomposite resist system that incorporates sub-nm fullerene C_{60} or C_{70} into a conventional resist material as a practical nanometer range resist system. Fullerene molecules show chemical and physical resistant characteristics, and their incorporation reinforces the original resist film, leading to substantial improvements in resist performance: etching resistance, pattern contrast, mechanical strength, and thermal resistance. A nanocomposite resist system based on an positive-type electron beam resist, ZEP520, and its application to the fabrication of X-ray masks, diffractive grating elements, nano-printing molds, and quantum dots are presented.

INTRODUCTION

With the advances in device miniaturization and integration, the fabrication of nanometer patterns with dimensions smaller than 0.1 μm is increasingly in demand not only for single-device study but also for ULSI research and development. However, it has been difficult to fabricate such ultrasmall patterns due to the poor resistant qualities of conventional resist materials, regardless of the means of exposure. First of all, there are the problems of contrast degradation and defects, which are illustrated in Fig. 1(a). Because resolution generally increases with decreasing resist film thickness, an extremely thin film is used for ultrasmall pattern fabrication (1). Such film is likely to result in contrast degradation or rounding at the upper corner region of the patterns during development and in the formation of defects during etching due to its poor etching resistance. Another problem is pattern collapse [Fig. 1(b)]. For substrate etching with a thicker resist film, the aspect ratio of a pattern, defined as the ratio of pattern height and pattern width, becomes higher, and this often results in pattern

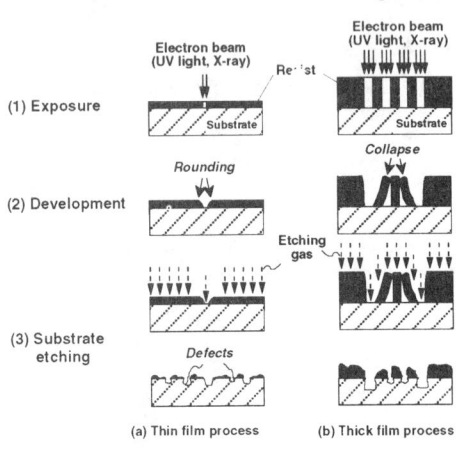

FIGURE 1. Problems in conventional resist processes.

collapse due to the poor mechanical strength of such thin and high patterns. Because pattern collapse practically determines resolution, it has become a serious concern in actual device fabrication. As a way to overcome these problems, we have proposed a novel approach called the nanocomposite resist system, which incorporates fullerene C_{60} or C_{70} into a conventional resist material to enhance resist performance (2), (3). In this paper, we first present the concept of the nanocomposite resist system, and then describe enhanced resist quality in a system composed of a C_{60}/C_{70} mixture and a positive-type electron-beam (e-beam) resist, ZEP520. We also discuss future developments and applications of this system.

CONCEPT

Figure 2 shows the basic idea of the nanocomposite resist system. A film of conventional resist material spin-coated on a substrate appears to be a closely packed film, but from a microscopic viewpoint, such a thin organic polymer film has spaces not occupied by polymer molecules and is porous in nature compared to metal or other inorganic materials [Fig. 2(a)]. Developer or etching species easily pass through the pores and induce pattern contrast degradation or etching defects. In addition, the porous organic film generally has poor mechanical strength. By filling the unoccupied spaces with highly etching resistant ultrafine particles, the intrusion of both the developer and etching species is blocked and pattern contrast and etching resistance are thereby enhanced [Fig. 2(b)]. It is also expected that close packing with such ultrafine particles could increase the rigidness or strength of the film by increasing its density. Since fullerene molecules are highly etching resistant (5) and ultrafine as shown in Fig. 2, we used C_{60} or C_{70}. These molecules have several advantages for incorporation: They readily dissolve in some aromatic solvents, and are chemically and physically stable.

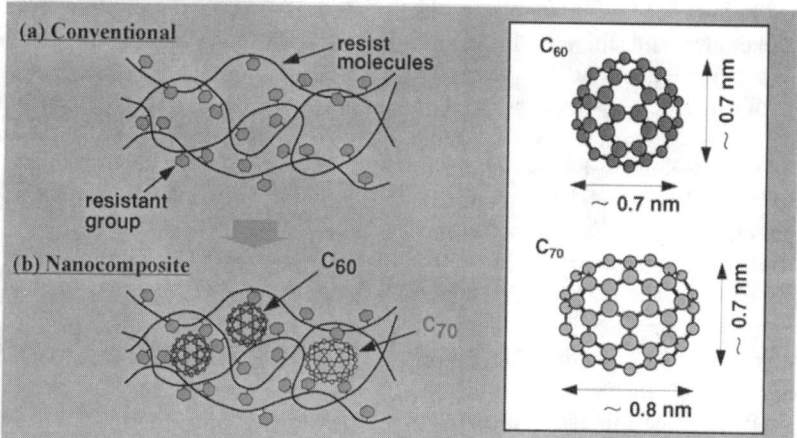

FIGURE 2. A schematic drawing of the nanocomposite resist system. C60 and C70 molecules fill up the spaces among the matrix resist molecules to form a closely packed and reinforced resist film.

RESIST CHARACTERISTICS AND APPLICATIONS

Commercially available fullerene C_{60}, C_{70}, and a C_{60}/C_{70} mixture (ratio: ~ 4/1), which is available at a much lower price than C_{60} or C_{70}, were incorporated into ZEP520 resist (C_{60}, C_{70}, and C_{60+70}@ZEP) by dissolving them with o-dichlorobenzene. The detailed preparation process is described in a previous report (4). Patterning experiments were done with an e-beam machine (25 kV), and exposed samples were processed following the conditions for pure ZEP.

The enhancement of etching resistance was evaluated using reactive ion etching (RIE) with C_2F_6 gas. As shown in Fig. 3, the etching rate of all the fullerene-incorporated ZEPs decreases in the same manner with increasing fullerene content. Under the present development conditions, we could not obtained good-quality patterns at higher contents due to heavy residue caused by the strong dissolution inhibiting effect of the fullerenes. The best-quality patterns were obtained in the range from 5 to 10 wt%, which corresponds to an etching resistant enhancement of 10 to 15 %. While the strong dissolution inhibiting effect degrades pattern quality, a moderate inhibiting effect improves pattern contrast. We observed a contrast enhancement effect at 5 wt%, as shown in Fig. 4. Fullerene is insoluble in the ZEP developer and thereby limits the dissolution of the resist at the upper corner of a pattern, which is weakly exposed by e-beam and is removed in the pure ZEP. Mechanical resistance or strength was also enhanced by fullerene incorporation. Figure 5 shows 90-nm-pitch and resist patterns of (a) pure ZEP and (b) 10 wt% C_{60}@ZEP formed in a 250-nm-thick film, which is relatively thick for nanometer pattern fabrication. During the development process, the pure ZEP patterns collapse due to their poor mechanical strength. On the other hand, the 10 wt% C_{60}@ZEP patterns

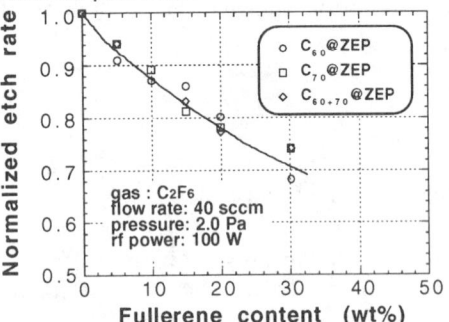

FIGURE 3. Enhancement of dry-etching resistance.

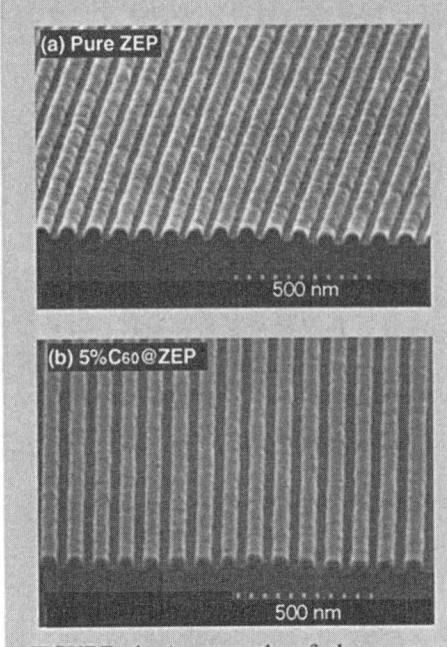

FIGURE 4. An example of the contrast enhancement of patterns. The pure ZEP patterns show rounded upper corners, while the 5 wt% patterns show steep profiles.

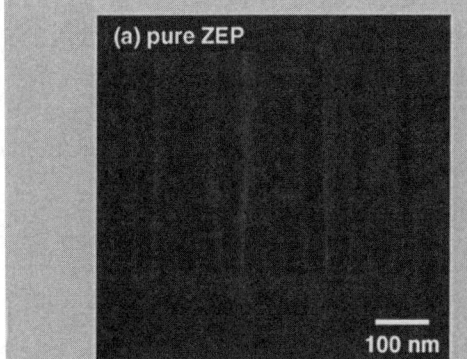 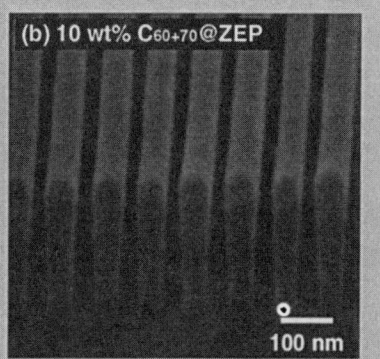

FIGURE 5. Enhancement in mechanical strength is evaluated by forming high aspect-ratio patterns: 90 nm pitch and 250 nm height. Patterns are collapsed in (a) pure ZEP, while no collapse is observed in (b) 10 wt% C_{60+70}@ZEP.

exhibit an extremely high aspect ratio of about 5.5 with no pattern collapse. We also observed an enhancement of thermal resistance in the nanocomposite systems. Patterns of 10 wt% showed practically no adverse swelling after heat treatment at 120 ℃ and only slight deformation even at 150 ℃, whereas those of the pure ZEP swelled at 120 ℃ and completely flowed at 150 ℃.

We have applied the new resist system to various device fabrications and obtained good results. Figure 6 shows a cross-sectional view of 50-nm Ta absorber patterns of a fabricated mask for x-ray lithography. The Ta thickness is 400 nm, so the aspect ratio is as high as 8. The mask was patterned with the 10 wt% C_{60+70}@ZEP. The mask composition and fabrication processes are described in detail elsewhere (6). This successful result can be applied to the fabrication of a Fresnel zone plate for x-ray microscopy. Grating patterns for semiconductor lasers were also fabricated by using the 10 wt% C_{60+70}@ZEP (Fig. 7). 80-nm-pitch patterns were etched into a 140-nm-thick SiN film through a 200-nm-thick resist by C_2F_6 RIE. The SiN patterns would be further transferred into a InP substrate in an actual device process. For deep etching of the substrate, the 10 wt% system has a great advantage over pure ZEP, since it provides

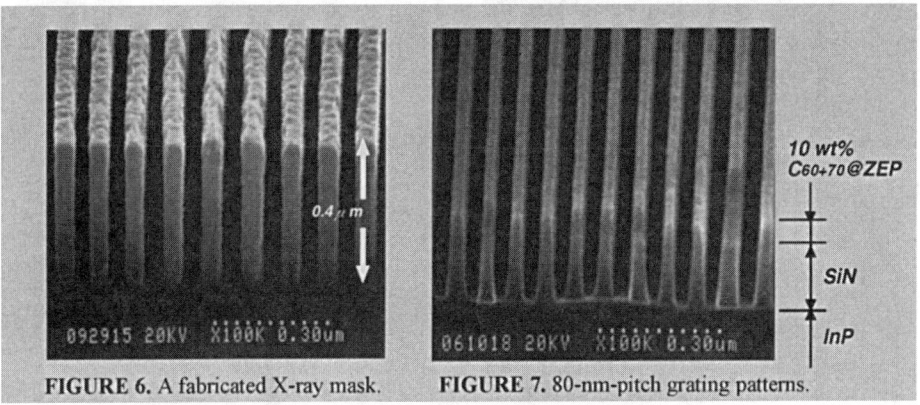

FIGURE 6. A fabricated X-ray mask.   **FIGURE 7.** 80-nm-pitch grating patterns.

high-aspect-ratio patterns strong enough to etch a thicker SiN film.

An array of dots with dimensions smaller than 100 nm is of great interest in the fabrication of quantum box lasers and nanocompact disks. Figure 8 shows an array of ~ 80 diameter and 125 nm period Ti dots formed by a lift-off technique using a single layer of the 10 wt% system. The enhanced thermal resistance of the system minimizes damage caused by increased temperatures during metal deposition by electron beam. We are currently working on the fabrication of quantum box lasers and nanocompact discs using this technique (7), (8). Further improvements in resolution will be possible by introducing a bilayer resist system composed of a fullerene-incorporated ZEP top layer and a pure ZEP bottom layer. In this new system, an inversely tapered pattern favorable for lift-off is formed by changing the fullerene content to optimize the sensitivity of the top layer.

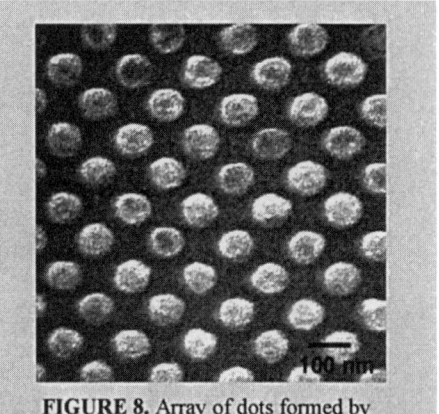

FIGURE 8. Array of dots formed by lift-off using the 10 wt% C_{60+70}@ZEP.

FUTURE DEVELOPMENTS

C_{60} and C_{70} are not soluble in common or nontoxic resist solvents like ethyl lactate. This limits the extensive application of nanocomposite resist from practical and environmental points of view. However, some fullerene derivatives have been and will be synthesized to enhance solubility in common organic solvents. Such derivatives can be readily incorporated into conventional resist materials at much higher contents than C_{60} or C_{70}. The higher degree of incorporation will make the quality of future nanocomposite systems much greater than that of the present ZEP system. In addition, the solubility enhancement has the potential to increase the sensitivity of nanocomposite resist by making it possible to combine it with a chemical amplification resist, in which sensitivity is greatly enhanced by the catalytic chain reaction of acids produced by high energy irradiation.

REFERENCES

1. Chen, W., and Ahmed, H., Appl. Phys. Lett. **62** (13), 1499-1501 (1993).
2. Ishii, T., Nozawa, H., and Tamamura, T., Appl. Phys. Lett. **70** (9), 1110-1112 (1997).
3. Ishii, T., Nozawa, H., Tamamura, T., and Ozawa, A., J. Vac. Sci. Technol. **B15** (6), 2570-2574 (1997).
4. Shibata, T., Ishii, T., Nozawa, H., and Tamamura, T., Jpn. J. Appl. Phys. **36** (12B), 7642-7645 (1997).
5. Tada, T., and Kanayama, T., Jpn. J. Appl. Phys. **35** (1A), L63-L65 (1996).
6. Ozawa, A., Ohoki, S., Oda, M., and Yoshihara, H., IEICE Trans. Electron. **ED77-C**, 255-262 (1994).
7. Kuramochi, E., Temmyo, J., and Tamamura, T., Appl. Phys. Lett. **71** (12), 1655-1657 (1997).
8. Pang, S. W., Tamamura, T., Nakao, M., Ozawa, A., and Masuda, H., "Direct Nano-Printing on Al Substrate using SiC Mold," in *Digest of Microprocesses and Nanotechnology'97*, 1997, pp. 14-15.

Cluster structure and elastic properties of superhard and ultrahard fullerites

Vladimir D. Blank*, Sergei G. Buga*, Nadejda R. Serebryanaya*, Gennadii A. Dubitsky*. Vaycheslav M. Prokhorov*, Michail Yu. Popov*, Natalia A. Lvova*, Vadim M. Levin† and Sergei N. Sulyanov‡

*Research Center for Superhard Materials,
7a Centralnaya St., Troitsk, Moscow Region, 142092, Russia
†Center for Acoustic Microscopy, Russian Academy of Sciences,
Moscow, 117334, Russia
‡Shubnikov Institute of Crystallography, Russian Academy of Sciences, Leninski Prospect 59,
Moscow, 117333, Russia

Abstract. Velocities of the longitudinal and shear sound waves are measured in ultrahard fullerites created by static high-pressure-high-temperature treatment under P=13 GPa and T= 1670-1870 K. The highest value of 26.0 km/s for the longitudinal waves is measured, that is about 40% more than in diamond. Bulk modulus of different ultrahard fullerites covers the range of about 600 -1700 GPa. The highest hardness is about 30×10^3 kg/mm^2. We ascribe these unique properties to formation of 20-30 atoms clusters by the walls of adjacent molecules under process of 3D-cross-linking. Most distinctly these clusters declare themselves in the cubic structure with the lattice parameter about 6 Å and 32 atoms per unit cell.

INTRODUCTION

Recent studies of the structure of high-pressure-high-temperature treated C_{60} fullerite have shown that beyond the stability of C_{60} molecule the materials created under pressure below about 8 GPa have a graphite-like features while under pressure above about 20 GPa - a diamond-like [1-3]. After the heat treatment in the pressure range 8-20 GPa the very interesting quenched fullerite states were created. Different 3D-polymerized crystal structures were obtained as well as the new disordered carbon states [3-5]. Physical properties of these metastable states are very different to the properties of other known carbon materials. Hardness of these new materials cover a wide range $(3 \div 30) \times 10^3$ kg/mm^2 [3,4,6,7]. The materials with the hardness higher than that of diamond were called "ultrahard" [4]. The very first practical application of such materials as ultrahard nanoindenters for correct measurements of the hardness of different faces of natural diamonds at room temperature is described in [6]. The elastic properties and structure of ultrahard fullerites are of particular interest. In this study we have measured the velocities of sound in the samples of ultrahard fullerites, evaluate elastic constants and discuss correlation between bulk modulus, hardness and structure of these materials.

SAMPLES AND ACOUSTIC MEASUREMENTS

Samples of pure pristine C_{60} (99.98%, C_{70}: 0.01%) were treated under pressure of 13 GPa and temperatures in the range 1500 - 1870 K for 1 min in a specially designed high-pressure apparatus [4]. After the thermobaric treatment the specimens were embedded into a steel cartridge clips; front and back faces were polished for acoustic microscopy study. The thickness of the samples with plane-parallel reflecting surfaces was 1.2 - 2 mm and their diameter was about 2 mm.

The wide-field pulse scanning acoustic microscope WFPAM was applied in a reflection mode with the small-aperture ultrasonic beam (half-aperture angle of lens is 11°) at a driving frequency f = 50 MHz [8]. Ultrashort probing ultrasonic pulses of 30 ns were used for measurements. They propagated into a specimen mainly as a beam of the longitudinal wave. A mean diameter of the acoustic spot on the specimen face was about 100 μm. The time interval t_l between the signals reflected at the front and back faces of a sample determines the magnitude of the longitudinal sound velocity C_L: $C_L = 2d / t_l$, where d is a specimen thickness. Due to the mode conversion at the faces of the samples a part of the incident beam propagates into a specimen as the transverse waves. In the particular values of the distances between an acoustic lens and the sample surface the echo-signal was formed by the beam which penetrates in the specimen as the pure longitudinal wave but after reflection at the backface propagates back as the transverse wave or vice versa. The transverse wave velocity magnitude was derived from measurements of the time intervals t_l for longitudinal wave and t_{tl} for mixed mode of propagation: $C_T = d /(t_{tl} - t_l/2)$.

Table 1. Density ρ, velocities of the longitudinal C_L and shear C_T sound waves, bulk elastic modulus B, shear modulus G, Young's modulus E and the Poisson's ratio σ for the fullerite samples synthesized under pressure P and temperature T. Data for synthetic polycrystalline diamond (our) as well as known data for monocrystal diamond, graphite and pristine C_{60} are presented for comparison.

N	P, GPa / T, K	ρ, g/cm³	C_L, km/s	C_T, km/s	B, GPa	G, GPa	E, GPa	σ
1.	12.5 / 1000	3.1	17.0	9.4	540	280	660	0.28
2.	13 / 1670	3.1	17.0	7.2	690	160	450	0.39
3.	13 / 1770	3.3	18.4	8.7	790	250	680	0.36
4.	13 / 1870	3.15	26.0	9.7	1700	300	850	0.42
	Polycrystal. synth. diam	3.74	16.0	9.6	490	350	850	0.22
	Monocryst. diamond	3.51	17.5-18.6	11.6-12.8	445	354-535	884-1144	0.1
	Graphite [9]	2.27	4.0÷21.6	0.3÷14.0	290			
	Prist. C_{60} [10]	1.68	3.0÷3.4	1.6÷2.0	10.8	4.85	12.6	0.31

Ultrasonic scanning of the samples have shown their acoustic inhomogeniouty. The areas with a minimum values of t_l were studied for a quantitative measurements. Elastic moduli were calculated on the basis of the measured velocities and densities of the samples. The results of measurements and calculations for ultrahard samples synthesized at pressure of 13 GPa and different temperatures are represented in the Table 1. We have measured acoustic properties of the synthetic polycrystalline "carbonado"-type diamond in the same manner. These data as well as the data for superhard fullerite sample obtained at P=12.5 GPa, T=1000 K and known data for monocrystal diamond, graphite and pristine C_{60} are presented for comparison. Values of velocities are averaged over 3-4 different points for each sample with the deviation of ± 0.1 km/s. Density of the samples was measured by flotation method using a mixtures of diiodmethane and acetone liquids of different concentrations.

As seen from the Table 1, the velocity of the compressional sound wave in the samples obtained at T = 1770 and 1870 K is higher than that in diamond and the bulk elastic modulus B of all studied samples is essentially higher than that for diamond and graphite. On the other hand, the velocity of the shear wave in the fullerites is less than in monocrystal diamond, thus their shear modulus G is less than that for diamond. The Poisson's ratio σ of the studied fullerites is much more than that of diamond.

It is interesting to note, that the values of $B = 540 \div 690$ GPa for the samples No.'s 1 and 2 obtained at relatively lower pressure and temperature correlate with the estimation of about 630 GPa calculated in [11]. We suppose that under these synthesis conditions the C_{60} cages in the nanostructure still prevail. A new sharp growth of B upwards 1 TPa was not predicted, but this phenomena may be also explained, at least partially, on the basis of theoretical approach [11] using the model of cross-linked fullerene-like carbon clusters for the structure of ultrahard fullerites.

STRUCTURE OF THE MATERIALS

Structure of the studied ultrahard materials highly disordered. X-ray study and Raman scattering [4,5] revealed only broad bands similar to those of amorphous diamond-like films with high ratio of sp^3- to sp^2-sites. However, very high hardness and particular elastic properties direct on the drastic difference in the structures of ultrahard fullerites as compare to known diamond-like materials. High bulk modulus, almost twice more than that of diamond was calculated for C_{60} molecule [11], but cross-linking of the molecules with sp^3-hybridization was not considered at that time. The experimental data show that 3D-polymerization of C_{60} in fcc and bcc lattices as well as in disordered C_{60} solid actually takes place. In this case formation of 20-30 atoms clusters by the walls of adjacent molecules happens [3]. Thus we suppose that the structure of 3D-polymerized C_{60} may be considered as a random 3D network of C_{60} and C_{20-30} clusters with conjugated interatomic bonds. Elastic properties of the material with such structure depend on the ratio of different clusters and type of interatomic bonds. Fig. 1 shows the x-ray diffraction patterns of the fullerite samples obtained in a narrow temperature-pressure range 1770-1830K; 12-13 GPa. Filtered Cu K_α radiation was used; the range of the scattering angle in Fig. 1 is up to $50°$ (about 2.0

Å of d-spacing). Top curve shows formation of the diamond-like peak in fullerite at very high temperature 2100 K and P=13 GPa. The value of pressure 12 GPa is not sufficient to preserve C_{60} structure at T=1770 K and the cross-linked layered carbon structure [3,5] forms at this conditions (bottom curve). Under P=13 GPa, T=1830 K only two distinct broad diffraction bands were observed, corresponding to 2.16 and 1.2 Å,. The diffraction pattern of the sample synthesized at medium value of pressure 12.5 GPa and T=1830 K contains more bands. Their maximums correspond to 3.0, 2.18, 1.8, 1.2 Å of d-spacings. We consider such diffraction pattern as a very disordered cubic structure with the lattice parameter of about 6 Å and 32 atoms per unit cell. This structure is built on the basis of the clusters corresponding to tetrahedral and octahedral emptiness in the 3D-polymerized fcc structure.

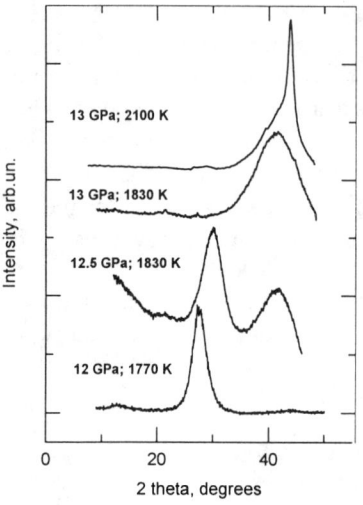

Fig. 1. X-ray diffraction patterns of the fullerite samples obtained at different P-T-conditions pointed out in the plot.

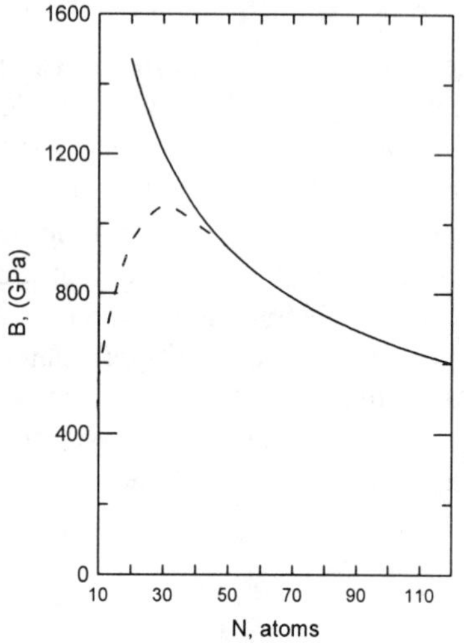

Fig. 2. Bulk modulus B of spherical carbon cluster versus number of atoms N according to relation used in [11] (solid line). Dashed line is discussed in the text.

Fig 2. displays the dependence of bulk modulus on the number of atoms N in a fullerene (or fullerene-like cluster). Solid line is plotted using the relation $B=2h/3(S_{11}+S_{12})$ from [11]. Here R- radius of a fullerene and we substitute it with $\sqrt{fN}$ for spherical shape (f-is a coefficient that we consider to be constant in the first approach). S_{ij} are elastic complaints of graphite. The value of $(S_{11}+S_{12})$ practically equal for graphite and diamond, thus we suppose in the first approach that sp^3-hybridization do not affect on bulk modulus of the molecules at cross-linking . Dashed curve $B(N)$ is drown up taking into account known value of B for diamond as a cluster of 10 atoms and our experimentally determined values of B for ultrahard fullerites assuming that they include clusters of about 20 to 60 carbon atoms.

MEASUREMENTS OF HARDNESS

The hardness of different fullerite samples including ultrahard was measured by the "NanoScan" (NS) measurement system based upon the principle of the atomic force microscope [6,7]. It was specially designed to carry out hardness tests by sclerometric (plowing) method. Ultrahard fullerite tips were used as a microindenter.

The results of measurements are presented in Fig. 3.

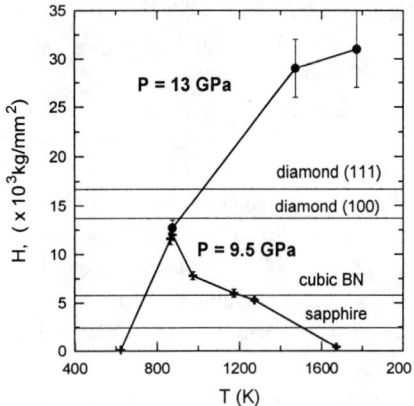

Fig. 3. Hardness H of the fullerite samples treated at pressures of 9.5 and 13 GPa as a function of temperature of synthesis T.

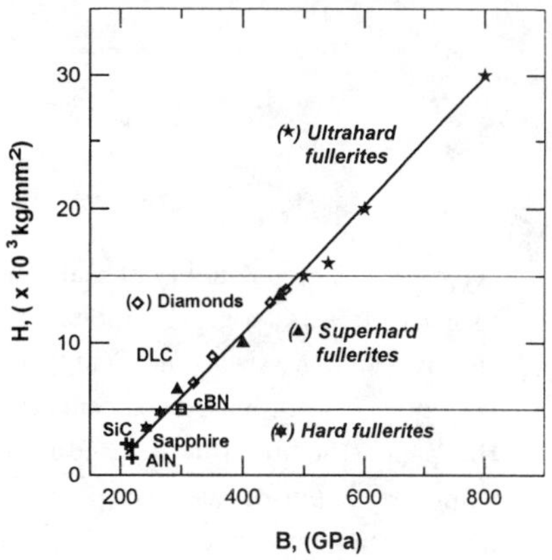

Fig. 4. Hardness H of different hard, superhard and ultrahard materials versus bulk modulus B.

The hardness of different faces of natural diamond was measured by the same method and it is pointed out in the Fig. 3 for comparison. Fig. 4 shows that the hardness of ultrahard fullerites with bulk modulus up to 790 GPa fits the same linear dependence on B as it is known for other hard and superhard materials [12]. Hardness of hard and superhard fullerites accords to this empirical rule too.

REFERENCES

1. Marques, L., Hodeau, J.-L., Nunez-Regueiro, M., Peroux, M., *Phys. Rev. B*, **54**, 12633-12636 (1996).
2. Hirai, H., Kondo, K., Yoshizawa, N., and Shiraishi, M., Appl. Phys. Lett., **64**, 1797-1799 (1994).
3. Blank, V., Buga, S., Dubitsky, G., Serebryanaya, N., Popov, M., and Sundqvist, B., *Carbon* (in press)
4. Blank, V.D., Buga, S.G., Serebryanaya, N.R., et al., *Phys. Lett. A*, **205**, 208-216 (1995).
5. Blank, V.D., Buga, S.G., Serebryanaya, N.R., et al., *Phys. Lett. A*, **220**, 149-157 (1996).
6. Blank, V., Popov, M., Lvova, N., Gogolinsky, K., Reshetov, V.,. *J. Mater. Res.*, **12**, 3109-14 (1997).
7. Blank, V.D., et al. *Diamond and Related Mat.* **7**, 427-431 (1998).
8. Blank, V., Levin ,V., Prokhorov, V., Buga, S., Dubitsky, G., Serebryanaya, N., *JETP* (1998), in press.
9. Blackslee, O., *J.Appl. Phys.*, **41** (8), 3373-3377 (1970).
10. Kobelev, N.P., Nikolaev, R.K., Soifer,Ya.M. et al., *Chem. Phys. Lett.*, **276**, 263-267 (1997).
11. Ruoff, R.S., and Ruoff, A.L, *Nature*, **350,** 663-664 (1991).
12. Prelas, M., Popovici, G., Bigelov, L.K., *Handbook of Industrial Diamonds and Diamond Films,* New York, Marcel Dekker, Inc, 1998, ch.3, p. 82

Electronic Properties of Nanotube Junctions

Ph. Lambin and V. Meunier

FUNDP, 61 Rue de Bruxelles, B 5000 Namur, Belgium

Abstract. The possibility of realizing junctions between two different nanotubes has recently attracted a great interest, even though much remains to be done for putting this idea in concrete form. Pentagon–heptagon pair defects in the otherwise perfect graphitic network make such connections possible, with virtually infinite varieties. In this paper, the literature devoted to nanotube junctions is briefly reviewed. A special emphasize is put on the electronic properties of C nanotube junctions, together with an indication on how their current-voltage characteristics may look like.

INTRODUCTION

The atomic lattices of two different single-wall nanotubes can be joined while preserving the threefold coordination of all the sites. Such a connection demands a few topological defects, a single pair of pentagon and heptagon in the simplest case [1]. The most promising junctions would be these ones connecting semiconducting to metallic nanotubes [2]. In this respect, hetero-junctions between C and $B_xC_yN_z$ nanotubes look promising [3] because the band gap of a BCN nanotube is only weakly dependent on its radius. A challenging issue would be to realize Schottky barriers in a controlled way which, through their rectifying properties, could act as fast switching nanodevices [4]. Although there is no direct experimental evidences that such junctions could work, there has already been a report of rectification of the current flowing across metal-semiconductor interfaces along a rope of single-wall carbon nanotubes [5]. The present paper provides a short survey of theoretical data obtained for junctions between C nanotubes, with a special emphasis put on their electronic properties. How the density of states varies locally in the junction region is illustrated on a few typical examples. The implications of these results on the current-voltage curves of the nanotube junctions are discussed.

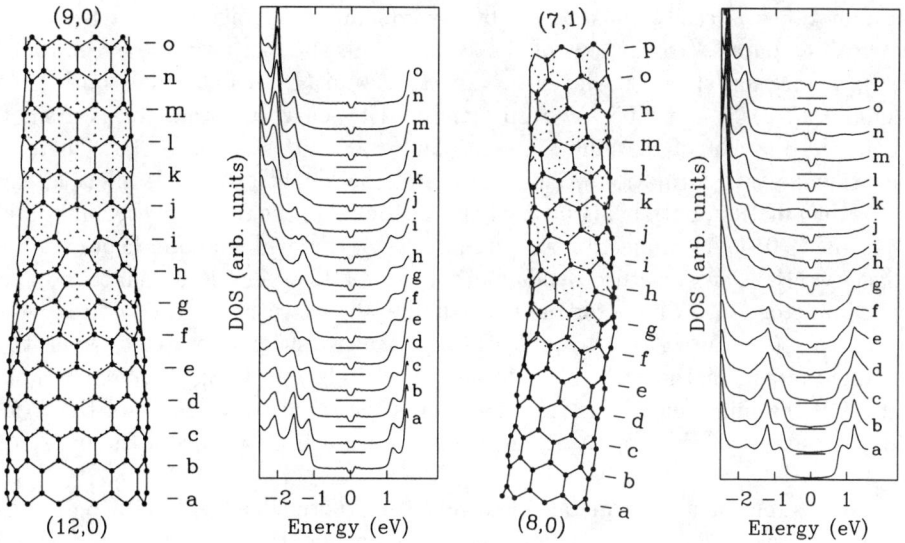

Figure 1(a) (left): (9,0)-(12,0) connection [9]. The curves a, b ... are local DOS averaged over the atomic sites around the sections indicated by the corresponding labels in the atomic structure. The heptagon and the pentagon are located at the sections f and i, respectively. The Fermi level is at zero energy, the horizontal bars indicate zero DOS.

Figure 1(b) (right): Same as in fig. 1(a) for a (8,0)-(7-1) junction [11]. There is a pair of adjacent pentagon-heptagon between the sections g and h.

LOCAL DENSITIES OF STATES

The density of states (DOS) of a carbon graphitic network around the Fermi level E_F is dominated by the π electron states, which are well described by a tight-binding Hamiltonian. In small diameter nanotubes, the curvature of the lattice has a significant effect on the π DOS [6]. The lattice curvature can be taken into account by setting the hopping interactions between first-neighbor π orbitals as

$$\langle \pi_i | H | \pi_j \rangle = V_{pp\pi} \hat{\pi}_i . \hat{\pi}_j - (V_{pp\sigma} - V_{pp\pi}) \cos \theta_i \cos \theta_j$$

with $V_{pp\sigma} \approx -V_{pp\pi} = \gamma_0$ (= 3 eV in this paper). In that equation, θ_i is the angle between the orientation $\hat{\pi}_i$ of the π-orbital at site i, normal to the atomic network, and the direction of the CC pair ij. In tight-binding, the DOS can easily be calculated on any atomic site of interest by the recursion technique. This method was used for computing the data presented below. A structural optimization of all the junctions considered in this paper was performed as described in ref [7].

Conical sections between straight nanotubes have often been observed in multi-wall systems [8]. An atomic model of such a conical junction between two zigzag nanotubes is easily constructed by introducing a pentagon and a heptagon along a line parallel to the axis of the structure [9, 10]. An example is the (9,0)-(12,0) connection shown in fig. 1(a). The (9,0) and (12,0) nanotubes both have a small gap (0.08 and 0.05 eV, resp.) induced by curvature that forms a small dip in the plateau of metallic states around $E_F = 0$ (see fig. 1(a)). There is a continuous transformation of the width and the height of the metallic plateau across the junction, both being inversely proportional to the radius of the nanotube.

In the (8,0)-(7,1) connection shown in fig. 1(b), the pentagon and heptagon are adjacent [11]. The structure bends at an angle of 12°. The (8,0) nanotube is a semiconductor (1.4 eV). As for the (7,1) tubule, the curvature of the lattice opens a small gap (0.1 eV). Fig. 1(b) indicates how the transition operates between the metallic plateau in the DOS of (7,1) and the band gap of (8,0). A remarkable feature of the junctions of figs. 1(a) and (b) is the rapid convergence of the DOS towards the one of the constituent nanotubes on moving away from the interface [11].

An example of a true metal-semiconductor junction is the (10,0)-(6,6) connection shown in fig. 2 [12]. Here the pentagon and the heptagon are located at diametrically-opposed positions, which bends the structure at an angle of 36°. Bends of that sort have been observed experimentally [13]. The DOS rapidly vanishes in the gap of the (10,0) semiconductor. In the upper part of the junction, the DOS does not converge monotonously towards the plateau typical of a metallic nanotube, but oscillates instead on both sides of the Fermi level. These oscillations are most likely due to interferences between incoming and reflected Bloch waves whose propagation is blocked by the semiconducting band gap. These interferences could form a $2k_F$-oscillation like standing pattern in front of the junction, and should be observable by STM. The corresponding wavelength $3a/2 = 0.37$ nm is characteristic of the armchair geometry.

CURRENT-VOLTAGE CHARACTERISTICS

A Bardeen perturbative formulation of the tunnelling conductance across a junction between two zig-zag nanotubes has been developed in ref. [9]. By simplifying further this approach, the tunnel current I in a biased junction can be written as

$$I = \frac{2e}{h} 2\pi N t^2 \int_{E_F - eV/2}^{E_F + eV/2} n_1(E + eV/2) n_2(E - eV/2) \, dE$$

where N is the number of bonds making the connection, with an average hopping interaction t, n_1 and n_2 are the DOS of the two nanotubes, assumed to have their Fermi levels at the same energy. In the absence of any charge transfer, the $I - V$ curve is symmetric with respect to zero bias. In a metal-semiconductor junction, $I = 0$ for $|V| < E_g/2$ where E_g is the band gap. There is no rectification unless the

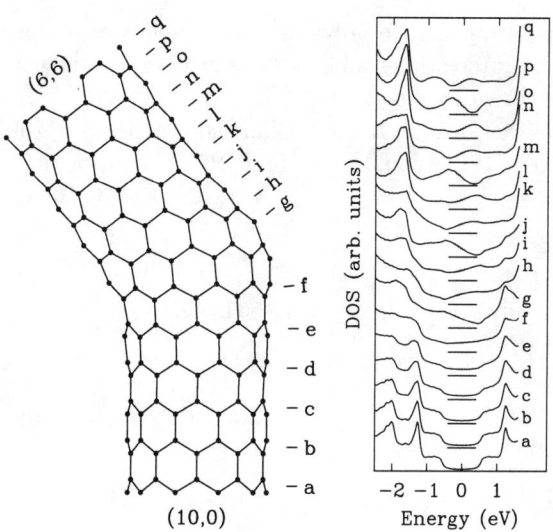

Figure 2: Same as in fig. 1 for a (10,0)-(6,6) junction [12]. The pentagon forms the protruding knee at location g, the heptagon is at the opposite border.

semiconducting part be doped. The differences between metal-semiconductor and metal-metal junctions gradually wash out by increasing the radii of the connected nanotubes. Indeed nanotubes with radii larger than $\sim$3 nm have their DOS already close to that of a graphite, which varies like $|E - E_F|$ near the Fermi level. The above convolution formula then gives I proportional to V^3 for all the connections. $I - V$ curves measured experimentally by scanning tunnelling spectroscopy follow this cubic law reasonably well, except at low bias where a parabolic law of the current was observed [5].

When the effects of curvature on the π-π interactions are neglected, all the nanotubes (L, M) with $L - M$ divisible by 3 are metallic. At this approximation, the conductance at zero bias of a junction between two of them, as given by the Landauer formula, becomes a purely topological property [14]. It is equal to $4e^2/h$ when the connected nanotubes are identical, and decreases rapidly by increasing the ratio R_1/R_2 of the radii. This is because a Fermi Bloch wave traveling on the largest nanotube cannot easily constrict to the smaller diameter of the other nanotube, and a large fraction of it is therefore reflected back. In certain circumstances, a junction between two metallic nanotubes can be *insulating* [15]. A gap of conductance takes place when the Fermi states on both sides of the junction belong to different representations of the rotational symmetry group of the connected system. This demands a conical connection based on several pairs of pentagon-heptagon that preserves the common rotational symmetry of the constituent nanotubes.

To summarize, it is hoped that this short survey of theoretical results on nanotube junctions was sufficient to prove their rich variety of properties. Whereas the exploration of these interesting properties has just started, new ideas are frequently proposed such as three-line junctions having the form of a T [16].

This work was performed under the auspices of the Interuniversity Research Project on Reduced-Dimensionality System (PAI P4-10) of the Belgian OSTC.

References

[1] Dunlap, B.I., Phys. Rev. B **49**, 5643-5649 (1994).

[2] Dresselhaus, M., Phys. World **9**, 18-19 (May 1996).

[3] Blase, X., Charlier, J.C., De Vita, A., and Car, R., Appl. Phys. Lett. **70**, 197-199 (1997).

[4] Service, R.F., Science **271**, 1232 (1996).

[5] Collins, P.G., Zettl, A., Bando, H., Thess, A., and Smalley, R.E., Science **278**, 100-103 (1997).

[6] Kane, C.L., and Mele, E.J., Phys. Rev. Lett. **78**, 1932-1935 (1997).

[7] Meunier, V., Henrard, L., and Lambin, Ph. . Phys. Rev. B **57**, 2591-2596 (1998).

[8] Iijima, S., Ichihashi, T., and Ando, Y., Nature **356**, 776-778 (1992).

[9] Saito, R., Dresselhaus, G., and Dresselhaus, M.S., Phys. Rev. B **53**, 2044-2050 (1996).

[10] Charlier, J.C., Ebbesen, T.W., and Lambin, Ph., Phys. Rev. B **53**, 11108-11113 (1996).

[11] Chico, L., Crespi, V.H., Benedict, L.X., Louie, S.G., and Cohen, M.L., Phys. Rev. Lett. **76**, 971-974 (1996).

[12] Lambin, Ph., Fonseca, A., Vigneron, J.P., B.Nagy, J., and Lucas, A.A., Chem. Phys. Lett. **245**, 85-89 (1995).

[13] Terrones, M., Hsu, W.K., Hare, J.P., Kroto, H.W., Terrones, H., and Walton, D.R.M., Phil. Trans. R. Soc. London A **354**, 2025-2054 (1996).

[14] Tamura, R., and Tsukada, M., Phys. Rev. B **55**, 4991-4998 (1997).

[15] Chico, L., Benedict, L.X., Louie, S.G., and Cohen, M.L., Phys. Rev. B **54**, 2600-2606 (1996).

[16] Menon, M., and Srivastava, D., Phys. Rev. Lett. **79**, 4453-4456 (1997).

Phthalocyanine-C_{60} Composites as Improved Photoreceptor Materials?

B.Kessler, C.Schlebusch, J.Morenzin, W.Eberhardt

Forschungszentrum Jülich, IFF, D-52425 Jülich, Germany

Abstract. With the methods of photoelectron spectroscopy (UPS), X-ray absorption near edge spectroscopy (XANES) and optical transmission spectroscopy we demonstrate that an electron transfer from light-induced excitonic states of different phthalocyanines (Pcs) towards unoccupied electronic states of C_{60} is energetically possible, thereby increasing the generation efficiency for free charge carriers. No indication for a ground state electron transfer is found for any of the Pcs in contact to C_{60}. The improvement of τ-H_2-Pc due to the admixture of C_{60} is directly shown by an enhanced photoconductivity.

INTRODUCTION

Organic photoconductor materials, especially metal phthalocyanines (M-Pcs) with M=TiO or VO and the metal-free H_2-Pc, are widely used in commercial photoreceptors which are the light sensitive parts in laser printers and xerographic devices (1). They are less toxic than anorganic components and their sensitivity reaches into the infrared region. Therefore infrared lasers can be used which are less expensive due to their universal application in mass technology products.

A crucial step for the performance of a photoreceptor is the generation of free charge carriers. The efficiency of the charge carrier generation process depends on the competitive recombination probability of the light induced excitonic state of the Pc. It can be enhanced when an acceptor is placed close to the Pc-molecules which is able to separate the electron and the hole and thereby reduces the recombination rate. C_{60} has already demonstrated its capability for this purpose with different organic materials (2-7). With our experimental analysis of the electronic states of several M-Pcs in contact to C_{60} we are able to determine whether such an electron transfer is energetically possible for the material combinations studied here. This process can only occur if the charge-separated state is energetically preferred compared to the intermediate excitonic state. An unwanted ground-state electron transfer between the M-Pc and C_{60} that might enhance the dark conductivity is unlikely due to the rather high chemical inertness of C_{60}.

EXPERIMENTAL

Two different kinds of samples are analyzed: we either use a dispersion directly from the production process of photoreceptor devices that contains a Pc with a polymeric binder (8) and an optional admixture of about 5% C_{60} ("technical" sample) which is cast onto a metal substrate and inserted into ultra-high vacuum after drying, or we prepare an *insitu* sublimed film of Pc with a thin overlayer of C_{60} ("model" sample). M-Pcs with M=TiO, VO, Cu, Ni, Fe and two different phases of metal-free H_2-Pc are analyzed.

For the analysis of the unoccupied density of states we use X-ray absorption near edge spectroscopy (XANES). Using tuneable synchrotron radiation from the HE-PGM 3 at BESSY (Berlin) a C-1s core electron is excited into the unoccupied states. Measuring the secondary electron yield as a function of the photon energy thereby gives the unoccupied density of states via the absorption cross section.

The occupied density of states is analyzed by photoelectron spectroscopy with ultraviolet radiation (UPS) using a He-resonance light source (hv=21.2eV) and an hemispherical electron analyzer.

Information on the excitation energy of the Pc is obtained by optical transmission spectroscopy in the energy region of hv=1eV to 4eV.

Measurements of the photoconductivity and the dark conductivity directly confirm the improvement of the Pc by an admixture of C_{60}. For these measurements a sandwich structure is prepared on a glass substrate that consists of the sample material between a transparent indium-tin oxyde (ITO) electrode and an evaporated Au electrode. Monochromatic light with a wavelength of 610nm is used for a measurement of the photocurrent as a function of an applied voltage between -0.1V and 0.1V.

RESULTS AND DISCUSSION

Using X-ray absorption near edge spectroscopy (XANES) we study the unoccupied density of states of different M-Pcs (M=TiO, VO, Cu, Ni, Fe and H_2-Pc) and M-Pcs with low coverages of C_{60} (0.5ML to 2ML, not shown). The spectral features of the overlayer samples are composed of structures from the particular M-Pc substrate and from the C_{60}. By subtracting the spectra of the pure M-Pc from the overlayer spectra a measured spectrum of pure C_{60} can be very well reproduced. Since no significant peak broadening or extra features occur we can conclude that none of these Pcs undergoes a chemical reaction with C_{60}. Therefore we do not expect a dramatic enhancement of the dark conductivity of these Pcs when they are doped with C_{60}.

Fig.1 displays our results for optical transmission spectra from two different samples of H_2-Pc, namely a "technical" and a "model" sample. Their spectra show slightly differing onset energies for the optical excitation that corresponds to the singlet excitonic state of the Pc. The singlet excitonic state can be populated with energies larger than 1.5eV and 1.6eV for the dispersion and the sublimed sample, respectively. We attribute this difference to different phases in which the organic molecules are stacked in order to build molecular crystallites in the solid (9). Such phases usually differ in the stacking angle and the stacking distance between the molecular subunits. Whereas the molecular crystals grow in a mixture of the thermodynamically stable α- and β-phases during the sublimation process, the dispersion-sample crystallites are in the metastable τ-phase. As a result of these spectra we see that singlet excitonic states of H_2-Pc have a minimum energy of 1.5eV.

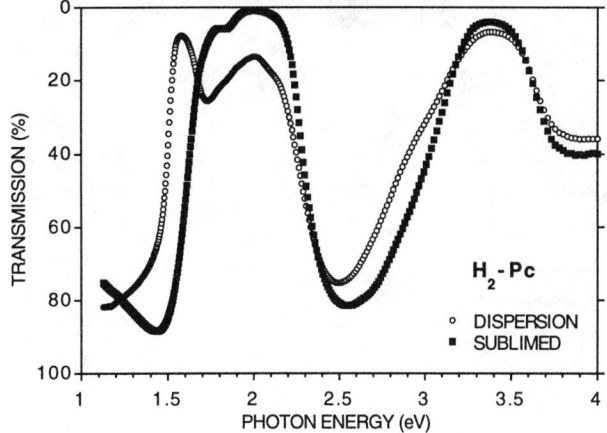

FIGURE 1. Optical transmission data from a technical (open dots) and a model sample (solid squares) of H_2-Pc.

As a next step we analyze the occupied density of states by photoelectron spectroscopy with ultraviolet radiation (UPS). As an example we discuss the results for a sublimed layer of H_2-Pc with a coverage of 2ML C_{60}. This UPS-spectrum can be decomposed into the single components of α,β-H_2-Pc and of C_{60} (see Fig.2). From this we determine the energies of the top of the valence band (VB) of α,β-H_2-Pc and of the highest occupied molecular orbital (HOMO) of C_{60}. Using the published value for the gap of C_{60} of 2.3eV (10) we obtain the position of the lowest unoccupied molecular orbital (LUMO) of C_{60}, and all 3 levels can be drafted in a model of the electronic states in the right part of Fig.2. From this model it is obvious at least 1.2eV are needed in order to transfer an electron from the VB of α,β-H_2-Pc into the LUMO of C_{60}. We therefore can conclude that the charge separated state is 1.2eV above the ground state, whereas the energy of the singlet excitonic level is at least 1.5eV as obtained from the transmission spectra. That makes a charge separation of the excited intermediate state via an electron transfer into the LUMO of C_{60} energetically possible. Since most of the M-Pcs have singlet excitonic levels at about 1.5eV to 1.8eV (1, 11) and our UPS-results show transfer energies $\leq$1.2eV for the other M-Pcs (not shown) we can conclude that for all the M-Pcs studied here an improvement of the charge-generation efficiency is expected when they are doped with C_{60}.

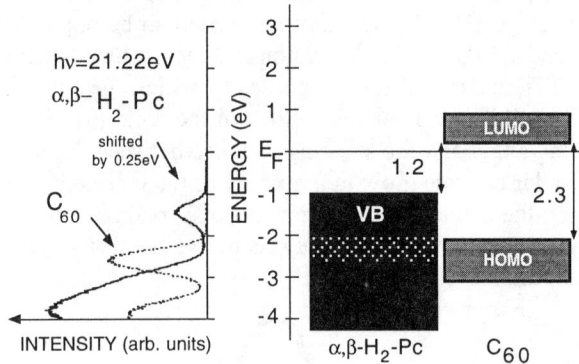

FIGURE 2.
left: Position of UPS-features of α,β-H_2-Pc and C_{60} in contact to each other (see text).

right: Model of the electronic levels of α,β-H_2-Pc and C_{60} in contact to each other; C_{60} gap from (10).

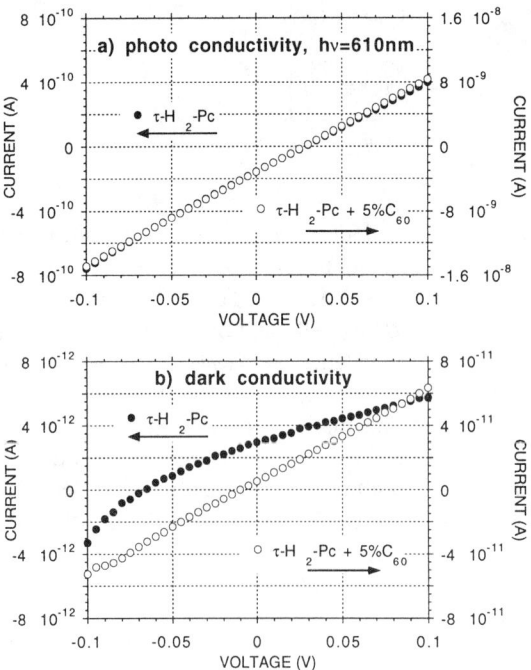

FIGURE 3. Photoconductivity (top) and dark conductivity (bottom) of technical samples of τ-H_2-Pc and binder with (open symbols, right scale) and without C_{60} (filled symbols, left scale).

Fig.3 displays the results of the photoconductivity (top panel) and the dark conductivity (bottom panel) for an undoped technical sample of τ-H_2-Pc (left scales) and a C_{60}-doped sample of this material (right scales). The spectra are displayed in such a way that the measured data fit into the same window for undoped and doped material. Therefore a comparison of the scaling factor at the top panel directly shows that the doping improves the photoconductivity by a factor of 20. The contrast between the photoconductivity and the dark conductivity of the undoped sample (left scales: top range divided by bottom range) corresponds to about 100. For the C_{60}-doped sample (right scales: top range divided by bottom range) the contrast improves to 200.

SUMMARY

Our results demonstrate that photoreceptors containing M-Pcs may show an enhanced charge-generation efficiency when they are doped with C_{60}.

ACKNOWLEDGEMENTS

We like to acknowledge the financial support by the BMBF-VDI under contract number 13N6906. B.K. acknowledges the financial support by the Ministerium für Wissenschaft und Forschung des Landes Nordrhein-Westfalen.

REFERENCES

1. K-Y.Law, *Chem.Rev.* **93** (1993) 449.
2. N. Sariciftci, A. J. Heeger, *Int. J. Mod. Phys.* **B8** (1994) 237.
3. A.Watanabe, O.Ito, *J.Chem.Soc., Chem.Commun.* **11** (1994) 1285.
4. K.Yoshino, T.Akashi, K.Yoshimoto, S.Morita, R.Sugimoto, A.A.Zakhidov, *Solid State Commun.* **90** (1994) 41.
5. C.Schlebusch, B.Kessler, S.Cramm, W.Eberhardt, *Synthetic Metals* **77** (1996) 151.
6. B.Kessler, C.Schlebusch, S.Cramm, and W.Eberhardt, "Organic Photoconductors and C_{60}", in *Physics and Chemistry of Fullerenes and Derivatives*, editors: H.Kuzmany, J.Fink, M.Mehring, S.Roth, World Scientific, Singapore (1995), p.283.
7. C.Schlebusch, J.Morenzin, B.Kessler, S.Cramm, and W.Eberhardt, "Unoccupied Density of States of Commercially used Organic Photoreceptors doped with C_{60}", in *Molecular Nanostructures*, editors: H.Kuzmany, J.Fink, M.Mehring, S.Roth, World Scientific, Singapore (1998), p.511.
8. M.Biermann, M.Lutz, AEG Elektrofotografie, D-59581 Warstein-Belecke, Germany.
9. A.Kakuta, Y.Mori, S.Takano, M. Saweda, I.Shibuya, *J. Imaging Technol.* **11** (1985) 7.
10. R.W.Lof, M.A.van Venendaal, B.Koopmans, H.T.Jonkman, G.A.Sawatzky, *Phys.Rev.Lett.* **68** (1992) 3924.
11. B.H.Schlechtman, W.E.Spicer, *J.Mol.Spectry.* **33** (1970) 28.

New Metallic Alloys Incorporating Fullerenes and Carbon Nanotubes

R. J. Doome, A. Fonseca and J. B. Nagy

Laboratoire de Résonance Magnétique Nucléaire,
Facultés Universitaires Notre-Dame de la Paix,
61, Rue de Bruxeeles, B-5000, Namur, Belgium

Abstract. In order to open new routes to fullerenes application, we have investigated the effect of fullerenes and nanotubes in metallic alloys. Fullerenes mixture and carbon nanotubes have been used as new carbon sources in the synthesis of metallic alloys (Al, Fe and Ni). After melting under inert atmosphere, macroscopic homogeneous alloys were obtained with iron and nickel but the aluminium based alloys looked rather inhomogeneous due to an incomplete melting. From the samples analysis by chemical reactions and XPS, it was concluded that the carbon is essentially located on the alloy surface as carbide and sp^2 structures. Except for the aluminium based alloy where some fullerenes were still detected, thermal treatment as well as metal catalytic effect led to the decomposition of the fullerenes in the alloys. Nevertheless, carbon nanotubes kept their structure and were trapped in the alloys. The hardness of these new alloys were determined and compared to values of common alloys incorporating graphite and norit-A as carbon sources. The preliminary results showed slightly higher hardness values for alloys incorporating fullerenes and weaker values for alloys incorporating carbon nanotubes.

INTRODUCTION

Since the publication in 1992 of a new way to use fullerene's structure to reinforce steel [1], we have developed new metallic alloys incorporating fullerenes and carbon nanotubes. New alloys were prepared by melting processes and samples obtained were analysed by different methods to determine whether the fullerene's structure was conserved or not. The reason to use fullerene in the synthesis of metallic alloys holds in the hope that fullerenes and especially carbon nanotubes could build a kind of carbon backbone in the metallic matrix leading to better mechanical properties.To evaluate the mechanical properties, we have measured the hardness of the samples, but other mechanical data were not available since the amount of sample was limited (about 50 g). All these results were compared with results obtained from metallic alloys made of graphite or Norit-A as carbon sources. The aims of our experiments were both to get new high tech material and to study the stability of fullerenic compounds under severe conditions (molten metal).

EXPERIMENTAL

Multi-walls carbon nanotubes were produced by catalytic decomposition of acetylene over cobalt supported on zeolite NaY as described in the literature [2, 3]. The support was removed by dissolution in hydrofluoric acid. Fullerenes soot was produced by mild oxidation in a premixed benzene/oxygen/argon flame at low pressure (70 torrs). This soot contains about 5% of fullerenes, mainly C_{60} and C_{70} in our experimental conditions. The fullerenes were separated from soot by Soxhlet extraction using toluene as solvent and analysed by HPLC. The latter showed the presence of C_{60} and C_{70} in a ratio 80/20, less than 1% of higher fullerenes and some PAHs which have been removed by petroleum ether washing. Powder mixtures have been prepared as described in table 1 and then, melt in an inductive oven (model PV 8910) working under argon atmosphere. This kind of oven allows the complete melting of the samples in less than one minute.

Table 1: Composition of initial powder mixtures before the melting process.

Sample name	Chemical composition	Weight of metal (g)	Weight of carbon (g)	Weight % of carbon	Atomic % of carbon
Fe1	Fe + fullerenes	39.5	0.8173	2.03	8.78
Fe2	Fe + nanotubes	39.4	0.4598	1.15	5.15
Fe3	Fe + norit-A	39.5	0.8417	2.09	9.02
Fe4	Fe + graphite	39.4	0.8223	2.04	8.85
Ni1	Ni + fullerenes	44.5	0.8825	1.94	8.84
Ni2	Ni + nanotubes	44.5	0.3961	0.88	4.17
Ni3	Ni + norit-A	44.7	0.8638	1.90	8.63
Ni4	Ni + graphite	44.7	0.8804	1.93	8.78
Al1	Al + fullerenes	13.5	0.2873	2.08	4.56
Al2	Al + nanotubes	13.6	0.3104	2.23	4.88
Al3	Al + norit-A	13.6	0.2755	1.99	4.35
Al4	Al + graphite	13.5	0.2961	2.15	4.70

Homogeneous melting were obtained for iron and nickel based alloys, but it was quite impossible to melt the aluminium by inductive heating because aluminium is diamagnetic and the very small granulometry could prevent the field lines propagation. So, we have used a resistive oven to melt the aluminium based alloys. This kind of oven works under vacuum, the temperature was set at 1200°C and the cooling was achieved by a nitrogen flow. However, the melting was incomplete and the samples looked inhomogeneous.

XPS measurements have been collected on a SSX-100 spectrometer calibrated with the value corresponding to the full width at half maximum of Au $4f_{7/2}$ signal. TEM observations have been performed on a Philips EM 301 with a lateral resolution of 3 Å.

RESULTS AND DISCUSSION
Chemical Analysis

The general principle of the caracterization by chemical analysis consists in taking a stuck of the alloy (about 3 g) and to dissolve the metal in aqua regia. Then, we analyse the residu by HPLC using a commercial Buckyprep column for fullerenes and by TEM for carbon nanotubes. HPLC results do not show the presence of fullerenes in the iron or nickel based alloys. Thus, fullerenes have been decomposed during the melting process. The thermal decomposition was enhanced by the catalytic effet of iron or nickel as it is the case in Cepek's experiments [4]. For the aluminium based alloy, we recovered 1.6 % of the initial fullerenes amount. This could be in agreement with a possible resistance of fullerenes in melting aluminium. The 98.6 % of missing fullerenes was decomposed into graphitic form or carbide, or has been sublimated during the one hour melting process. TEM observations of the alloys incorporating nanotubes show that the structure of the nanotubes is partially conserved. Only some nanotubes have been eaten away from the tips or imperfect sections (connections between two tubes, unwell graphitic parts,...). We also observed helicoidal tubes after the metal dissolution and this could confirm the perfect and continuous turbostratic structure for this kind of nanotubes. The reaction time of the melting was found to be a very important parameter: The shorter it is (inductive heating) the more nanotubes do survive. The catalytic destruction of fullerenes compounds seems to be stronger for the case of nickel than for iron.

XPS Characterization

The iron and nickel based alloys were analysed by X-ray photoelectron spectroscopy to collect more information about the chemical environment of both metal and carbon. Thus, the XPS experiments allowed us to determine the bond energy of the different carbon species at the surface of the sample. Before each measurement, the metal was cleaned by ionic sputtering for 20 minutes (± 1 mm) . The XPS results are listed in table 2. The carbon was observed to be localised at the surface of the alloys except for the "Fe + fullerenes" alloy which looks more homogeneous in its mass. Only traces of carbide were observed in the iron based alloy incorporating nanotubes. This is in complete agreement with the conservation of the sp^2 structure of the nanotubes. The percentage of carbide found in the alloy incorporating nanotubes could come from the amorphous carbon produced together with the nanotubes.

Hardness Determination

The hardness of the iron based alloys has been determined by the depth of the diamond print at the surface of the polished samples. The results obtained are presented in table 3.

Table 2: Summary of XPS results.

Metallic alloy composition	Atomic % of carbon			
	in initial mixture	at surface of alloy	in sp^2 form	in carbide form
Fe + fullerenes	8.78	12.27	34.8	65.2
Fe + nanotubes	5.15	49.96	82.2	17.8
Fe + norit-A	9.02	27.99	35.7	64.3
Fe + graphite	8.85	17.01	39.9	60.1
Ni + fullerenes	8.84	29.18	64.5	35.5
Ni + nanotubes	4.17	46.69	48.8	51.2
Ni + norit-A	8.63	60.87	58.8	41.2
Ni + graphite	8.78	/	/	/

Table 3: Hardness of the iron based alloys and depth of the diamond print on the surface of the alloy.

Type of alloy	(im)print d (mm)	Hardness (Vickers)
Fe + fullerenes	0.089	469
Fe + nanotubes	0.113	290
Fe + norit-A	0.091	450
Fe + graphite	0.091	450

We observe that the alloy incorporating fullerenes is slightly harder than the reference alloys (Fe+graphite, Fe+ norit-A). This could be explained by a better distribution of the carbon in the alloy as the XPS results suggest. The alloy incorporating nanotubes is softer than the references. This is likely due to the fact that the samples were polished and since the nanotubes are essentially located at the surface, the polishing have removed nearly all the carbon leading to a very poor carbon alloy.

CONCLUSIONS

We have used fullerenes and carbon nanotubes as carbon sources to synthesise new metallic alloys. XPS measurements and chemical analysis show that fullerenes decompose into graphitic form or carbide, but the structures of the nanotubes are partially conserved. As far as iron or nickel alloys are concerned, the hardness of the new alloys do not show any advantages compared with conventional carbon sources.

REFERENCES

1. R. Job; High Tech Materials Alert (1992), 9 (2)
2. V. Ivanov, J. B.Nagy, Ph. Lambin, A. A. Lucas, XB Zang, D. Bernaerts, G. Van Tendeloo, S. Amelinckx and J. Van Landuyt; Chem. Phys. Lett (1994), 223, 329
3. K. Hernadi, A. Fonseca, J. B.Nagy, D. Bernaerts, A. Fudala and A. A. Lucas; Zeolites (1996), 17, 416
4. C. Cepek, A. Goldoni and S. Modesti; Phys. Rev. B-Condensed Matter (1996), 53, Iss 11, 7466

Realization of Large Area Flexible Fullerene - Conjugated Polymer Photocells: A Route to Plastic Solar Cells

C. J. Brabec[1], F. Padinger[1], V. Dyakonov[1], J. C. Hummelen[2], R. A. J. Janssen[3], and N. S. Sariciftci[1]

[1]*Christian Doppler Laboratory for Plastic Solar Cells, Physical Chemistry Linz, J. Kepler University Linz, A-4040 (Austria)*
[2]*University of Groningen, 9747 AG Groningen (The Netherlands)*
[3]*Laboratory of Organic Chemistry, TU Eindhoven, 5600 Einhoven (The Netherlands)*

Abstract. Various interesting photophysical phenomena in composites of fullerenes and non degenerate ground state conjugated polymers with highly extended π-electrons in their main chain can be explained by the ultrafast electron transfer from the conjugated polymer (donor) to the fullerene (acceptor) upon illumination. This photoeffect was utilized for the production of small area (mm²) photovoltaic devices which show energy conversion efficiencies $\eta_e > 1\%$ and carrier collection efficiencies $\eta_c > 20\%$. In this work we present efficiency and stability studies on large area (6 cm by 6 cm) flexible solar cells based on a soluble alkoxy PPV (3,7 - dimethyloctyloxy methyloxy poly(phenylenevinylene)) and a highly soluble fullerene derivative, 1-(3-methoxycarbonyl)propyl-1phenyl [5,6]C_{61} (PCBM). The enhanced solubility of PCBM compared to C_{60} allows a high fullerene - conjugated polymer ratio and strongly supports the formation of donor - acceptor bulk heterojunctions.

INTRODUCTION

The utilization of organic materials for photovoltaic devices has been investigated intensely during the last couple of decades (for a summary of the early reports see for example (1,2,3)). Because of the ultrafast photoinduced electron transfer (4) with long lived charge separation, the conjugated polymer/C_{60} system offers the special opportunity to produce thin film photovoltaic devices from solution. The photoinduced charge separation happens with quantum efficiency near unity. The performance (5) of such bulk heterojunction devices is remarkably enhanced compared to devices made from the single components. Conjugated polymers seem to fulfill all properties acquired for photovoltaic energy conversion: strong light absorption and the possibility of charge separation in presence of a strong electron acceptor like fullerenes.

RESULTS AND DISCUSSION

The photovoltaic devices have been produced by spin casting from solution, yielding a typical film thickness around 100 - 200 nm. For the high work function electrode, transparent ITO substrates, either on glass or on polyester, have been used. The low work function electrode, Al, was evaporated onto the spin cast film.

Fig. 1 shows the intensity of the photoluminescence (dots, right axis) as a function of the fullerene concentration in alkoxy PPV/PCBM composites. Luminescence quenching already happens at very low fullerene concentrations (below 1 mol% fullerenes). Percolation of fullerenes to a connected path (around 17 mol% for small molecules) is not necessary for luminescence quenching. Even at low fullerene concentrations the very effective electron transfer takes place. To obtain an efficient photovoltaic response it is further necessary to collect these photogenerated charges. The squares in Fig. 1 (left axis) show the short circuit current I_{sc} as a function of the fullerene concentration.

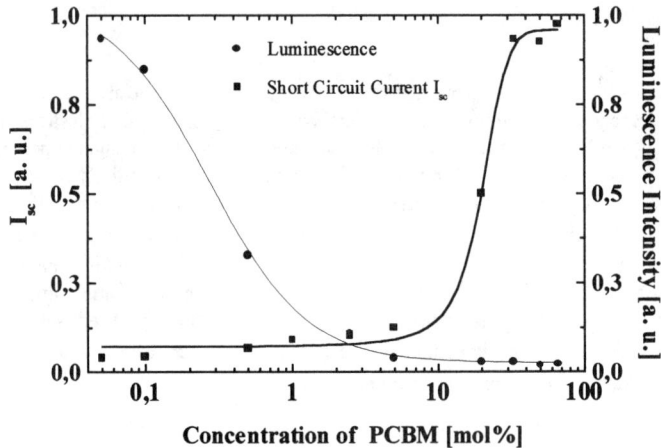

FIGURE 1. Luminescence quenching (full dots, right axis) and short circuit current I_{sc} (full squares, left axis) vs. molar fullerene concentration in the composite.

The lines are fits to sigmoidal logistic distributions. One can see clearly, that charge carrier transport occurs at higher fullerene concentrations than luminescence quenching. At 20 mol% of fullerenes I_{sc} already reaches 50 % of its saturation value, while at concentrations below 15 % I_{sc} is app. 10 % of the saturation value. At 33 mol% saturation is reached and no enhancement of the current is found by further addition of fullerenes. Microscope studies on these compounds show, that at higher fullerene concentrations (>33 mol%) PCBM tends to aggregate and/or phase separate from the conjugated polymer, leading to inhomogeneous films. The quality and homogeneity of the composite film strongly influences the efficiency of the solar cell. Inhomogeneous films with pin holes and large serial resistivities lead to small fill factors (FF), lower rectification and decreased open circuit voltages (V_{oc}). A further

parable effects are found for fullerenes. In Figure 3b the V_{oc} of oxygen protected
ic solar cells is monitored. One can see, that the shelf life of the sealed devices is
: than 150 days.

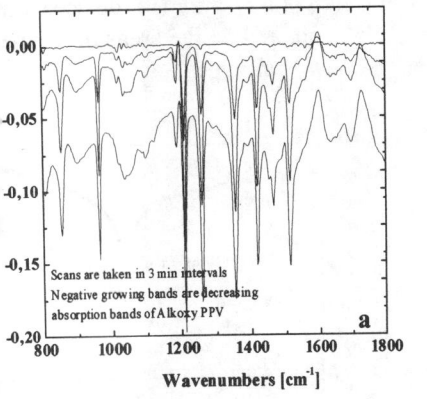 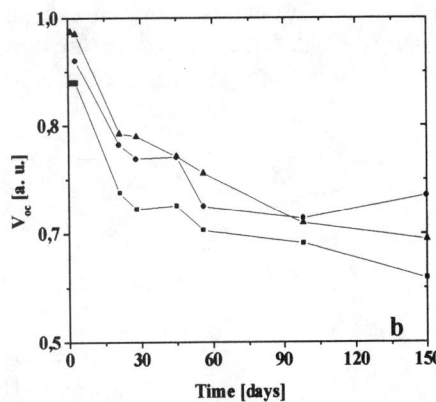

URE 3. (a) In situ FTIR degradation study in ATR geometry. Difference spectra are gained by
ding the reference spectrum (dark, in Argon) through spectrum taken under illumination
$_nW/cm^2$) and oxygen. Scans are taken in 3 minutes intervals. (b) V_{oc} vs. shelf life time [days] of
ected plastic solar cells.

CONCLUSION

In plastic solar cells fullerenes act in a double role – as highly efficient e^- acceptors
well as e^- conductors. The power efficiency of plastic solar cells (> 1.2 %) is limited
charge transport. Plastic solar cells can be upscaled to large flexible substrates
hout losing efficiency. Oxygen protection of plastic solar cells increases the shelf
time over 150 days. Further improvements in device efficiencies are expected by
timizing the composite composition, the network morphology and the charge
nsport properties of the single components.

REFERENCES

. Simon and J. J. Andre, Molecular Semiconductors, Springer, Berlin, (1985).
J. B. Whitlock, P. Panayotatos, G. D. Sharma, M. D. Cox, R. R. Sauers, and G. R. Bird, Optical
gineering, 32, 1921 (1993).
C. W. Tang, Appl. Phys. Lett., 48, 183 (1986).
N. S. Sariciftci, L. Smilowitz, A. J. Heeger, and F. Wudl, Science, 258, 1474 (1992).
G. Yu, J. Gao, J. C. Hummelen, F. Wudl, and A. J. Heeger, Science, 270, 1789. (1995).

important parameter, determining the quality of the thin film is ther(
the substrate. We studied the I/V characteristics of the fullerene/c(
thin films on two different substrates and in two different geometr
ITO glass substrates with active areas around 15 mm², and (ii
polyester substrates with typical areas of 6 cm by 6 cm and active ar
mm². Figure 2 shows a comparison between these two cell types.

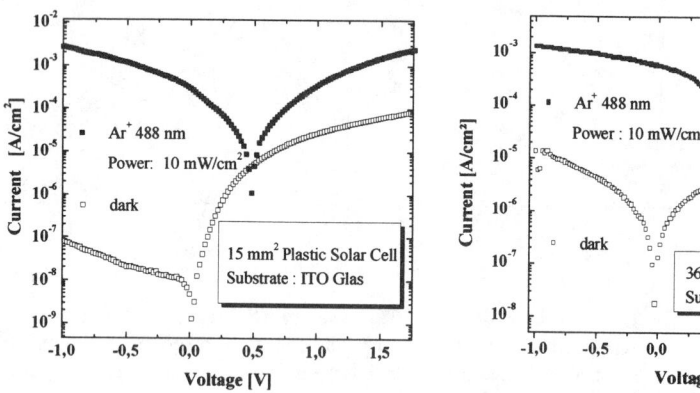

FIGURE 2. (a) I/V characteristics of small area glass cell (□ dark - ■ illumina
mW/cm²) and (b) large area polyester cell (□ dark - ■ illuminated with 488 nm,

Although dark negative currents are higher for the flexible substr;
show comparable V_{oc} and I_{sc}. For both cells we calculated a filling
0,35. The overall efficiency of the cells is calculated with ap
monochromatic illumination (488 nm) with 10 mW/cm². These data
upscaling to flexible substrates does not influence the efficiency
electron/phonon efficiency η_c of the conjugated polymer/fulle
estimated to be higher than 20% for thin films (5). Obviously
inefficient charge transport processes, probably due to the hopping c
fullerenes.

A very important parameter for industrial applications is the lifetir
Conjugated polymers are well known for their sensibility to phot
oxygen containing environments. We studied the degradation beha
components of the plastic solar cell by in situ FTIR measurements i
(attenuated total reflection). Fig. 3a shows the difference spectra for
Negative growing bands are vanishing absorption bands, positive g!
interpreted as enhanced absorption features. Illumination with lig
atmosphere bleaches the polymer, seen by the vanishing absorpt
between 800 and 1500 cm⁻¹. After degradation the polymer film is
mutual growing of absorption bands around 1700 cm⁻¹ is interpreted
of carbonyl bands, which is in accordance with the bleaching of

Fullerenes and nanostructured plastic solar cells

Joop Knol and Jan C. Hummelen[Ψ]

Stratingh Institute, Materials Science Centre,
University of Groningen, Nijenborgh 4, 9747 AG Groningen, The Netherlands

Abstract. We report on the present on the present status of the plastic solar cell and on the design of fullerene derivatives and π-conjugated donor molecules that can function as acceptor-donor pairs and (supra-) molecular building blocks in organized, nanostructured interpenetrating networks, forming a bulk-heterojunction with increased charge carrier mobilities. Finally, we report on the first and basic steps towards the preparation of such molecular building blocks.

INTRODUCTION

The need to develop inexpensive renewable energy sources continues to stimulate new approaches to the production of efficient, low-cost photovoltaic (PV) devices. The recent discovery of photoinduced electron transfer[1] in composites of conducting polymers (CP's) as donors and suitable fullerene (F) derivatives as acceptors provided a molecular approach to high-efficiency photovoltaic conversion. Because the time scale for photoinduced charge transfer is ultrafast as compared to charge recombination, the charge separated state can be described as being meta-stable (lifetime ~10^{-4}s) with a quantum efficiency close to unity (Scheme 1).

$$\underset{\text{(ground state)}}{CP + F} \xrightarrow{h\nu} \underset{\text{(excited state)}}{CP^{*} + F} \underset{\text{recombination}}{\overset{e^{-}\ \text{transfer}}{\rightleftharpoons}} \underset{\text{(charge separated state)}}{CP^{\cdot +} + F^{\cdot -}}$$

<u>*Scheme 1.*</u> *Photoinduced electron transfer in composites of conducting polymers and fullerenes.*

In contrast to single junction devices (e.g. *n*-Si/*p*-Si), PV devices consisting of blends of CP and suitable fullerene derivatives (CPC blends) behave like bulk-heterojunction materials with interpenetrating networks of donors and acceptors in which the whole blend is photoactive.

Recently, it has been shown that such "all plastic" (except for the electrodes) PV devices consisting of blends of alkoxy-poly(phenylene vinylene) (alkoxy-PPV) and

CP442, *Electronic Properties of Novel Materials:* XII International Winterschool
edited by H. Kuzmany et al.

the soluble fullerene derivative[2,3] 1-(3-methoxycarbonyl)propyl-1-phenyl [6,6]C_{61} (denoted as [6,6]-PCBM) can be made reproducibly with active areas of 4 cm^2 without loss of efficiency[4] (see Figure 1). Device manufacturing includes a single spincast procedure of the alkoxy-PPV/[6,6]-PCBM blend from solution upon a flexible substrate coated with ITO followed by deposition of the upper aluminum electrode.

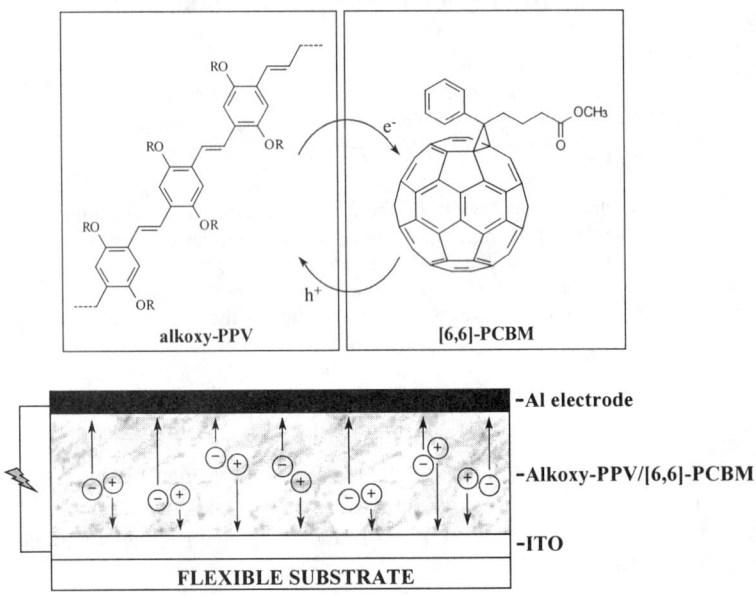

Figure 1.   Schematic representation of PV device.

Due to the presence of electrodes with different work functions an internal electric field is generated, along which the separated charges (i.e. hole and electron) move in opposite directions.

There are two major points of interest for further development of this type of PV cell: lifetime (stability) and charge carrier mobility. Since the bulk-heterojunction is a rather disordered assembly on the nanometer scale, it is expected that the overall charge carrier mobility within the PV cell is still low. With respect to the fullerene phase, optimum charge carrier mobility (i.e. electron transport) would be achieved if the independent fullerene units could be assembled in a more or less straight alignment between the two electrodes.

Towards assembly of novel functionalized fullerene derivatives

Through the rich chemistry of addition reactions to fullerenes[5], a variety of functional groups can be introduced that affect macroscopic properties (e.g. solubility).

We have studied the introduction of complementary functional groups ("lock" and "key") that can induce head-to-tail assembly of fullerene units on the molecular level. The connection of complementary functional groups to the fullerene core yields a fullerene derivative **1** that is expected to give a sponteneous head-to-tail assembly via multiple "lock"-"key" interactions (Scheme 2).

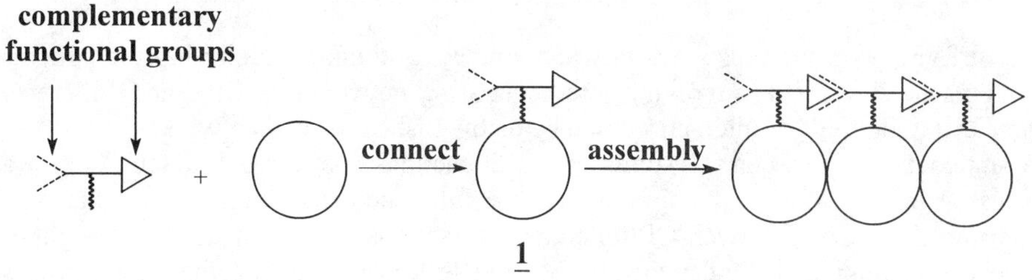

Scheme 2. Assembly of fullerenes through complementary functional groups.

For this reason we have prepared methanofullerenes **2a** and **2b** which are structurally related to [6,6]-PCBM (see Figure 2).[6] Both compounds have been characterized by several spectroscopic techniques (IR, UV/VIS, NMR). Especially **2b** shows remarkable solubility in different solvents such as carbon disulfide, toluene, tetrahydrofuran and *o*-dichlorobenzene which is an important factor during the manufacturing of PV devices. Both **2a** and **2b** meet the criterion of bearing complementary functional groups including the carboxylic- (acidic "key") and the *N,N*-dialkylamino- (basic "lock") functionalities. We have found that recrystallization of **2b** under carefully controlled conditions leads to the formation of polycrystalline bowl-shaped superstructures of ~0.4 mm (!) outer-diameter as identified by microscopy (see Figure 2).

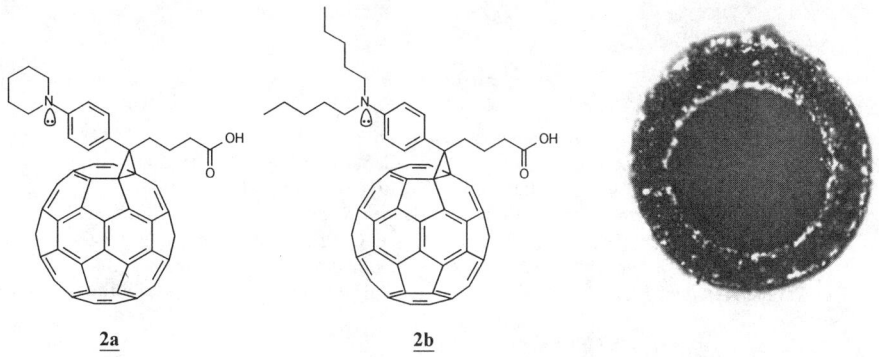

Figure 2. Structures of methanofullerenes **2a** and **2b** and top-view of bowl-shaped polycrystalline superstructures of **2b**.

We speculate that the actual assembly involves a poly hydrogen bonded array of fullerene units. A further study of the assembly features of both **2a** and **2b** in the solid state and in solution is needed to unambiguously answer this question.

Nanostructured interpenetrating CP/fullerene networks

There are numerous ways in which one could think of modifying the structure of a bulk-heterojunction involving interpenetrating networks of CP's and fullerenes. Three different types of interactions can be distinguished in such networks:
(a) Interactions between CP-units; (b) interactions between fullerene-units (as discussed in the previous paragraph); (c) CP/fullerene interactions within the bulk-heterojunction. Control over CP/fullerene interactions on the nano-level might be achieved by linking the independent CP- and fullerene-units together. This process could involve either linking of CP and fullerene through covalent bonds and/or linking through a spontaneous assembly process of functional groups (i.e. the principle of "lock" and "key"). Currently we are selecting and investigating suitable molecular building blocks materials for this particular purpose. Ultimately, the impact of the linking process on the architecture of the bulk-heterojunction and physical PV device parameters such as exciton transport, electron transfer efficiency, charge recombination, charge carrier mobilities etc. is to be verified.

Acknowledgments

Prof. N.S. Sariciftci, Dr. C. Brabec and Mr. F. Padinger of the Johannes Kepler University Linz (Austria) are acknowledged for the work on PV devices. Dr. R.A.J. Janssen and Ir. B. de Waal of the Technical University Eindhoven (The Netherlands) are acknowledged for the work on CP's. This work was financially supported by The Netherlands' Agency for Energy and the Environment (Novem).

References

Ψ   Present e-mail address: J.C.Hummelen@chem.rug.nl.
1. N.S. Sariciftci, L. Smilowitz, A.J. Heeger, F. Wudl, *Science* **258**, 1474 (1992).
2. J.C. Hummelen, B.W. Knight, F. LePeq, F. Wudl, *J. Org. Chem.* **60**, 532 (1995).
3. G. Yu, J. Gao, J.C. Hummelen, F. Wudl, A.J. Heeger, *Science* **270**, 1789 (1995).
4. For more technical details about this PV device, see: C. Brabec, F. Padinger, V. Dyakonov, J.C. Hummelen, R.A.J. Janssen, N.S. Sariciftci, *"Realization of large area flexible fullerene-conjugated polymer photocells: a route to plastic solar cells"*, contribution in these proceedings.
5. A. Hirsch, *Synthesis* 895 (1995).
6. Experimental details: J. Knol, J.C. Hummelen, to be published.

Semiconductor Device Structure Based on Fullerene: Ag/ C60 Thin Film Schottky Barrier

E.A. Katz[*], D. Faiman[*+], S. Shtutina[+], B.Mishori[§], and Yoram Shapira[§]

[*]*National Solar Energy Center, Ben-Gurion University of the Negev, Sede Boker Campus, 84990 ISRAEL, keugene@bgumail.bgu.ac.il*
[+]*Department of Physics, Ben-Gurion University of the Negev, Beersheba, 84105 ISRAEL*
[§]*Department of Physical Electronics, Tel-Aviv University, Ramat-Aviv, 69978 ISRAEL*

Abstract. A possibility to produce a Schottky barrier at the Ag/C_{60} thin film interface has been demonstrated. The device structure exhibited rectifying behavior in the dark and photovoltaic properties.

INTRODUCTION

Recently we demonstrated a possibility to produce a Schottky barrier between a C_{60} single crystal and silver paste [1]. We also reported on photovoltaic properties of C_{60} thin film - silicon heterojunction [2]. However, we didn't successfully produce any stable Schottky barrier with C_{60} thin films. Therefore, we decided to use an Ag layer as a substrate and to improve the C_{60} film crystallinity in order to produce a Schottky barrier between the C_{60} and Ag layers. This paper presents a study of the crystalline structure and Surface Photovoltage (SPV) spectra of such C_{60} thin films as well as the dark and light current-voltage (I-V) curves for a C_{60}/Ag interface.

EXPERIMENTAL DETAILS AND RESULTS

C_{60} thin films were evaporated on the glass substrates partially predeposited with an Ag layer (Fig. 1, a), using a vacuum deposition technique. The starting powder was commercially obtained C_{60} (99.98%).
The crystalline structure of the C_{60} films were determined by X-ray diffraction. C_{60} films and C_{60}/Ag interfaces were also studied by SPV spectroscopy [3]. Dark and light I-V measurements of the C_{60}/Ag interfaces were performed with an HP 4140 pA Meter / DC voltage source. Light measurements were performed using a solar simulator (Solar Constant 575).

Using special conditions for C_{60} evaporation we have obtained films with homogeneous thickness for both the Ag and clear glass parts of the substrate (Fig. 1, b). However the structure of the C_{60} film grown onto the Ag part of the substrate differs principally from that deposited onto the glass part of the same substrate. Fig. 2 displays the x-ray diffraction pattern for such a film with a thickness of 100 nm. For the film grown on the Ag layer, one can observe narrow and intensive peak (111) and its higher harmonics: (222) and (333). In other words this C_{60} film has a (111)-texture, i.e. a well-aligned growth of the C_{60} molecules, with (111) planes parallel to the substrate surface, takes place under these conditions. The sizes of crystalline domains for textured C_{60} films are sufficiently high: values of a coherence length along <111> direction were up to 630 Å.

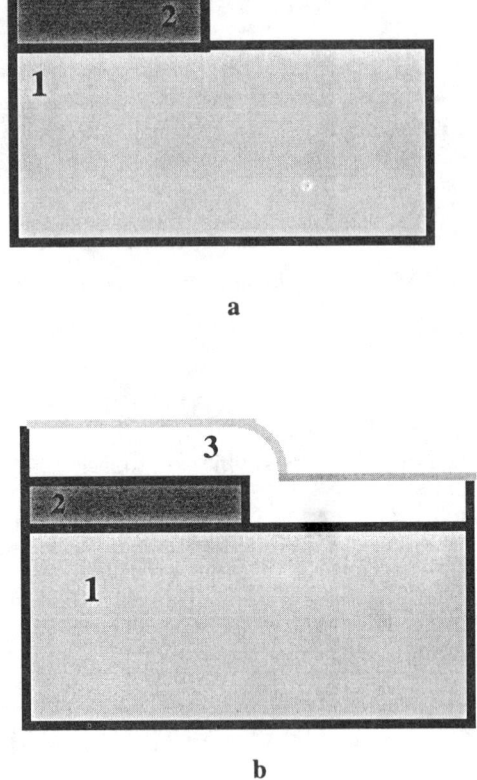

FIGURE 1. Schematic cross-section of the substrate (a) and deposited C_{60} thin film (b). 1- glass substrate, 2 - Ag layer, 3 - C_{60} thin film.

Fig. 3 represents the SPV spectra for similar C_{60} film with a thickness of 300 nm. The spectrum for the film grown on clear glass (Fig. 3, a) was very similar to that we observed previously for non-textured C_{60} films grown on glass substrates [3].The negative sign of the SPV indicates the n-type of photoconductivity of C_{60} films. On the other hand the C_{60} film grown on the Ag-covered part of the same substrate and having well-ordered textured structure exhibits a spectrum with the similar characteristic energies but with a positive sign (Fig. 3, b).

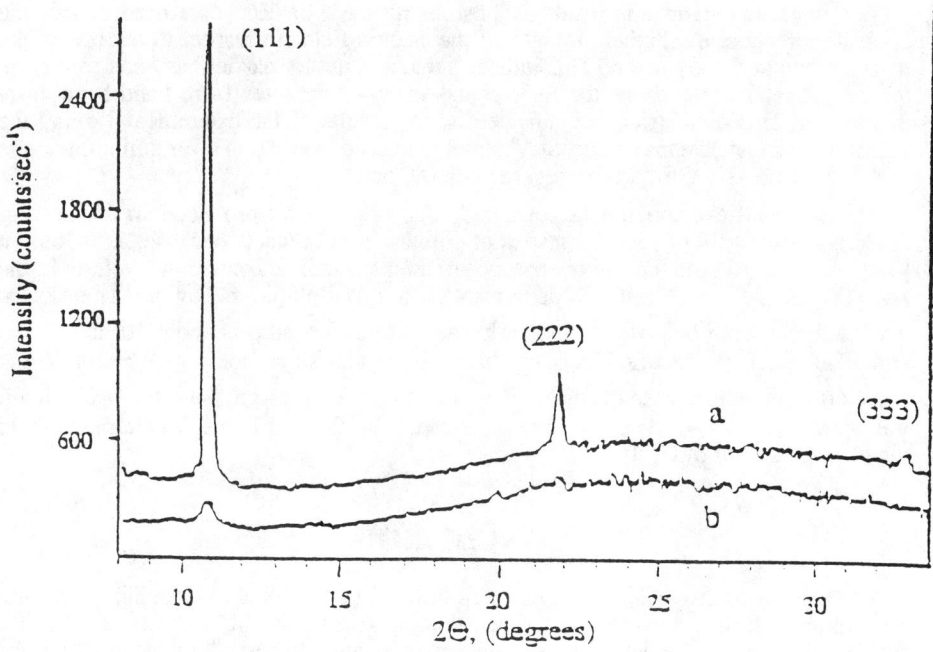

FIGURE 2. X-ray diffraction pattern of the C_{60} film grown on the Ag (a) and glass (b) part of the same substrate. Broad line belongs to the glass substrate.

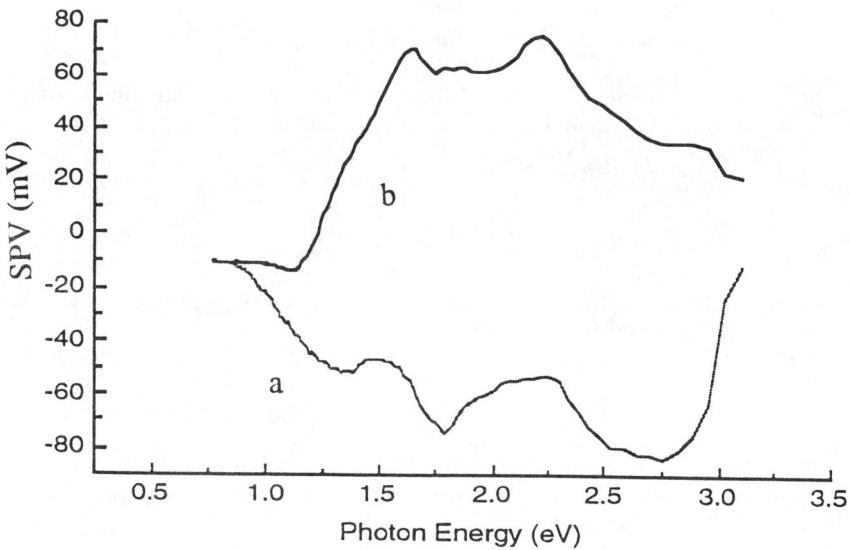

FIGURE 3. SPV spectra for the C_{60} film on the glass (a) and Ag (b) parts of the substrate.

We may understand this result by postulating the existence of a barrier with the opposite direction and higher strength of the electric field (in respect to the one at the front surface of the C_{60} layer). This indicates that a Schottky barrier has been formed at the C_{60}/Ag interface and that the SPV signal mainly originates from band-bending at this barrier. This conclusion is confirmed by the results of the experiment varying the C_{60} film thickness: the maximum SPV signal increased from 60 mV for a film thickness of 400 nm to 300 mV for a film thickness of 180 nm.

To study the I-V characteristics of the C_{60}/Ag interface we produced an Al/ C60/Ag device structure using Al semi-transparent dots as front contacts because, as is known [4], the Al/ C_{60}/Al structure (in the absence of an intermediate insulating layer, as in the Al/AlO$_x$/ C_{60}/Al configuration) demonstrates a quasi-ohmic behavior. The device exhibited rectifying behavior in the dark (the rectification ratio is about 10^2 at ± 2 V) and photovoltaic properties. Under irradiation by a solar simulator with P = 700 W·m^{-2} the short-circuit current density was found to be J_{sc} = 22.3 µA·cm^{-2} and the open-circuit voltage was V_{oc} = 180 mV. The low value of the fill factor FF = 0.3 is attributable to the high resistivity of C_{60} films [5].

CONCLUSIONS

A possibility to grow C_{60} films with well-ordered (111)-textured crystalline structure on Ag-predeposited substrates has been demonstrated. The SPV spectra for such a film grown on an Ag layer have been reported and discussed by postulating the existence of a Schottky barrier at the C_{60}/Ag interface.

The I-V characteristics of the C_{60}/Ag interfaces have been presented. The device structure exhibited rectifying behavior in the dark and photovoltaic properties.

ACKNOWLEDGMENTS

This work was funded by the Israel Ministry of National Infrastructures. One of us (E. A. K.) acknowledges the financial support of Ormat Industries, Ltd and the Sally Berg Foundation.

REFERENCES

1. M. Koltun, D. Faiman, S. Goren, E. A.Katz, E. Kunoff, A. Shames, S. Shtutina and B. Uzan, *Sol. Energy Mater. Sol. Cells* **44**, 485-491 (1996).
2. E.A. Katz, D. Faiman, S. Goren, S. Shtutina, B. Mishori and Yoram Shapira, *Full. Sci. Tech.* **6**, 103-111 (1998).
3. B. Mishori, E.A. Katz, D. Faiman, and Yoram Shapira, *Solid State Commun.* **102**, 489-492 (1997).
4. H. Yohehara and Ch. Pac, *Appl. Phys. Lett.* **61**, 575-576 (1992).
5. A. Hamed, Y.Y. Sun, Y.K. Tao, R.L. Meng and P.H. Hor, *Phys. Rev. B*, **47**, 10873-10876 (1993).

APPENDIX

Scientometrics on Fullerenes and Nanotubes

Werner Marx, Michael Wanitschek and Hermann Schier
Max Planck Institut fuer Festkoerperforschung, Heisenbergstr. 1, D-70569 Stuttgart
(e-mail=marx@and.mpi-stuttgart.mpg.de or wani@and.mpi-stuttgart.mpg.de)

Introduction

The availability of computers, networks, and extended data bases make statistical evaluations of published literature fairly easy. One problem of using this method in condensed matter physics and material sciences is the definition of the search field by appropriate choice of search terms (key words). E. g. in many papers on „nanotechnology" this word will neither appear in the title nor in the abstract. In addition, the term is used for a large variety of subjects. The situation is quite different in the field of fullerenes and carbon nanotubes. This is a closed area with about 15 000 publications within the last few years, all of which can be found by using the chemical names.

A statistical analysis of published literature is attractive for
- active researchers
- industry
- funding agencies

Active researchers will be able to select the journal in which most of the papers in their field of specialization are published (e.g. „Synthetic Metals" or the „Kirchberg Proceedings") or they will learn to improve their impact on the scientific community by choosing the periodical which is most likely to be quoted by scientists in general („Science"? „Nature"? „Condensed Matter News"?) or by peers in their own field („Synthetic Metals"? „Advanced Materials"?). Managers in industry might find help in deciding when to switch from observation of a field to active engagement in research and development. Furthermore they will be able to localize the most active groups for co-operations, for instance, if they intend to participate in research networks sponsored by the European Communities. Funding agencies might be interested in the regional distribution of research activities (Why are there more fullerene publication from China than from Russia?), they might spot out deficiencies in particular fields of research and development (There is very little nanotube research in Germany), and they might like to use scientometrics to control the efficiency of certain funding actions (Has the number of multinational publications increased by the establishment of European TMR Networks?)

Of course, there is an important CAVEAT: *Scientometrics measures the impact on the scientific community but not scientific quality!* (Just as a library of 20 000 000 volumes does not necessarily have to have a copy of the Guthenberg Bible or 20 000 paintings art gallery might not have a single Rembrandt).

The following graphs and tables summarize a scientometric evaluation of the fullerene and nanotube literature, which was carried out on the host STN International in the context of the International Winterschool on Electronic Properties on New Materials: Molecular Nanostructures - IWEP 98, Kirchberg, Tyrol, Austria, February 28th to March 7th, 1998 [1].

Number of Publications

On January 19th, 1998 there were 10 331 „fullerene species" registered in the chemical compounds file of Chemical Abstracts. The total number of publications on these species was 14 920, out of which 1 564 were on nanotubes. The files compile data from more than 10 000 journals, reports, conference proceedings, patents, and books.

Figure 1 shows a plot of number of publications versus year of publication both for fullerene and for nanotubes. Of course, the sudden rise in the fullerene literature is due to Kraetschmer's discovery of a new route of fullerene synthesis, which made this material easily accessible to almost any laboratory in the world. Table 1a compiles the journals with the largest numbers of fullerene publications, and Table 1b those with publications on nanotubes. The Kirchberg proceedings[2] are at rank 5 and 9, respectively. This information tells us that these proceedings have a well established position in the field. The reader will get a good survey without having to select the relevant papers from a plethora of other material.

Figure 2 shows the regional distribution of research activity, Fig. 2a for fullerenes, and Fig. 2b for nanotubes. These graphs are based on the nation assignment of first authors by Chemical Abstracts. A surprising result is that China ranks third, after America and Japan, and well before the European countries and Russia. Germany has a good position in fullerenes (rank 4), but in nanotubes she is at the rear end (rank 10, after India).

Number of Citations

According to the files of the Science Citation Index (SCI) there are 7 479 publication out of the total of 14 920 which have been cited at least once, i.e. every second paper has not (yet) been quoted at all. There are 18 802 publications with citations on fullerens, and the total number of fullerene citations is 140 158. Consequently a citing publication has 7.5 references on fullerenes on the average. The 1 564 publications on nanotubes have been cited 7 545 times in 3 179 publications. For a reliable evaluation the files of Chemical Abstracts and Science Citation Index had to be combined and many technical details [3] had to be taken care of (of which the treating of umlaute as in „Krätschmer" is just one example). The reader should be aware that *citation lists are not ranking lists*. Depending on their age the various publication had time intervals of different lengths during which they had the chance to accumulate citations!

Table 2 compiles the most cited publications on fullerenes. The „star" certainly is Kraetschmer with his method of mass production of fullerenes, followed by Kroto, whose first synthesis of fullerene was honored by the 1997 Noble price (Curl, Kroto, Smalley). Figure 3 shows the time dependant citation of the two most cited publications from Table 2.

Table 3 compares the „field-specific impact" of those journals which have published at least 50 papers on fullerenes. The field-specific impact differs from the „Journal Impact Factor" (JIF) of the Institute for Scientific Informations (ISI). The field-specific impact considers only the fullerene literature and should help researchers to select those journals which have the biggest influence in the community specialized in fullerenes. As can be seen from Table 3, the high-JIF journals „Nature", „Science", and „Physical Review Letters" also rank first in fullerene-specific impact.

Acknowledgement

A preliminary version of this survey has been submitted to „Condensed Matter News". We thank the editor for the permission to use some of the material.

References

[1] *Molecular Nanostructures*, ed. by J. Fink, H. Kuzmany, M. Mehring, S. Roth, AIP New York, in print

[2] *Electronic Properties of Novel Materials: Progress in Fullerene Research,* ed. by J. Fink, H. Kuzmany, M. Mehring, S. Roth, World Scientific, Singapore 1994
Physics and Chemistry of Fullerenes and Derivatives, ed. by J. Fink, H. Kuzmany, M. Mehring, S. Roth, World Scientific, Singapore 1995
Fullerenes and Fullerene Nanostructures, ed. by J. Fink, H. Kuzmany, M. Mehring, S. Roth, World Scientific, Singapore 1996
Molecular Nanostructures, ed. by J. Fink, H. Kuzmany, M. Mehring, S. Roth, World Scientific, Singapore 1998

[3] W. Marx: Cogito 4-96, 35-38 (1996)

Table 1a: Compilation of the 30 journals with the largest number of publications on fullerenes

	Publications	Journal
1	723	CHEM PHYS LETT
2	719	PHYS REV B: CONDENS MATTER
3	503	PROC - ELECTROCHEM SOC
4	390	J PHYS CHEM
5	380	KIRCHBERG PROCEEDINGS" (UP TO IWEPNM 1996)
6	331	J AM CHEM SOC
7	326	FULLERENE SCI TECHNOL
8	325	SYNTH MET
9	247	SOLID STATE COMMUN
10	228	PHYS REV LETT
11	200	MATER RES SOC SYMP PROC
12	160	PROC SPIE-INT SOC OPT ENG
13	149	J CHEM PHYS
14	142	APPL PHYS LETT
15	142	J CHEM SOC., CHEM COMMUN
16	128	MOL CRYST LIQ CRYST SCI TECHNOL., SECT C
17	105	PHYSICA C (AMSTERDAM)
18	104	SCIENCE (WASHINGTON, D C)
19	100	TETRAHEDRON LETT
20	95	MOL CRYST LIQ CRYST SCI TECHNOL., SECT A
21	88	J ORG CHEM
22	86	NATURE (LONDON)
23	84	ANGEW CHEM
24	79	Z PHYS D: AT., MOL CLUSTERS
25	78	CARBON
26	68	J PHYS CHEM SOLIDS
27	66	APPL PHYS A
28	66	NUCL INSTRUM METHODS PHYS RES., SECT B
29	65	IZV AKAD NAUK, SER KHIM
30	63	J PHYS.: CONDENS MATTER

Table 1b: Compilation of the 30 journals with the largest number of publications on nanotubes

	Publications	Journal
1	63	PHYS REV B: CONDENS MATTER
2	56	CHEM PHYS LETT
3	52	CARBON
4	37	PROC - ELECTROCHEM SOC
5	31	APPL PHYS LETT
6	30	NATURE (LONDON)
7	30	SCIENCE (WASHINGTON, D C)
8	28	PHYS REV LETT
9	23	KIRCHBERG PROCEEDINGS" (UP TO IWEPNM 1996)
10	23	MATER RES SOC SYMP PROC
11	21	SYNTH MET
12	17	FULLERENE SCI TECHNOL
13	17	PCT INT APPL
14	15	J PHYS SOC JPN
15	14	SOLID STATE COMMUN
16	13	J MATER RES
17	12	ADV MATER (WEINHEIM, FED REPUB GER)
18	12	ELECTRON MICROSC 1994, PROC INT CONGR ELECTRONMICROSC
19	12	JPN J APPL PHYS., PART 2
20	11	J PHYS CHEM
21	9	J PHYS CHEM SOLIDS
22	9	MOL CRYST LIQ CRYST SCI TECHNOL, SECT C
23	8	CHEM MATER
24	8	EUROPHYS LETT
25	8	NATO ASI SER., SER E
26	8	PHYS LETT A
27	8	Z PHYS D: AT., MOL CLUSTERS
28	7	J APPL PHYS
29	7	J VAC SCI TECHNOL., B
30	7	PROC SPIE-INT SOC OPT ENG

Table 2: The 50 most-cited publications on fullerenes

	Citations	First Author		Journal			
1	2859	KRAETSCHMER W	1990	V347	P354	NATURE	
2	2657	KROTO H W	1985	V318	P162	NATURE	
3	1210	HEBARD A F	1991	V350	P600	NATURE	
4	874	IIJIMA S	1991	V354	P56	NATURE	
5	724	HEINEY P A	1991	V66	P2911	PHYS REV LETT	
6	684	HAUFLER R E	1990	V94	P8634	J PHYS CHEM-US	
7	587	KRAETSCHMER W	1990	V170	P167	CHEM PHYS LETT	
8	557	AJIE H	1990	V94	P8630	J PHYS CHEM-US	
9	533	KROTO H W	1991	V91	P1213	CHEM REV	
10	531	ROSSEINSKY M J	1991	V66	P2830	PHYS REV LETT	
11	522	SAITO S	1991	V66	P2637	PHYS REV LETT	
12	475	HADDON R C	1991	V350	P320	NATURE	
13	461	TAYLOR R	1990	P1423		J CHEM SOC CHEM COM	
14	454	ALLEMAND P M	1991	V253	P301	SCIENCE	
15	444	HOLCZER K	1991	V252	P1154	SCIENCE	
16	410	STEPHENS P W	1991	V351	P632	NATURE	
17	399	EBBESEN T W	1992	V358	P220	NATURE	
18	384	TAYLOR R	1993	V363	P685	NATURE	
19	380	DAVID W I F	1991	V353	P147	NATURE	
20	378	CHAI Y	1991	V95	P7564	J PHYS CHEM-US	
21	377	DIEDERICH F	1991	V252	P548	SCIENCE	
22	371	TANIGAKI K	1991	V352	P222	NATURE	
23	364	HAWKINS J M	1991	V252	P312	SCIENCE	
24	348	HADDON R C	1986	V125	P459	CHEM PHYS LETT	
25	348	KROTO H	1988	V242	P1139	SCIENCE	
26	347	ARBOGAST J W	1991	V95	P11	J PHYS CHEM-US	
27	345	BETHUNE D S	1991	V179	P181	CHEM PHYS LETT	
28	343	YANNONI C S	1991	V95	P9	J PHYS CHEM-US	
29	337	ALLEMAND P M	1991	V113	P1050	J AM CHEM SOC	
30	333	UGARTE D	1992	V359	P707	NATURE	
31	329	RAO A M	1993	V259	P955	SCIENCE	
32	328	WEAVER J H	1991	V66	P1741	PHYS REV LETT	
33	325	DAVID W I F	1992	V18	P219	EUROPHYS LETT	
34	307	ZHANG Q L	1986	V90	P525	J PHYS CHEM-US	
35	302	KROTO H W	1987	V329	P529	NATURE	
36	300	MINTMIRE J W	1992	V68	P631	PHYS REV LETT	
37	292	FLEMING R M	1991	V352	P787	NATURE	
38	289	CURL R F	1988	V242	P1017	SCIENCE	
39	289	OBRIEN S C	1988	V88	P220	J CHEM PHYS	
40	285	BETHUNE D S	1990	V174	P219	CHEM PHYS LETT	
41	285	HARE J P	1991	V177	P394	CHEM PHYS LETT	
42	274	SCHMALZ T G	1988	V110	P1113	J AM CHEM SOC	
43	267	VARMA C M	1991	V254	P989	SCIENCE	
44	258	WUDL F	1992	V25	P157	ACCOUNTS CHEM RES	
45	252	CREEGAN K M	1992	V114	P1103	J AM CHEM SOC	
46	248	LOF R W	1992	V68	P3924	PHYS REV LETT	
47	247	TYCKO R	1991	V67	P1886	PHYS REV LETT	
48	244	SCHLUTER M	1992	V68	P526	PHYS REV LETT	
49	240	TYCKO R	1991	V95	P518	J PHYS CHEM-US	
50	238	NEGRI F	1988	V144	P31	CHEM PHYS LETT	

Table 3: Fullerene-specific impact of the journals having published at least 50 papers on fullerenes

Citations per Publication	Citations	Publications	Journal
210.14	18072	86	NATURE (LONDON)
89.32	9289	104	SCIENCE (WASHINGTON, D C)
47.68	10870	228	PHYS REV LETT
42.33	14011	331	J AM CHEM SOC
39.36	3306	84	ANGEW CHEM
30.33	11828	390	J PHYS CHEM
25.89	3676	142	J CHEM SOC, CHEM COMMUN
24.42	3639	149	J CHEM PHYS
23.88	1361	57	EUROPHYS LETT
22.69	16405	723	CHEM PHYS LETT
17.75	1207	68	J PHYS CHEM SOLIDS
17.08	1503	88	J ORG CHEM
15.10	10856	719	PHYS REV B: CONDENS MATTER
15.04	135	142	APPL PHYS LETT
13.08	693	53	J CHEM SOC, FARADAY TRANS
10.61	605	57	J CHEM SOC, PERKIN TRANS 2
10.04	542	54	CHEM LETT
9.83	2429	247	SOLID STATE COMMUN
9.18	716	78	CARBON
8.48	848	100	TETRAHEDRON LETT
8.20	541	66	APPL PHYS A
6.87	419	61	SURF SCI
6.76	338	50	THIN SOLID FILMS
6.45	387	60	JPN J APPL PHYS, PART 2
6.26	657	105	PHYSICA C (AMSTERDAM)
6.08	480	79	Z PHYS D: AT, MOL CLUSTERS
6.00	318	53	PHYS LETT A
5.29	312	59	INT J MASS SPECTROM ION PROCESSES
5.21	328	63	J PHYS: CONDENS MATTER
4.38	289	66	NUCL INSTRUM METHODS PHYS RES, SECT B
3.55	710	200	MATER RES SOC SYMP PROC
1.76	574	326	FULLERENE SCI TECHNOL
1.75	570	325	SYNTH MET
0.87	83	95	MOL CRYST LIQ CRYST SCI TECHNOL, SECT A
0.77	46	60	THEOCHEM
0.55	209	380	"KIRCHBERG PROCEEDINGS" (UP TO IWEPNM 1996)
0.55	29	53	CHIN PHYS LETT
0.52	34	65	IZV AKAD NAUK, SER KHIM
0.46	59	128	MOL CRYST LIQ CRYST SCI TECHNOL, SECT C
0.03	13	503	PROC - ELECTROCHEM SOC
0.02	1	59	GAODENG XUEXIAO HUAXUE XUEBAO
0.01	2	160	PROC SPIE-INT SOC OPT ENG
0.00	0	57	CHEM COMMUN (CAMBRIDGE)

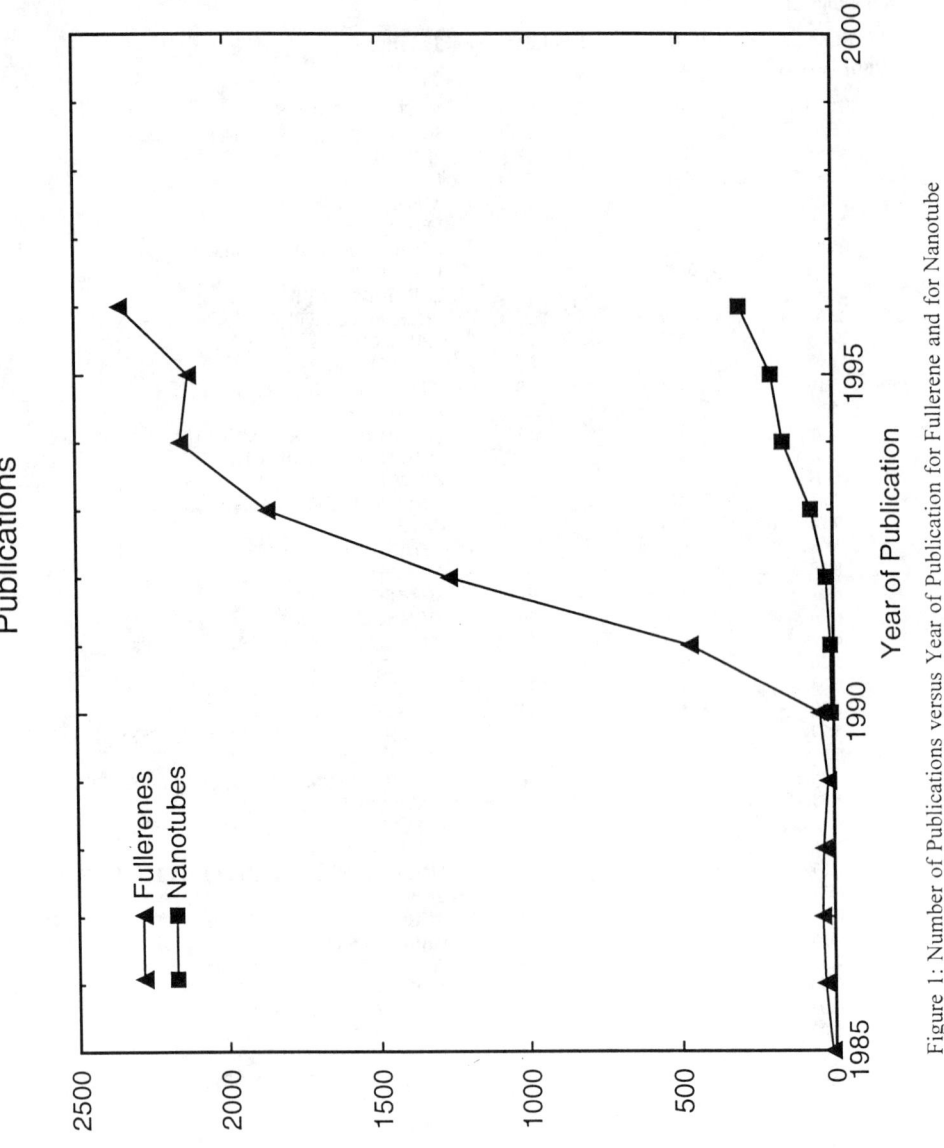

Figure 1: Number of Publications versus Year of Publication for Fullerene and for Nanotube Literature

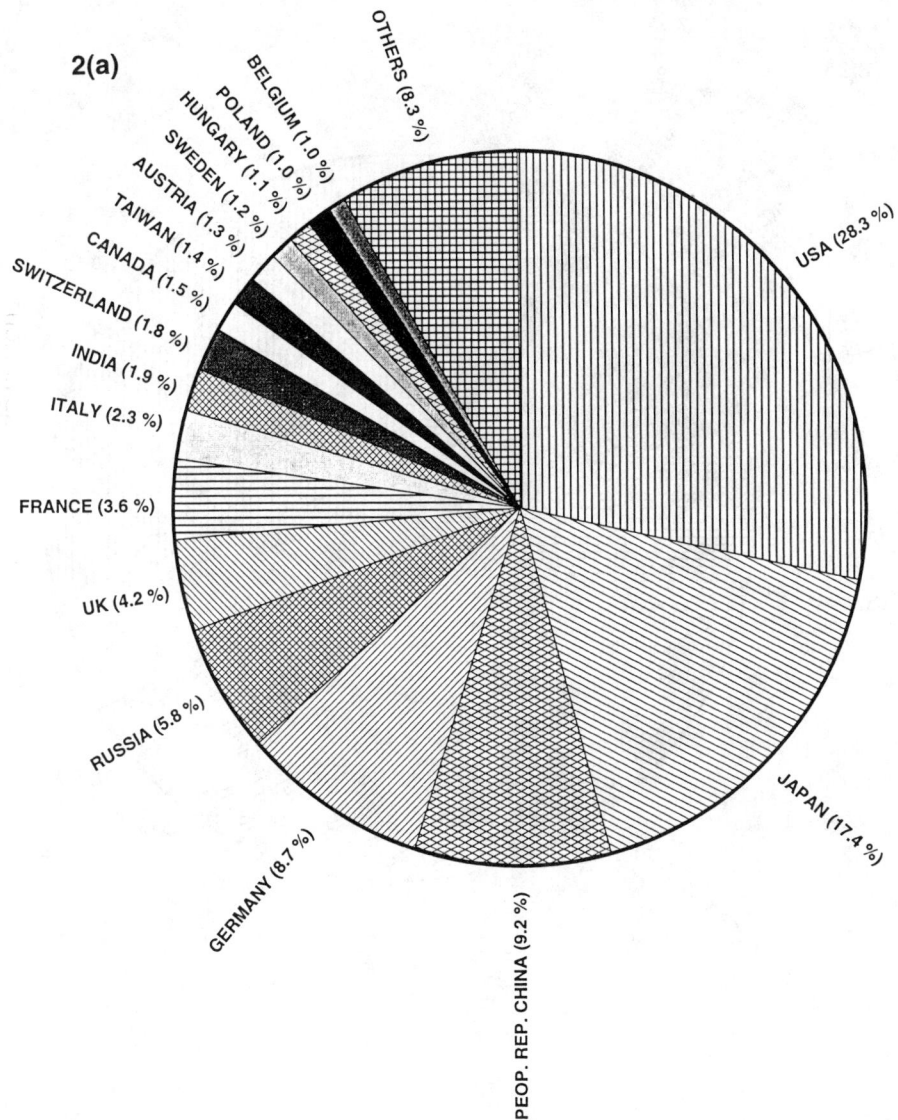

Figure 2: Regional Distribution of Publications. a) Fullerenes b) Nanotubes

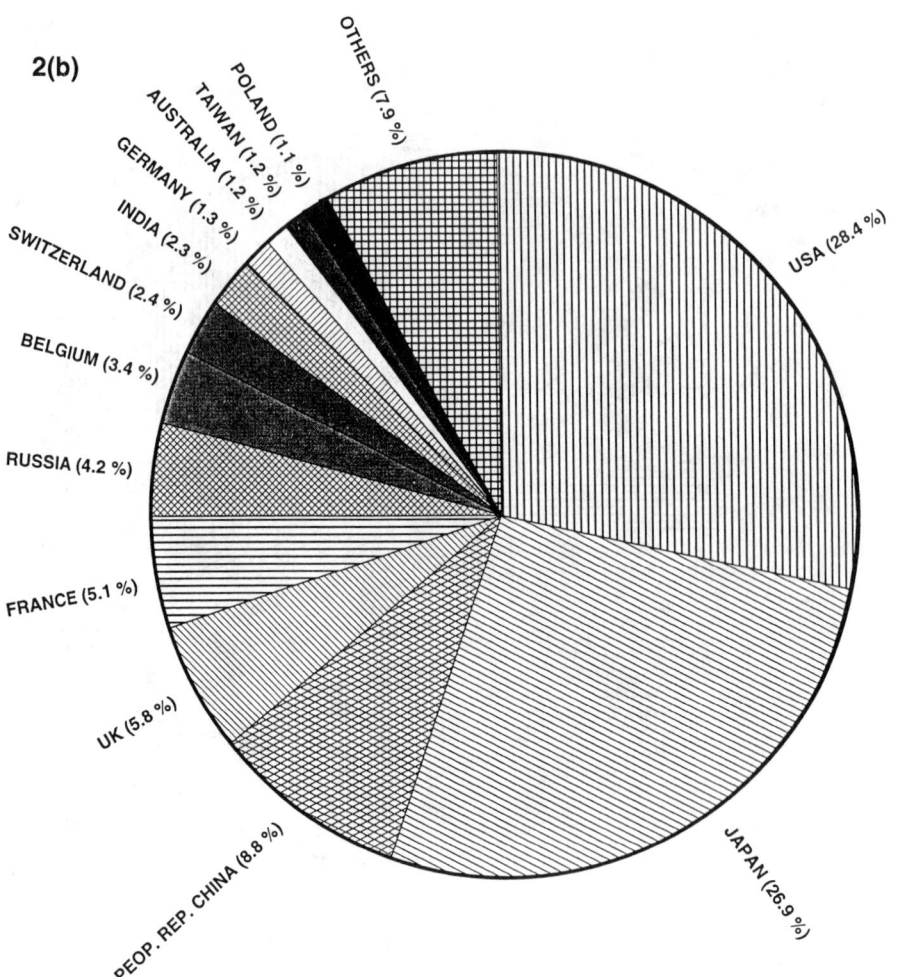

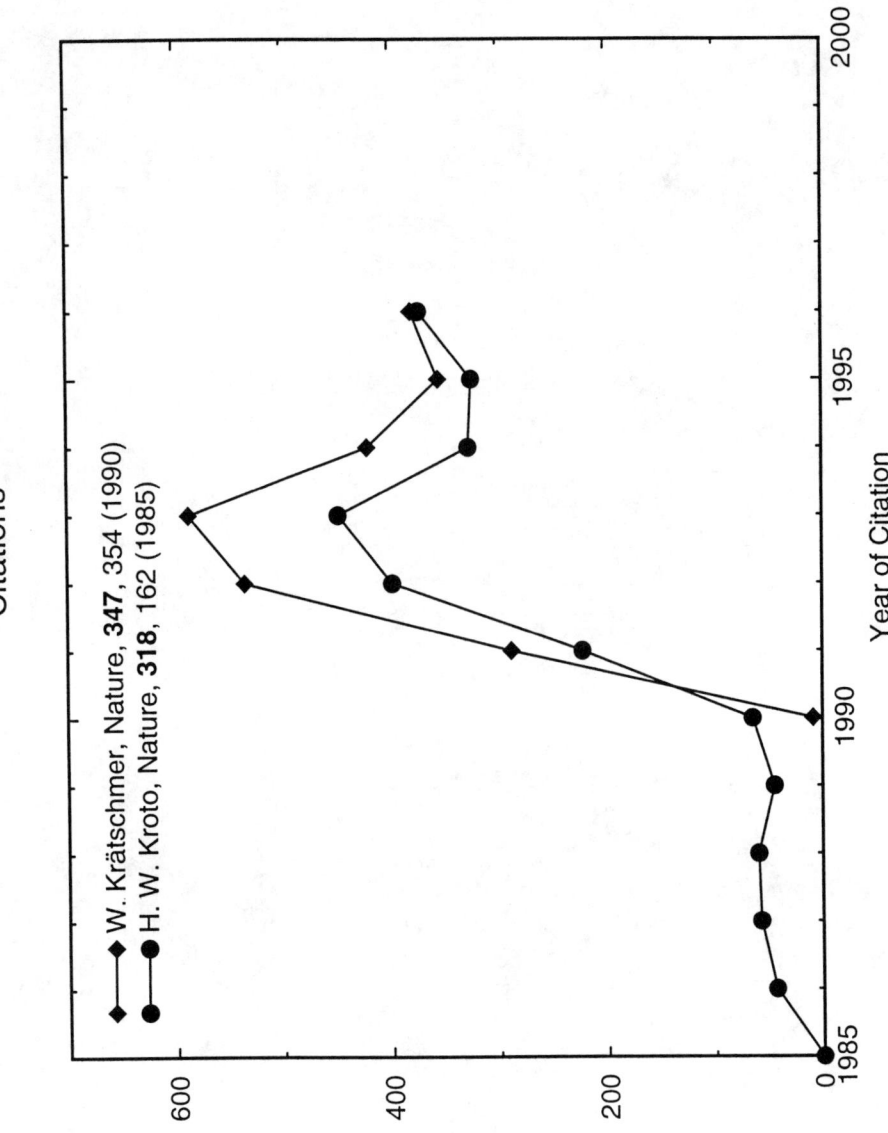

Figure 3: Time dependant citations of the most cited publications of W. Krätschmer and H. W. Kroto

Author Index

A

Achey, R. M., 412
Adla, A., 383
Ago, H., 87
Alessio, E., 232
Alloul, H., 310
Alvarez, L., 116
Alvarez, M. M., 403
Andrievsky, G. V., 172
Anglaret, E., 116, 318
Apih, T., 412
Arčon, D., 340, 412
Ashey, R. M., 407
Astashkin, Y. A., 172
Auban-Senzier, P., 314
Azamar, J.-A., 241

B

Bachtold, A., 65
Barra, A.-L., 340
Basca, R., 467
Baumgartner, G., 296, 314
Beeli, C., 489
Beguin, L., 51
Benito, A. M., 116, 128
Bennati, M., 300
Bergeron, D. J., 92, 456
Bernier, P., 3, 57, 116, 123, 128, 156
Bezmelnitsyn, V. N., 261
Bietsch, W., 344
Bigioni, T., 403
Biró, L. P., 164
Blank, V. D., 499
Blau, W. J., 97
Blinc, R., 340, 407, 412
Blumentritt, S., 460
Bonard, J.-M., 132, 439, 467, 489
Bondarenko, A., 348
Bonnamy, S., 51
Böske, T., 265
Bozhko, A. D., 34
Brabec, C. J., 257, 519
Brouet, V., 310
Brown, C. M., 327

Buga, S. G., 499
Bulavin, L. A., 172
Bulina, N. V., 227
Burger, B., 101, 211
Burghard, M., 39, 44, 74, 460
Buzaneva, E. V., 172
Byszewski, P., 348, 353

C

Cabioc'h, T., 430
Campbell, E. E. B., 368
Carano, M., 232
Čermák, I., 425
Čermáková, I., 425
Ceroni, P., 232
Cevc, P., 340
Chabanenko, V., 348
Châtelain, A., 439, 489
Choi, J-P., 241
Chouteau, G., 340
Christensen, M., 335
Chugreev, A. V., 253
Chupina, O. V., 227
Churilov, G. N., 227
Churilova, Ya. N., 227
Clark, T., 392
Clauss, W., 92, 456
Clérac, R., 241
Colomer, J.-F., 16
Coulon, C., 241

D

Dalal, N. S., 407, 412
Da Ros, T., 232
Delafond, J., 430
de la Fuente, G. F., 116, 128
Delaney, P., 69
Denanot, M. F., 430
Dennis, J., 241
Diduszko, R., 348
Dietel, E., 396
Ding, A., 12
Dinse, K.-P., 383, 396

Dolinšek, J., 412
Doome, R. J., 515
Drichko, N. V., 357
D'Souza, F., 241
Dubitsky, G. A., 499
Duclaux, L., 51
Duesberg, G. S., 39, 44, 61, 74, 97
Dunsch, L., 207, 373
Durov, S. S., 172
Dutton, P. L., 456
Dyakonov, V., 257, 519

E

Eberhardt, W., 509
Egger, R., 147
Eisler, H.-J., 215
Eletskii, A. V., 261
Ellwood, I., 79
Enderle, M., 237

F

Faiman, D., 527
Fally, M., 277
Fefelova, V. V., 227
Fink, J., 83, 265, 271
Fischer, J. E., 3, 34
Fonseca, A., 16, 515
Förderer, M., 425
Forró, L., 65, 136, 296, 310, 314, 467
Fowler, P. W., 420
Frentrup, W., 265
Fuhrer, M. S., 69

G

Gaucher, H., 51
Girard, J. C., 430
Gmeiner, J., 344
Gogolin, A. O., 147
Golden, M. S., 83, 265, 271
Goze, C., 156
Graja, A., 357
Griffin, R. G., 300
Grobert, N., 25, 29
Gromov, A., 368

Grupp, A., 300, 340, 447
Gruß, A., 396
Grushko, Yu. S., 253
Guillard, T., 116
Guldi, D., 232
Gunnarsson, O., 287, 292
Gyulai, J., 164

H

Happ, M., 253
Hare, J. P., 25, 29
Havlik, D., 237
Hayashi, M., 87
Heid, R., 322
Hellwig, C., 265
Henneberger, F., 253
Hennrich, F. H., 215
Hernadi, K., 20
Hernández, E., 156
Herwig, P., 447
Hirsch, A., 223, 363, 392, 396
Holmes, W., 69
Hone, J., 79
Hsieh, Y. Y., 241
Hsu, W. K., 25, 29
Huang, Yu., 283
Huber, P., 237
Huczko, A., 348
Hulman, M., 101, 379
Hummelen, J. C., 257, 519, 523

I

Inakuma, M., 379
Ishii, T., 494
Ito, A., 283
Ivanov, V. A., 451
Iwasa, Y., 219, 296
Iyer, V. S., 447
Izuoka, A., 219

J

Jánossy, A., 296
Janssen, R. A. J., 257, 519
Jantoljak, H., 123, 136

Jaouen, M., 430
Jerome, D., 314
Jiao, J., 489
Johnson, A. T., 92, 456
Jörissen, L., 481
Jouguelet, E., 57
Journet, C., 3, 57, 116, 123, 128
Jung, Ch., 265

K

Kalhofer, S., 425
Kane, C. L., 143
Kanzow, H., 12
Kappes, M. M., 215
Karakassides, M. A., 416
Kasuya, A., 107, 112
Kataura, H., 87
Katz, E. A., 527
Kessler, B., 509
Khoury, J. T., 403
Kim, G.-T., 61
Kim, H. J., 34
Kiricsi, I., 20
Klizting, K. V., 74
Klos, H., 481, 485
Knapp, C., 383, 396
Knechtel, W. H., 97
Knol, J., 523
Knorr, K., 237
Knorr, S., 300, 340, 447
Knupfer, M., 83, 265, 271
Koch, E., 287, 292
Kochkanjan, R., 348
Komatsu, K., 194, 211
Konarev, D. V., 357
Kordatos, K., 327
Koretz, Ya., 227
Kowalska, E., 348, 353
Krätschmer, W., 425
Krause, M., 373
Krawez, N., 368
Kroto, H. W., 25, 29
Krstic, K., 44
Krstic, V., 39
Kucharski, Z., 348, 353
Kuhlmann, U., 123
Kumazawa, Y., 87
Kuran, P., 207, 373

Kurita, N., 186
Kürti, J., 101
Kusakabe, K., 331
Kutner, W., 207

L

Laforge, Ch., 203
Lambin, Ph., 168, 203, 504
Lamp, P., 481
Lamy de la Chapelle, M., 3, 128
Lange, H., 348
Laplaze, D., 116, 128
Lappas, A., 327
László, I., 435
Lauginie, P., 51
Lebedkin, S., 194
Lee, R. S., 3, 34
Lefrant, S., 3, 128
Levin, V. M., 499
Lips, K., 363, 388
Liu, K., 61
Loiseau, A., 3
Louie, S. G., 69
Lvova, N. A., 499
Lyubovskaya, R. N., 357

M

Maniwa, Y., 87
Margadonna, S., 327
Márk, G. I., 164
Martin, R. M., 292
Martinez, M. T., 116, 128
Marx, W., 533
Maser, W. K., 3, 116, 128
Mathis, C., 57
Mauser, H., 392
Mehring, M., 300, 305, 340, 447
Meingast, C., 194
Mele, E. J., 143
Melenevskaya, E. Yu., 253
Metenier, K., 51
Meunier, V., 168, 504
Micholet, V., 3
Milia, F., 407
Mishori, B., 527
Mitani, T., 219

Morenzin, J., 509
Müllen, K., 447
Muno, M., 79
Muñoz, E., 116, 128
Muster, J., 39, 44

N

Nagel, P., 194
Nagy, J. B., 16, 20, 515
Nalimova, V. A., 34
Nazarenko, A. V., 172
Neugebauer, H., 249
Nishina, Y., 107, 112
Noworyta, K., 207
Nozawa, H., 494
Nuber, B., 223, 363

O

Obraztsova, E. D., 132
Ogitsu, T., 331
Ogul'chansky, T. Yu., 172
Okada, S., 198
Okun, M. V., 261
Omerzu, A., 340
Ono, Y., 87
Ōsawa, E., 179, 186
Osborne, A. J., 25
Ovchinnikov, S. G., 227

P

Padinger, F., 519
Paolucci, F., 232
Pasler, V., 194
Paulsson, M., 190
Pénicaud, A., 241
Petit, P., 57
Petrakovskaya, E. A., 227
Petridis, D., 416
Philipp, G., 44, 74
Pichler, T., 83, 265, 271
Pidduck, A. J., 29
Piedigrosso, P., 16
Pietzak, B., 223, 363, 383, 388, 396
Pilloud, D. L., 456

Piskoti, C., 79, 183
Pogorelov, V. E., 172
Popov, M. Yu., 499
Pouhova, Ya. I., 227
Prassides, K., 327, 331
Prato, M., 232
Prilutski, Y. I., 172
Prokhorov, V. M., 499

Q

Quéré, F., 310

R

Rachdi, F., 305
Radomska, J., 348
Razbirin, B. S., 253
Reeves, C. L., 29
Reichenauer, G., 481
Renker, B., 322
Reuther, U., 223
Richards, P. L., 69
Rinzler, A. G., 34, 83, 101
Rivière, J. P., 430
Roffia, S., 232
Rojik, I., 20
Rols, S., 116
Roth, S., 39, 44, 61, 74, 460
Rubio, A., 156

S

Saito, S., 198
Saito, Y., 107, 112
Salisbury, B. E., 403
Salvetat, J.-P., 51, 65, 136, 467
Sariciftci, N. S., 249, 257, 519
Sauvajol, J. L., 116, 318
Savchenko, A. A., 227
Sawatzky, G. A., 152
Schaaff, T. G., 403
Scharber, M., 257
Scharff, P., 172
Schier, H., 533
Schilder, A., 344

Schlebusch, C., 509
Schmalz, A., 12
Schmid, M., 61
Schober, H., 322
Schönenberger, C., 65
Schranz, W., 237
Schütz, W., 485
Schwoerer, M., 344
Seifert, G., 420
Semkin, V. N., 357
Senet, P., 203
Seraphin, S., 489
Serebryanaya, N. R., 499
Shafigullin, M. N., 403
Shapira, Y., 527
Shibata, T., 494
Shimoda, H., 296
Shinohara, H., 241, 379
Shriner, K., 241
Shtutina, S., 527
Shul'ga, Y. M., 357
Simon, F., 296
Simovic, B., 314
Sing, M., 265
Siska, A., 20
Sklovsky, D. E., 34
Slanina, Z., 179
Smalley, R. E., 34, 83, 101
Solovyov, L. A., 227
Sommerhalter, C., 223
Stadelmann, P. A., 439, 489
Stafström, S., 190, 335
Starukhin, A. N., 253
Steinmetz, M., 237
Stöckli, Th., 439, 467
Strunk, C., 65
Sugano, M., 107, 112
Sugawara, T., 219
Sulyanov, S. N., 499
Sundqvist, B., 194, 322
Surján, P. R., 443
Szekeres, Z., 443

T

Takahashi, H., 107
Tamamura, T., 494
Tanaka, K., 87, 283
Tanaka, T., 194

Tani, Y., 107
Tanigaki, K., 327, 331
Tanoue, K., 219
Tellgmann, R., 368
Terrones, H., 25, 29
Terrones, M., 25, 29
Thier, K.-F., 305
Thomsen, C., 123, 136
Tohji, K., 107, 112
Tománek, D., 159
Tou, H., 87
Trapp, V., 481
Trasobares, S., 25, 29
Tsuneyuki, S., 331

U

Ugolkova, E. A., 451

V

Valli, L., 232
van den Brink, J., 152
Varadarajan, U., 69
Vezmar, I., 403
Vietze, K., 420
Vizard, C., 29

W

Waiblinger, M., 363, 383, 388, 396
Wallis, D. J., 29
Walton, D. R. M., 25, 29
Wang, Z. L., 439
Wanitschek, M., 533
Wehmeier, M., 447
Weiden, N., 383, 396
Weidinger, A., 223, 363, 383, 388, 396
Whetten, R. L., 403
Whitney, M., 79
Wright, P. J., 29

Y

Yagi, T., 219
Yaguzhinski, S. L., 34

Yamabe, T., 87
Yashchuk, V. N., 172

Z

Zaritowskij, A., 348
Zettl, A., 69, 79, 183
Zgonnik, V. N., 253

Zhao, X., 179
Zhu, Y. Q., 25, 29
Zhuravlev, M. Ye., 451
Zoriniants, G., 257